Hardback ISBN 9780957679139

Published by
Llyfrau Cambria Books, Wales, United Kingdom.
Cambria Books is a division of
Cambria Publishing Ltd.
Discover our other books at: www.cambriabooks.co.uk

Formatting by
Carolyn Michel
Design Elements

Photo Verity Johnson

Understanding The Bird of Prey

Nick Fox

Contents

Preface

A lot has happened since the first edition of this book in 1995. Following the pesticide crisis, we've seen a change in public attitudes toward birds of prey, a blossoming of international research and conservation work, a long and successful struggle to breed raptors in captivity resulting in a major increase in supply of birds for falconry, and we've had a technological revolution in telemetry, computers, and genetic research. On top of that we now have robotic prey, the development of falcon racing and recognition by UNESCO of the cultural significance of birds of prey, but widespread erosion of habitats and biodiversity. So this second edition has become a major re-write.

Birds of prey turn people into fanatics, and, like all good fanatics, they cluster into factions, most of whom do not see eye to eye with one another! A kinder person than me would probably call these factions 'disciplines'. People come to raptors from many viewpoints; we have field biologists, ecologists, ethologists, taxonomists, birders, artists, falconers, breeders, showmen, re-habilitators, game-keepers, conservationists, preservationists, educators, anthropologists, philosophers, politicians, administrators, law enforcers, historians - all with their own viewpoints. It can be hard to see things from each other's perspectives, and harder still to write for them.

This is not a theoretical book; application is the acid test of theory and this book is firmly based on real life and first-hand applied science as far as possible. Theory can take you only so far. There comes a point where you have to get in there and 'just do it, mate'. And that's when you start to realise that the water is deeper and muddier than you had thought…

It is called '*Understanding* the Bird of Prey' because I am trying to provide basic principles. Once you understand the underlying principles, you can apply them to your own particular circumstances. There are a number of books on the market which are based on process: 'This is how I do it and it works'. These writers have spent years perfecting systems that work for them, more or less by trial and error. But when others try to copy them, things fall apart because everyone's situation is slightly different, or quite likely they are unable to devote the time and effort needed. Small details are key to everything, and I have included detail in places where year after year I have seen people get into difficulties.

Publishing technology has also moved on. Some people say 'You can't learn falconry from books'. I also used to teach people through courses, usually a group of five or six each week. This is very inefficient, so we distilled this book, and some of the course material, into 10 Bird of Prey Management films, each one modular and lasting 90 minutes. Now this new edition will also be available as an e-book and as a paper book produced as Print on Demand, linking to the films, creating a package which you can dip into or out of as you wish.

Many people have helped with the final production. My initial peer reviewers were Neil Forbes MRCVS, Greg Simpson MRCVS and Nigel Harcourt Brown MRCVS (Structure and function), Prof. Colin Pennycuick (Flight), Prof. David Bird and Mr. Bob Berry (Breeding), Dr Frederic Launay and Mr. Steve Layman BSc (Development and behaviour), Mr. Steve Layman BSc (Training and conditioning), Dr Ian Newton (Hunting strategies of wild raptors), Prof. Tom

Cade, Dr Mike Nicholls, Dr Ken Felix, Mr Hal Webster, Mrs Jemima ParryJones, Dr Walter Bednarek, Mr. Paul Llewelyn and Dr Cecilia Lindberg (Ethics), and Ms. Renatta Platenberg MSc (Rehabilitation). Earlier drafts have been commented on by Dr Tom Bailey MRCVS, Dr Steve Sherrod, Dr Nigel Barton, Mr. David Hancock, Miss Katie O'Neal, and Mr. Seth Layman. Dr Chris Eastham undertook some of the last minute data processing to ensure the data are as recent as we could make them.

Since then Dr Andrew Dixon, Dr Eugene Potapov, Dr Robert Kenward, Helen Nakielny, Jo Oliver, Angel Garcia-Rojo, Jim Chick, Colin Wynn, Lynn Oliphant, Tom Spink, Jevgeni Shergalin, Dr Fernando Feás, Patrizia Cimberio, Remy van Wijk, Neil Davies, Roy Lupton, Mark Upton, Iolo Williams, Mike Calvin, Noel Hyde, Debbie Stewart, Steve Chindgren, Mark Williams, Verity Johnson, Alessandra Oliveto, Alan Gates, Martyn Standley, Kurien Koshy Yohannan and many other people have helped with information, images, discussions, and comments. I thank them all for their support and patience, and apologize for not being able to name everyone.

Foreword

Birds of prey have stirred human emotions since time immemorial because of their fierce beauty, great strength, commanding presence, and superlative skill as hunters. Who knows when the first primitive man looked skyward and decided to adopt the eagle as his totem? Who first saw a peregrine stoop and thought to take her for his hunting companion? Surely it happened well before recorded history.

Today, birds of prey are arguably the most popular group of birds. Certainly they attract the most ardent and zealous devotees, whether they be falconers, rehabilitators, breeders, banders, scientists, or birdwatchers. There is something about these winged predators that evokes the strongest possible feelings in the breasts of their admirers and causes such people to do all kinds of strange things to maintain a closeness with their feathered idols, as Bil Gilbert detailed in a most revealing account for *Audubon* some years ago. To have such a magnificent creature as a hawk, falcon, or eagle accept you as its companion and to allow you to enter its space is to gain a rare perspective on life.

At a time in our social and cultural development when most human beings have lost direct contact with nature and living things and are separated more and more from the natural world, there is a great need for nurturing avocations that give people the opportunity for intimate, eyetoeye relationships with other kinds of animals. Deep in the human psyche there is a longing to know 'other bloods.' It is a part of what the Harvard biologist, E. O. Wilson, has termed 'biophilia' the love of life that stems from the ancestral and mystical connectedness that human beings have and feel for other forms of life. For many it is mainly a subconscious yearning, but it becomes a vibrant, overt passion for a few, special people. An opportunity to experience this kinship with other animals and with nature helps people to understand their dependence on the earth's natural processes and on the web of interconnected relationships among all living things.

Unfortunately, all too many of us now receive most of our information and knowledge of animals and the living world from television or the glossy pages of magazines and coffee table books. More and more human attitudes about nature are shaped by these representational, often distorted pictures rather than by real-life experiences.

It was not always so. When I was a boy growing up in the 1930s and 1940s in Texas, and later in California, I had a constant succession of wild pets ranging from jars full of red ants to 'horny toads,' ground squirrels, deer mice, opossums, crows, owls, and my first hawk when I was nine years old. This venerable tradition of pet-keeping no doubt goes back to the very origins of humanity and was the basis for the domestication of animals. The list of great naturalists and biologists whose careers sprang from such simple pursuits is too long to enumerate but includes such names as Aristotle, Frederick II of Hohenstaufen, John James Audubon, Louis Agassiz, Theodore Roosevelt, William Beebe, Aldo Leopold, Roger Tory Peterson, Fran Hamerstrom, and at least two Nobel prize winners, Niko Tinbergen and Konrad Lorenz, whose classic 1935 paper, *Der Kumpan in der Umwelt des Vogels*, was based largely on his personal experiences with hand-tamed birds.

One wonders how many potentially great biologists are being lost to society now because opportunities for hands-on experiences with wild animals are so limited. Gone are the days when a young boy or girl could go freely into the woods or field and bring back a wild pet as a companion. It is now against the law to do so in most places. There are international treaties, national laws, state and provincial laws, municipal laws, and societal disapprobation of all sorts. It is, of course, true, and we all know that many of these wild creatures that used to be taken so lovingly into childhood embrace ended up either dead or leading miserable lives in confinement. That is part of the reason for the laws and social disapproval. Still, on balance, there is much to be said in favour of keeping wild animals as companions, and today the grosser forms of unintended mistreatment can be avoided with modern knowledge of veterinary practices, husbandry, and caretaking.

It is of more than passing interest that falconry, raptor rehabilitation, and captive breeding are among the few legal and acceptable ways in which people can still have personal relationships with non-domesticated animals. The practitioners of these pursuits are privileged members of society, and because birds of prey are so highly esteemed and basically rare creatures, they have a special responsibility to make certain that the birds which come into their ken and care receive the most enlightened and humane treatment possible. Many new innovations in the care and handling of birds of prey have come about in the last 30 years: captive breeding, telemetry, greatly increased knowledge of nutrition and treatment of diseases and injuries, understanding of behaviour and how it can be modified, especially by imprinting, and new equipment such as Aylmeri jesses, loop leashes, shelf perches, and other things.

As a practicing falconer and raptor biologist, Nick Fox has been in the thick of many of these developments. He received his Ph.D. degree from the University of Canterbury for original research on the little-known New Zealand falcon, he has propagated raptors for many years, done pioneering telemetry studies on the home range and movements of the once endangered Mauritius kestrel, worked on the reintroduction of the red kite in Great Britain, and most recently been involved with the National Avian Research Centre of Abu Dhabi raising and training captivebred falcons for Arab falconry. This book is a distillation of the knowledge and understanding of raptors that he has acquired through these diverse activities over the span of time in which the practices involved in handling birds of prey have undergone such radical and improved change.

Understanding the Bird of Prey will serve as an advanced primer for the many devotees of raptors coming on the scene as the 21st century approaches and as a stimulating refresher for the old hands, who always need to be challenged by new ideas and methods. Together with an up-to-date manual on avian veterinary practices, the reader of this book has all the basic scientific and managerial underpinnings needed to keep birds of prey physically and psychologically fit in the captive environment and the means to enter that very special and inexhaustibly fascinating world of the eagle, the hawk, and the falcon.

TOM J. CADE

The Peregrine Fund, Inc.

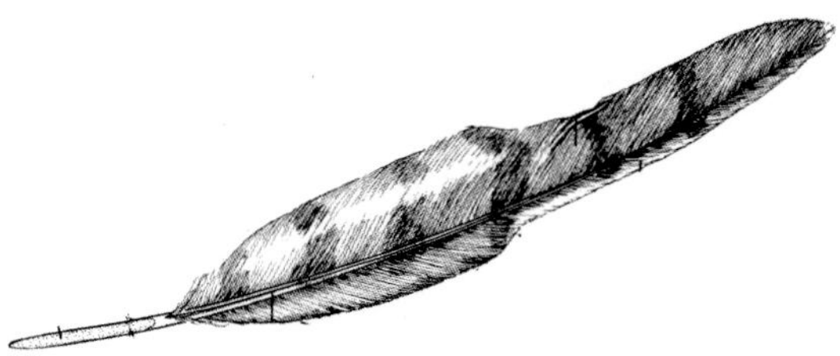

C1 Structure and Function

This chapter is accompanied by the Bird of Prey Management film *Anatomy.*

A wild Mongolian saker. *Photo Andrew Dixon*

1.1 Introduction

The first scientific study of bird structure was made by a falconer, Frederick II, Holy Roman Emperor and King of Sicily, Germany, Italy and Jerusalem, in his thirteenth-century manuscript *The Art of Falconry*. His work stood virtually alone for 300 years until 1572 when Coitier, the Dutch father of avian anatomy, published his studies. Unlike Aristotle before him, Frederick insisted on seeing the evidence for himself. He reared children in isolation to investigate speech and he cut men open to observe the process of digestion. He wasn't always right; he thought birds had diaphragms, but his approach to falconry through natural science was a sound one which we can only update.

The aim here is to describe structure in relation to function, particularly those aspects of interest to falconers, breeders, and rehabilitators. It is based primarily on dissections of Common Buzzards *Buteo buteo,* which are available here as road casualties, but also on various other raptor species that are mentioned where they differ.

Raptors rely on flight for their survival and their skeleton, feathers, skin, and flight muscles make up about 60 percent of the payload (figure 1.2.1). The contribution of each part of the body toward survival is balanced against the cost of carrying it. As you work through this book, you will see how everything about a raptor, its structure, flying ability, and behaviour patterns, are honed to a peak of refinement by that moment of truth when a hawk actually hunts its prey. Most raptor populations are controlled through mortality, mainly at the stage when juveniles are trying to survive on their own. Those that survive to breed have come through an uncompromising test of their structure and function.

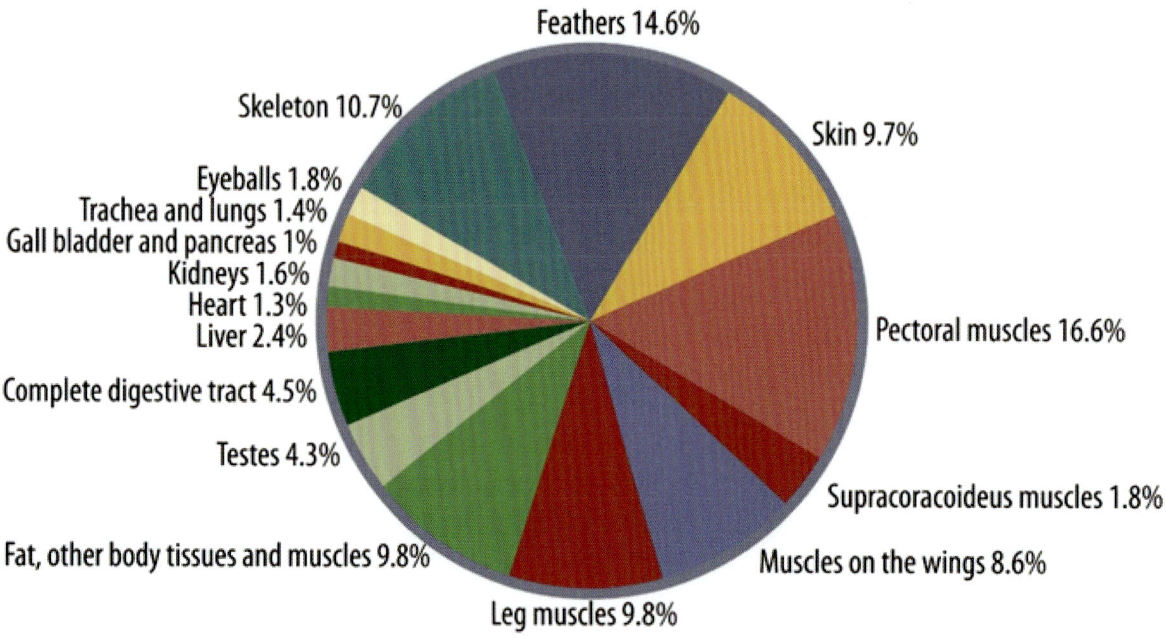

Figure 1.1.1 A 129 gram male European Sparrowhawk *Accipiter nisus*, showing the proportions of his body parts by weight.

1.2 Taxonomy

The taxonomy of diurnal birds of prey and owls is complicated and with continuing genetic studies the classification is constantly being tweaked.

Class Aves (Bird)

Sub-class Neornithes (Modern birds)

Order Falconiformes (Raptors)

Family Sagittariidae (Secretary Bird)
Family Pandionidae (Osprey)
Family Cathartidae (New World Vultures)
Family Falconidae (Falcons and Caracaras)
Family Accipitridae (Hawks, Eagles, Kites, Harriers and Old World Vultures)

Order Strigiformes (Owls)

Family Tytonidae (Barn Owls)
Family Strigidae (All other Owls)

Early taxonomists based their 'species' on what the birds looked like – their phenotypes. They looked at variations in plumage and morphology. Often working from museum skins, this method can become self-fulfilling. You put all the red ones in one pile and call them one species, and all the black ones in another pile and call them another species. This led to all sorts of inconsistencies. For example the juvenile New Zealand Falcons (*Falco novaeseelandiae*) which are dark brown, were called a different species from the adults, which are streaked. If the taxonomists had actually gone to a nest and seen the adults and young together, the misunderstanding would have been quickly resolved. Confusions like this continue to this day and each author copies the one before.

Various cladograms have been published postulating relationships between birds of prey. Michael Wink, using molecular nucleotides, placed Storks as more closely related to Falcons than are Accipiters, and Accipiters more closely related to Harriers than to Buzzards, Eagles and Kites.

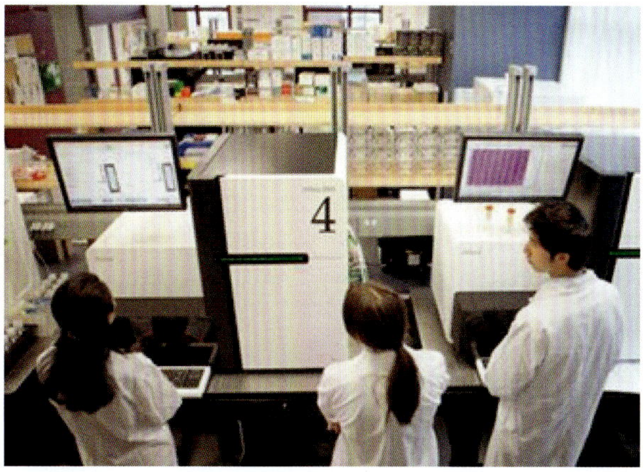

Analysing the peregrine genome at the Beijing Institute of Genetics.
Photo Andrew Dixon

Our genetic studies (Xiangjiang Zhan *et al*) show that Peregrines *Falco peregrinus* and Sakers *F. cherrug* diverged from one another about 2.1 million years ago. Nittinger*et al* and Wink *et al* concluded that the Hierofalcon group (Lanners *F. biarmicus*, Luggers *F. jugger*, Sakers, Gyrfalcons *F. rusticolus* and Black Falcons *F. subniger*) probably radiated only in the last interglacial period and, because they all interbreed freely (except for the Black falcon which hasn't been tried), are probably best regarded as a super-species *Falco hierofalco*.

We have analysed the genomes of the peregrine and saker falcons and since then more raptor species have been analysed by Cho *et al*. But in doing this, we stumble into a sampling problem. When we took the DNA sample for the peregrine of course we didn't take samples from every single peregrine in the whole world, or even from every 'sub-species', we just took it from one individual. If we could afford to sample as many birds as we wished, we could ripple out from the individual and include her siblings and parents, then all the peregrines in a population within a set area, then all the peregrines in the local so-called 'sub-species', and so on. It all starts to get extremely messy! At what points during this rippling out of your sample do you put a ring around a group and declare it to be a 'species'? Most of the species definitions, based on gene flow between groups, or on the ability to breed together, do not hold up in modern classifications. The four main groups within Hierofalco can interbreed indefinitely and probably continue to have some gene flow when individuals meet.

On top of this, what we conclude happened a million years ago, or ten thousand years ago, or even a hundred years ago, may well be overtaken by recent events. I remember Jerry Anderson, a British falconer living in Iran, showing me photos of *wacheris* there in the 1960s. They were essentially intermediate forms between the lanner to the west and the lugger to the east. Nowadays they are mostly all gone, as are the lanners of the Arabian peninsula, (together with the *jebelia*, the little mountain Red-naped Shaheens *Falco peregrinus pelegrinoides*), so it is just academic. The Altai falcon too was heavily trapped out in the 1980s and 1990s and whatever is there now has probably been genetically swamped by adjacent populations.

Thus although lawyers treat 'species' as if they are hard categories, biology is much more complicated and vague; we have many definitions. Falconry, which is a form of applied biology, needs to bear this firmly in mind especially in matters of taxonomy and genetics and how they relate to ethology. Species are often more like braided rivers, with strands separating and sometimes coming back together, rather than closed legal boxes.

Long before Karl von Linnaeus and Charles Darwin came on the scene, falconry had its own names for all sorts of different raptors going back into the mists of time. Delicious names like *bahari, nebli, chargh, jerudi, alethe, piqu, tarlan, baloban, alphanet, ossifrage, tagaroti, boreal,* and *turumti.* I will leave those alleyways for you to explore at your leisure.

Taxonomy places most emphasis on the phenotype - what the animal looks like - and this is simply an expression of its genotype. But remember the genotype also dictates each animal's *ethotype*, the suite of instincts it has, which is the starting point for its behaviour. We'll be looking at this in Chapters 4 and 6, but for now remember that although we tend to lump groups of raptors together, their genotypes may be very different and therefore their ethotypes have a very different starting point. Thus the accipiters are from a completely different family of birds from the falcons. Their nearest common ancestor was about 75 million years ago. Falcons are closer to crows and passerines than to hawks. Everything about them is fundamentally different. These deep differences can be seen for example in their bone structure, such as in their nares. or in their moult sequences. All the similarities we see are not due to close genetic relatedness but to convergent evolution. They look and behave alike because they are performing similar roles. We can assume that their ethotypes, or suites of instincts, have many parallels with each other wherever their ecology demands it, just as their phenotypes do. But there may be differences too. Accipiters have notoriously hair-trigger responses. They need this for the kind of fast, violent attacks they do, often in wooded cover. They are also very spooky and fearful. They are as 'wild as hawks'. Even hard imprints can revert to wildness if left out for a few days. Thus their fear response and imprinting may be slightly different from the falcons. Perhaps their fearfulness has more of a genetic component which can be masked by intensive early imprinting or later habituation but more easily reasserts itself if not constantly reinforced. By and large they need more prolonged and closer contact with humans than falcons do. Some of these hard-to-get-at aspects can be traced right back to fundamental taxonomic differences.

Accipiters and falcons are probably more closely related to chickens than to each other. But you can train chickens too. This old re-habbed battery hen in 1974 became an excellent mouser!

1.3 The skeleton

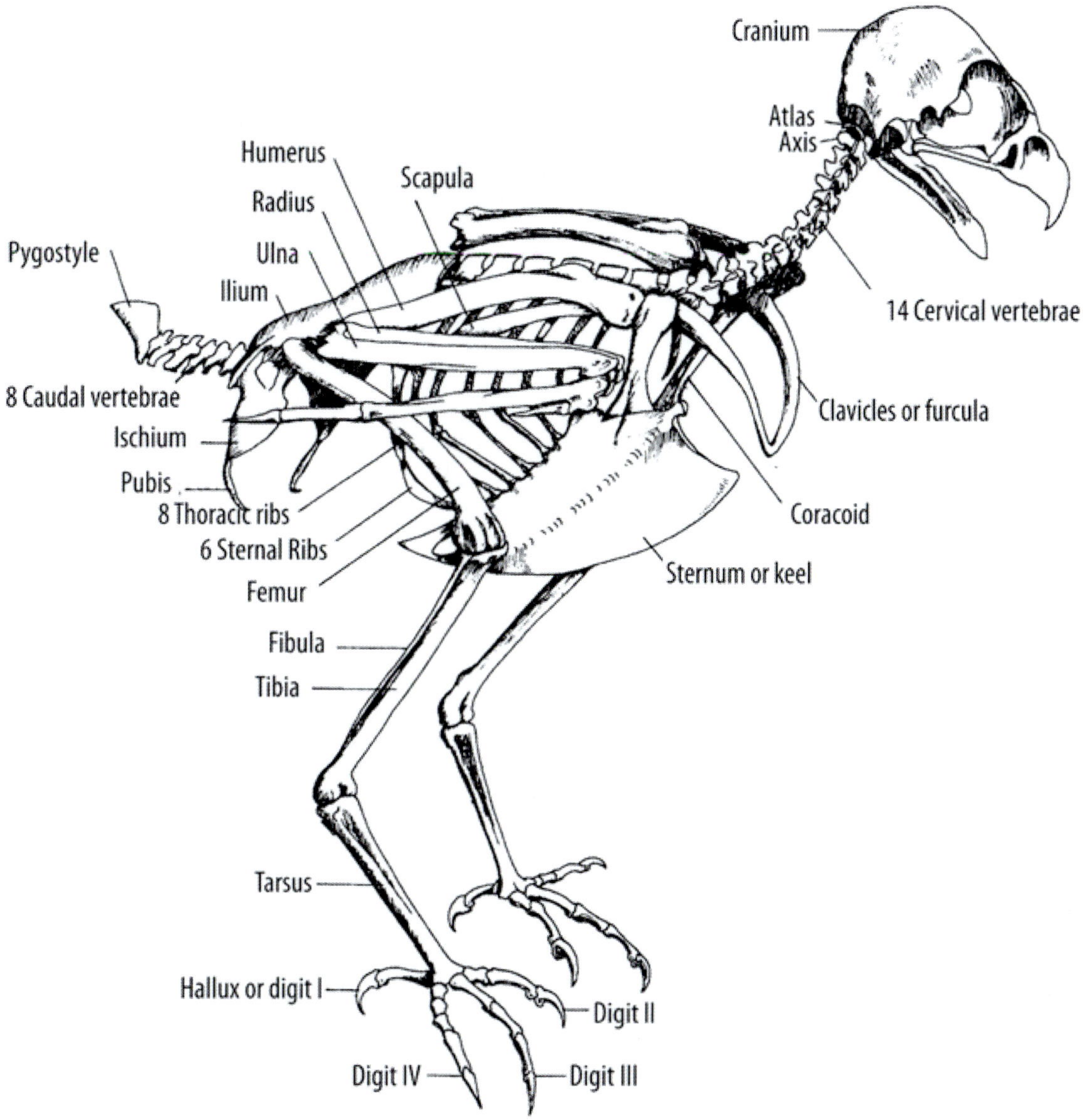

Figure 1.3.1 The skeleton

The demands of flight have stripped down the hawk skeleton to combine lightness with strength. Most of the bones are hollow and filled with light spongy bone or internal cross-braces and marrow. The pelvis, or synsacrum, and the back, or notarium, are both fused structures separated by one free vertebra in the midback. This helps give rigidity to the bird's body. The neck is very flexible allowing free movement for the head and compensating for the rigidity of the body and the relative immobility of the eyeballs.

The lower part of the body is made up of the large, shield-like breastbone or sternum. This, with its deep keel and the fused clavicles or wishbone below the crop supports the powerful pectoral muscles which drive the wings. To prevent the sternum from pulling up and crushing the head there are two stout braces, the coracoids, which run up to the shoulders. Being so exposed to impact, the clavicles can often be broken in head-on collisions. When this happens the hawk droops the wing on the injured side and the pectoral muscles may waste slightly on that side.

Not only are some of the bones air-filled, but many of them are pneumatic or connected to the system of air sacs (see section 1.6). Should the hawk break its upper wing it can actually breathe through the broken bone. The lower leg bones

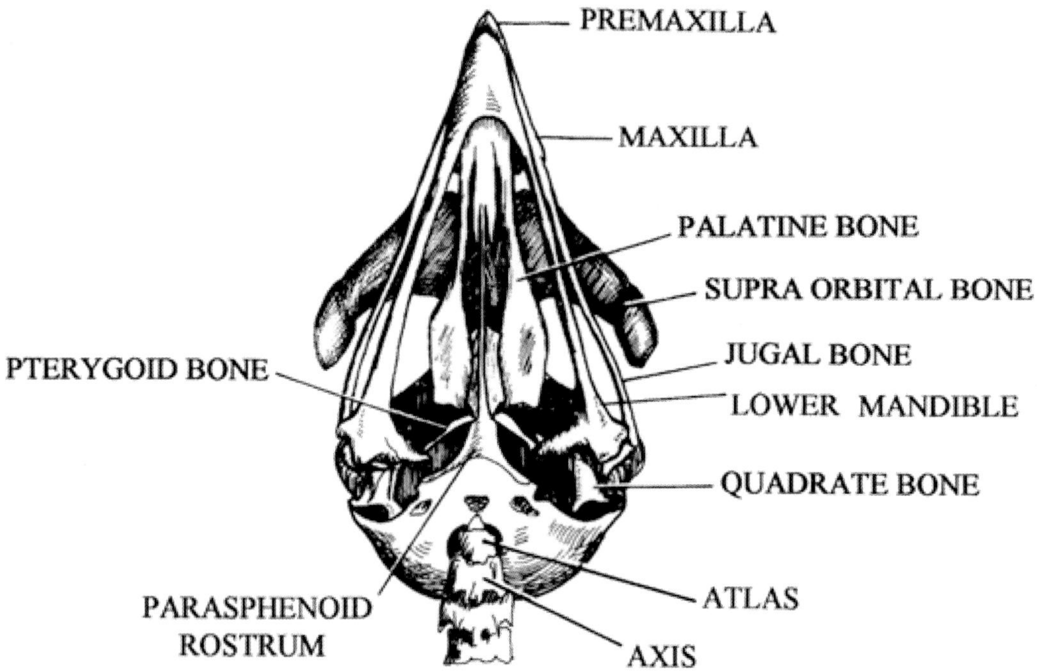

Figure 1.3.2. The skull of the common buzzard from below.

contain bone marrow, which makes red blood corpuscles.

The limb bones are thin-walled tubes and the delicate internal cross-braces, which give them their girder-like rigidity, are aligned with the forces exerted on the bone in response to the forces on the whole animal. Thus if a growing hawk is not given the opportunity to exercise adequately in its first months of life the cross-braces will be randomly oriented and the limb bones will be weak.

The raptor skull (figures 1.3.2 and 1.3.3) is highly adapted and is much lighter than those of the archosaurian reptiles from which it evolved. The beak is light and covered with horny keratin. In the falcons and falconets there is an extra notch called the tomial tooth just behind the tip of the beak. This is thought to aid in breaking the necks of prey, the normal falcon method of killing. The beak grows down from the waxy cere, which is usually gray or yellow and is sensitive to the touch. Inside the outer horny layer of the beak is the main growing zone, which is well supplied with blood vessels. Below this is the bony core, the premaxilla. Should the beak become overgrown or damaged (figure 1.3.3) it is a relatively simple matter to 'cope' or clip the tip and reshape the sides with a sharp knife, rat-tail file, emery board or Dremel. Splitting or flaking at the side of the beak is best treated by trimming back any loose snags and then cutting back the tip of the beak as far as is safe. This reduces the leverage on the split. If the hawk is provided with plenty of bony joints or tirings to pull at, and some abrasive surfaces to feak or wipe her beak on, it will stay in good order. The beak should never be allowed to get too overgrown otherwise the 'quick' or soft interior will extend, preventing reshaping.

The nostrils or nares are circular in the falcons, with a central pin or tubercle. The rear portion of the aperture is shielded by a thin plate of bone. In accipiters, buteos, eagles, and owls, the nostrils are oval with a much-reduced tubercle.

The eyes are large, occupying about two thirds of the skull and almost touching each other inside the head. They are protected and supported by a ring of bony sclerotic plates embedded around the rim of each eyeball and by a supra-orbital ridge forming a tough bone and cartilage eyebrow, shading the eye and giving the hawk its fierce appearance. Behind this ridge, toward the centre of the skull, the eyeball lies unprotected immediately below the skin and is vulnerable to damage by thorns or barbed wire.

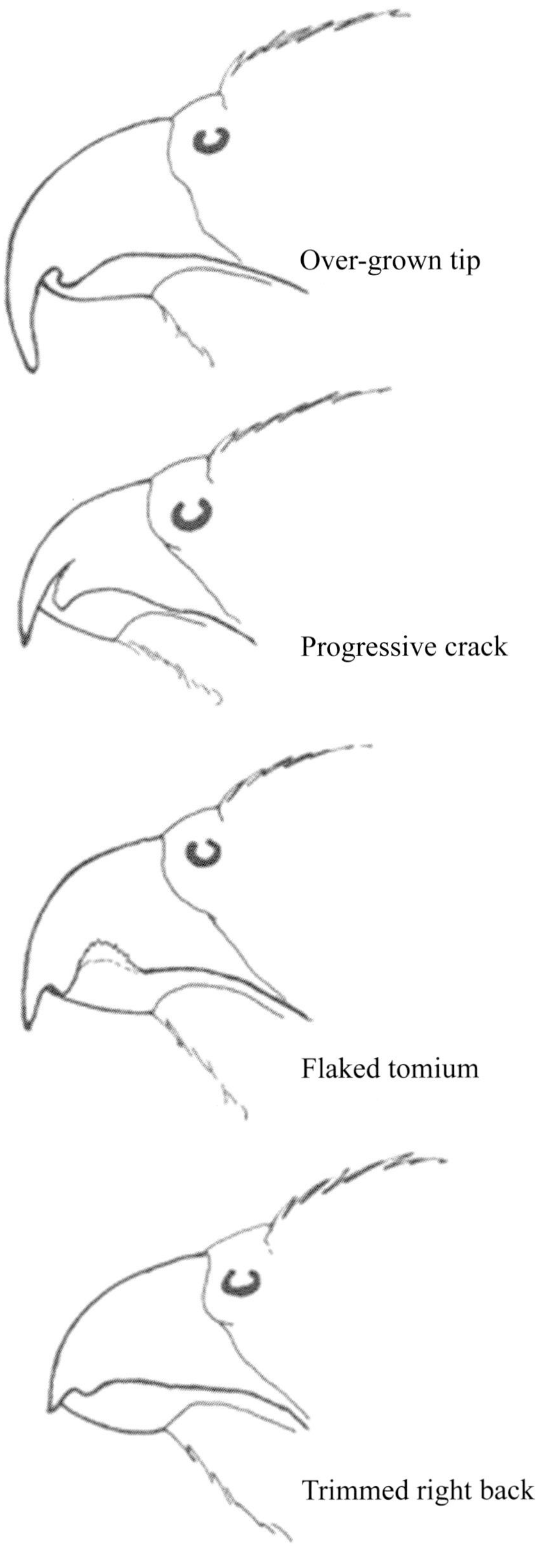

Figure 1.3.3. Coping falcon beaks.

The jaws are double hinged like those of snakes. When the hawk opens its mouth, two small quadrate bones, one in each angle of the jaw, swing forward two or three millimetres (figure 1.3.4). The lower jaw hinges down from the quadrate bones and the upper jaw has a flexible bony hinge behind the cere. As the quadrate bones swing forward, they push the jugal arch or cheekbones forward and these push the beak upward. At the same time, two small rodlike pterygoid bones inside the throat are also pushed forward by the quadrate bones and they in turn push forward the palatine bones lining the roof of the mouth. The palatine bones are fused to the interior of the beak. Thus, when the hawk opens its mouth, the beak, the cheeks, entire palate, and lower jaw move forward and outward. In some of the owls, such as the Barn Owl *Tyto alba*, opening the mouth also extends the jaw outward on each side so that the gape increases by about six millimetres, enabling prey to be swallowed whole.

For the falconer the significance of these bone movements can be seen when fitting the hood. As the cheek bones and lower jaw move forward when the mouth opens, they act as wedges. A leather mouthpiece that crosses them will inevitably either rub the delicate skin that covers them or will let in light. The only way to fit the hood so that the hawk can open its mouth sufficiently to cast, and which is light tight and nonabrasive, is to cut the mouthpiece so that it seals the eyes at the lores (the soft skin in front of the eyes) and then follows just above the jugals to the rictus or corners of the mouth (see figure 3.9.1). Meanwhile, the bulge in the back of the hood caused by the braces, fits snugly under the occiput at the back of the skull, preventing the hood fitting from being hooked off. At the sides, the hood also curves inward sufficiently under the quadrates to prevent removal. In this way the hood can be locked on and yet gives very little restriction to the gape.

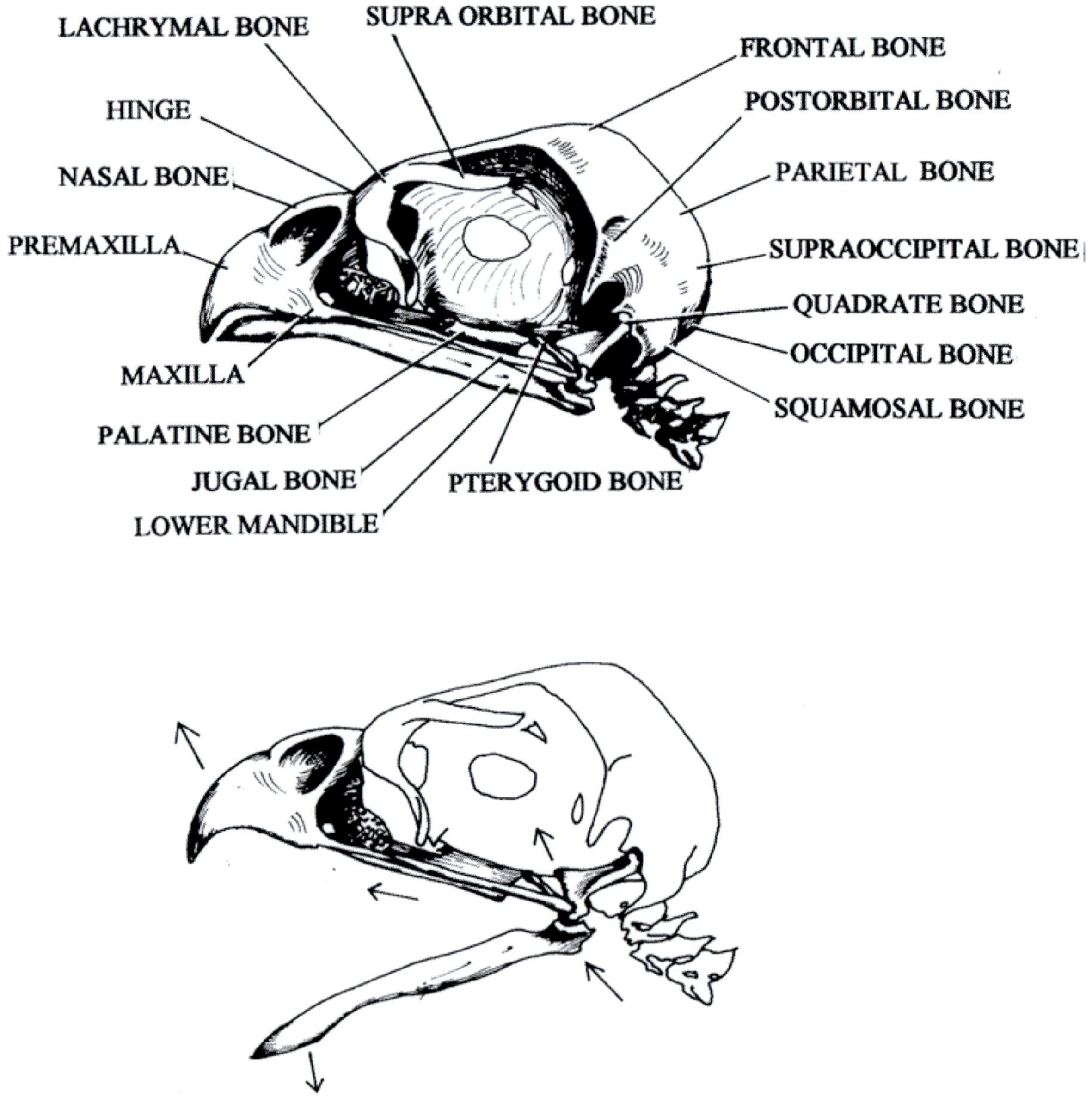

Figure 1.3.4. The skull of the common buzzard, showing the movement of the jaws.

1.4 The senses

'He has eyes like a hawk.' We all know that hawks can see well, but what are their eyes really like and do they even live in the same visual world as us?

The eyesight of birds of prey is very finely tuned and evolved for their lifestyle. Eyesight is critical to high speed hunting. Like humans and related species, but unlike most mammals, birds see in colour as well as in black and white. The light-sensitive retina at the back of the eye contains colour-sensitive cones as well as tone-

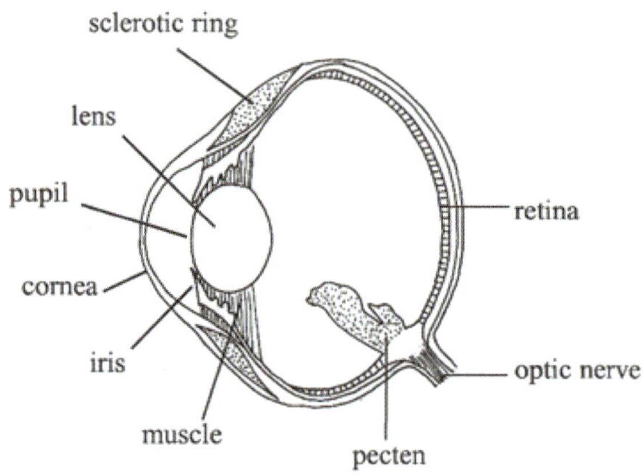

Figure 1.4.1. The anatomy of the eye.

sensitive rods for vision in poor light. They also have a wider spectrum of colours than ours. While we have only one part of the visual field in focus, the buzzard has two focal spots or foveas in each eye and these are joined by a horizontal strip of extra sensitive area (figure 1.4.2). One of these foveas, the central or 'deep' fovea, faces outward and is adapted for detecting small movements. Sharp vision is energy expensive; only about 1.5° of the field of vision provides maximum detail. When a hawk cocks her head slightly to examine the sky, she is using one of her deep foveas (figure 1.4.3). The other, temporal, fovea faces forward and is best at resolving detail, in conjunction with binocular vision. The horizontal strip is used for scanning. Even outside the foveas, the buzzard has visual acuity about twice ours. Using its temporal foveas and binocular vision, together with head bobbing, the buzzard can build up a detailed stereoscopic image and a good perception of distance.

The red-tail *Buteo jamaicensis* has a binocular vision of 33°, Cooper's hawk *Accipiter cooperii* 36°, the saker 37° and the Harris hawk 47°. The Cooper's best binocular vision is above the plane of the bill, an adaptation for attacking from below, whereas the redtail's best binocular vision points more downwards. Searching raptors such as the Short-toed Eagle *Circaetus gallicus* are better adapted for scanning, having only about 20° of binocular vision. Searching raptors also tend to have less eyeball mobility: American kestrels *Falco sparverius* can move their eyes only 2°

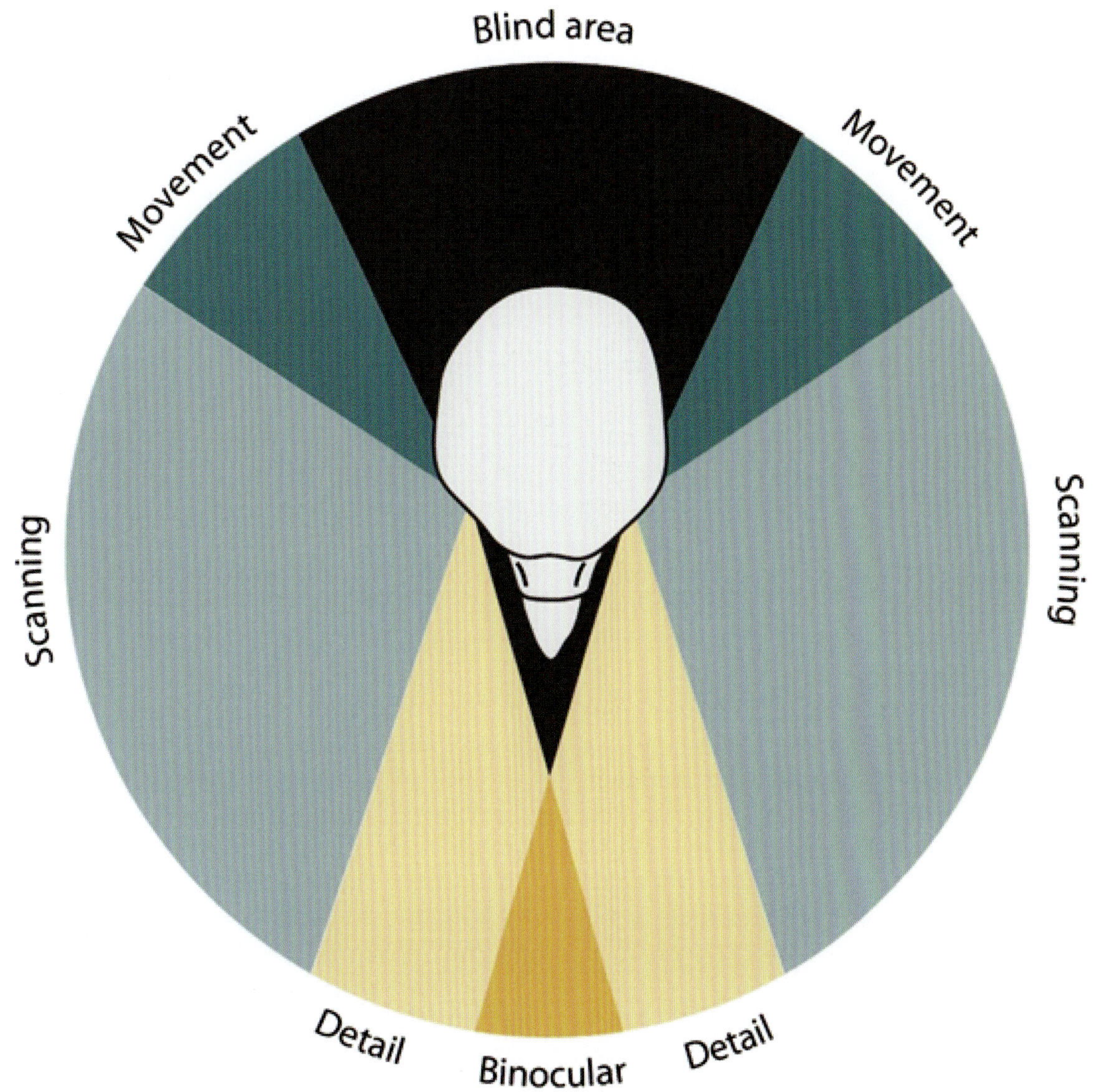

Figure 1.4.2. The field of vision of a common buzzard.

Figure 1.4.3. This saker has noticed an eagle. She cocks her head to use her central fovea.

whereas Coopers and redtails can move 10-12°. Owls, which have primarily binocular vision and immobile tubular bony eyeballs, have little lateral vision and no central foveas.

At the top of each cone cell is an oil droplet - usually red, orange, yellow, clear, or green. These heighten the contrast of coloured objects and also act as haze filters. Protruding into the eye from the retina is the pecten, a large, elongated body well-supplied with blood vessels. It is thought to supply nutrients to and remove waste from the retina by diffusion across the eye fluids. It may also act as a movement detector by casting a shadow across the retina.

In contrast to mammals, most of the muscles controlling the hawk's lens and pupil are striated muscle and can be contracted voluntarily. Thus a peregrine, unhooded into bright sunlight, can promptly adjust her pupils to avoid being dazzled. Similarly, she can fly well toward the sun. To accommodate close focusing, the pupil and the Crampton's muscle around it, are able to squeeze the lens and cornea and this can be seen in accipiters. When they examine very near objects, such as food in their feet, their pupils shrink noticeably.

The third eyelid, or nictitating membrane, is translucent or transparent (some waterbirds use them as spectacles for focusing under water). The membrane cleans and protects the eye and the hawk seldom uses her other eyelids during the daytime. When she grapples with prey, her third eyelids constantly flick across to protect her eyes from sharp feathers or thorns. She cleans her eyes by rubbing them on her scapulas; it would be too risky to use claws for this, as she does for scratching the rest of her head. Most raptors have prominent eye ridges, providing protection during collisions and acting as sun shades.

Our human vision, at 30 cycles/degree, is about twice as sharp as a dog's, at 12 c/d. A peregrine's visual acuity is up to about 140 c/d, around five times better than ours. Many wild falcons hunt well after dusk, and some hunt in moonlight or by streetlights. They have better discrimination in low light than accipitrid raptors. Accipiters sleep soundly and I have caught sparrowhawks in our barn at roost using a small torch. But visual acuity depends on many factors: light levels, depth of field, distance, movement, and latency; it is a complex feature.

Pigeons can detect the polarization plane of light and use this for navigation. It is likely that raptors too can detect the polarization of light and thus sunlight, the sky, and reflections to them show polarization patterns.

Raptors can see things that are much closer together in time than we can. For example, the old cine film is designed for human vision. It flickers at 25 frames per second which the human eye translates as one moving picture, because we can only see up to 20 events per second. This is called the Flicker Fusion Frequency or FFF. Raptors have FFFs of about 120-150 Hz, or events per second, and thus have difficulty in interpreting cine film as pictures. Raptors which rely on high-speed chases and have to avoid branches or catch agile prey, need a high FFF. A dragonfly, for example, has an FFF of up to 300 events per second with a high-speed brain and reactions to match. A Hobby *Falco subbuteo,* trying to catch one, needs similar fast reactions otherwise it would have no chance. These very fast sensory systems need very fast nervous systems to work them and in accipiters, for example, it is likely that some of these nerve pathways from the eyes and ears, via sensory neurons, to the motor neurons which control the muscles, have only minor links with association neurons in the brain. Given the time taken for impulses to traverse the nervous system and the latency periods between impulses, the accipiter would not have much time to 'think' consciously about its reaction to ensure that it does the appropriate thing. We thus say that the bird is 'nervous;' it seems to overreact in the wrong way to harmless stimuli. We can train out some of this nervousness but we cannot reconstruct the bird's entire sensory and nervous systems. It is these fast reactions that make it such an efficient predator. Thus it is actually living in a world moving about 5-6 times faster than ours and it is important to remember that although physically we may be standing side by side, mentally we are in different worlds. If you try piloting a robotic prey you will soon notice that your human brain works at an inferior speed compared to the raptor chasing you.

The same high-speed nervous system, with fast impulses, and short refractory times, which serves the hawk's eyes, also supports its hearing. The hearing of birds is better than ours. A seabird can arrive back at a nesting colony and still identify its own chick by its calls, through a cacophony of similar birds calling. Human hearing ranges from infrasound at 20 Hz to ultrasound at 20,000Hz. Birds hear over a similar range but peak in the 1000 - 4000Hz range and can discriminate between closer frequencies than we can. They have only one ear ossicle, the *columella auris*. A thick tegmentum or fold of tissue rapidly dampens vibrations in the eardrum enabling birds to detect sounds only 0.6 - 2.5 milliseconds apart. Their FFF for hearing is thus 2,500 - 6,000. We have to slow down the speed of bird calls to appreciate their complexity. Hawks can easily recognize one another by their individual voices even though they may look the same. Parent falcons can pick out their offspring and their mates from among similar falcons even after some weeks separation, and show it in no uncertain terms! Hawks can readily discriminate between different footfalls and different car engines. Just as with eyesight, raptors live in a completely different sound world to us. Forest and nocturnal raptors are particularly receptive to sounds, as are the harriers. The soft, silent plumage of owls is probably as much an adaptation to enable the owl to hear the prey as it is to prevent the prey hearing the owl, and of course their asymmetrical ears allow a high degree of directionality.

It is likely that the nervous system works at different speeds according to hormonal influence. The drowsy animal reacts only slowly, whereas when the animal is under the influence of adrenalin, its reactions speed up and its FFF for both vision and hearing markedly increase, as if it is experiencing events in slow motion. This neurohormonal response curve would enable the hawk to match the high metabolic cost of achieving a high FFF, with its immediate needs.

The ability of a hawk to hold its head still when on a moving perch or hovering is not just done through eyesight. If a hawk is hooded and placed on a rotating perch it is capable of holding its head stationary. The organs of balance are well developed in birds; imagine standing on one foot on a waving branch at night in a storm with your head tucked under your shoulder, fast asleep!

It used to be thought that most birds, with a few notable exceptions such as the kiwis and the turkey vulture, cannot smell. But this is probably an oversimplification. It is true that few can detect airborne scents and the olfactory lobes of the brain that deal with smell are small, but examination of the nostrils does show scent-detecting cells in many species. Each nostril has three cavities or conchae separated by delicate bony baffles like the interior of a small snail shell. The first cavity acts as a baffle and thermoregulator. The second is covered with mucus-producing ciliated cells and warms and moistens the air. Only the third cavity contains scent-detecting cells. This cavity opens directly into the roof of the mouth and it is thus possible the bird is able to smell, as well as taste, the food actually in its mouth. I have noticed on several occasions various hawks pick up and reject meat while holding it only by the horny tips of the beak where it seems no taste buds could be functioning.

The third cavity also opens into the infraorbital sinus. This sinus, which lies in front of and slightly below the eye, links up with the sinus on the opposite side and also with various small cavities throughout the skull. An infection here of sinusitis may infect one or both sides and can be difficult to eradicate.

The nasal glands are a flattened ovoid in falcons and lie half in the orbit and half extending into the orbital sinuses. In accipiters they are thinner and more elongated, lying tightly against the front of the orbit. In kites and buteos the glands lie in a depression in the frontal bone.

Birds of prey produce a salty solution from the nares thought to be secreted by the nasal glands in much the same way that some marine birds use salt glands. This is an extra-renal route for the excretion of excess sodium and chlorine ions. Although there is some experimental evidence supporting this, it does not explain why the nasal secretions should coincide with eating.

The nasal secretions appear to be more of an overflow of fluids from the buccal cavity, which are sneezed away when they block the nares. Females feeding young chicks drool large quantities of liquid onto the food for the chick to ingest and it is not clear if this fluid is salivary or nasal, or both, in origin. Research by Claus Fentzloff has revealed high levels of calcium in this liquid, which presumably is of benefit to young chicks.

The central tubercle in the nostril of falcons remains a puzzle. Kessel *et al* built a 3D model

of a peregrine nostril and results indicate that the tubercle has a dampening effect on air flow. For a falcon needing to breath during a stoop, the headwind air pressure could otherwise blow up the entire air sac system. Yet other species that lack tubercles, such as eagles, can travel just as fast. In the dynamic gliders such as the albatrosses, petrels, and fulmars, part of the nostril is thought to act as air velocity sense organs; these species are also able to smell air scent. In the falcons, the second nasal chambers on each side interconnect. Should air strike one nostril with more pressure than the other, then air should pass between the nostrils through the passageway. Whether or not this passageway is lined with sensory cells awaits further investigation.

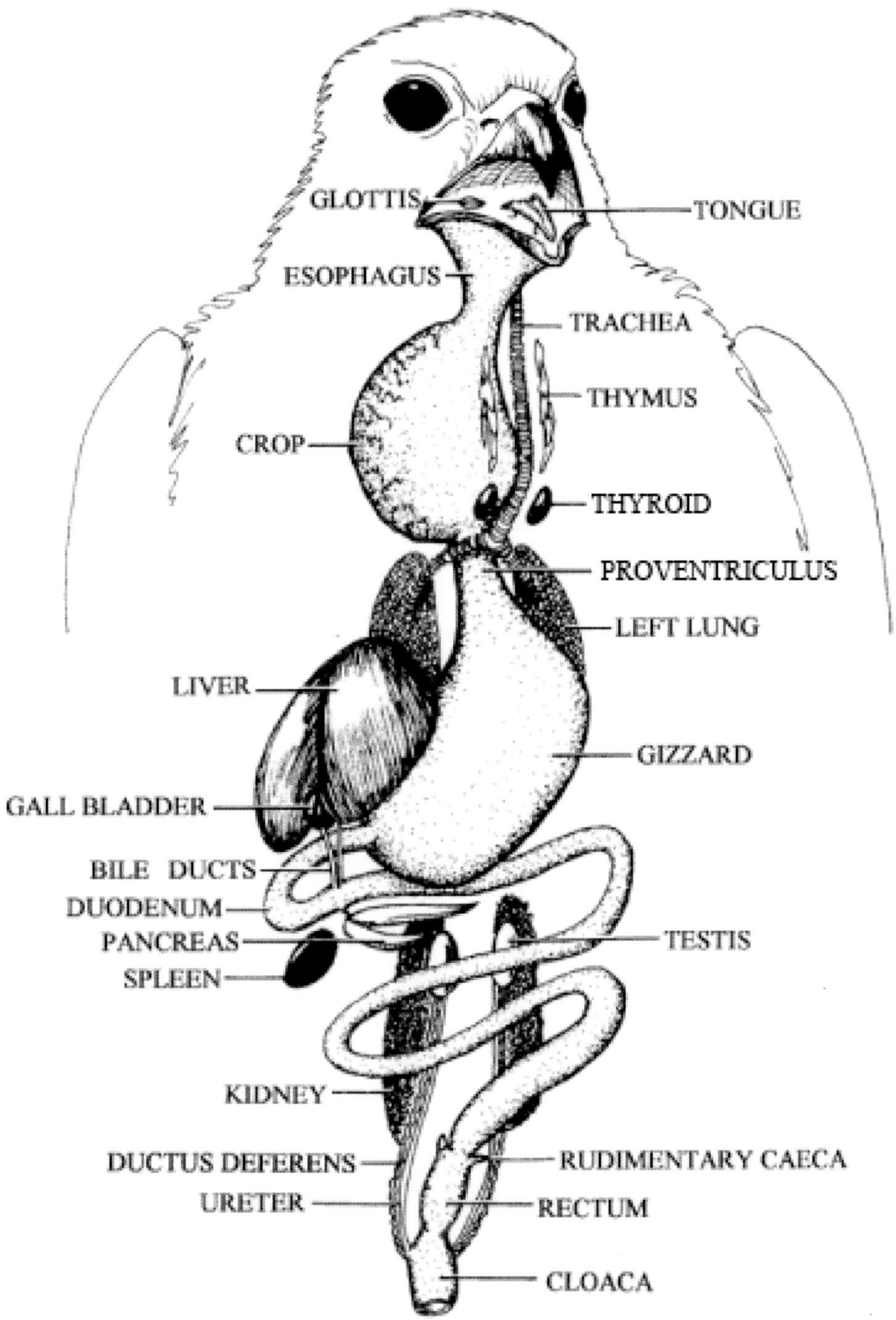

Figure 1.5.1. The digestive system of the common buzzard.

1.5 The digestive system

The back of the tongue and the palate have backward-pointing spines, or papillae, which help the hawk gulp down food. As the beak closes, the entire upper palate moves backward, pushing the food before it. The esophagus is slightly muscular and opens into the crop, which is extremely elastic. It lies to the right side of the windpipe, or trachea, and is therefore asymmetrical, although this does not become obvious until the crop is almost full. No digestion, apart from that caused by saliva, occurs in the crop; it is mainly a storage organ. Bit by bit the food is 'put over' into the proventriculus or glandular stomach. To do this the hawk first stretches up its head and then pushes it down chin first in a flattening movement, often wriggling its head far to one side (figure 1.5.2). It all looks a little alarming but the hawk is perfectly well. Owls have no crop; instead the food passes directly to the stomach.

Figure 1.5.2. 'Putting over' the crop.

In birds of prey the muscular stomach, or gizzard, is relatively thin walled. Unlike chickens, hawks do not need to grind up hard seeds. The proventriculus and the gizzard thus merge together into a pear-shaped organ. The walls of the proventriculus are lined with epithelial cells that secrete strong peptic enzymes and hydrochloric acid, making a pH of 2.0 - 3.5. Diurnal raptors can digest most of the bones of their prey although keratinous structures, such as beaks, claws, and feathers, tend not to be affected. Owls are less able to dissolve bones and their pellets often contain near-complete skeletons.

The gizzard is lined with a one-millimetre-thick, slightly cheesy, yellow layer called the cuticle or koilin which is secreted by a layer of mucus cells. The cuticle protects the gizzard from the acid secretions of the proventriculus and provided that the bird has a natural diet the cuticle is constantly worn away, or sometimes is shed completely (see section 6.5). The outlet of the gizzard contains a grooved valve that prevents indigestible food going further. The soft parts of the prey are leached out by the chemical secretions and travel on down the intestine for further digestion and absorption. The feathers and fur are slowly compacted by the contractions of the gizzard to form a pellet. These contractions occur every 20 - 30 seconds and in the common buzzard have a pressure of 8 - 26 millimetres of mercury. When the gizzard is almost full of unwanted matter the pellet or pellets are regurgitated. Normally this occurs before midmorning, either at the night perch or sometimes after killing, but usually before feeding on the first kill of the day. Captive hawks can often be induced to cast an hour or two early by showing them food.

The liver is large in birds of prey. The liver and gall bladder of a buzzard together weigh about 20 grams - 2.5 percent of its total body weight. It is in two lobes overlying the gizzard, protected by the pectoral muscles and sternum and cushioned from impact by the thoracic airsacs that surround it. The liver stores excess sugars, fats, and vitamins (especially vitamin A), synthesizes certain proteins, makes bile which is stored in the gall bladder, and excretes waste products from the blood, detoxifying it.

The gall bladder has two ducts which empty into the duodenum about 20 centimetres beyond the gizzard, about one-sixth of the way along the intestine. Bile is normally secreted continuously at about 0.5 - 1 milliliters per hour and its job is to emulsify fats. Although normally important in carnivorous birds, the gall bladder was absent in several peregrines recorded by Gorham and Ivy. The bile pigment, biliverdin, stains the mutes green when the gut is empty; it can also pass back up, staining the castings of an empty hawk.

The pancreas lies in a loop of the duodenum and has one to three ducts into the intestine. Its enzymes digest fats, proteins, and carbohydrates that are then absorbed in the intestine. The pancreatic secretions are alkaline to neutralize the stomach acids.

Digestion and absorption are the chief functions of the intestine and the food is carried along by peristaltic waves. The walls are lined with tiny finger-like villi to increase the surface area for absorption. The intestine is often host to parasitic infections such as coccidia, roundworms, and tapeworms which, although relatively harmless at low levels, can, at higher densities, ulcerate the delicate lining and lead to anaemia and expose the bird to secondary infections. During the first few days of life the yolk sac feeds via a duct into the small intestine between the jejunum and the ileum. Once the yolk is used up, the duct atrophies, leaving a small papilla, the vitelline diverticulum.

Gamebirds have a pair of large blind-ended tubes, the caeca, coming off the lower portion of the intestine at the sphincter with the rectum. These are used for the bacterial decomposition of crude fibrous foods, but raptors, which have a high protein diet, have only rudimentary caeca about four millimetres long. These usually contain lymphoid tissue.

Unlike mammals, birds have only one urinogenital opening, the cloaca. The upper part of the cloaca, the coprodeum, receives faeces from the intestine; this faecal material is black or brown. The middle part of the cloaca, the urodeum, receives semi-solid urine from the kidneys via the two ureters which constantly move the urine down by peristalsis; the urine is extremely concentrated and also contains insoluble white uric acid crystals. Urine in the urodeum is moved by retroperistalsis back up into the rectum where more water is reabsorbed. Urine and faeces are then excreted from the cloaca together, the coprodeum being semi-everted so that faeces do not come into contact with the urodeum and the proctodeum. On the dorsal or upper surface of the proctodeum is a small lymphatic pocket, the cloacal bursa or Bursa of Fabricius. This helps to combat infections, particularly when the bird is young.

The length of the gut, from mouth to cloaca, is about 135 centimetres in the buzzard. The rates at which food travels along it are discussed in section 6.3. Work by Dr Nigel Barton has shown how the 'attacking hawks' (8.1), which rely mostly on agile prey, have shorter, lightweight, less efficient guts than the 'searching hawks,' such as the buzzards, which are more efficient at digesting their food. Thus some accipiters have gut lengths only half as long as buteos of the same size, and need more food, but digest it more quickly.

The gut itself hosts huge quantities of bacteria, the 'biome', some of which aid digestion. In herbivorous birds some of these bacteria break down cellulose releasing the plant cell contents, but in raptors their role is still little researched. The spectrum of bacteria is important and the question is – how do they get there in the first place? When the mother feeds the young chicks she drools saliva which helps make the meat digestible and presumably contains some bacteria, but not necessarily the same ones found lower in the gut. Fecal material of course contains huge quantities of bacteria and studies on other species have shown benefits in 'inoculating' young animals with limited quantities of faeces. Faeces from prey animals do not necessarily contain the same bacteria as needed by the predator. Probiotics and live yoghurt may be useful but it is unknown if they have bacteria suited to birds of prey. Antibiotics may destroy many of the gut bacteria, leading to an unbalanced biome which needs time to recover. The biome that a young chick first establishes may set the pattern for the rest of its life, rather like a newly planted wood or pond, the species that get in first can predominate for a long time. Does a strong hawk with a good appetite and immune system have a good biome that she can pass on to her progeny? Or is it mainly chance how the biome develops? This is an exciting and potentially rewarding area for future research.

1.6 The respiratory system

The respiratory system in birds has two main functions—to exchange carbon dioxide for oxygen and to regulate body temperature. Whereas we humans have a rather inefficient tidal flow of air in and out of our bag-like lungs, resulting in a permanent stagnant residue, birds have a unique circular flow system so that air goes in at one end of the lungs and through to the other, rather like a radiator. In addition, birds have a complex system of airsacs that extend throughout the body cavity and, in many birds, even into the hollows of the main bones and vertebrae (figure 1.6.1). Birds do not have a large moveable diaphragm. Instead, the intercostal muscles between the ribs raise and lower the sternum, pumping the airsacs like bellows. This is often synchronized with wing beats during flight. In many species, breathing and wing beating are in time during cruising flight. The action of the wings beating helps the movement of pulling air through the lungs. As the bird needs more oxygen, it doesn't necessarily breath faster, it breathes more deeply.

The airsacs themselves are very like flimsy polythene bags, they cannot be used for gas exchange (the lungs have to do this) but as they ramify so widely throughout the bird's body they are extremely effective in the transfer of heat, particularly in cooling the bird during hard exertion.

As the bird expands its thorax by lowering the sternum, air is first drawn into the lower airsacs, the posterior thoracic, and the abdominal airsacs.

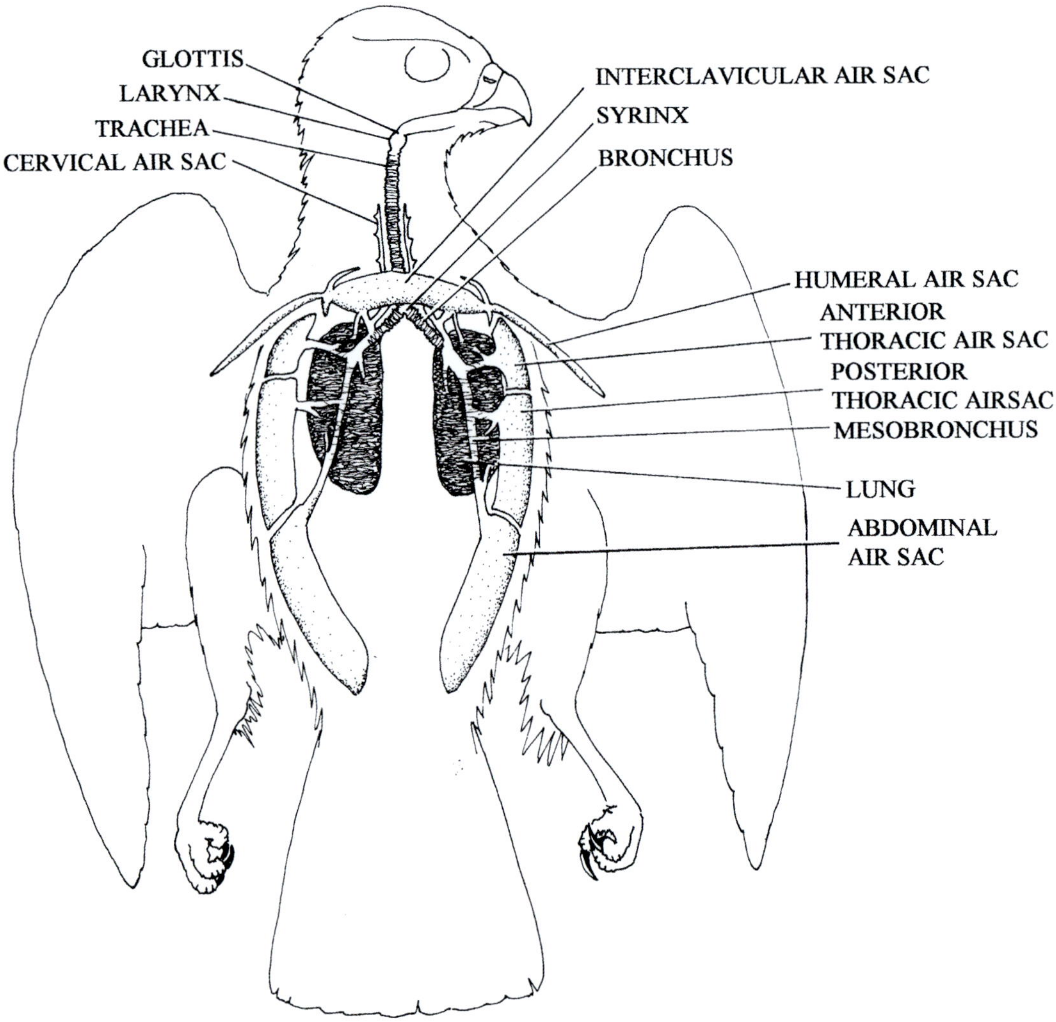

Figure 1.6.1. The respiratory system of the common buzzard.

When the sternum is raised, the air is then squeezed forward into the lungs. The next time the chest is expanded, air is drawn out of the lungs into the forward airsacs. Then, when the sternum is raised for the second time, the air is expelled back up the trachea and out. So it takes two complete breathing movements of the chest for any given 'piece' of air to go through the whole system and

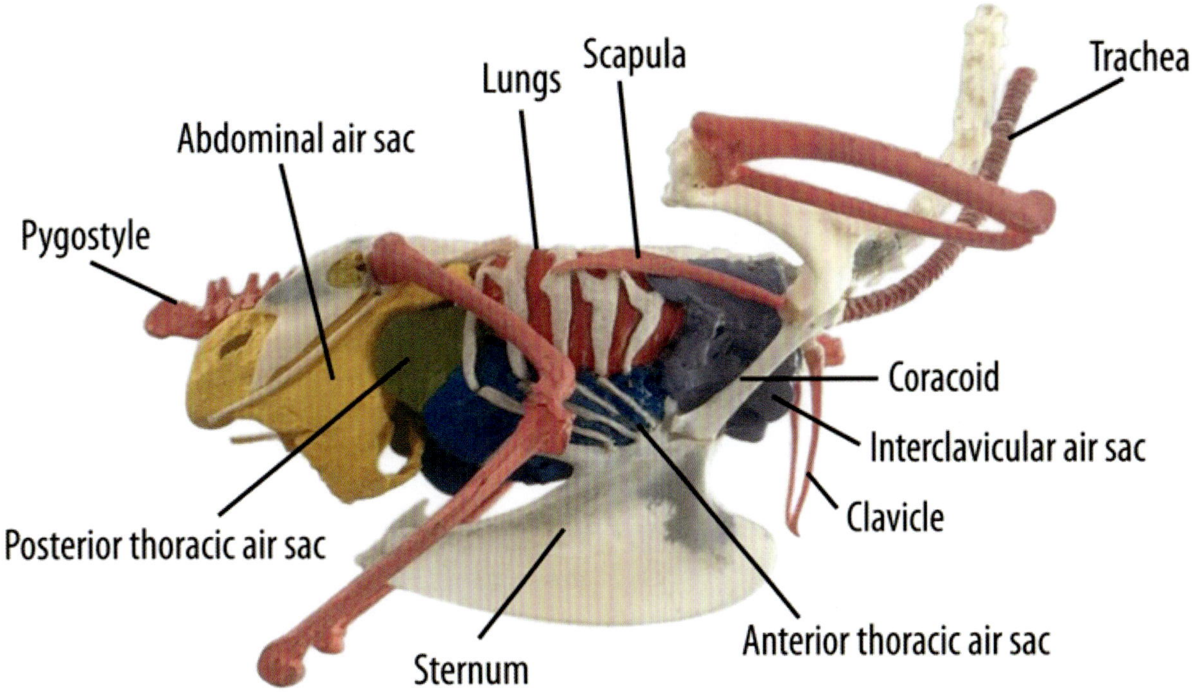

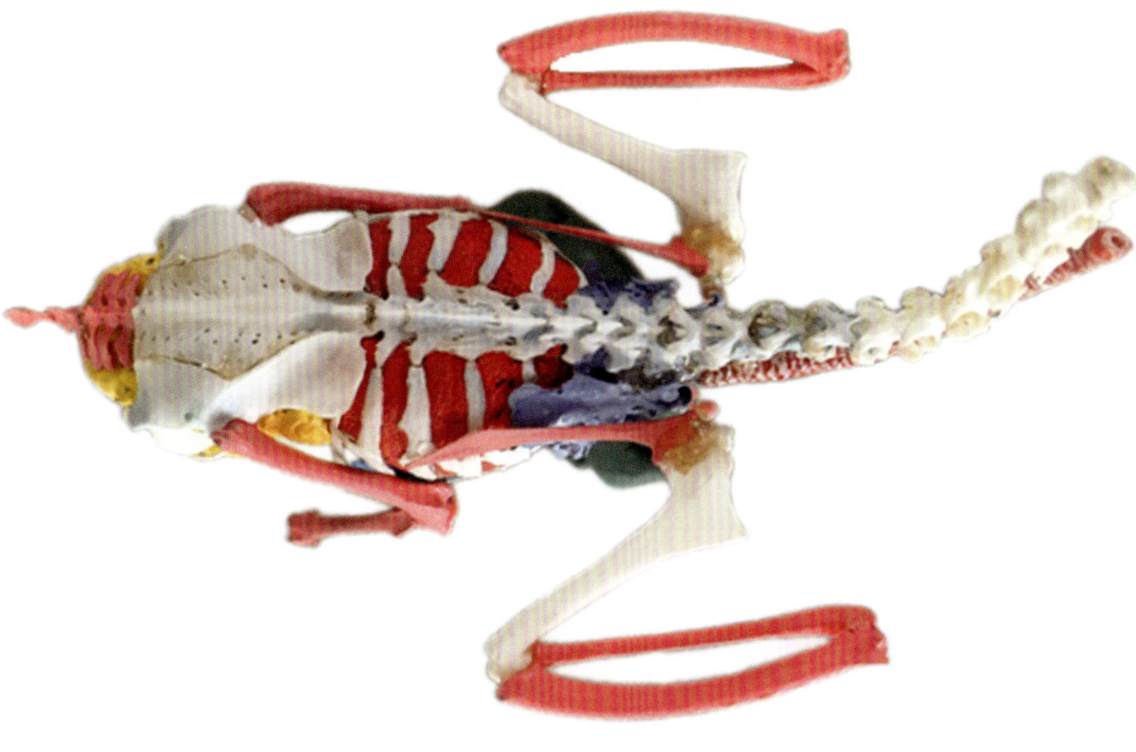

Figure 1.6.2 In this preparation I have infused the airsacs with latex, then dissolved the soft tissues away. The bones coloured pink contain bone marrow, but the humeri, pelvis and even some of the vertebrae (in white) link up to the respiratory system.

thus two 'pieces' are in transit at any one time. Because of this, when the bird is expanding its chest, it is in fact emptying its lungs, unlike us. Only when it lowers its chest again is air pushed forward into the lungs. However, because the airsacs buffer the sharp differences in air pressure, the overall effect is that air suffuses more or less continuously through the lungs.

Birds live life at a faster pace than we do. Flight requires a higher body temperature and a faster metabolism. So birds require proportionately more oxygen. But because bird blood haemoglobin is about twice as efficient as ours at shuttling oxygen and carbon dioxide, and because of their free-flowing breathing system, a buzzard needs to breathe at only 15 - 30 breaths per minute; half the rate of a mammal of the same weight.

Birds cannot sweat because they have no sweat glands. A peregrine dripping with sweat after a hard flight would not only look dreadful but would scarcely be able to fly at all, just like one that has bathed. Despite this, birds do lose some water by passive evaporation from the skin. When they get too hot, they mainly lose heat by the evaporation of water from the internal surfaces of the airsacs and panting, or in owls, gular fluttering, further increases the rate of evaporation. Overheating is one of the main limiting factors on flight performance for many birds. In temperate climates, an exercising falcon, such as a peregrine, will start panting to cool itself long before it needs to pant for increased gas exchange.

The voice of birds of prey is reasonably simple although piercing. It is produced in the syrinx where the trachea divides to go to the lungs. The evocative cries of raptors have to carry long distances.

The airsacs and lungs are the seat of various illnesses including aspergillosis and air sacculitis or inflamed airsacs. Aspergillosis is a fungal infection and looks like bread mold growing in the airsacs. Should the interclavicular airsac become inflamed, or infected, the hawk may cast its food or refuse a large crop. This is because the weight of the crop pinches the airsac against the clavicles and causes discomfort.

The airsacs of free-living raptors often become choked with long, thin *Serratospiculum* nematode worms. These can be relatively harmless unless there are too many or if they cause a secondary infection. It can be tricky to treat these because they must be physically removed. Their intermediate hosts are ground beetles.

Sometimes, especially in young raptors, an airsac will leak air under the skin and the bird blows up like a balloon. Although alarming, it can usually be treated by pricking the skin with a sterilized needle, deflating the skin, and holding in the hand for a while, maintaining a light pressure over the skin. The leak usually seals spontaneously in the fast-growing bird.

1.7 The urinogenital system

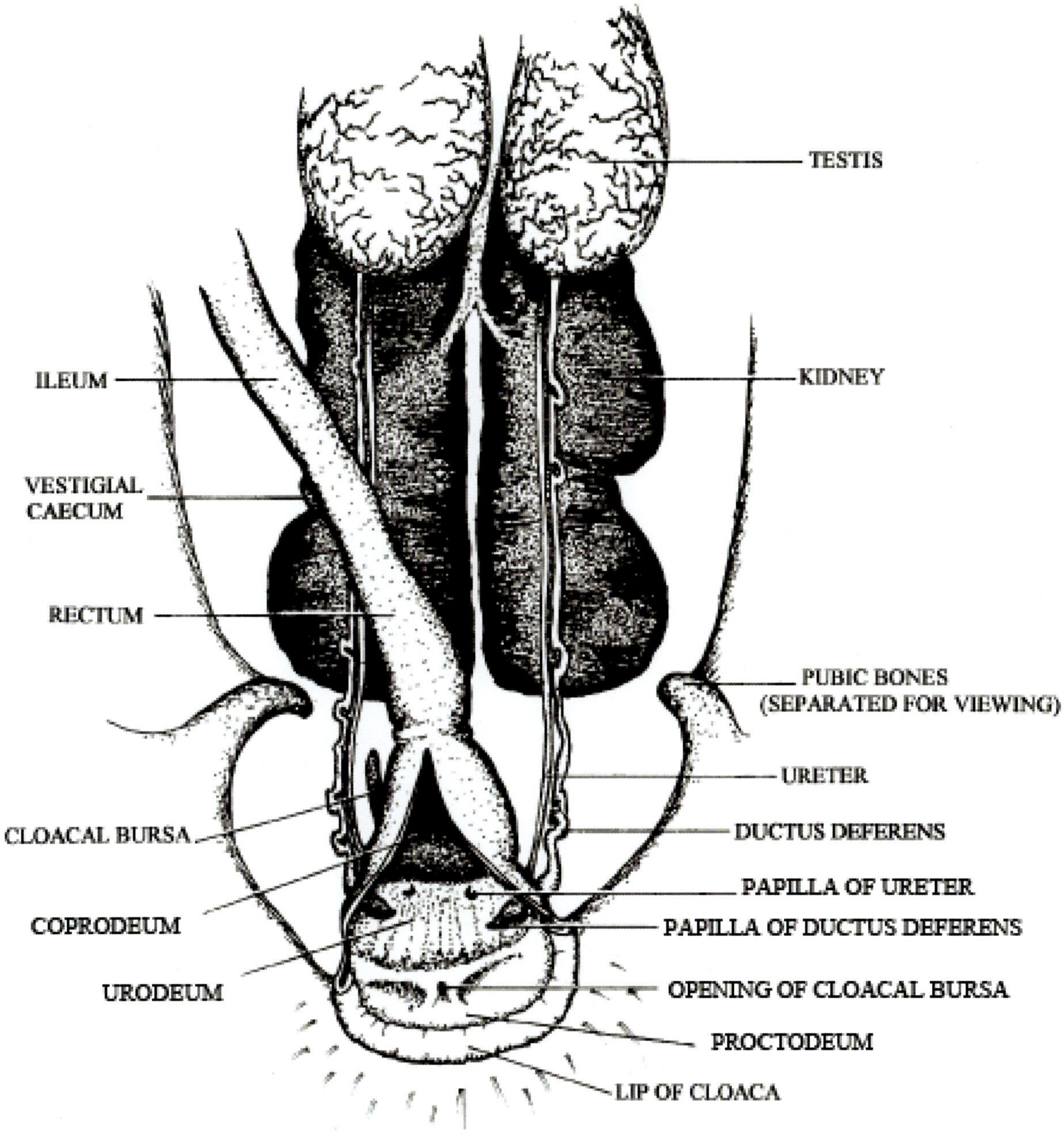

Figure 1.7.1. The urinogenital system of a male New Zealand falcon in breeding condition.

The kidneys lie in a hollow in the pelvis called the renal fossa (figure 1.8.1). The ureters carry waste products from the kidneys to the urodeum, the middle chamber of the cloaca. Birds' kidneys are much more efficient at conserving water than are those of most mammals, about 75–80 percent of the urine is in the form of insoluble uric acid which forms the white part of the mutes. This is probably an adaptation for survival during the time the chick is confined inside the eggshell.

The reproductive system of birds is extraordinary in that, when not in use, it shrivels away almost to nothing. This gives the bird less weight to carry around all winter. With the lengthening days of early spring the pituitary gland in the brain releases hormones - follicle stimulating hormone, luteinizing hormone, and prolactin - which stimulate the sex organs to grow. Soon they have grown to 200–300 times their dormant size and start to produce hormones of their own. Some of these feedback and control the hormone secretions of the pituitary and sex organs and stimulate further development, courtship, and nesting behaviour.

Males have two functional testes just above the kidneys. These enlarge in early spring and start to produce spermatozoa that pass down into the ductus deferens where they are stored. After insemination the semen is stored in the oviduct of the female and may fertilize eggs for several days afterward. Although spermatozoa can swim using their long tails, their progress is slow and erratic. The main movement of the semen is accomplished by the oviduct itself undergoing a reverse peristalsis. During copulation oxytocin released from the female's pituitary gland induces

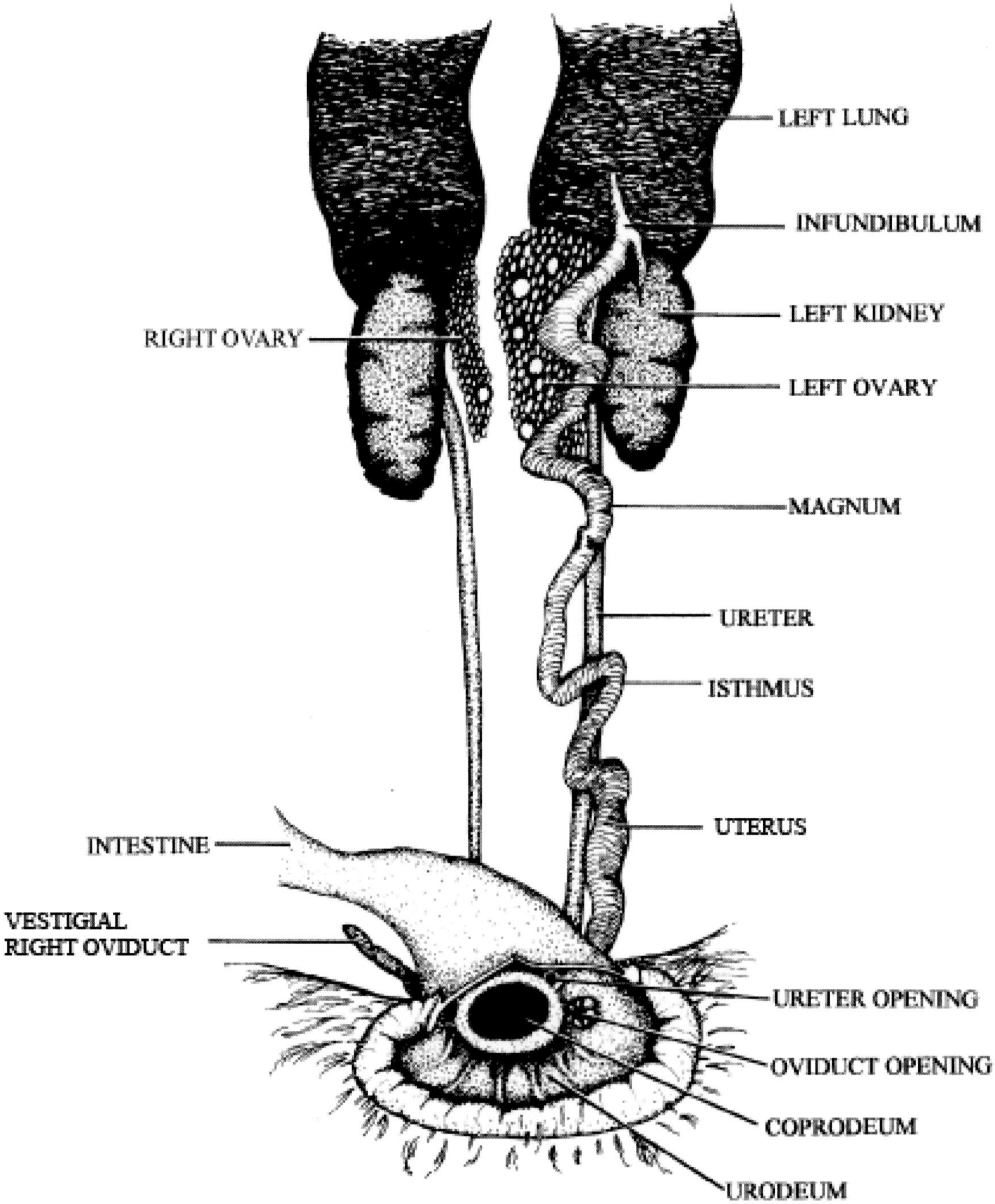

Figure 1.7.2. The urinogenital system of a female common buzzard in winter.

waves of contractions that transport the semen high up the oviduct within about three minutes of copulation. In the wild, falcons frequently maintain the copulatory position for two or three minutes after the male has left and, if well placed, one can see the cloaca contracting and expanding rhythmically. Often the female wails softly at this time.

In female birds, often only half of the reproductive system, the left side, is functional. However, in birds of prey, the right side is often present and may be partially active. Early in the breeding season the ovaries enlarge and start to produce small, yellow, yolk-like ova. When mature, an ovum leaves the follicle in the ovary and is swept into the funnel-like infundibulum, which has also enlarged. This is called ovulation and in chickens occurs 15–75 minutes after the preceding egg has been laid. In raptors this interval is usually longer. Fertilization must take place at this point before the outer layers of the egg are laid down during the egg's progression down the oviduct. Curiously, unlike mammals, in birds it is the female that decides the sex of the offspring. Some of the eggs contain a male chromosome and some do not. All the sperm cells contain a male chromosome.

The details of reproduction and development of the egg is discussed in Chapter 2.

1.8 The blood system, the lymphatic system, and the nervous system

The heart in the common buzzard has four chambers and lies immediately inside the sternum. It weighs about seven grams, 0.9 percent of the total body weight. The heart rate increases from a resting rate of 80-100 up to 250 beats per minute during exercise. In more active species in steady flight, the heart rate can stabilise at 600-700 beats per minute, six times the resting heart rate. The breathing rate increases too, by twenty times, to over 400 breaths per minute.

Tests on the blood are a useful diagnostic tool. They can reveal systemic infections, high cholesterol levels, parasitic diseases, pesticide levels, the DNA profile, and so on. A variety of tests evaluate the red blood cells or erythrocytes (packed cell volume and total count, haemoglobin concentration, mean corpuscular values, reticulocyte count, and erythrocyte morphology), the white blood cells or leucocytes (total count, differential, and morphology), the thrombocytes, and the coagulation rate.

The lymphatic system is difficult to identify with the naked eye. Only one organ, the spleen, is clearly visible (see figure 1.5.1). This is a dark red, oval body 10-15 millimetres in diameter lying beside the gizzard. It is concerned with the antibody response to infection and produces white blood cells. It also dismantles old red blood cells.

The nervous system of birds is similar to that of mammals. The cerebral cortex, which probably controls conditioned behaviour and learning, is large but not as well developed as in mammals, and the corpus striatum controlling perception, locomotion, and instinctive behaviour is dominant. Birds probably have less direct conscious control of body movements than do mammals. Portmann in 1950 found that a kestrel whose cerebral hemispheres had been removed was able to capture and hold a mouse but did not know what to do with it next. Bird behaviour thus tends to be more instinctive and stereotyped than in mammals.

In some ways they can be thought of as going from one sequence of behaviour to the next, each sequence being controlled by the corpus striatum and the overall picture set by the bird's physiological state as governed by its hormones. Thus when a bird is 'ready' it starts to build a nest. The nest building is mainly instinctive; first timers usually build almost as good a structure as experienced ones. The same urge drives the bird to collect the nest material and to achieve this it may need to show some intelligence and learning—for example in finding out where suitable material is and how to bring it back when one of the items is unwieldy. The sight of a clutch of eggs in the nest is the signal for the next sequence—incubation—and so on. Many actions which are quite complicated and appear to be intelligent, including most of the search and attack strategies (see Chapter 8), seem to be instinctive and appear for the first time in a complete and near-perfect form, rather than being developed through many learned stages.

Although when things go wrong, such as a chick falling out of the nest, birds often behave stupidly because their instinct program is not set up to cover such a contingency, the system by and large works well. For a sparrowhawk which may live for only two breeding seasons there is more survival value in doing it all by instinct than by learning. It might otherwise take several seasons before it succeeded by trial and error in building an adequate nest. Obviously its parents could not teach it how to build a nest because the eggs are not laid until after the nest has been built.

Raptors probably rank among the more intelligent birds and are certainly brighter than chickens, but they are not as clever as the opportunists such as the crow family. In general, the more specialized a hawk is, the less it is able to adjust to different circumstances and the less intelligent it appears to be. Those with a wide repertoire of hunting techniques (see section 8.3) are able to survive in a variety of habitats and are all-rounders, with intelligence to match. Where the specialist relies on superior physical ability and suitable habitat, the all-rounder is more able to adapt and to use its brain to outsmart the prey.

1.9 The wing structure

The wings are based on the pectoral girdle, which consists of the shoulder blades or scapulars, the coracoids, the fused clavicles or wishbone, the humerus, and the wing bones. The main tendons, which control wing movements, are powered by the strong pectoral muscles, which attach to the keel and to the clavicles. This system keeps weight off the wings and below the bird's centre of gravity, thus making the bird more stable. The breast muscles immediately under the skin are the massive pectorals, which pull the wing downward and provide thrust. Between these and the sternum are the supracoracoideus muscles, which raise the wings using tendons, which run through a pulley hole in each shoulder, the triosseal canal. As raising the wings is easier than the downstroke the supracoracoideus muscles are only 5-10 percent of the size of the pectorals.

The pectoral muscles are made up of red muscle fibres and white muscle fibres. These are discussed more fully in 7.1. The pectorals muscles have about twice as many mitochondria as the supracoracoideus, and about 1.5 times

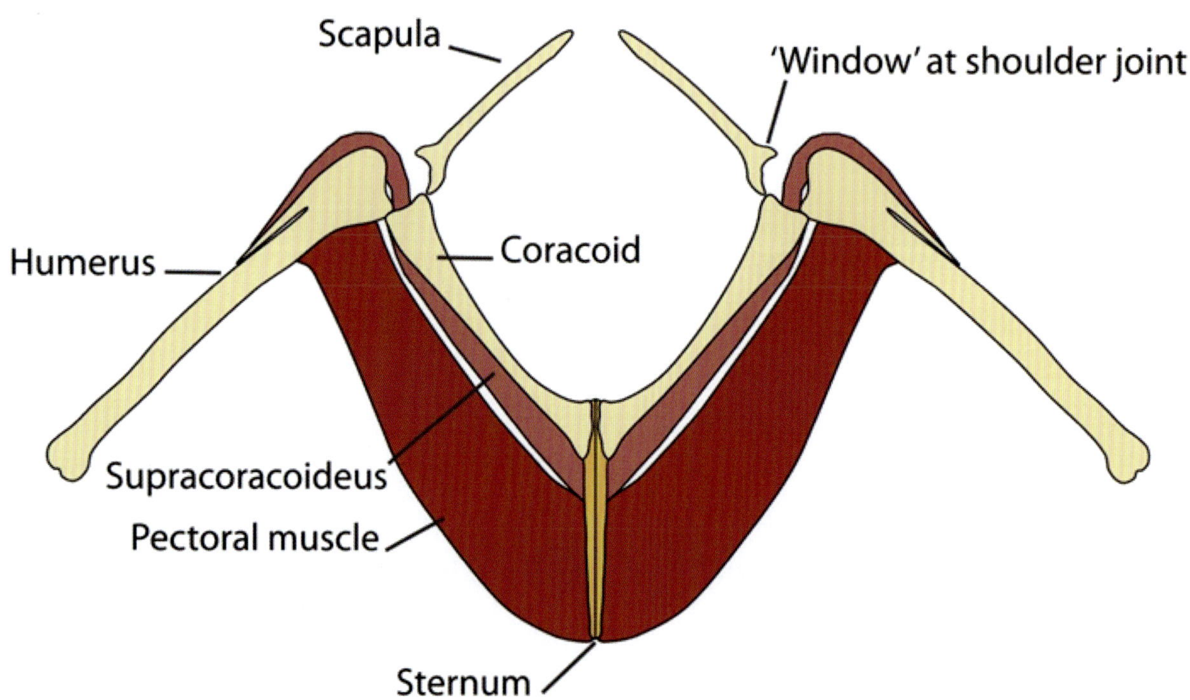

Figure 1.9.1. The pectorals pull the wings down. The supracoracoideus muscles pull them up via a pulley through the shoulder, to the top of the humerus.

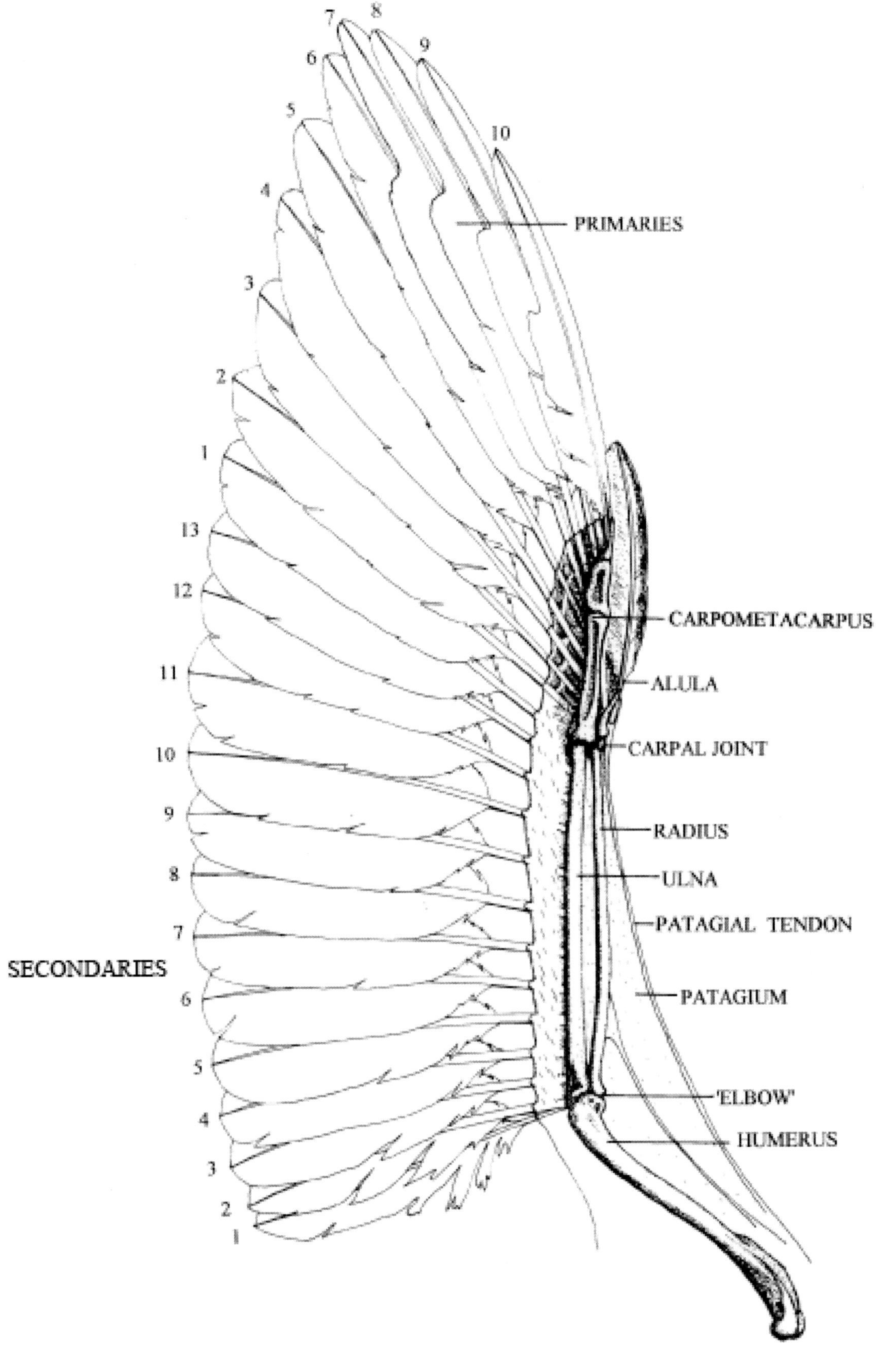

Figure 1.9.2. The structure of the wing of the common buzzard

the oxidative activity. My data on a European Sparrowhawk, a Merlin *Falco columbarius*, a Common Kestrel *F. tinnunculus*, five New Zealand falcon falcons, two common buzzards, a Red Kite *Milvus milvus*, a saker, a Harris Hawk and a Griffon vulture *Gyps fulvus*, show pectorals of 11.3-17.6 percent of the total body weight, and supracoracoideus of 0.9-1.5 percent of the total body weight. The Griffon had proportionately the heaviest pectorals, reflecting the scaling effects of such a large (9.25 kilograms) bird, but it also had proportionately the smallest supracoracoideus (see 1.16).

The accipiters, as well as having red fibres for general flying, also have white fibres for sprinting. This enables them to explode from the fist with the same vigor as a rocketing pheasant. During acceleration, and when climbing, accipiters develop propulsion on the upstroke as well as on the downstroke (see 1.17). The shoulders rotate, producing a backward flick of the slotted primaries that store energy in them and straighten during the upstroke. The supracoracoideus muscles, which raise the wings, have a high proportion of white fibres and are noticeably paler. They give some power to the upstroke during the sprint.

The contractions of the breast muscles lower the upper wing bone or humerus (figure 1.9.2). This is air filled and connected to the airsac system. It is strengthened by small cross-bracing structure structures inside the cavity. The humerus carries the small tertiary feathers, but no flight feathers. From the humerus, the radius and ulna carry the secondaries, each feather being attached by two ligaments to small bony knobs on the ulna. The secondaries provide lift and they number from ten in accipiters to thirteen in the common buzzard and as many as twenty-five in the Bateleur eagle

As this saker brings in prey, she flares her alulas for landing.
Photo Stig Olsen

Terathopius ecaudatus. Between secondaries 4 and 5 is an extra tectrix or covert feather, which makes it appear as if a secondary is missing. The radius, a long, slim bone along the leading edge of the wing, acts as a brace and, should the hawk hit an obstacle, is first in line for damage.

Between the humerus and the radius is a large flap of skin called the propatagium which makes an aerodynamically 'clean' edge to the aerofoil shape of the wing. It is held in place by two elastic tendons that run up to small muscles at the shoulder. Should these slacken, the propatagium fails to contract fully when the wing shuts and a fold remains visible. This condition is prevalent in some bloodlines of peregrines. It does not noticeably affect the bird's flight, but nonetheless, birds with this defect should not be used for breeding. If the elastic tendon is severed completely in an accident it needs very careful repair if the bird is to regain full powers of flight with a proper aerofoil.

The radius and ulna connect to the wrist or carpal joint, which, like our wrist, is complex both in structure and in movement. A blow or injury to the joint can cause the joint capsule to swell in a condition known as the 'blaine,' a bursitis similar to tennis elbow or housemaid's knee. Like so many joint problems it may, with rest and warmth, clear up entirely, or it may reappear under stress and persist, in which case the hawk will have to be retired from strenuous flying.

Two structures arise from the carpal joint: the alula and the manus or hand. The alula is the remains of the thumb and supports three small stiff features called the bastard wing. As air speed over the wing falls below a certain level, the alula erects and acts as a Handley Page slot, smoothing the airflow over the wing and, by putting off turbulence a little longer, enables the bird to fly more slowly without stalling. This can be clearly seen when the bird lands or brakes.

Occasionally, for reasons that are as yet unclear, a young falcon may develop with one or both alulas drooping. While this is unsightly and regarded as a defect, it doesn't seem to affect their performance and some race trainers simply trim them off. Dropped alulas are fixable if treated when the bird is just fledging. Take some dissolving suture thread and tie it tightly around the outer alula feather so that it cannot slip down. Then bring the two thread ends under the other two alula feathers to the base of P5 or P6 and thread them around the shaft as close to the socket as possible. Now tighten the threads until the alula is pulled well back behind the leading edge of the wing, then knot it off. Thus the thread is attached at both the alula and at P6, but there is a little slack between them; they are not actually lashed together. The stitch is invisible and can removed in a month or so if it hasn't already dissolved.

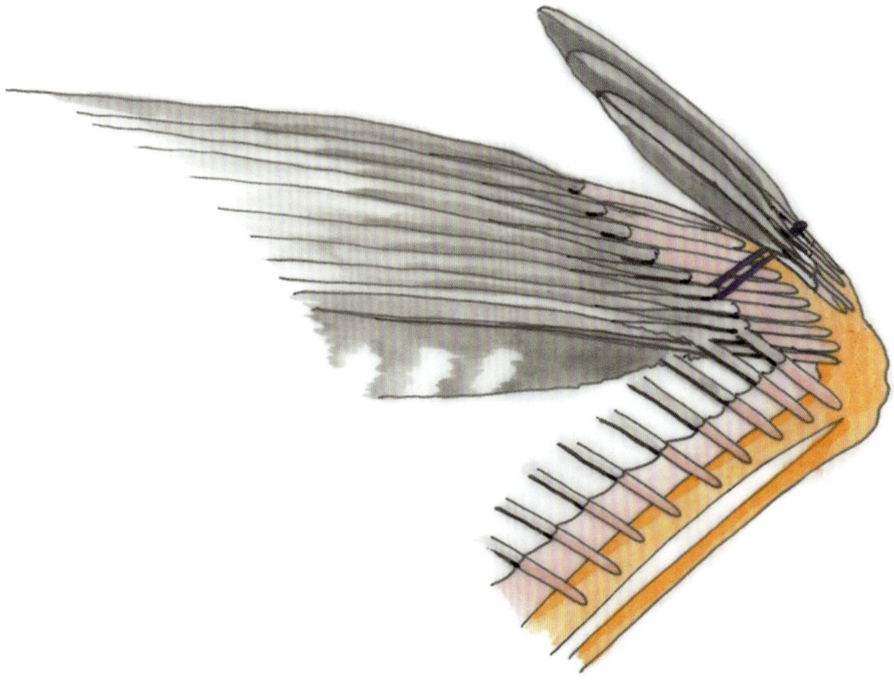

Figure 1.9.3. Repairing a dropped alula.

The manus consists of the fused remnants of the fingers and supports ten primary or flight feathers. The primaries are the source of power in flight and they fold away like scissors under the secondaries when the wing closes. The way in which they work is complex, as is the whole action of the wing. The primaries and secondaries are fixed in sockets attaching them to the manus and the ulna, and these sockets are connected together by elastic membranes. These are essential to maintain even feather spacing under pressure.

One should be very skeptical of the claims of some rehabilitators that a bird's flight is normal just because it can stay in the air for a few hundred metres. A rehabilitated accipiter or large falcon may be capable of apparently normal cruising flight and yet be significantly impaired in power, agility, and endurance during an attack. Whereas many species of birds which rely on their wings mainly just for transport can survive quite drastic wing injuries, raptors, particularly the rapacious species which rely on their attacking ability, cannot.

1.10 The tail structure

The tail is made up of two rows of five feathers placed each side of the pygostyle or tailbone. Two additional feathers, the deck feathers, lie centrally, just above the others. The pygostyle has a raised keel to support the tail feathers embedded about 17 millimetres into their sockets. Above the root of the deck feathers is the preen gland with a small wick supplying oil for preening.

The tail is used for steering and braking, particularly by those species that hunt in forests. The Long-tailed Hawk *Urotriorchis macrourus* and the Micrastur forest falcons show the most extreme development. Because the tail creates drag it is dispensed with as far as possible by the high-speed aerial species. Some species that fly fast and accurately, such as swallows and swifts, have taken up both options—long outer tail feathers for steering and the rest of the tail shortened to reduce drag. These V-shaped tails help to smooth the two vortices that spill off the edges of the tail during flight.

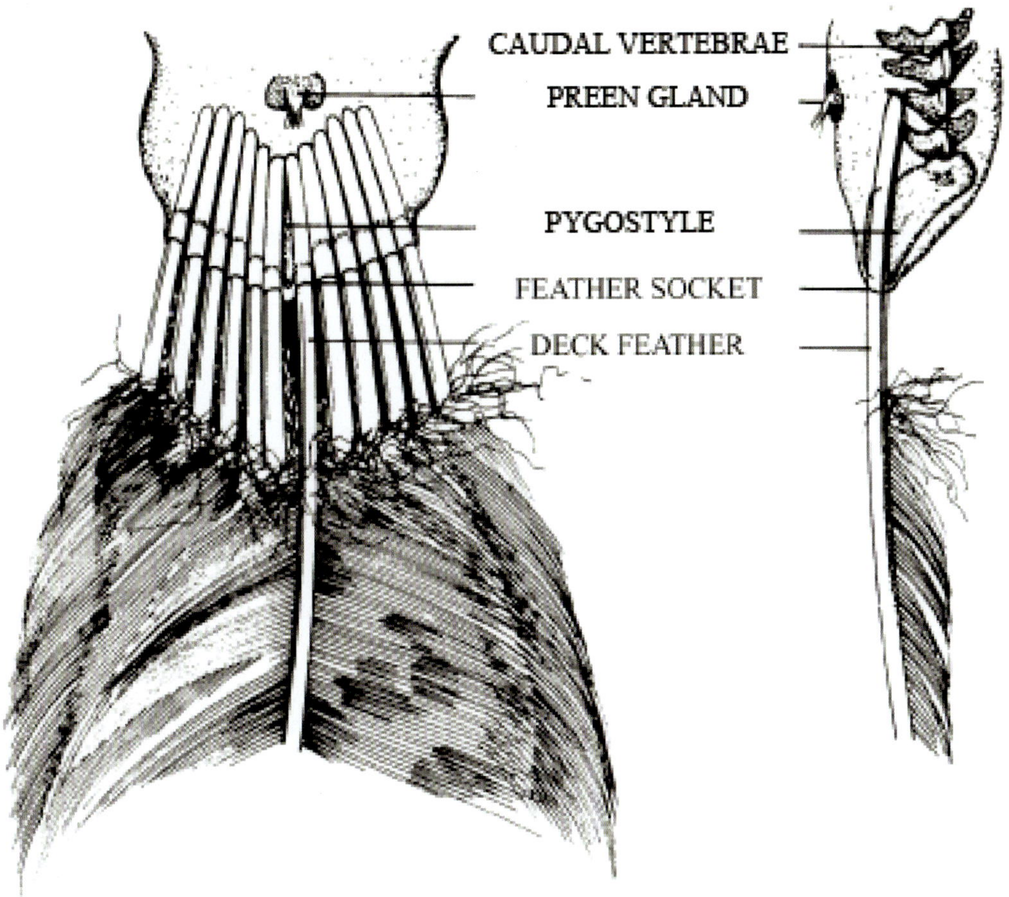

Figure 1.10.1. The structure of the tail of the common buzzard.

The small amount of extra lift provided by the tail is offset by the extra drag and only when soaring in rising air is the tail opened for added lift. The tail also tends to stabilize pitching in flight; the up and down movement like a boat heading into waves. But a long tail, with its extra drag, induces turbulence more easily and thus large birds tend to have short tails. This reduces stalling speed and makes it easier for them to land. A young hawk tends to constantly adjust and flare her tail, compensating for slight unbalances in movement; as she becomes a more confident flier she keeps her tail more closed.

1.11 Feather structure

Feathers are made of keratin, the same protein as hair, horn, and fingernails. They are dead once complete and can thus be trimmed or repaired without the bird feeling anything; only the sockets are alive. They are not distributed all over the body but grow mainly in well-defined zones (figure 1.11.1). The body feathers are erectile and are each controlled by four muscles. On a cold day the bird fluffs out for extra insulation; if warm or stressed, it sleeks its feathers down. Birds such as gyrfalcons have more down on the feather-free areas for better insulation in cold climates. Some of this down can be removed when flying them in hot climates, to prevent undue heat build up. Overheating is the main performance-limiting factor for gyrs in warm climates.

Raptors have five main types of feathers, the flight feathers, the contour feathers covering the body (some of which have after-feathers), down feathers for insulation (especially in the chick), small hair-like filoplumes scattered over the body, and tiny hair-like crine feathers around the beak and nostrils which shed off dried blood. Some raptors, particularly the harriers, have powder-down feathers that flake into talc-like powder giving a fine bloom to the feathers and perhaps a degree of waterproofing.

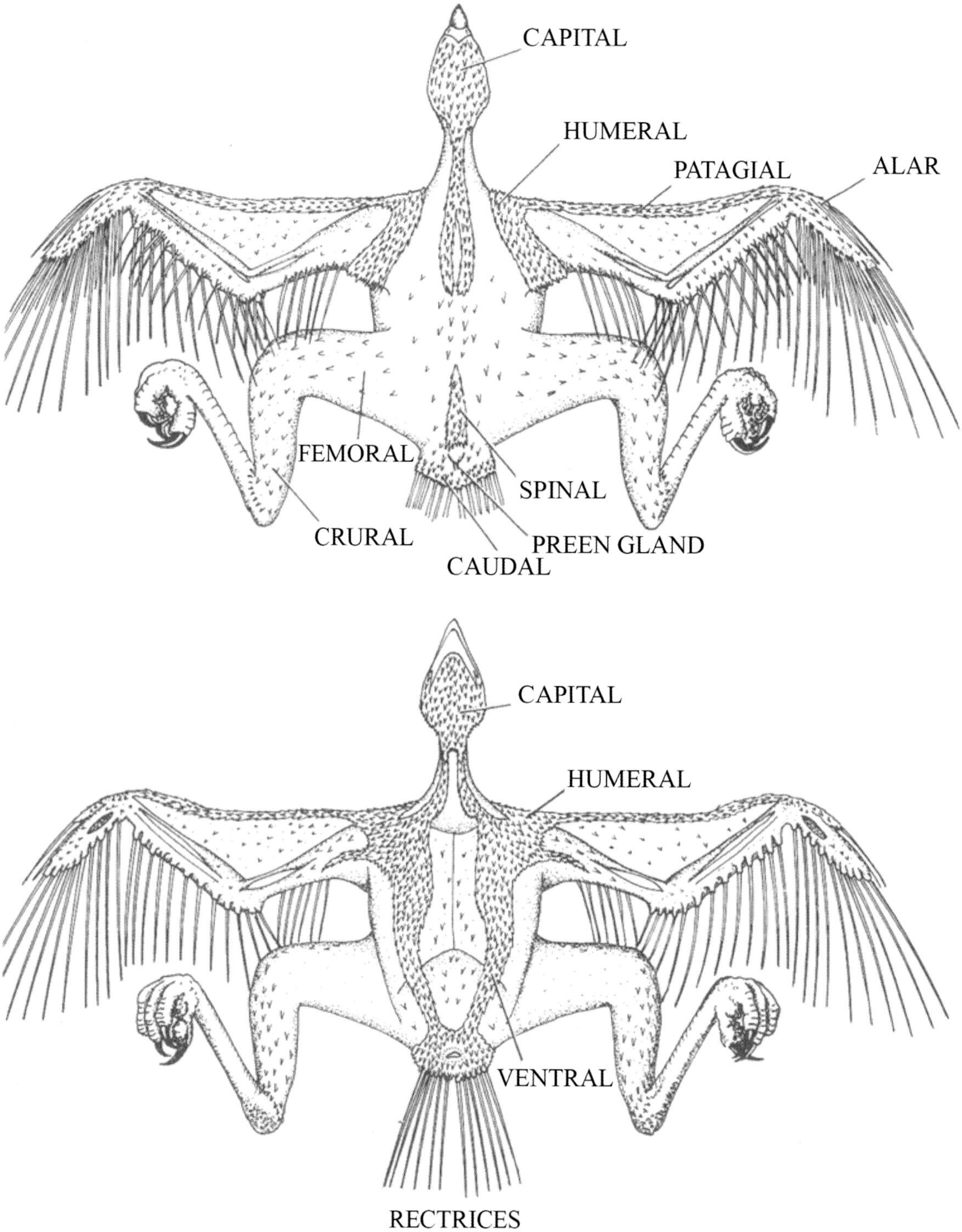

Figure 1.11.1 The feather tracts of the common buzzard.

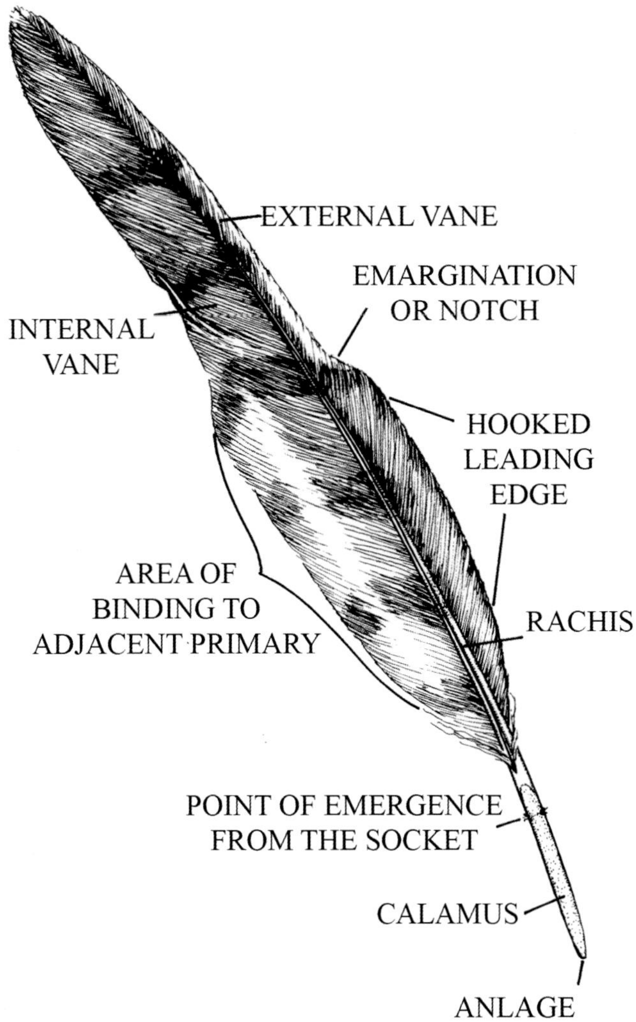

Figure 1.11.2. A primary feather from the right wing of a European sparrowhawk, viewed from below.

The flight feathers consist of a hollow quill or calamus and a central shaft or rachis with vanes on each side made up of hundreds of barbs (figures 1.11.2 and 3). The barbs are held together by millions of barbules which have hooks on the lower leading edge locking the whole structure together. When the bird preens these are settled back in place like a zip fastener. Some of the barbs hook adjacent feathers together so that, for example when the bird is soaring, the adjacent wing feathers 'stick' together to form a continuous sheet, that can slide up and down as the wing closes.

Different feathers have different degrees of stiffness and elasticity. For example, the leading primaries are stiffer than the trailing ones and the primaries of an adult peregrine are stiffer than those of a juvenile. Across the species we have measured using the Pennycuick and Lock technique, we found that the peregrine had the stiffest primaries, followed by the northern goshawk, gyrfalcon, saker, Harris, and common buzzard. There is more work to be done on this; feather elasticity is tricky to measure.

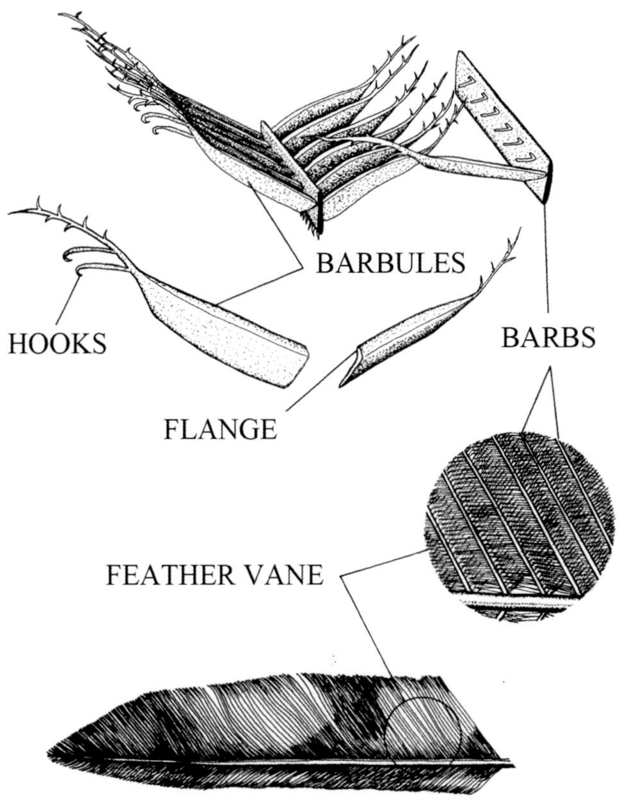

Figure 1.11.3. The structure of a European sparrowhawk primary feather.

1.12 Feather maintenance

The hawk looks after its feathers by preening using light waterproofing oil from the preen gland at the base of the tail, and by bathing in water or dust. Many species also sunbathe; the ultraviolet light acting on the provitamin S in the oil on the skin produces vitamin D^3, which is essential for calcium absorption and without which growing chicks may develop rickets. I frequently use silicon fabric waterproofing spray on hunting falcons during periods of wet weather with beneficial effects. In one disastrously wet grouse-hawking season we used a whole can on three falcons in three weeks. The desert falcons particularly, are quickly drenched in our British climate, and hawks taking ground game soon get sodden by the wet grass.

Raptors are susceptible to feather parasites especially in the summer in warm climates, or when taking prey such as rooks and crows which harbour a lot of parasites. A heavy infestation may indicate a concurrent disease or injury and may also be implicated in the spread of avian pox and avian malaria. The commonest parasites are the pale grey lice that crawl on the feathers and are often visible when the hawk is hot. Tiny feather mites can damage the feathers by stripping the barbules and the repulsive large, flat, sideways scuttling hippoboscid flies are often seen on young birds. Some of these black hippoboscids are capable of short flights, usually up one's sleeve. All these parasites can be swiftly eradicated using Frontline or other similar preparations specifically produced for birds. It is wise not to offer a bath for a day or two after application and to keep the hawk hooded and head shielded during spraying or dusting. Spray under the wings, not on the upper surfaces.

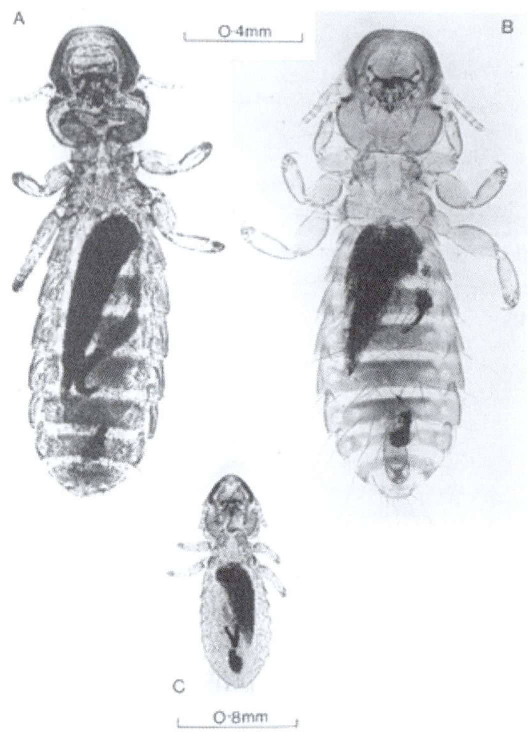

Falcon lice *Degeeriella rufa*
A. Adult female, B. Adult male, C. Nymph.

If a feather is slightly bent, it can be restored to shape by warming it up. Vultures use the same method; after many hours of soaring, their primaries become set in a curve, rather like a bow that has been left strung. By sunbathing on a perch for a while the vulture uses the heat to straighten out the feathers.

Broken feathers can be repaired by splinting, imping, plugging or sewing in depending on where they are damaged. Flight feathers must be able to twist and bend naturally and therefore repairs on the outer two thirds of the feather are seldom successful for long. Proper flexion is reduced and this puts extra strain on the ends of the reinforcement, which break after a while. We use very fine fretsaw blades like springy wire. A flexible glue keeps the two butts together. Repairs to the inner third are better, assuming an exactly matching replacement is available. Carbon or bamboo plugs in the hollow quill are probably the easiest way of achieving this and it should be borne in mind that the hollow shaft of the feather continues for some distance inside the skin socket. In buzzard-size hawks this distance varies from 15–30 millimetres according to the feather concerned. About one half of this distance can safely be used for plugging and therefore even if only a small stump of feather is showing, it is repairable. Don't try to push the plug right in to the end of the calamus; this could be painful. Instead, make the plug slightly loose so that it slips in easily, but before putting it in, inject the hollow stump with epoxy. This effectively extends the reach of the plug inside the socket without hurting the bird.

Broken feathers should always be repaired promptly otherwise neighbouring feathers will break under the extra strain. Therefore when putting the bird away to moult it is important to ensure that all feathers are complete even if the replacements are from the wrong species, otherwise the newly growing feathers, lacking support, may break off in the blood. Rehabilitators should become skilled at feather repairs to ensure that birds can be released back to the wild, with a good serviceable set of feathers, rather than be held for a protracted period to moult, which raises further complications.

Falconers' birds tend to break their feathers much more than do wild raptors and most of this is traceable to bad management. Of course, genuine accidents do occur in the field. Inexperienced hawks often catch quarry so that they fall into a tree rather than on to the ground. The hawk, sprawled or wedged in the branches, is unable to kill the quarry and yet is unwilling to let it go. One is lucky if feather damage is the only result. Fortunately, hawks soon learn not to do this. Ground game, such as hares, are also a fruitful source of broken feathers. If your goshawk by the end of the season looks as if she has survived a trip through a jet engine then you have been over-facing her and are lucky if she hasn't ruptured an airsac. Although it is always tempting to try for the largest quarry that the hawk can catch, it is much harder to take the smaller, more agile species.

Most feathers are broken or weakened through bad management. Poor handling on the fist, badly designed equipment, and general carelessness and incompetence all take their toll. If your hawk is breaking feathers while in captivity than a stringent review of technique is called for.

1.13 Feather growth

Raptors go through a series of plumages in their lives. They hatch with a short natal down. After about 7-14 days their second, thicker, down coat grows through. This allows them to control their own temperature to a large extent, thus freeing up the female to go hunting to meet their high food demands. As the growing chicks approaches full body weight, the juvenile plumage grows through. For a time, the chicks are harlequins of feathers and down before emerging in full juvenile plumage. Many juveniles undergo some body moult in their first autumn, but their main moult starts in their next spring. Then they gradually grow their adult plumage, being clean moulted by the autumn. Flight feathers are replaced gradually over the easier summer months, to minimize any impairment to the bird's hunting ability. Many larger species, such as eagles, are unable to complete the moult in one year and it drags on so that the bird always has a mix of at least two plumages.

Moulting is triggered by a variety of environmental and hormonal factors. An increasing day length triggers the moult via the pituitary gland, and starvation or illness can retard it, delay it, or stop it altogether. Females usually moult heavily while they are incubating whereas the males, which are busy hunting, delay their moult until the females are ready to hunt too. Migratory species tend to stop moulting during the migratory period, the migratory tundra peregrine, for example, starts its moult in the summer in Alaska or northern Canada then interrupts it until it reaches South America in September or October. Once on the wintering grounds the moult is resumed.

Birds that have developed with a minimal fear response (see section 4.10), such as hand-raised birds, are better adapted to captivity and do not suffer from the persistent low-level stress that wild-caught or some parent-reared individuals do. Low-level stress entails constant secretion of adrenalin which is antagonistic to the sex hormones; such birds are therefore harder to bring into breeding condition and do not usually moult rapidly or well. They cannot be flown at such high weights as the better-adjusted birds.

At higher levels, adrenalin causes constriction of the peripheral blood vessels such as those to the skin and to the feather buds or papilli. A human being, when badly frightened, goes 'white as a sheet.' So too does a bird, but the skin cannot be seen through the feathers. Any feathers which were growing when the bird was frightened have their blood supply restricted and hence receive less nutrients for a period. Adrenalin also reduces the nutrient levels in the blood because it reduces

This Mongolian saker has just hatched her chicks and has started her wing moult.
Photo Andrew Dixon.

the blood supply to the intestine. The tail feather of a peregrine grows at about 0.2 millimetres per hour. When frightened, the bird secretes adrenalin in seconds and the effects may take one to two hours to subside. During this time, the rapidly growing epidermal cells in the follicle may be starved of nutrients and a stress mark or fault bar may grow down about 0.5 millimetres wide.

Similarly, if the bird is starved or ill for a period, the blood-sugar levels and nutrients fall and again the deprived follicles produce a 'hunger trace' or, in extreme instances, the whole growing shaft may be pinched off. Similar weaknesses develop in the horns and hooves of cows when calving or in human hair or fingernails when you are seriously ill. All these structures are made of keratin and all depend on a good blood supply for normal growth.

Reduction in blood pressure to the follicles resulting in reduced nutrients and oxygen reaching the feather germ also occurs naturally. The metabolism of birds is lowest during sleep (see also section 6.3). The normal temperature of about 41.6 to 41.9 °C drops to around 41.1° C and the metabolic rate, and hence circulation rate, is reduced. When a bird is in poor condition, or if its metabolic demands are high, as in winter or during illness, this can result in fault bars every night, or in owls, every day. If you take a feather from a small bird and hold it up to the light, you can often see stripes across the vanes showing the different growth between night and day. If the bird is starved or debilitated, as well as fault bars forming, after three to four days the growth rate of the feather slows down too. Because of the way the feather germ is positioned in the follicle, the barbule-forming region is the first to suffer when nutrients are reduced.

Although feathers have an extremely high strength to weight ratio, they are still quite heavy. A bald eagle *Haliaeetus leucocephalus* measured by Brodkorb had feathers and down weighing 677 grams - 16.6% of its total body weight, whereas its skeleton weighed only 272 gram - 6.7%. Pigmented feathers, such as black areas containing melanins, tend to be tougher because more pigments are secreted in the black areas. Thus pale feather tips usually wear off rapidly until a dark area is reached. Pale marks in the middle of a dark feather can wear away differentially or be more susceptible to the nibblings of feather mites or lice. Some falcon species, such as the gyrfalcon, Sooty Falcon *Falco concolour*, and Eleonora's Falcon *F. eleonorae* are genetically dichromatic; that is they have a dark phase caused by a preponderance of eumelanin, a true black pigment. Other genera, such as the buteos, also show melanistic phases, some of which probably involve phaeomelanins which give dark browns and red browns.

Feathers are dead once complete and, when moulted, do not drop loose but are pushed out by the growing new feathers. If a feather follicle is damaged by injury, it may grow twisted or stunted feathers. Or, if the follicle is normal but the blood supply is reduced, for example by scarring or infection, it may grow normal feathers which, once they reach a certain length, make heavier demands on the follicle than the blood supply can cope with; feather growth continues for a little longer but the vanes do not unfurl and finally the shaft pinches off. Repeated feathers attempt to grow but suffer the same fate. The rest of the feathers on the bird grow normally. This may be due to an infection. One way to treat it is to cut off the unfurled feather every two weeks just beyond the sheath. This reduces the physical strain on the injured socket while the feather is growing. After six or eight weeks the quill forms, the feather hardens off and the socket is less delicate. A replacement feather can then be fitted into this quill and with any luck the socket will take the strain. If the socket has healed completely, the bird should moult normally next time. If the dermal papilla or the blood supply is affected, the prognosis is less cheerful.

Prey species, such as pigeons and gamebirds, have very loose feathers which come out in bunches when a hawk tries to grab one. Such feathers are regrown and pigeon racing enthusiasts therefore pull out a broken feather rather than repair it. Raptors, on the other hand, have much tougher follicles and if a feather is forcibly wrenched out, the socket may be permanently damaged, particularly if the feather was in the blood.

> You can plug the empty socket with wax or a short length of quill to prevent it closing up and distorting the new feather. To do this, select a length of quill slightly narrower than the original feather and cut off at the point where the vanes start. Tie a thread tightly round the shaft, preferably using dissolving stitch material. Smear the plug part with a disinfectant cream. Gently winkle the quill into the socket as far as it will comfortably go, then tie the ends of the thread around the adjacent feather. If the thread is still intact after two weeks, snip it but leave the plug in place until it is pushed out naturally. The commonest feather to come out is a deck feather bearing a transmitter.

Very rarely, raptors show a continual moult; as soon as the feathers are grown or half grown they all drop out again. This appears similar to French moult in budgerigars, which is caused by a virus but in raptors may be due to a malfunction of the thyroid, pituitary, pineal gland, or gonads. Stress on a feather, such as caused by an over-heavy tailmount, can also stimulate follicle development, causing the feather to be shed. If a juvenile feather is shed prematurely, it is replaced by an adult feather and this can be used to sex some species, such as merlins, by pulling out a body feather from a bird that cannot be sexed on the basis of measurements. The jack merlin will grow a nice blue replacement.

This is a donut-shaped moulting pen in Abu Dhabi.

All falconers hope for a clean and rapid moult and great is the gnashing of teeth in late summer when a few lingering old feathers stubbornly refuse to budge. On the first hawking trip of the season they oblige you by dropping out on the cadge. So what can be done to ensure a smooth moult?

First of all it is essential to reduce stress as much as possible, especially for any nervous birds. Personally, I dislike having any hawk tied up when it is not being flown and from January until July all my birds are in breeding pens or, if juvenile, in moulting pens. There they can get a little exercise, bath, sunbathe, play, and watch the world go by, up out of the way of damp and mutes.

Other than a good varied diet, including additional vitamins and essential amino acids, and perhaps some honey, it is best not to use any moulting agents unless you know you have a real problem bird. Injections of the female sex hormone progesterone can cause early moulting, as can thyroxine, but further research on these is needed, especially as thyroxine can be toxic close to therapeutic levels. The nearer to the normal date that treatment is given, the less hormone is required to trigger the moult. They may thus be useful to make the moult start early but they do not make the moult period itself substantially shorter. The activities of the thyroid appear to affect moulting the most, then the sex hormones, then the pituitary.

Another approach is to alter the photoperiod by bringing the apparent year forward a month or two. One regime used successfully in America is a twenty-hour day length in very early spring until the first feathers drop, then eight hours winter photoperiod. This has resulted in all the primaries being moulted at once and the birds unable to fly. I tried a regime similar to this, but slightly less drastic, and one of the four falcons was unable to fly for three weeks while moulting. Essentially, you are bringing the summer equinox forward and getting the bird as quickly as possible onto a shortening autumn photoperiod. Others have been successful by simply using 24 hour light from the winter solstice onwards. One can buy light bulbs that give a spectrum similar to natural day light but it is not clear how necessary this is. Even street lights can mess up photoperiods.

Raptors show two main patterns of moult. The accipitridae, which include all the accipiters and buteos, have a descending moult. The primaries are dropped one by one, starting in the angle of the wing and working outward so that the outermost primary (P10) is the last to drop. The secondaries also start from the angle of the wing and work in toward the body.

In contrast, falcons start their wing moult from the centre of the primaries, usually P4, outward in both directions, and from the centre of the secondaries, usually S5, in both directions outward.

Both groups of birds normally drop the centre (deck) feathers first from the tail, the remaining tail feathers being shed in a wide variety of sequences. Usually no feather is dropped until its neighbour is three-quarter to half grown although occasionally a falcon will drop all its remaining ten tail feathers as soon as its new deck feathers are complete. Because feathers are dead, the body has no way of detecting if they are broken. Broken feathers are moulted no sooner or later than whole ones.

The whole process of moulting in birds of prey is spread out over the summer months avoiding stress periods such as rearing young or migrating. A first-year bird which drops its first feather in mid-March is doing quite well to be clean moulted by mid-August. Migratory raptors such as the *calidus* peregrines that we worked on in Siberia experience a complex photoperiod. They arrive back on the nesting grounds in May and soon have 24 hour daylight during breeding. By August there is a brief dark night period and the falcons start to head south across the taiga forests

The outer primaries and the rectrices of this *calidus* peregrine in July are very faded compared to her new feathers coming through. The overlying feathers have left dark, unfaded shadows.

of Russia. Some will winter in latitudes across north Africa, the Middle East, to Indonesia, but some carry on further south, crossing the equator and 'winter' in the 'summer' there. Trying to breed or even just moult these arctic falcons is therefore quite difficult and their migratory behaviour is hard-wired in their genome.

1.14 Wing shape, aspect ratio, dihedral, and tail variations

Above all else, the shape and structure of a raptor's wings and tail reflect its potential as a flying machine, and in turn a reflection of the ways it searches for and catches its prey. When you consider that a majority of young raptors die in their first year, primarily because they cannot catch enough food, then you appreciate the massive selection pressure finely tuning individuals for their habitats. This applies not just to the individual but also to the splitting up of each species into two groups, male and female, each adapted by size to exploit subniches within the same ecosystem.

This fine-tuning can be seen in many features related to flight. Many hawks, such as buzzards, rely on searching for easy prey so that the main selection pressure is on their searching ability (see 8.1). The attack, when it comes, is just a simple one. These birds tend to have low wing loadings, and are specialists at gliding and soaring. They are usually fairly stable in the air, with a large dihedral and medium to high aspect ratio.

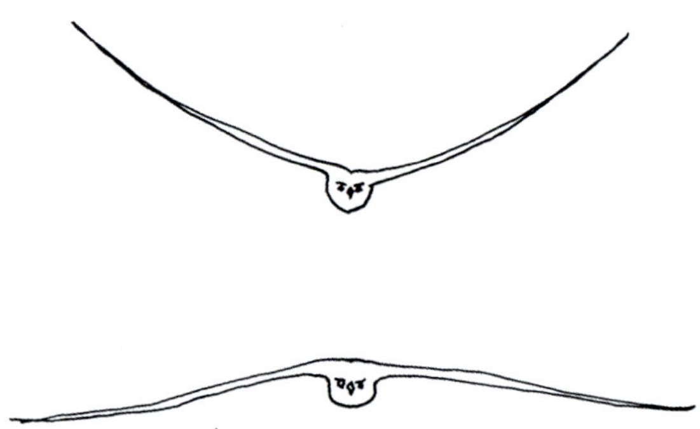

Figure 1.14.1. Top: A Red kite has a large dihedral which gives lateral stability against rolling.
Below: An accipiter has a negative dihedral making it laterally unstable.

This male gyr/peregrine is gliding at an optimum glide angle. If he was soaring he would open his primaries to optimise lift over distance. If he wanted to go faster he could half close his wings, but would descend more steeply.
Photo Stuart Grimwood.

The dihedral angle measures how high the wings are held above horizontal, making the bird self-righting and less prone to rolling (figure 1.14.1). The soaring birds tend to be relatively stable so that they can almost fall asleep and yet still remain level. The attacking hawks, on the other hand, are selected during the attacking phase of the hunt and need to be fast and agile. They tend to have a negative dihedral, or anhedral, making them prone to roll easily and be very unstable. A mere flick sends them over, which is what they need to be manoeuvrable. It is the difference between a white-water canoe and a punt. Of course birds can change their wing configuration at will and can increase stability, not only by changing the dihedral, but also by sweeping back the wings.

The searchers, while soaring, have open, slotted wing tips, which are ranged from the front of the wing tip to the back at decreasing angles of attack (figure 1.16.4). During fast airspeeds the tips bend up further and twist, increasing the angle of attack into lift. The bird soars up, but slows down. The primaries straighten a little, maintaining an optimum angle of attack for the airspeed. This appears to be a passive mechanism, but nerves in the feather bases presumably indicate the extent of bend on the primaries and tell the bird when the airspeed is so low that it must resume flapping. The effect of the slotting is to maintain adhesion of the air passing over the top of the wing, and thus reducing the stall speed. This effect is much less developed in the attacking hawks, which, because of their higher wing loadings, glide at higher speeds. They achieve it by altering their whole wing profile (see figure 1.16.3).

The aspect ratio is the ratio between the wingspan and the average wing width. Because birds' wings are not exact rectangles, we calculate this as Span^2/Wing area. The high-speed gliders, such as the peregrine, have long, narrow high aspect ratio wings of about 7.7 (figure 1.14.2). This reduces the induced drag, but makes it harder for them to manoeuvre, especially at high speeds when the strain on the open wing is at its greatest. The slow-speed soarers, such as the buzzards, vultures, kites, and some eagles, have a medium aspect ratio of about 6.3, and quite a cambered wing. They do not attempt agile manoeuvres at any speed. The accipiters, traditionally known as 'shortwings' also have a medium aspect ratio, but their wings are a different shape. Their wings are not primarily adapted for gliding and soaring, but for supplying thrust. The key to the accipiter wing is acceleration, and this is discussed more fully in 1.17. Thus it is possible for two wings to have the same wing area, but have different functions. For example, the female New Zealand falcon wing and the male peregrine wing are the same area, but they have different aspect ratios.

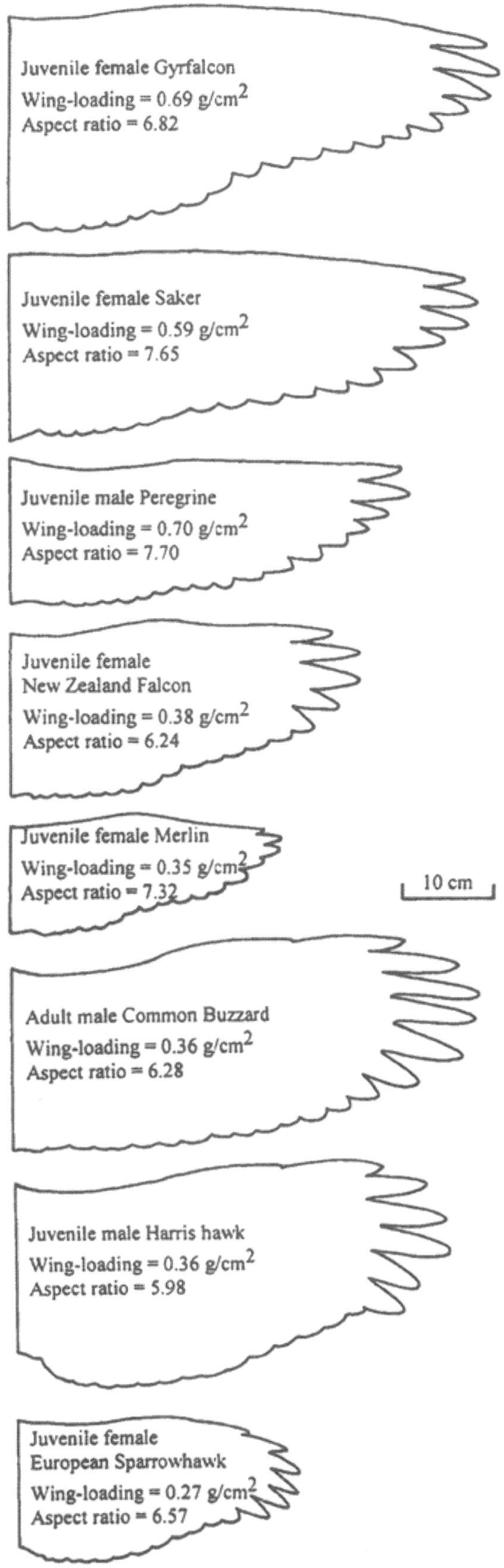

Figure 1.14.2. Tracings of some raptor wings showing the different aspect ratios.

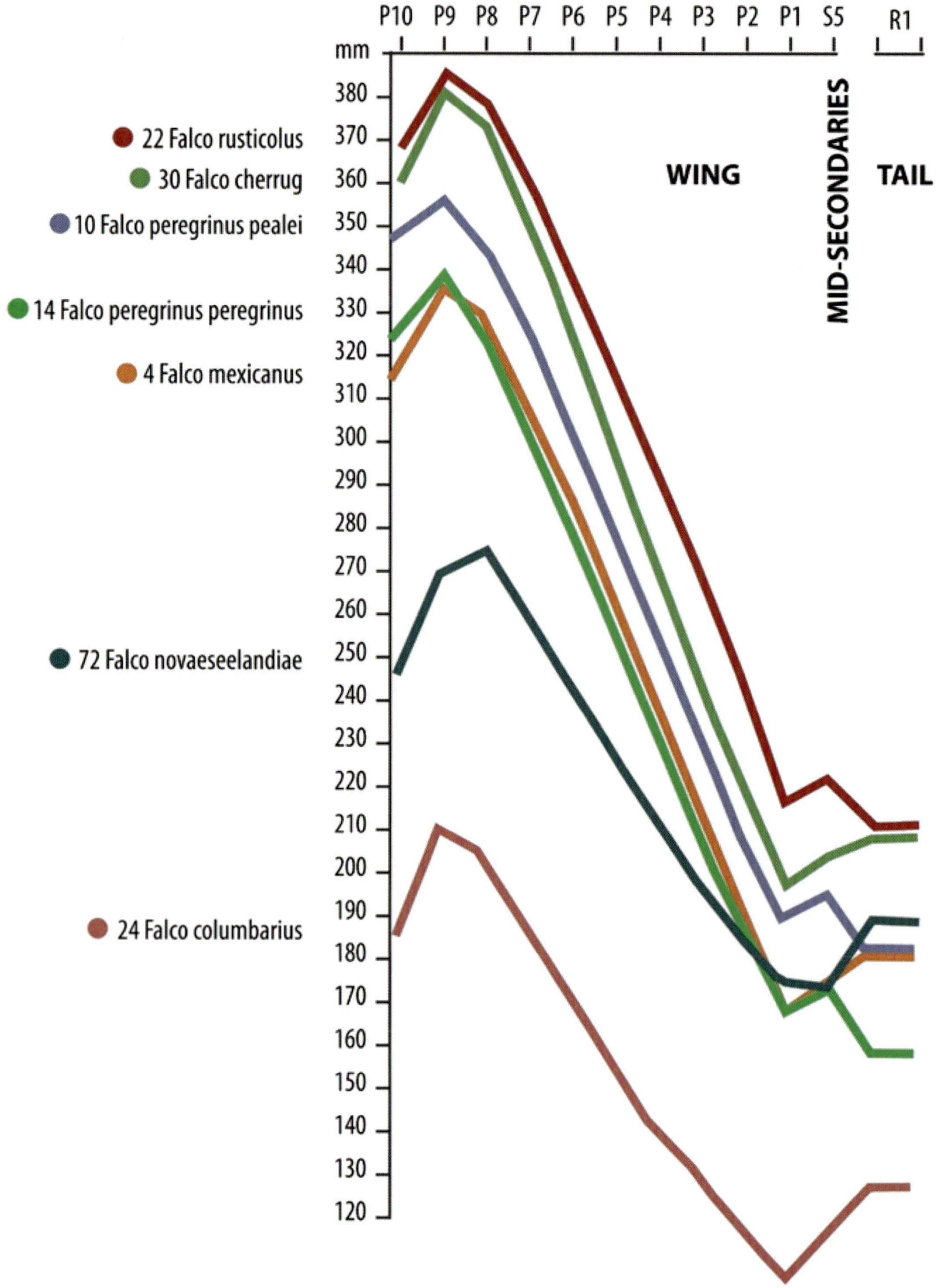

1.14.3 Primary, mid-secondary and tail lengths for six falcon species (including two sub-species). Both sexes and all ages combined to weighted means. Sample sizes as shown.

By measuring raptors (see 2.4) it is possible to see the specialist differences in the shape of the wing and tail (figures 1.14.3 and 4). The peregrines have the most pointed, narrow wings and short tails. They are the high-speed gliders. The desert falcons, such as the gyr, saker, and prairie have less pointed, wider wings and longer tails. They have a more general-purpose format, switching from bird to mammal prey as the need arises. The New Zealand falcon has a more accipitrine shape

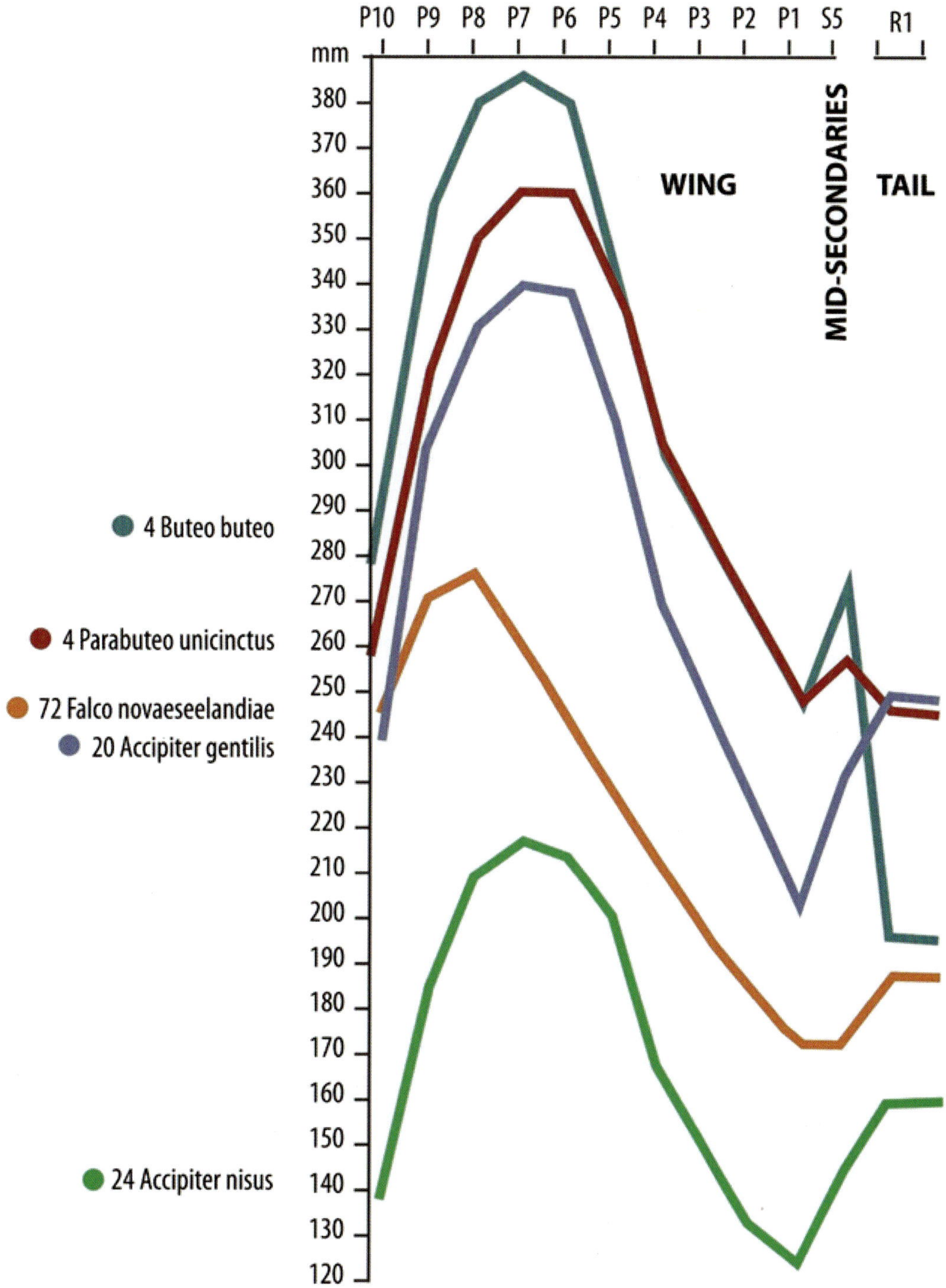

1.14.4 Primary, mid-secondary and tail lengths for five raptor species. Both sexes and all ages combined to weighted means. Sample sizes as shown.

and I have shown it again among the accipiters for a comparison. The merlin too, is somewhat accipitrine, chasing small agile birds and needing to be very manoeuvrable.

The common buzzard (figure 1.14.4) has a long, broad wing and a short tail. The goshawk has a shorter, narrower wing tip and a long tail. The Harris wing lies between the two but also it has a long tail. The sparrowhawk is a shorter-winged version of the goshawk. From these figures you can also see that the secondaries are enlarged in the accipiters, the Harris hawk, the New Zealand falcon and the gyrfalcon, or, putting it another way, the inner primaries are reduced.

There is plenty of room for more research on these topics. The important thing to understand is that there are many physical differences between the species, often small, but normally significant. To appreciate them, they must be viewed within the whole lifestyle of the bird: its skeletal and muscle systems, its physiology, its psychology, and the details of the way it hunts. Everything is interrelated.

1.15 Variations between juveniles and adults

Young birds are afraid of flying fast. This is very sensible of them. Have you ever driven your car as fast as it can possibly go? Speed kills! Every year we watch the falcons develop their flying skills and it is quite obvious that, to start with, they are not happy about flying fast because they cannot handle the landings. Fortunately, at this stage their feathers are still in sheath and they tend to flop without any real damage. But if you confine a hawk, such as a saker, in a breeding pen and don't let her fly until she is hard-penned, she may well have a panic attack. She takes off, develops quite an airspeed before she realizes what is happening, and unable to brake, land, or even turn, she may fly a kilometre or two before crashing. If she crashes into a tree and scrabbles onto a branch she may refuse to try flying again for one or two days. As you sit below the tree during this time, you can reflect on why it has happened.

The wild young raptors that develop naturally learn over their first four weeks of flying to cope with the manoeuvres of normal flight. They play together and snatch leaves off trees, gradually developing their coordination. But their parents are still supplying the food, so that they do not have to attempt hazardous flight manoeuvres chasing prey. This is when the selection pressure starts. One way around the problem is not to try to catch fast agile prey. Many juvenile raptors, such as red-tailed hawks, sakers, and prairie falcons, in their first autumn behave more like 'searchers' than 'attackers' (see Chapter 8). They catch easy things like worms, insects, voles, and other small prey. When winter comes, many of these prey disappear. They get locked into the frozen ground, hidden by snow, or just die back. The juveniles must turn to more difficult prey and become more like 'attackers.' Many don't make it. Those that do can handle higher-speed attacks and are able to survive on lower prey numbers. By the next spring, when prey is plentiful, they have proven themselves as hunters and are able to provide not only for their own needs but for a family too.

All this is reflected in the wings. The juveniles tend to have longer wing and tail feathers than the adults. This gives them a lower wing-loading and a lower stall speed. They can fly more slowly than their parents, more like 'searchers.' Their feathers also tend to be softer, less easily broken, but not as efficient for fast flight.

We have a lot of data on this but it is very complicated because the different genera change in different ways. The peregrine and merlin, for example, develop shorter narrower wings and shorter tails in their adult plumage. Accipiters shorten their tails but get longer, more pointed, wings. They become more falcon-like. The only way to study this is to obtain data on both juvenile and adult plumages for the same individuals, rather than single plumages for random individuals. This is another rich area for someone looking for a research project.

1.16 Gliding and soaring

Gliding and soaring are forms of unpowered, fixed-wing, flying dependent on three forces: lift, weight, and drag. The bird moves forward through the air under the influence of gravity, in a sinking, angled descent. The forward movement generates lift through the aerofoil of the wing, but the climb rate is less than the sink rate through the surrounding air. If the whole mass of air rises faster than the bird sinks, then the bird can still gain altitude; this is soaring. In gliding, the bird sinks through the air in order to travel horizontally; it trades height for distance, whereas in soaring, the bird is raised up by the air going up faster than the bird is sinking. Thus soaring

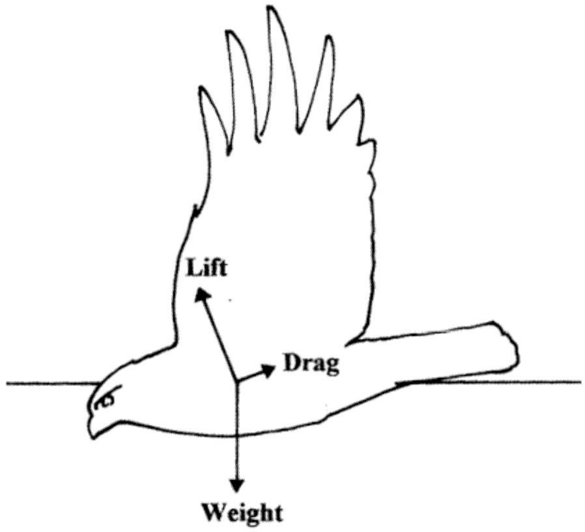

Figure 1.16.1. The forces acting on a soaring buzzard.

is a form of gliding which uses rising air to gain altitude. When gliding, the hawk slowly loses height as it gains distance and the ratio of height to distance travelled is called the glide angle. A reasonably capable gliding bird with long narrow wings, such as an adult peregrine, achieves a glide angle of 5-6° in still air, a ratio of about 1 in 10, which means that from a height of 300 metres (1000 ft) it reaches a horizontal distance in still air of 3 kilometres (about 2 mi), or even better. This is achieved by reducing drag through having a hard, streamlined contour, long, narrow wings and a short tail, not by reducing weight. Adjustments in weight alone alter the speed of descent but do not affect the glide angle. A heavy peregrine still achieves the same glide angle as a light one, but travels faster during the glide. The smooth contours minimize the disturbance to the surrounding air. The air rolls smoothly over the wings, spinning off the trailing edge of the wing tips as vortices (figure 1.16.2).

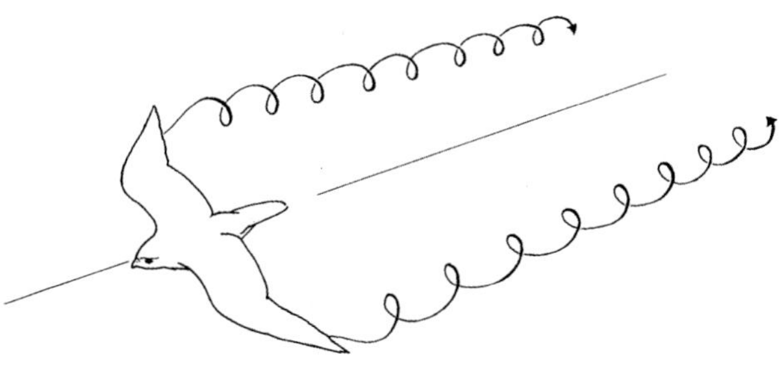

Figure 1.16.2. Vortices spiral off the wingtips of a peregrine as she glides on fixed wings.

As the glide angle gets steeper, the bird folds her wings in closer to the body (figure 1.16.3) and travels faster. This reduces the leverage strain on the pectoral muscles but increases drag slightly.

It is very common to see a wild falcon, in her home area, circle up on open wings with tail fanned in a column of rising air, perhaps near a hill face, until she is a speck in the sky, and then glide in a straight line toward her chosen hunting area. Using these two flying techniques very little energy is needed to cover her whole home range. Once you get used to being with the falcons, you can 'see' these invisible lifts, almost as clearly as an elevator in a department store. If you pilot a robotic bird you can find them for yourself, and if you have a barometric altimeter on board you can easily see the difference.

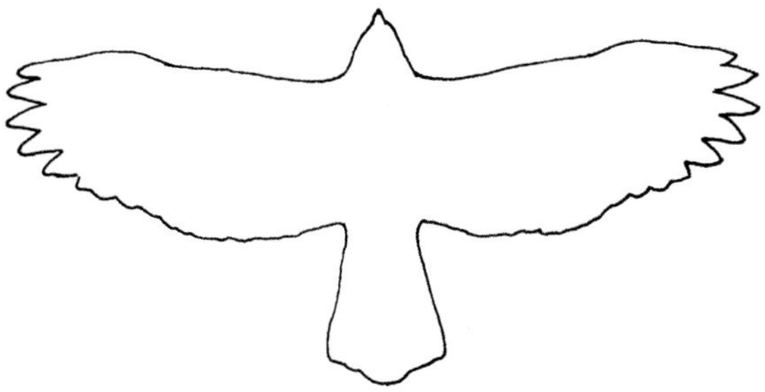

Harris gliding at 6.0° at 7.0 metres a second.

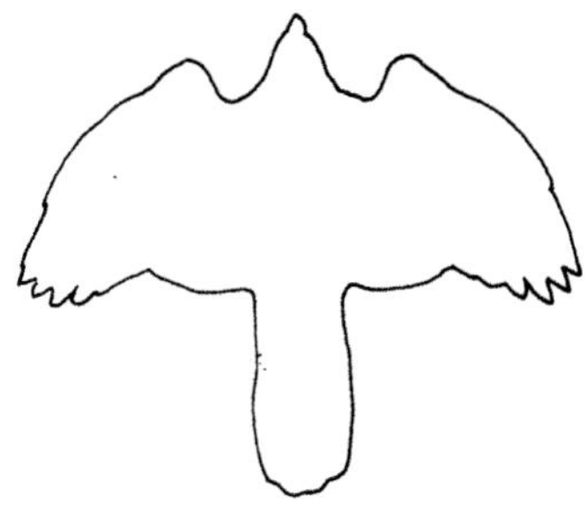

Harris gliding at 8.5° at 16.2 metres a second. (After Tucker and Heine, 1990)

Figure 1.16.3. The profile of a Harris hawk alters as the glide becomes steeper and faster.

Soaring is a form of gliding in that the bird is sinking through the air all the time. When soaring, the bird is not necessarily trying to cover a big horizontal distance but rather to stay in the same locality gaining height, cooling off, displaying or searching for prey. The emphasis therefore is not so much on travel but on staying up. This means

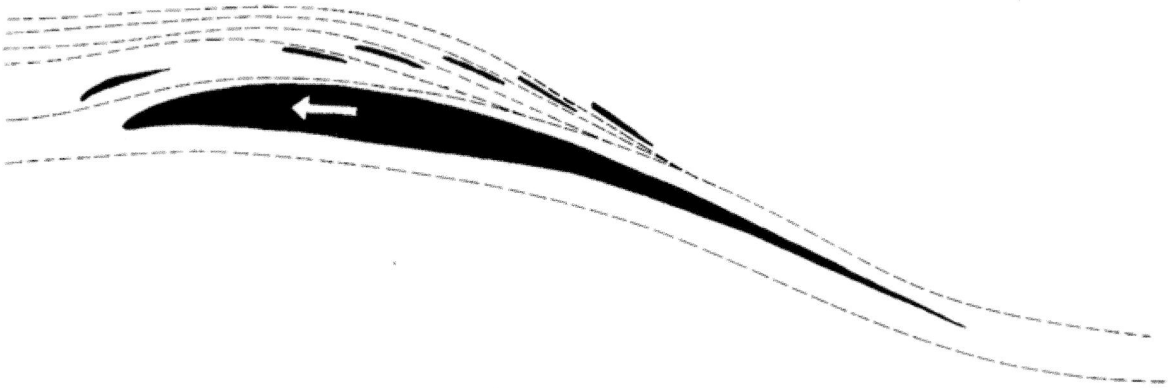

Figure 1.16.4.The airflow over a buzzard's wing tip before the stall.

the bird's airspeed can be sacrificed so that for any given glide angle she sinks less per minute. Therefore, the specialist soarers have reduced weight as much as possible. At normal flight speeds, drag is proportional to airspeed, so at lower airspeeds it is less important, and therefore the bird can have big lifting surfaces—big wings, deep secondaries, big-fanned tail. The extra drag these give is more than compensated for, at low speeds, by the extra lift they give.

Also in soarers, the outer primaries are emarginated so that the wing tips are deeply slotted or 'fingered.' Each primary tip thus acts as a separate narrow aerofoil giving lift with minimum drag and also lowering the stalling speed. To achieve this, the primaries are progressively stiffer toward the leading edge of the wing. The wing stalls when the airflow over the top of the wing, instead of being smooth and sucking the wing up, becomes turbulent. Slotting redirects air over the top of the wing, reducing the turbulence, lowering the stalling speed so that the bird can stay airborne at lower forward speeds.

Other species, such as accipiters and game birds, have slotted wings for an entirely different reason. Pheasants do not soar! These birds use their slotted wings for a power sprint and this is discussed in 1.17.

The specialist soarer, having reduced its sink rate to a minimum, needs a source of air which is rising faster than the hawk is sinking. This is supplied either by forced updrafts at hill slopes or by air warming up to form thermals. Obviously in flat areas the hawk must rely on thermals alone. In cool climates these are unreliable and specialist soarers, and some eagles and buteos, cannot survive there. Their range is restricted to places where lifting air is available often enough for them to find food. They usually must wait each morning until the air heats sufficiently to rise. Frederick II in Sicily forbade his falconers to fly their falcons after 9:00 a.m. because of the risk from eagles, which start to soar then.

Although most raptors can glide and soar, the most rapacious species, such as the large falcons and accipiters, are primarily adapted for the attacking phase of the hunt (see 8.1), and this is usually a powered, flapping flight. But, unlike aeroplanes, birds are alive and can alter their flight profiles from moment to moment. Thus a peregrine, by adjusting her wings and feathers, can optimize herself for high-speed gliding and then change to a soaring profile. You can read her intentions by her shape: narrow, slightly flexed wings and closed tail indicate that she is gliding and is intent on covering a big horizontal distance—perhaps she has seen some distant prey (see 8.13). Fanned wings and tail show that she is striving for height rather than distance. She may have found a column of rising air that will take her right up to the cloud base.

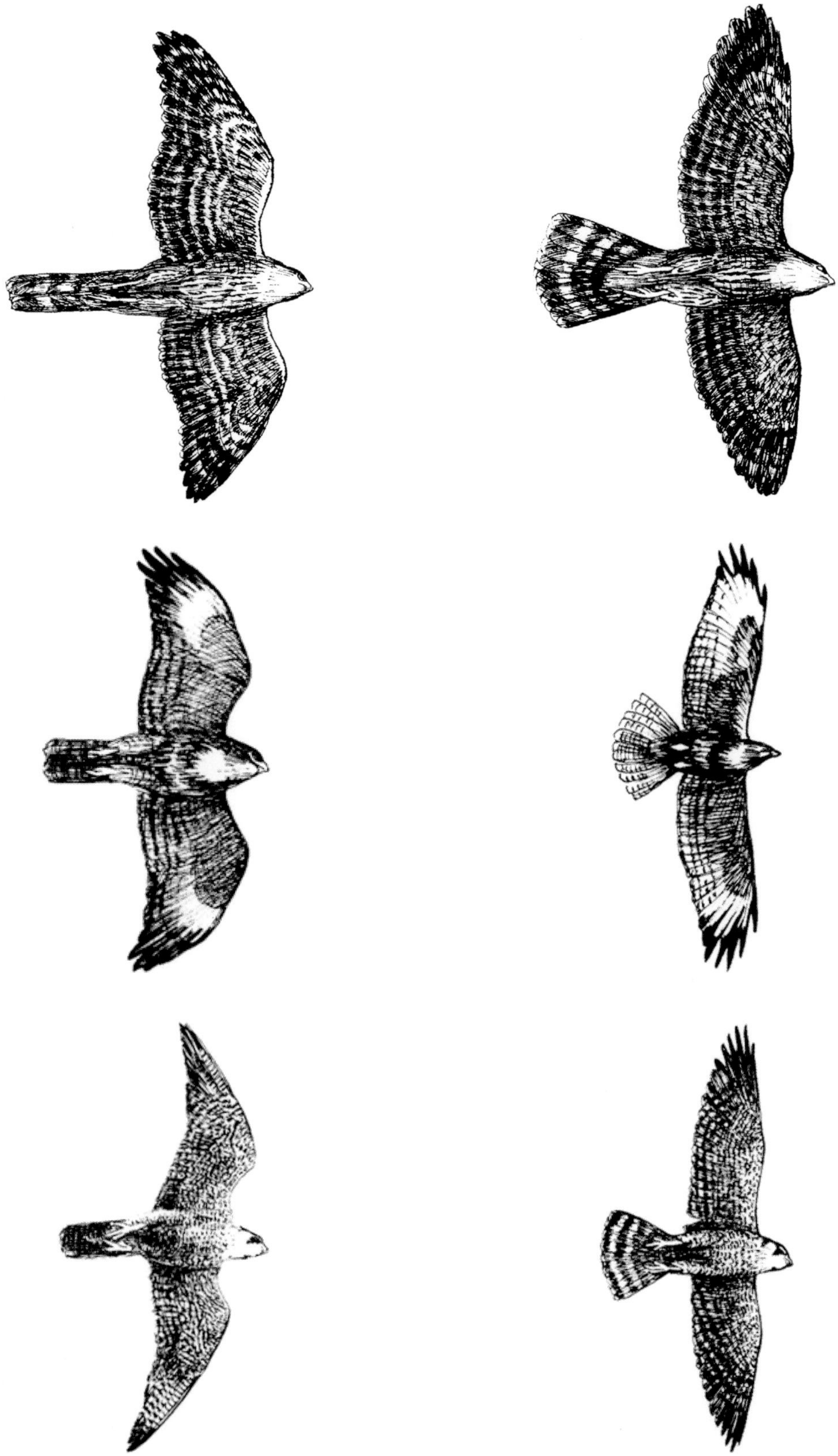

Figure 1.16.5.
Top: Juvenile northern goshawk gliding and soaring.
Middle: Common buzzard gliding and soaring.
Bottom: Adult peregrine gliding and soaring.

Figure 1.16.6 Adult peregrine in fast glide, about to stoop.

Species with low wing-loadings, such as the buteos, mainly search for quarry that is easy to catch. They are masters of soaring and gliding. They cannot manage a fast agile sprint and their favourite attack is the glide attack (8.13).

1.17 Powered flight

Ignoring for a moment all aerial manoeuvres such as turns, take offs and landings, powered flight, in which the hawk propels itself by flapping its wings, has several recognizable gaits. First is steady flight, more or less equivalent to our walking. Studies by Spedding, Pennycuick, and Rayner have shown that when a falcon, such as a kestrel or peregrine, is flying very slowly, at less than 3 m/s, the upstroke is passive, gives no lift, and no longer generates a vortex. The result is a pulsing doughnut-shaped vortex ring caused by the thrust of the downstroke (figure 1.17.1). During the upstroke the wing is flexed quite closely into the body, and, as it were, draws in and then shrugs off the trailing vortex of air.

The buzzards and eagles fly in a similar way except that their wingbeat is deeper, slower, and folds up more. Their upstroke is faster than the downstroke. During strenuous flight it is likely that the upstroke is active, giving some lift, but in slower flight it is probably passive.

As the falcon speeds up to above 7 m/s, the airspeed is sufficient to raise the wings, reducing the workload of the supracoracoideus muscles. Now the upstroke is active and generates trailing vortices (figure 1.17.2). Instead of folding closely to the body, the wing is held stiffer and straighter in the upstroke, resulting in the typical flickering flight of the cruising peregrine (figure 1.17.3).

When the raptor wants to go faster than cruising speed it has three choices: the buzzards and big eagles can't really sprint, so to make a faster attack they must come in from a height in a fast glide (see 8.13). The falcons and the accipiters can actively fly faster but they do it in two different ways. At this point I seem to have reached the edge of known science, so I can only

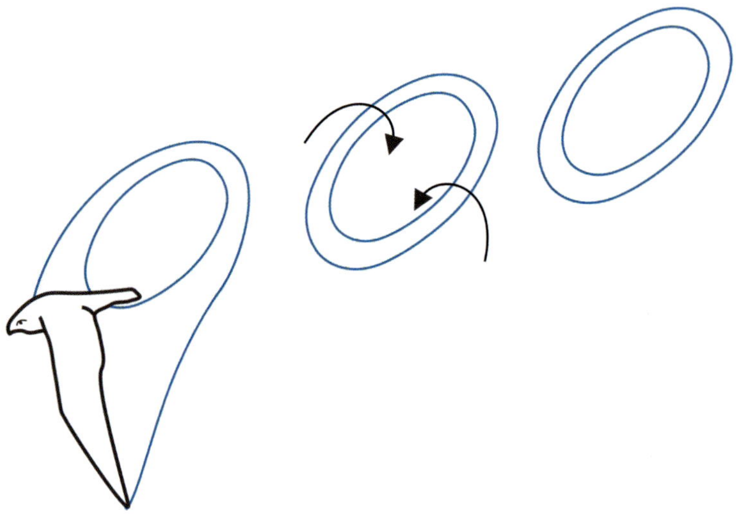

1.17.1. A vortex ring is formed by each downstroke of the falcon wing. The upstroke is passive.

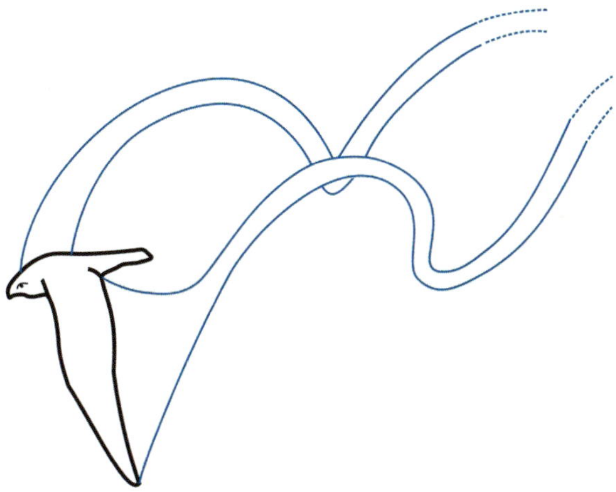

Figure 1.17.2. The falcon in normal flapping flight with continuous trailing vortices.

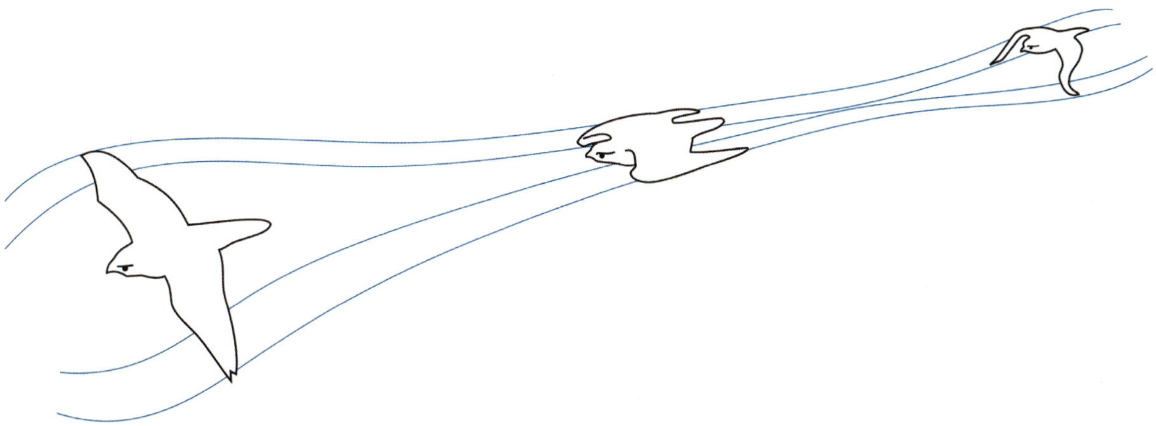

Figure 1.17.3. A peregrine with pumping, high speed wingbeat.

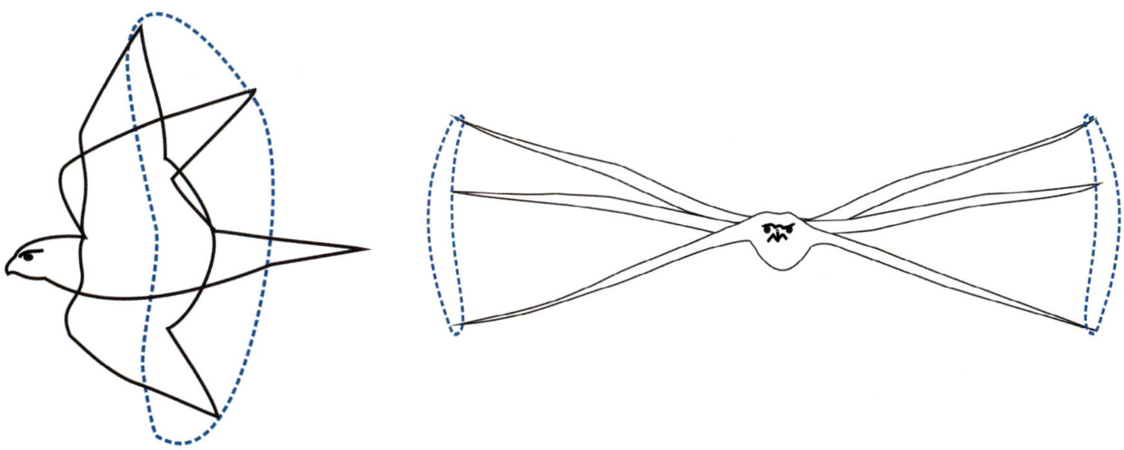

Figure 1.17.4. The stiff, shallow wingbeat of the cruising peregrine.

Young female gyr Bella in power flight attack. *Photo Stuart Grimwood.*

tell you what I think is happening. The problem of course is that it is not easy to study a sprinting raptor in detail, with or without a wind tunnel.

The falcon, already traveling very fast, at such a speed that her normal cruising wingbeats would slow her down, is like a cyclist who cannot pedal fast enough for it to make much difference. So the falcon glides with wings half shut, the secondaries providing lift, but otherwise minimizing leverage strain on the pectorals. The vortices trail smoothly behind (centre, figure 1.17.4) as in a steep glide. Then the falcon makes a series of deep pumping wingbeats in which it puts as much power as it can into some big fast downstrokes, faster even than the speed at which it is traveling, and angled to provide thrust, rather than lift. The result, which is very noticeable in a sprinting gyrfalcon, is that each wingbeat visibly pushes the bird forward in the air with a spurt of speed. It is almost as if it is being kicked from behind by an invisible boot. Only a short burst of these pumping beats is needed for the falcon to reach such a speed that further flapping can no longer help. By then something will have happened.

The accipiter also sprints but has a slightly different problem to solve. Whereas the big falcon needs to travel very fast over long distances, the accipiter needs to travel very fast over short ones. Her battles are won or lost in seconds. Therefore her problem is not so much one of top speed, but of acceleration. At the start of her sprint effort, the accipiter is either still, or moving slowly. So her sprint has to start from near zero. Whereas the gyrfalcon is like a cyclist who is winning a long race by sheer top speed, the accipiter is like a cyclist waiting at the starting line to win a race which finishes after 50 metres. At the finish she may not have even reached her top speed, but may still be accelerating. Therefore she has to cram as much effort as she can into a short period of time, and she cannot waste half of that time on a useless upstroke that does not contribute toward the thrust. She gets round this by using a slotted elastic wing, which not only stores energy, but also reduces drag in the upstroke.

During the down stroke (figure 1.17.5 a-e), thrust and lift are generated, but, because the wing is so short and the pectorals have such a mechanical advantage, the primary tips are bent backward to the point where their tips twist 90° from the plane of the wing, in both vertical and horizontal planes (figure 1.17.5 e). She then starts to draw in her wings, actively using her supracoracoideus muscles (1.17.5 f). The elastic primaries start to recover their normal shape, thrusting the air both downward and backward. This provides lift and thrust and also helps the supracoracoideus muscles. The wing is now halfway back up and semi-folded. The blades of the primaries are now edgeways on, in line with the angle of incidence. At the top of the upstroke they all come together, seal up and start to provide thrust again as the wing comes down.

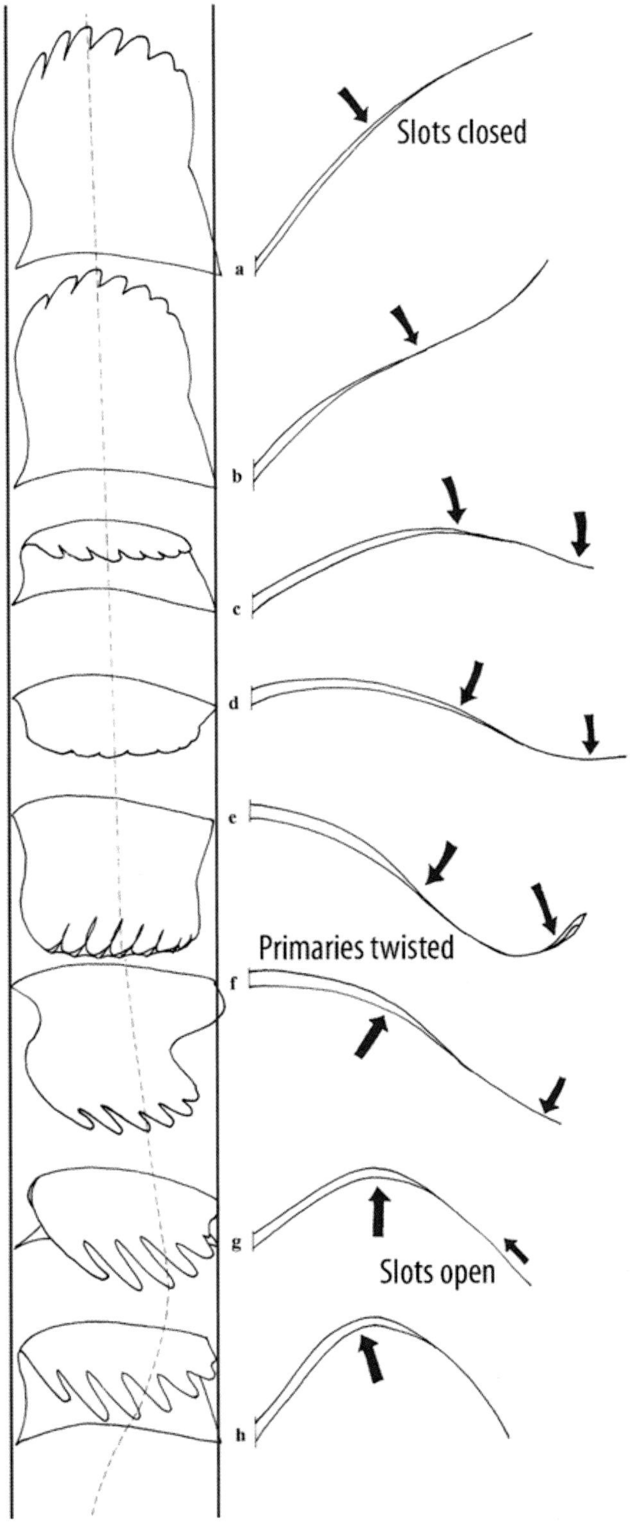

Figure 1.17.5 (a-e) The downstroke of a sprinting goshawk and (f-g) the upstroke.

Assuming still air and level, powered flight we can plot approximate acceleration curves for a variety of raptors. Lacking directly comparable data from fit hawks, I have compiled these curves from several sources, particularly from works by H. J. Slijper and T.A.M. Jack and from my own observations. While they may not be very accurate for absolute speeds, they give an idea of the differences between the species, although of course individual birds vary considerably.

Common buzzards (figure 1.17.6) are relatively weak at powered flight; slow to accelerate and with a slow top speed, they soon burn out. One seldom sees a buzzard sprint more than 100 metres and rarely 200 metres. She quickly starts to rest by gliding between beats, shown in the graph at 40 metres. During level flight she can manage about 10 metres per second, beating her wings at about 5.0-6.5 beats per second. There is very little quarry that travels slowly enough to tempt a buzzard into much more than an 80-metre effort. Most prey is either caught quickly (for example a vole) or quickly outdistances the buzzard in a tail-chase (for example, a partridge). Then the buzzard will tail off and perch, in this case, at 45 metres.

On the other hand, small accipiters, such as the sparrowhawk, develop full speed in under one second, within the first few metres. This explosive sprint gives them the edge over all other raptors in cover. But few maintain a flat out sprint after 100 metres; normally the flight has been concluded by then one way or the other, and if unsuccessful, the sparrowhawk will probably perch after 150 metres or so.

Slijper's work on goshawks showed that the male starts off faster but is overtaken by the female at around 70 metres. Once she gets going, the

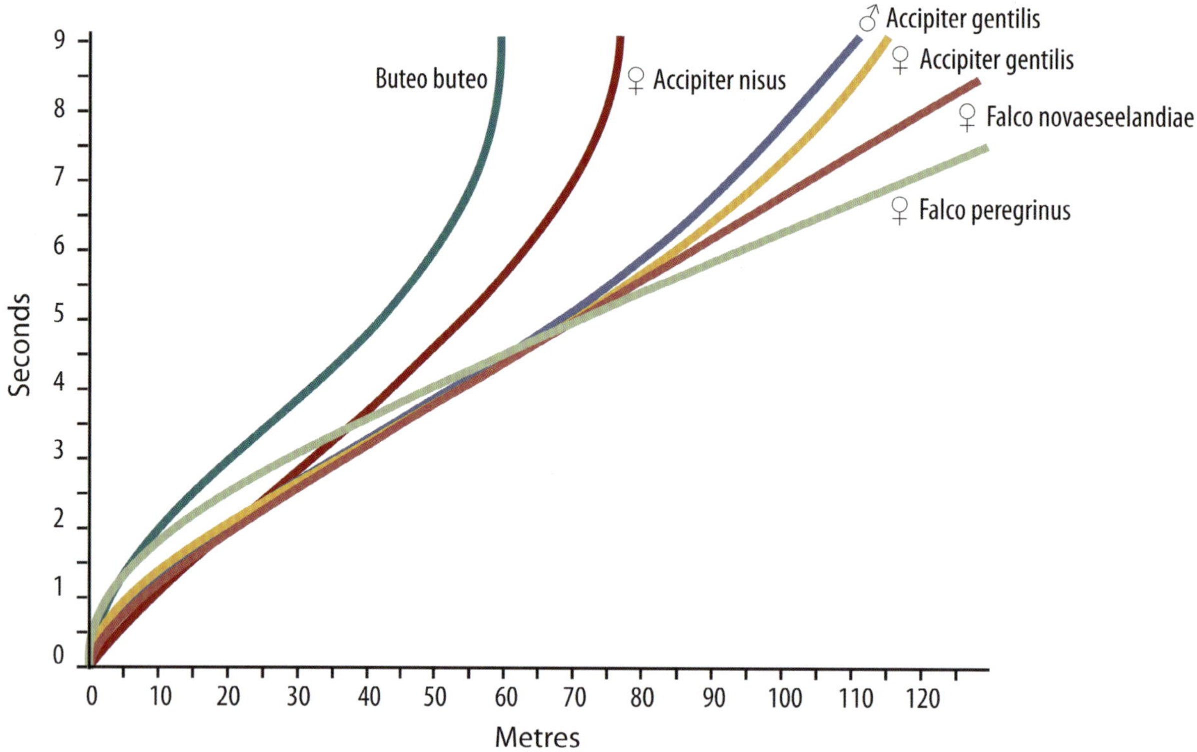

Figure 1.17.6 Acceleration curves for common buzzard, female sparrowhawk, male goshawk, female goshawk, female New Zealand falcon and female peregrine in a level sprint.

female is slightly faster. After about 130 metres or so goshawks usually slacken off somewhat; having failed to kill in the initial sprint, they either give up or cruise along gaining height to mark where the quarry puts in.

Over the first 20 metres, the peregrine is not much faster than the buzzard but by 50 metres she is building up speed, overtaking the goshawks by about 130 metres and keeping up a good pace for several hundred metres. Over a long distance she is probably outdistanced in level flapping flight only by the gyrfalcon. In the Flat Race to the Telwah, large falcons cover 400 metres in under 20 seconds, aided by ground effect, compressing the air between their wings and the surface of the ground. This represents 20 metres per second (45 mph).

The New Zealand falcon, with its accipitrine flight profile but falconine physiology, starts off much like a small male goshawk. By the time a female sparrowhawk has covered 80 metres the falcon has reached 100 metres. By about 130 metres, as the goshawks start to fade, she is going strongly but is overtaken by the peregrine which has got into her stride. By the time the peregrine has gone 280 metres the New Zealand falcon lags about 40 metres behind and both disappear over the horizon with no slackening.

The acceleration and top speed of the predator must parallel those of the prey in a tail-chase attack. Quail have similar flight patterns to sparrowhawks, pheasants to goshawks, and pigeons (with a slightly faster take off) to peregrines. Strong, long-flying prey are soon ignored by accipiters unless they can be caught in a sprint or ambush.

The effect of the rocketing ability of accipiters can be seen in their acceleration over the first 40 metres. Cooper's hawks, for example, normally beat their wings at 4.0–5.5 beats per second, but during take-off fly at 7–8 beats per second, using pectoral muscles which are about 17 percent of total body weight. They cover this distance when the buzzards and most falcons have barely gone 20 metres. Their performance over this distance is critical in making direct flying attacks (see 8.10), and in estimating the maximum distance at which it is feasible to attack a particular quarry species and have some hope of overhauling it. Large falcons tend not to attempt short direct attacks but prefer more calculated long ones. Buzzards do not sprint at all if they can possibly avoid it; instead they use height to give them an impetus for a glide attack, or a dive.

Apart from the infinite permutations of acceleration curves and attack distances, there is the question of agility. Here the accipiters, the merlins, and the New Zealand falcon score highest, the larger falcons and Harris hawks are somewhat less able, and finally, lumbering at the rear, come the buzzards. The quarry, too, vary a great deal in their agility. Agility tends to be bought at the expense of top speed; a long tail gives agility but creates drag.

1.18 The foot structure

The leg and foot of a raptor has seven functions: to support the weight of the body, to act as a rudder during some flight manoeuvres, to cushion the impact of landing, to catch, to hold and in some species, to kill prey, to help regulate body temperature, to preen those parts of the body inaccessible to the beak, and, in some species, to act as courtship signals. To perform some of these tasks effectively the foot must be light in weight for maximum agility and with a surface that is easy to keep clean of blood, dirt, and bacteria. To meet these requirements the foot has evolved so closely to the physical limitations of living tissues that there is very little margin for error and it is

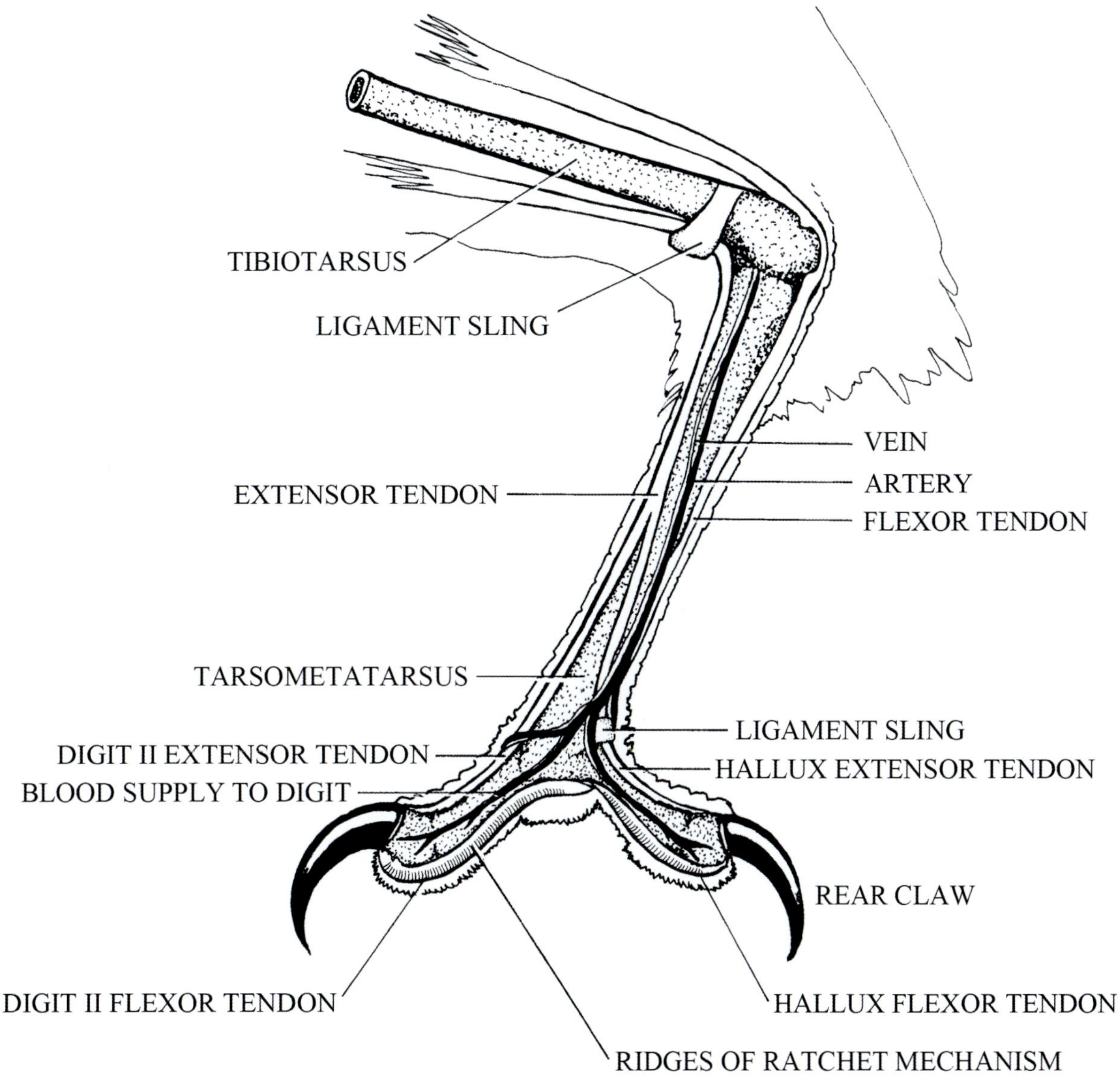

Figure 1.18.1 The ligaments and blood supply in the leg of the common buzzard.

not surprising therefore that the foot is so prone to disorders.

The foot basically is a series of bones and sheathed tendons. Behind the tibiotarsus and tarsometatarsus run the very strong flexor tendons. Muscles higher up in the thigh control these. The flexor tendons run right down through the ball of the foot along the underside of each toe to the furthest pads. When the muscles contract, the foot closes tightly. The tendons slide in grooves in the underside of the toe bones and are held in place by tough sheaths. The sheaths are lined with fine ridges rather like the grooves of a fingerprint and engage with rough ridges on the tendons themselves (figure 1.18.2). When the foot tightens, these ridges lock together like a ratchet mechanism on a handbrake. Possibly you will have noticed a stiff, jerky, creaky effect when loosening the grip of a live hawk; this is the ratchet being forcibly overridden. The ratchet has two useful purposes for the hawk: it enables her to 'lock' her foot closed on a branch while sleeping, and it means that once she has got a tight grip on her prey it does not require much muscular effort to maintain the grip. A human attempting to grip hard with his hand can only do so for a few seconds before the muscles lose power; if the victim stays quiet at first and then suddenly jerks away he finds the grip has relaxed. Of course the ratchet, like so many devices, is subject to technical hitches; not a few ospreys have drowned by being unable to unlock from an oversize fish. The use of a thin glove can encourage a falconry hawk, especially an excitable one, to lock onto the fist. These birds are known as 'sticky-footed.'

The flexor tendons run through a groove at the top of the tarsometatarsus in such a way that as the weight of the sleeping bird tends to buckle the legs, so the flexors tighten the grip on the perch. Conversely a goshawk, whose tendons are set at a fairly tight adjustment, is uncomfortable on a flat surface. She cannot open her feet completely without straightening her upper legs into a position that is tiring to maintain and, if kept on a flat surface for several weeks, her flexor tendons may collapse.

The extensor tendons run down the front of the tibiotarsus and the tarsometatarsus, one of them branching back through a ligament sling, or

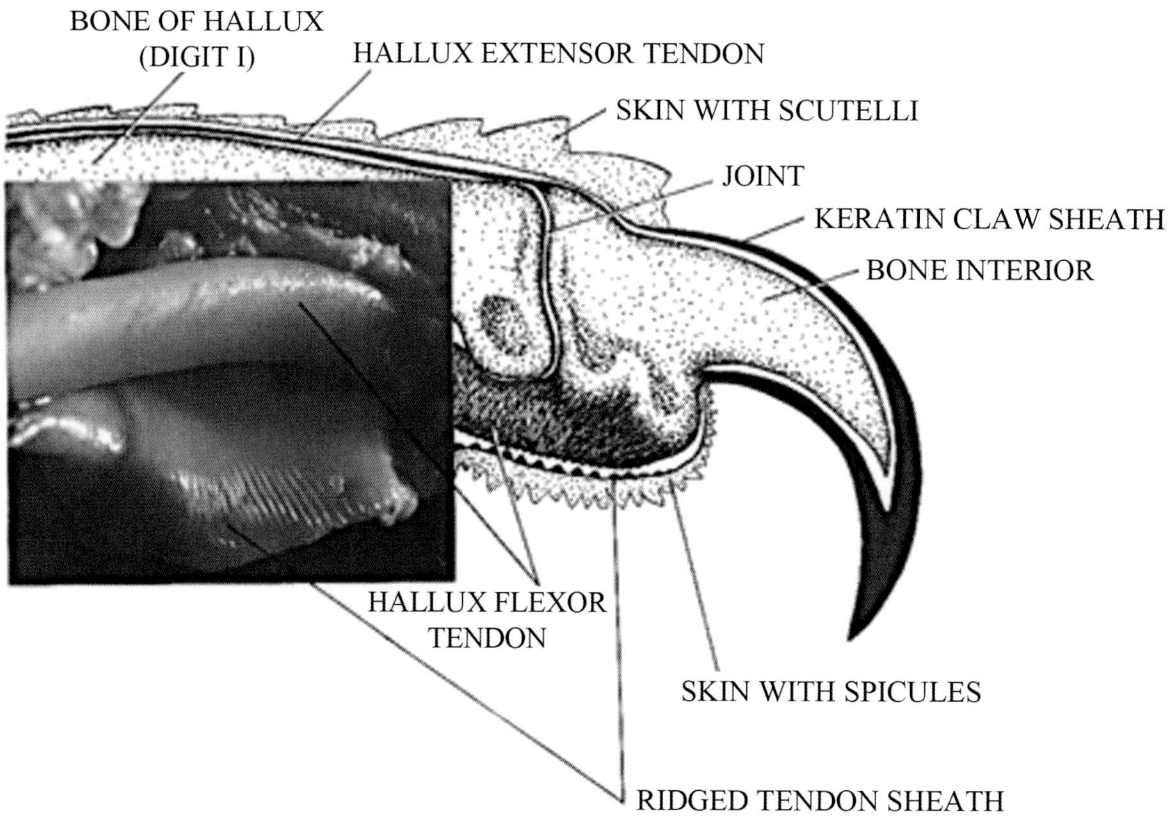

Figure 1.18. 2 The ratchet mechanism.

retinaculum, to a rear toe. Their effect is to extend the toes and of course they are weaker than the flexors. The two ligament slings act like pulleys and keep the tendons in place, otherwise they would pull away from the bone like a bowstring.

The blood vessels lie just under the skin thus avoiding being pinched by the tendons. Both artery and vein run alongside each other for most of their course. This means that the warm outgoing blood in the artery is cooled by the returning blood in the vein, which is itself warmed by the artery. This conserves heat in cold weather and during exertion or hot weather the vessels dilate allowing the bird to lose quite a lot of heat this way. Some vultures even excrete urine onto their feet to increase evaporative cooling. Fortunately, falconers' birds are usually better behaved!

As the foot consists mostly of bone and tendon, the blood circulation within it is meagre, especially during long periods of inactivity. This means that any injury—a sprain, torn ligament, bite, thorn, or scratch, can take a long time to heal. An infection is hard to overcome. Prolonged pressure can lead to degeneration of the underlying tissues or the formation of a core of dead tissue like a corn. The undersides of the feet have lump-like grips. If the hawk is kept on a smooth surface, these become the pressure points. First they become polished and often pink, then they slowly die, allowing bacteria to enter. The surrounding tissue reacts by building up layers of cells, resulting in a type of bumblefoot.

Some falconry species, such as peregrines, are very heavy for their size and so their feet have to carry more weight than usual. The large falcons are often very phlegmatic in captivity, increasing their risk of corns. A beef diet seems to make raptors more prone to foot trouble and if the bird is kept fat and fed in the evening an extra 25 percent weight is on the feet all night, just when the bird is inactive and with one foot up.

Preventive husbandry can reduce these risks. Birds are best kept on perches with sufficient curvature to prevent undue tension of the flexor tendons but of sufficient width to spread the weight evenly over the whole foot area. A rippled surface, such as astroturf, prevents pressure points without causing abrasion and a drained surface stops the soles of the feet being continually damp and soft and reduces bacteria. A balanced diet with tirings encourages good foot tissues and a good circulation, discouraging 'filled' legs, and reduces the likelihood of the hawk picking at its feet in boredom. The pen should be designed to prevent the bird landing heavily and bruising its feet. Feeding in the morning and a careful watch on the bird's weight reduces the constant pressure on the feet because the food is digested more rapidly during the daytime (see section 6.3). Careful coping of the claws, combined with proper handling to prevent self-inflicted injuries, and prompt attention to bites and thorns, will reduce the risks of infection. Although treatments for foot conditions are constantly being improved, prevention is still better than cure.

The claws are of different sizes and shapes. The hind claw and inner claw are primarily used

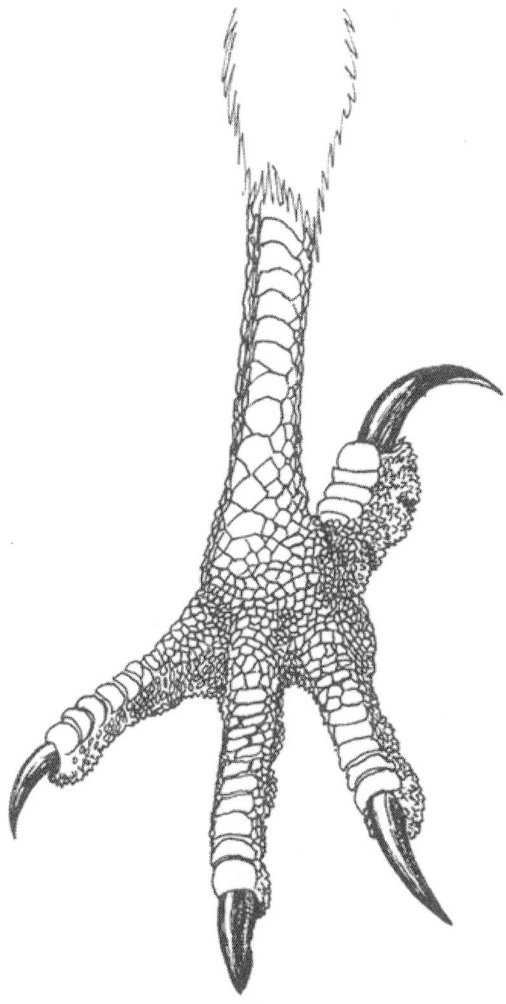

Figure 1.18.3 The right foot of the common buzzard.

for killing and gripping prey and are therefore the largest. The claw on the centre toe (digit III) is short with a lateral flange on the inner side. The hawk uses this to preen and scratch the delicate areas of the head that could be damaged if scratched with the point of a sharp claw. The outer toe's claw is also short. These two toes give the hawk stability when perched and increased reach for catching prey.

The skin of healthy adults of most raptor species is bright yellow due to carotene pigments. In the large falcons the breeding adult males have noticeably deeper yellow feet than their mates, whose feet have lost colour, possibly owing to the uptake of carotene for the developing eggs. The males show their bright legs during the tippy-toed walk of courtship display (figure 2.6.2). Many of the large falcons have pale blue or green feet when young, slowly turning yellow when they become sexually mature. But foot colour is also genetic; I have seen peregrine nests in the Arctic with one chick with blue feet and another with yellow feet. Also a domestic peregrine/saker with bilateral differences: one yellow foot and one blue foot, with one half of the tail longer than the other half.

The proportions of the feet vary according to the prey normally taken. Raptors that catch birds in the air, such as sparrowhawks, peregrines, merlins, Orange-breasted Falcons *Falco deiroleucos*, and New Zealand falcons, have long toes with pronounced grips underneath. The peregrines have short tarsi to withstand the impact of the stoop, whereas the sparrowhawks and New Zealand falcon have long tarsi to enable them to snatch birds from cover and during a twisting tail-chase.

Mammal specialists, such as the kestrels and many species of buteos, have shorter stronger toes; often these mammals are killed by the strength of the grip rather than by a bite.

Intermediate forms, which catch both birds and mammals, such as the goshawks and the desert falcons, have an intermediate foot, reflecting the dual-purpose role. Others are specialists, such as the Secretary Bird *Sagittarius serpentarius*, which has very long legs and small, strong feet for dealing with snakes. Careful examination of the feet shows that they are just as finely tuned for their lifestyle as are the wings. If some clumsy human comes along and puts rings, tags and anklets onto this triumph of evolution, don't be surprised if performance is handicapped.

1.19 Size

'I want the biggest one!' Some people are obsessed by size, but size is no measure of performance. You wouldn't choose a race horse just because it is the biggest one. And anyway, what do we mean by 'biggest'? How can we actually measure size?

We can choose a fixed element of the bird that can be measured. Beak to tail (body length) is no good because the neck can stretch. Measurements can vary by 15%. Similarly, girth, as measured with a tape around the widest part of the body also has a measuring error of 15%. On most raptors the longest fixed measurement is the wing length and I show how to measure this in 2.4. This has an accuracy of about 1.5%. between different people making the measurements. The next longest dependable measurement is tail length, with an accuracy of 3%.

Weight is another measure of size and this varies by 30% at least, as shown in 6.2. It is hugely unreliable. But if we try to make our measurements as consistent as possible: same age, same time of day, same fitness level, same food in the gut and so on, we still find that the correlation between physical size (using wing length as an index) and body mass is variable. We can find large lightweight hawks and small heavy ones. This will be reflected in their relative wing-loadings, which in turn affects their flight characteristics. So, rather than simply saying 'I want the biggest one', you might say, I want a small, heavy one with a high wing-loading who flies like a fighter jet. We have a male gyr here who initially trained at 950g but weighs 1450g when fat. You could pay a lot of money for 500g of fat!

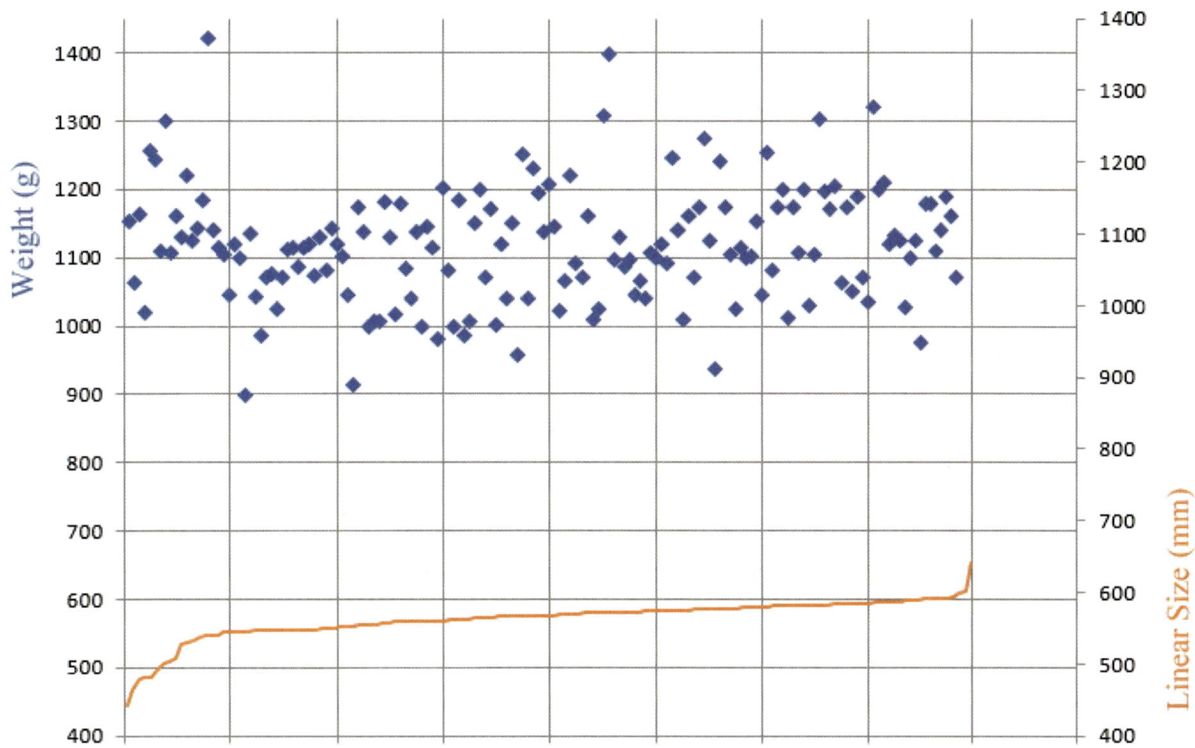

Figure 1.19.1 Weight scatter of male gyrs and gyr/sakers against linear size (wing plus tail length).

1.20 Sexual dimorphism and variations with lifestyle

Although size differences are very noticeable to the casual observer, size is just one of many parameters which should be looked at together to understand what is going on. Sexual dimorphism is when the sexes are of different sizes or shapes. In raptors the females are usually larger and some people call this Reverse Sexual Dimorphism (RSD). Sexual dichromatism on the other hand, is when the two sexes are of different colours. This is usually in the adults. Juveniles often also have a different colour plumage...you can see that the water is already getting muddy!

Colours are usually to do with the way hawks react with each other, for example in courtship. Many species, such as the saker and the common buzzard, are very variable but are hard to differentiate as juvenile or adult, male or female, by colour. So dichromatism is more of an evolutionary luxury rather than a necessity in raptors.

Sexual dimorphism is most pronounced in 'attackers' and least in the 'searchers' (see Chapter 8). Many other characteristics come in the same package. The attackers tend to have high wing loadings, an unstable, anhedral profile, and catch agile prey in an active attack. They have a lower hunting success but usually one kill is sufficient for the day and they have short, lightweight, less efficient guts. The main selection pressure acts on them during the attacking phase of the hunt.

Searchers tend toward the opposite on all these features. In between are a host of intermediate species making up a spectrum. So it seems likely that sexual dimorphism is selected for most strongly during the attacking phase of hunting agile prey. Raptors try to evade this selection pressure in two ways. As juveniles, some attackers behave temporarily more as searchers, as discussed in 1.15. As adults they can reduce the selection pressure by widening their prey base. This is achieved by one sex taking the smaller prey and the other taking the larger ones. In this way one species can virtually fill the niches of two species. Doing this may trespass on the niche of

The male New Zealand falcon (right) at 300g is 53% of the 570g female.
Photo Colin Wynn.

another species above or below in the size scale. This can be seen in the North American accipiters: the Sharpshin *Accipiter striatus*, the Cooper's hawk, and the northern goshawk, partitioning six niches between them. This could limit the extent to which one of them can diversify up or down the size scale. If you take away close competitors, the species could be more dimorphic, and this is seen in the New Zealand falcon, until a point is reached at which the female is so large that she might too easily intimidate the male.

Why should the female be the sex that is larger? Why not the male? The female is the one who lays the eggs and has to stay at the nest for six or seven weeks. She is dependent on the hunting ability of the male, and later, her chicks are also dependent on him. This period is timed to coincide with the young prey animals leaving their nests. The male has access to the young of all his normal prey species, plus the young of the female's prey species, which at that time are smaller. He must find prey that is worth the energy cost of ferrying back to the nest (he is better off consuming any insects he may catch because they are not worth ferrying home). He also has to find prey within the size range that he can easily carry. In the absence of neighbours, his hunting range is thus limited by how far away it is worth going to catch something and bring it back. By the time his chicks are getting large enough to be safely left alone for a while so that the female can also hunt, the prey have also grown larger. She can cash in on this more easily and economically than can the male. Thus it is all a balance between the method of attack, the size of the prey, and the timing of the development of young raptors and young prey.

There are other factors that arise from this situation. Theoretically the large female can more easily produce eggs, but on the other hand a female sparrowhawk is just as effective an egg producer as a female goshawk four times her size. A big female may be better at defending the nest. She may also be aggressive toward or dominate the male. I believe these are all outcomes of sexual dimorphism, rather than causes of it. There are plenty of predators in which the male is larger than the female. If the male was really such a rapacious fiend as he is painted, why doesn't he

Mobile vet clinic at the races in Riyadh.

just kill his defenceless young? In reality, raptors are strongly inhibited against killing anything at the nest and there are many instances in which raptors have brought young prey back to the nest still alive and then adopted them and reared them.

The most severe selection pressure and attrition in raptors operates during searching for, and most particularly during attacking prey. This is the moment of truth in a raptor's life and the time when poor performers are weeded out. The driver for sexual dimorphism operates through mortality of poor hunters, and the driver for the female being larger operates through productivity during nesting.

1.21 Health

Throughout this book I am assuming the raptors are healthy. But of course this is a massive assumption to make. All of us are hosts to many species of micro-organisms or have a health issue of one kind or another. Even genetically we are prone to various conditions. We may for the time being appear A1 healthy, but if physically or mentally stressed, underlying conditions may flare up. On the other hand it is easy to become paranoid and imagine diseases that are not there.

I am not a veterinarian and this book is about healthy raptors, not sick ones. You need to learn about all the common health issues. If you have to treat your bird, for example dress wounds, tube feed, administer medicines or collect samples, our film *Health Care* will guide you through basic nursing procedures. Careful consideration of the housing you use, hygiene protocols and nutrition will pre-empt sickness flaring up, but if you see some symptoms you are not sure about, get the bird to your vet, and do it quickly.

1.22 Further Reading

Arent, L.R. 2007. *Raptors in Captivity: Guidelines for Care and Management*. Hancock House. 978-0-88839-613-6.

Andersson, M and R.A. Norberg. 1981. Evolution of reversed sexual size dimorphism and role partitioning among predatory birds with a size scaling of flight performance. Biological Journal of the Linnean Society 15:105-130.

Bailey, T.A., N.C. Fox, A. Mukhtar, A. Toosi and J.H. Samour. 2000. Reconciling conservation interests with health and medical issues of the Laggar Falcon (*Falco jugger*) in Pakistan. In Raptor Biomedicine III 273-279. J.T. Lumeij*et al* (Eds).

Barton, N.W.H. and D.C. Houston.1993. The influence of gut morphology on digestion time in raptors. Comparative Biochemistry and Physiology Part A: Physiology. 105A. 571-578.

Baumel, J.J. (ed). 1979. Handbook of Avian Anatomy: *Nomina Anatomica Avium*. Academic Press, London. ISBN 0 12 083150 3.

Baumgart, W. 1991. *Der Sakerfalke*. Ziemsen Verlag. ISBN 3-7403-0258-5.

Beynon, P.H. *et al* (Eds). 1996. *Manual of Raptors, Pigeons and Waterfowl*. British Small Animal Veterinary Association Ltd. ISBN 0 905214 29 3.

Boström, J.E., M. Dimitrova, C. Canton, O. Håstad, A. Qvarnström and A. Ödeen. 2016. Ultra Rapid Vision in Birds. PLoS ONE 11(3): e0151099.

Brown, L.H. 1976. *Birds of Prey: their Biology and Ecology*. Hamlyn. ISBN 0600 313 069.

Brown, L.H. and D. Amadon. 1968. *Eagles, Hawks and Falcons of the World*. Country Life.

Burton, R. 1990. *Bird Flight. Facts on File*. Oxford. ISBN 0 8160 2410 3.

Butler, P.J., N.H. West and D.R. Jones. 1977 Respiratory and cardiovascular responses of the pigeon to sustained, level flight in a wind tunnel. Journal of Experimental Biology 71:7-26.

Cade, T.J. 1982. *Falcons of the World*. Cornell University Press. ISBN 0 8014 1454 7.

Campbell, T.W. 1988. *Avian Hematology and Cytology*. Iowa State University Press. ISBN 0 8138 0064 1.

Carruthers, A. C., Thomas, A. L. R. and Taylor, G. K. (2007). Automatic aeroelastic devices in the wings of a steppe eagle *Aquila nipalensis*. J. Exp. Biol. 210, 4136-4149.

Cheney, J. A. ,J. Stevenson, N.E. Durston, M. Maeda, J. Song, D. Megson-Smith, S.P. Windsor, J.R. Usherwood, and R.J. Bomphrey. 2021. Raptor wing morphing with flight speed. Journal of the Royal Society Interface, 18(180): 20210349.

Cho, Y.S., Jun, J.H., Kim, J.A. *et al.* 2019. Raptor genomes reveal evolutionary signatures of predatory and nocturnal lifestyles. Genome Biol. 20, 181.

Cooper, J.E. 1978. *Veterinary aspects of captive birds of prey*. Standfast Press. Glos. ISBN 0904 602 044.

Cramp, S and K.E.L. Simmons. 1980. *The Birds of the Western Palearctic* vol. II. Oxford University Press. ISBN 0 19 857505X.

Duriez, O. et al. 2014. How cheap is soaring flight in raptors? A preliminary investigation in freely-flying vultures. PLoSOne 9, e84887.

Durston, N. E., Wan, X., Liu, J. G., & Windsor, S. P. 2019. Avian surface reconstruction in free flight with application to flight stability analysis of a barn owl and peregrine falcon. The Journal of experimental biology, 222(Pt 9), jeb185488. https://doi.org/10.1242/jeb.185488

Eastham, C.P. 1999. Species concepts and their relevance to the taxonomy of the Desert Falcons. Falco 13: 18-20.

Eastham, C.P. 2000. Morphological studies of taxonomy of the saker (*Falco cherrug* - Gray 1833) and closely allied species. PhD thesis, University of Canterbury, Kent.

Eastham, C.P., Nicholls, M.K. & Fox, N.C. 2002. Morphological variation of the saker (*Falco cherrug*) and the implications for conservation. Biodiversity and Conservation 11, 305–325.

Farner, D.S. and J.R. King. 1974. *Avian Biology*. Academic Press, New York.

Fischer, D.; Van-Waeyenberghe, L.; Failing, K.; Martel, A. and Lierz, M. 2018. Single tracheal inoculation of *Aspergillus fumigatus* conidia induced aspergillosis in juvenile falcons (*Falco* spp.). Avian pathology, 47(1): 33-46.

Fowler, D.W. *et al.* 2009. Predatory Functional Morphology in Raptors: Interdigital variation in talon size is related to prey restraint and immobilisation technique. PLoS ONE 4.11. 2009

Fox, N.C., A. Mukhtar, C.P. Eastham and H.J. MacDonald. 1996. Research and Conservation of the Laggar Falcon (*Falco jugger*) in Pakistan. 2nD International Conference on Raptors, Urbino, Italy.

Fox, R., S.W. Lehmkuhle and D.H. Westendorf. 1976. Falcon visual acuity. Science. 16;192(4236):263-5.

Frederick II of Hohenstaufen. 1250. *De Arte Venandi Cum Avibus: The Art of Falconry*. Translated by C.A. Wood and F. M. Fyfe. 1943. Stanford University Press. ISBN 10: 0804703744 .

Frost, B. J., L.Z. Wise, B. Morgan and D. Bird. 1990. Retinotopic representation of the bifoveate eye of the kestrel (*Falco sparverius*) on the optic tectum. Vis.Neurosci. 5: 231-239.

George, J.C. and A.J. Berger. 1966. *Avian Myology*. Academic Press.

González-Martín-Moro, J, J.L. Hernández-Verdejo and A. Clement-Corral. 2017. The visual system of diurnal raptors: updated review. Arch. Soc. Esp. Oftalmol. 92(5):225-232.

Gorham, F.W. and A.C.Ivy. 1938. *General function of the gall bladder from the evolutionary standpoint.* Chicago: Field Museum of Natural History.

Hackett, S. J. *et al.* 2008. A phylogenomic study of birds reveals their evolutionary history. Science 320:1763-1768.

Heidenreich, M. 1997. *Birds of Prey: Medicine and Management*. Blackwell Science. ISBN: 0632041862.

Hildebrand, M., D.M. Bramble, K.F. Liem and D.B. Wake (eds.) 1985. *Functional Vertebrate Morphology*. Harvard University Press, Cambridge, Massachusetts.

Inzunza, O., H. Bravo, R.L. Smith and M. Angel, M. 1991. Topography and morphology of retinal ganglion cells in Falconiforms: a study on predatory and carrion-eating birds. Anat. Rec. 229, 271-277.

Jones, M. P., K.E. Pierce, Jr and D. Ward. 2007. Avian vision: a review of form and function with special consideration to birds of prey. J. Exot. Pet Med. 16: 69-87.

Kermode, A.C. 1970. *Flight without Formulae*. Pitman. ISBN 0 273 41680 4.

Kessel, A., F. Hoffman, A. Baars and L. Danter. 2017. Do bony tubercle inside falcon nostril effect breathing during high speed diving? Conference for the Society of Experimental Biology.

King, A.S. and J. McLelland (eds.). 1979 & 1981. *Form and Function in Birds*, vols. 1 & 2. Academic Press, London.

King, A.S. and J. McLelland. 1984. *Birds: Their Structure and Function*. Bailliere Tindall, London. ISBN 0 7020 0872 9.

Lumeij, J.T *et al.* (Eds). 2000. *Raptor Biomedicine III*. Zoological Education Network. ISBN 0-9636996-1-X

Lord, R. D., Jr 1956. A comparative study of the eyes of some falconiform and passeriform birds. Am. Midl. Nat. 56, 325-344.

Marshall, A.J. 1961. *Biology and Comparative Physiology of Birds*. Academic Press, New York.

Martin, G. R. 2012. Through birds' eyes: insights into avian sensory ecology. J.Ornithol. 153, S23-S48.

Martin, G. R. and Katzir, G. 1999. Visual fields in short-toed eagles, *Circaetus gallicus* (Accipitridae), and the function of binocularity in birds. Brain Behav. Evol. 53,55-66.

McLelland, J. 1990. *A Colour Atlas of Avian Anatomy*. Wolfe Publishing. ISBN 0 7234 1575 7.

Mills R, H. Hildenbrandt, G.K.Taylor, and C.K. Hemelrijk. 2018. Physics-based simulations of aerial attacks by peregrine falcons reveal that stooping at high speed maximizes catch success against agile prey. PLoSComputBiol 14(4): e1006044.

Mitkus, M., S. Potier, G.R. Martin, O. Duriez and A. Kelber. 2018. Raptor Vision. Oxford Research Encyclopedia of Neuroscience.

Muller, M.G. *Practical Handbook of Falcon Husbandry and Medicine*. Nova Science Publishers. 978-1607416081.

Naisbitt, R and P. Holz. 2004. *Captive Raptor Management and Rehabilitation*. Hancock House. ISBN 0-8839-490-X.

Nittinger, F, E. Haring, W. Pinsker, M. Wink and A. Gamau. 2005. Out of Africa? Phylogenetic relationships between *Falco biarmicus* and the other hierofalcons (Aves: Falconidae). JZS 43(4), 321–331.

O'Connor, R.J. 1984. *The Growth and Development of Birds*. John Wiley and Sons. ISBN 0 471 90345 0.

O'Rourke, C. T., M.I. Hall, T. Pitlik and E. Fernández-Juricic. 2010. Hawk eyes I: diurnal raptors differ in visual fields and degree of eye movement. PLoS ONE 5,e12802.

Pendleton, B.A.G., B.A. Millsap, K.W. Cline, and D.M. Bird (eds.). 1987. *Raptor Management Techniques Manual*. National Wildlife Federation scientific and technical series, no. 10. Washington, DC. ISBN 0 912186 82 8.

Pennycuick, C.J. 1972. *Animal flight*. Studies in Biology 33. Edward Arnold, London.

Pennycuick, C.J. 1989. *Bird Flight Performance, A Practical Calculation Manual*. Oxford University Press. ISBN 0 19 857721 4.

Pennycuick, C.J. and A. Lock. 1976. Elastic energy storage in primary feather shafts. J. Exp. Biol. 64: 677 689.

Pennycuick, C.J. and M.A.Rezende. 1984. The specific power output of aerobic muscle, related to the power density of mitochondria. Journal of Experimental Biology 108:377-392.

Piggott, D. 1977. *Understanding gliding: the principles of soaring flight*. A. & C. Black. ISBN 0 7 136 1640 7.

Reymond, L. 1985. Spatial visual acuity of the eagle Aquila audax: a behavioural, optical and anatomical investigation. Vision Res. 25(10):1477-91.

Reymond. L. 1987. Spatial visual acuity of the falcon, Falco berigora: a behavioural, optical and anatomical investigation Vision Res. 27(10):1859-74

Rayner, J.M.V. Form and function in avian flight. In Johnston, R. F. (Ed). *Current Ornithology*, vol. 5. Plenum Press, New York.

Redig, R.T. *et al.* (Eds). 1993. *Raptor Biomedicine*. Chiron Publications.

Reynolds, K.V., A.L. Thomas and G.K. Taylor. 2014. Wing tucks are a response to atmospheric turbulence in the soaring flight of the steppe eagle *Aquila nipalensis*. Journal Royal Society Interface 11, 20140645.

Samour, J. 2016. *Avian Medicine*. Elsevier Health Sciences. ISBN 9780723438328.

Schulz, R. 1982. Der Karyotyp von Habicht (*Accipiter gentilis*) und Sperber (*Accipiter nisus*), zugleichein Beitragzur Geslechtbestimmung in der Ordnung Falconiformes. Inaugural Dissertation, University of Giesen.

Skutch, A.F. 1976. *Parent birds and their young*. University of Texas Press. Austin & London. ISBN 0 292 76424 3.

Snyder, N.F.R. and J.W. Wiley. 1976. *Sexual size dimorphism in Hawks and Owls of North America.* Ornithological monographs no. 20. American Ornithologists' Union.

Spedding, G.R. 1982. *The Vortex Wake of Birds: An experimental Investigation*. PhD thesis, University of Bristol, U.K.

Stewart, D. 2018. *The Hunters. The Precarious Lives of New Zealand's Birds of Prey*. Random House, New Zealand. ISBN 978-00-14-377274-3.

Sturkie, P.D. 1965. *Avian Physiology*. Comstock Publishing Associates, Ithaca, New York.

Torre-Bueno, J.R. 1976. Temperature regulation and heat dissipation during flight in birds. Journal of Experimental Biology. 65:471-482.

Tucker, V.A. 1987. Gliding birds: the effect of variable wing span. Journal of Experimental Biology 133:33-58.

Tucker, V.A. and C. Heine. 1990. Aerodynamics of gliding flight in a Harris Hawk, *Parabuteo unicinctus*. J. Exp. Biol. 149: 469-489.

Voitkevich, A.A. 1966. *The feathers and plumage of birds*. Sidgwick & Jackson, London.

Wattel, J. 1973. Geographic Differentiation in the Genus Accipiter. Nuttall Ornithological Club 13:1:231.

Welty, J.C. 1975. *The Life of Birds*. W.B. Saunders Co. ISBN 0 7216 9231 1.

Wernery, U. *et al.* 2004. *A colour atlas of falcon medicine*. Schlutersche. ISBN 9783899930078.

Wink, M, H. Saur-Gurth, D. Ellis and R.E. Kenward. 2004. Phylogenetic relationships in the Hierofalco complex (Saker-, Gyr-, Lanner-, and Lagger Falcon.) in Chancellor, R.D and B. -U. Meyburg. Raptors Worldwide. WWGBP.

Wu, T.Y.T., C.J. Brokaw, and C. Brennen (eds.). 1975. *Swimming and Flying in Nature*. Plenum Press, New York.

Xiangjiang Zhan, Shengkai Pan, Junyi Wang, Andrew Dixon, Jing He, Margit G Muller, Peixiang Ni, Li Hu, Yuan Liu, Haolong Hou, Yuanping Chen, Jinquan Xia, Qiong Luo, Pengwei Xu, Ying Chen, Shengguang Liao, Changchang Cao, Shukun Gao, Zhaobao Wang, Zhen Yue, Guoqing Li, Ye Yin, Nick C Fox, Jun Wang & Michael W Bruford. 2013. Peregrine and saker falcon genome sequences provide insights into evolution of a predatory lifestyle. Nat Genet 45, 563–566.

Zeigler, H.P. and H.J. Bischof. (eds.) 1993. *Vision,Brain, and Behavior in Birds.* MIT Press, Cambridge, Massachusetts. ISBN 026224036.

C2 Managing a breeding programme

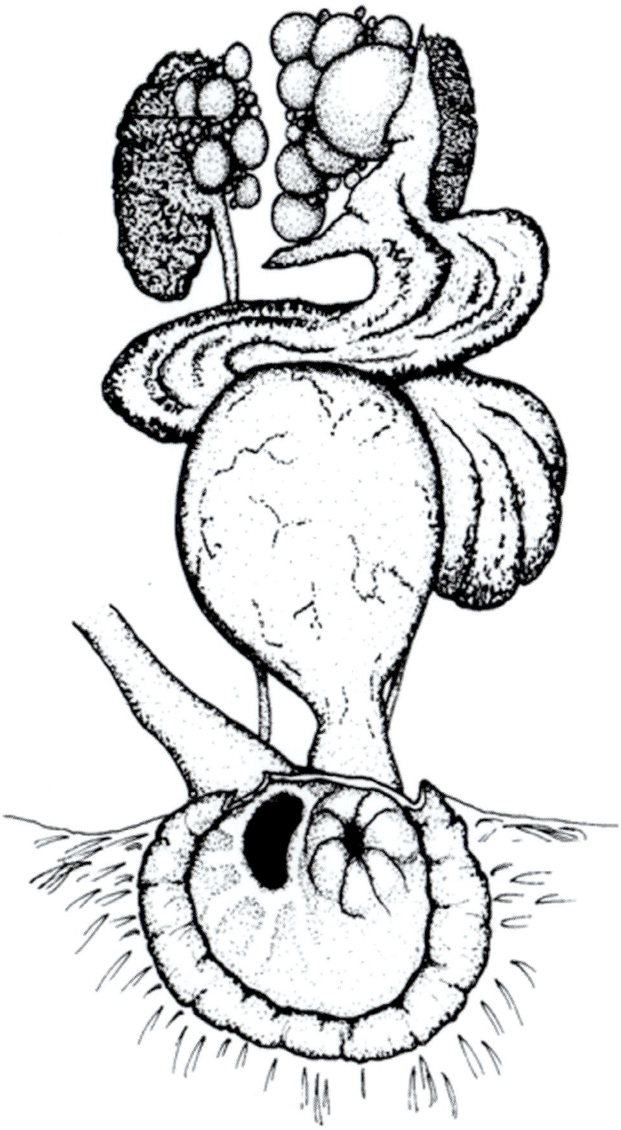

Figure 2.17.2 The reproductive tract of a female saker during egg-laying.

This chapter is supported by four films from the Bird of Prey Management Series,
Captive Breeding:
Part 1. Preparations for Breeding.
Part 2. Imprints and Insemination.
Part 3. Incubation and Hatching.
Part 4. Rearing.

Wild sakers copulating.
Photo Gombobaatar.

2.1 The process of domestication

The remarkable progress of domesticating many raptor species in the 1970s and 1980s is a process that will continue to unfold. Some state-run projects are deliberately breeding only indigenous subspecies or populations because the offspring are wanted for release to the wild. Some breeders are content to breed what they have; the offspring are only wanted for falconry and, once a particular pair are breeding, it is a pity to split them up. Other breeders are selecting for particular traits. For example, certain lines of peregrine which are renowned for waiting on are selected for game hawking. Physically and genetically they appear standard, but when flown they have a marked tendency to go up. Some breeders are looking for size; they put together large Aleutian Peale's peregrines or big Finnish goshawks. Others are selecting for speedy bloodlines. This process is bound to continue and in the next twenty years certain bloodlines will come to prominence for these reasons.

Some breeders are taking matters one step further by making inter-subspecies hybrids, inter-species hybrids and, in some cases even inter-generic hybrids, such as a Cooper's hawk crossed with a Harris hawk. Most of the wider outcrosses are very different genetically and in the long term will be sterile. Others, such as the gyr/saker, are very closely related, appear perfectly fertile and perhaps throw new light on our understanding of the taxonomy of these groups. Among falcons, multi-generation hybrids are now common, for example 31/32 gyr/saker falcons. The Domestic Falcon *Falco domesticus* is just around the corner. This is falconry's equivalent of the racing pigeon and the domestic poultry breeds, except that it can involve multiple species.

Many purists will hold up their hands in horror at all this, but the story is far from finished. Once the 'breed' stage is reached, there is then the question of open and closed stud books. This stage happened in Europe over the last two centuries with many of our horse and dog breeds. It was a somewhat haphazard and very interesting process developing, for example, the thoroughbred horse and the English pointer. Our forebears added a little of this and a bit of that to the recipe until they reached the type they wanted, then they closed the stud book and started breeding 'true' lines. Interestingly, mitochondrial DNA research into the bloodlines of the thoroughbred horse, reveals that their pedigrees are not as 'pure' as claimed. Various substitutions have gone on over the years! On balance, this was probably a good thing.

Nowadays one can only alter the English pointer by selecting from registered animals. If one decided to add some setter blood for example, the offspring would be classed as mongrels and would not be registerable by the Kennel Club. Insofar as this maintains the genetic integrity of the breed, the closed stud book is a good thing. On the other hand it means that the main evolutionary development of the breed is now over and that the breed is restricted now to the limited gene pool it originally started with. If this was large, well and good. But if it was small, the breed will gradually become more inbred through the generations. When that happens, all sorts of congenital abnormalities such as hip dysplasia and ingrowing eyelids start to crop up.

It will be some time before this stage is reached with raptors. It has been exciting meeting the challenge of domesticating birds of prey and it will certainly be interesting to watch the bloodlines emerge. For this to happen there has to be selection pressure. In race horses this is competitive racing. In hunting dogs this is fieldwork and competitive Field Trials. In show animals it is Show Standards (some of which are really grim). Field Trials are in many ways a good idea but they suffer from two problems; one is that they measure the training of the dog more than its genes. The other is that they have closed down the uses of the dogs into very tight stereotyped patterns, such as retrieving, based largely on shooting (they are now called 'gun' dogs rather than game dogs). This neglects other aspects of the dog such as how long their potential working lives are, how easy they are to train, how good they are with children, how easy they are to live with, and so on.

In raptors, the selection pressures are a little different. The first one is whether they will breed at all. For example we have a pair of gyr/sakers here which laid twelve eggs this year and produced twelve chicks. Every fertile egg this female has ever laid has hatched. We have other pairs which have never laid at all. Once the chick has been reared, it then must be evaluated in various ways before the decision is made to breed from it. We measure it, we train it and evaluate its hunting performance for the type of flight it is intended for, we assess its mental suitability for breeding, we look at its pedigree, and sometimes its DNA profile, then we look for a suitable mate for it and calculate the coefficients of relatedness to see if they will maximize the genes we are selecting for. Most breeders are unable to do this because of the economies of scale. They will be led to some extent by market forces. What sells well is obviously desirable. Unfortunately it is also sold and therefore out of the breeder's hands and not available for breeding. We get round this in two ways: one is that we train and hawk with a lot of our own birds each year, sometimes twenty or more. The other is that we lend out birds to other falconers and then call them back in once sexually mature. Now we can train and compete using roprey we can measure some of the parameters such as level speed and climb rate and assess our birds scientifically.

At present we don't have Breed Societies. Falconry clubs tend to be based on geographical area rather than breed. But a Harris Hawk Society may not be far away. Harris hawkers tend to be a sociable lot. There will probably never be a Goshawk Society. People who fly goshawks end up antisocial, introverted, stubborn, sensitive, and have generally used up all their patience on their birds....

Having looked at breeding from an overall perspective, we are now in a position to look at evaluating the bird, breeding from it and raising good youngsters.

2.2 The genetic assessment of pairs

You should have a clear idea of your general breeding plan from the start. You should know whether you are trying to maintain the widest variety of genes or whether you plan to intensify certain genomes by line breeding. Therefore you need some information about the genes which your birds carry.

The traditional way to do this is to use a pedigree (figure 2.2.1). This will show the ancestry back several generations, perhaps even to founder stock. It may also give additional information which is useful, such as how the bird was reared and who bred it. The pedigree may also link up to a computer stud book such as SPARKS. Pedigrees are simple and basic to a breeding plan. Surprisingly few people buying a hawk show the remotest interest in its pedigree. This will change in time as purchasers become more discerning.

Fewer people still know how to use a pedigree. It is not enough just to look at the pedigree of your new hawk and say, 'Oh look! Her grandmother was 'The Galloping Gourmet' she must be good! What if you are planning to put two birds together as a pair, and you can see that they have one or two shared ancestors in their pedigrees? How do you calculate how related they are?

Genetics unfortunately tends to be 90 percent theory and only 10 percent reality. Let's look first at what relatedness means. Let's start with two gyrfalcons, one from Alaska and one from Greenland. So they must be unrelated. Wrong. Of course they are related or they wouldn't both be gyrfalcons. The question is: how related are they? Human beings and chimpanzees for example share 95 percent of their genetic material. They are almost the same! We are all related to some extent and so we need to assess what this background level of relatedness is.

When two parents breed together, each donates half its genes. We can say therefore that, on average, the relatedness between parent and child is half (or 0.5 or 50 percent). The same applies to siblings: a brother and sister are 50 percent related, (except identical twins, which are 100 percent related.). The background level of relatedness lies somewhere between this close relationship of 50 percent and zero. In a genetically diverse population the background level of relatedness may be as low as 15 percent, but in a bottlenecked population it could be over 40 percent. Therefore we cannot say that our founder birds are unrelated, that they have a coefficient of relatedness of zero.

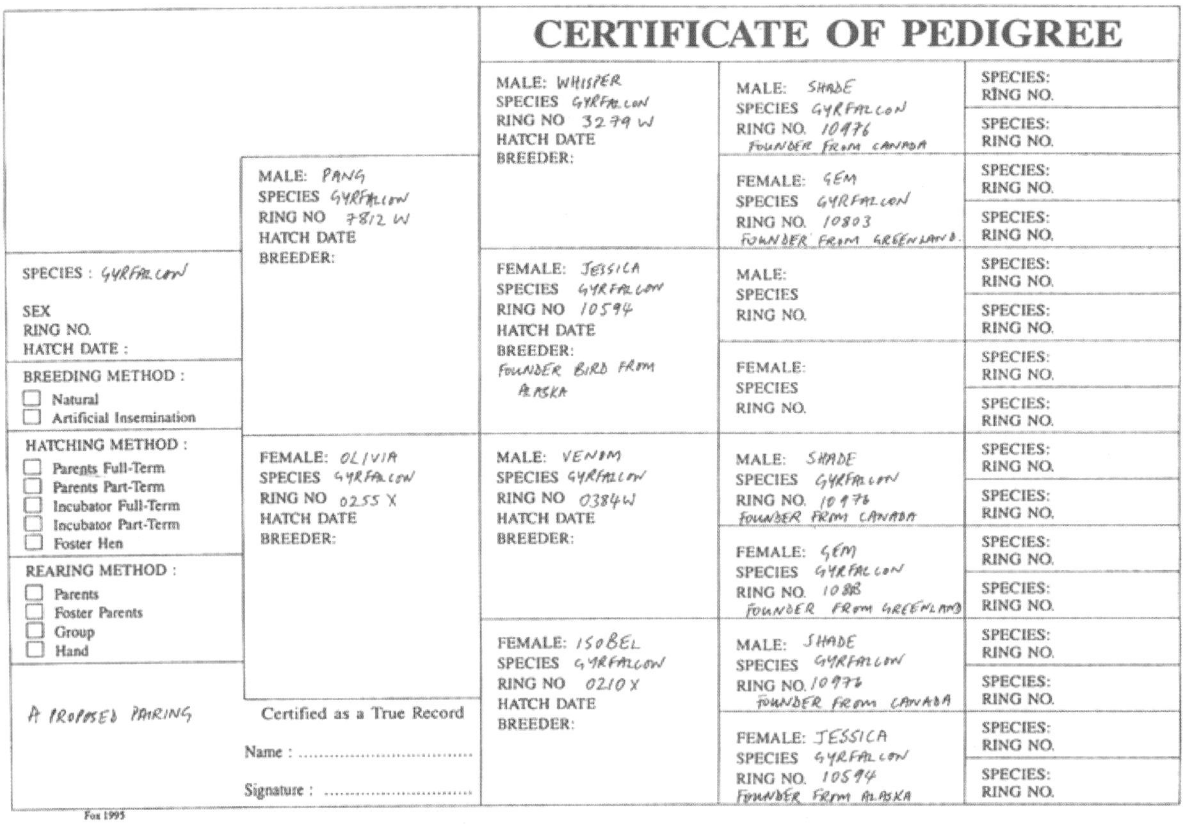

CERTIFICATE OF PEDIGREE

SPECIES : GYRFALCON
SEX
RING NO.
HATCH DATE :

BREEDING METHOD :
☐ Natural
☐ Artificial Insemination

HATCHING METHOD :
☐ Parents Full-Term
☐ Parents Part-Term
☐ Incubator Full-Term
☐ Incubator Part-Term
☐ Foster Hen

REARING METHOD :
☐ Parents
☐ Foster Parents
☐ Group
☐ Hand

A PROPOSED PAIRING

MALE: PANG
SPECIES GYRFALCON
RING NO 7812 W
HATCH DATE
BREEDER:

FEMALE: OLIVIA
SPECIES GYRFALCON
RING NO 0255 X
HATCH DATE
BREEDER:

Certified as a True Record
Name : ....
Signature : ....

MALE: WHISPER
SPECIES GYRFALCON
RING NO 3279 W
HATCH DATE
BREEDER:

FEMALE: JESSICA
SPECIES GYRFALCON
RING NO 10594
HATCH DATE
BREEDER:
FOUNDER BIRD FROM ALASKA

MALE: VENOM
SPECIES GYRFALCON
RING NO 0384W
HATCH DATE
BREEDER:

FEMALE: ISOBEL
SPECIES GYRFALCON
RING NO 0210 X
HATCH DATE
BREEDER:

MALE: SHADE
SPECIES GYRFALCON
RING NO. 10976
FOUNDER FROM CANADA

FEMALE: GEM
SPECIES GYRFALCON
RING NO. 10803
FOUNDER FROM GREENLAND

MALE:
SPECIES
RING NO.

FEMALE:
SPECIES
RING NO.

MALE: SHADE
SPECIES GYRFALCON
RING NO. 10976
FOUNDER FROM CANADA

FEMALE: GEM
SPECIES GYRFALCON
RING NO. 1088
FOUNDER FROM GREENLAND

MALE: SHADE
SPECIES GYRFALCON
RING NO. 10976
FOUNDER FROM CANADA

FEMALE: JESSICA
SPECIES GYRFALCON
RING NO. 10594
FOUNDER FROM ALASKA

SPECIES: RING NO. (×16)

Fox 1995

Figure 2.2.1. A pedigree form showing the proposed pairing calculated in the text.

For the sake of simplicity, SPARKS, and similar studbooks assume that founder birds are unrelated. The only way to find out what the background level of relatedness is, is to profile the DNA and calculate it. This is what we had to do with our New Zealand falcon program. We had to trap wild falcons from all over the breeding range, take blood and look at the DNA. This gave us a starting point. We discovered that some individuals trapped over 300 kilometres apart were related by up to 50 percent. This means they were as close as brother and sister! So much for founders being unrelated! What this meant was, not that they were brothers and sisters, but the population as a whole was closely related, having at some point dwindled to low numbers and then made it back up again. This is not necessarily a bad thing. A severe bottleneck weeds out the bad genes and thus further inbreeding is less likely to be harmful. Inbreeding in a more varied population is more likely to reveal 'genetic skeletons in the closet.'

A species, such as the ferruginous hawk, might have a very varied population in the wild but when imported to another country for breeding, maybe only ten individuals might come in. Of these maybe only six birds breed. Of these maybe one pair produces 50 percent of the total young. This is not too drastic at first, but go a decade and a few generations down the line, as we are now with some of our species, and problems can arise.

Calculations of coefficients of relatedness, and covariance charts, get very difficult. But it is not hard to follow relatedness in a simple three or four generation pedigree. This is what you do:

Take the pedigrees of the two birds you are planning to mate and put them side by side. If they have no common ancestors then you can assume for practical purposes that they are 'unrelated' or more properly, their relatedness is close to background level. Let's choose a pairing with three common ancestors, Jessica, Gem, and Shade which are wild founder birds:

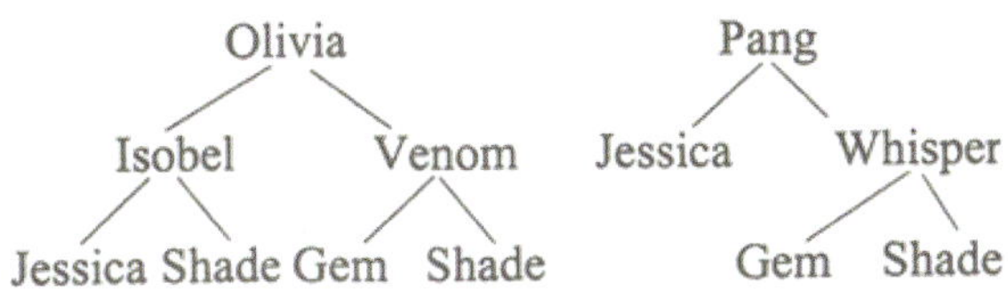

Each generation is related by 50 percent to the next. So first we trace Jessica up to Olivia. They are 0.5 X 0.5 = 0.25 related. Pang is also 0.5 related to Jessica. So for this one common ancestor, Olivia and Pang are related by 0.25 X 0.5 = 0.125.

For Gem, their relatedness is 0.5 X 0.5 X 0.5 X 0.5 = 0.0625.

For Shade it is more difficult. Both Olivia's grandfathers are Shade. This doubles the genetic contribution for that generation level. So for Olivia, Shade is 0.5 X 0.5 + 0.5 X 0.5 = 0.5. Pang's complement is 0.5 X 0.5 = 0.25, thus 0.5 X 0.25 = 0.125 for Shade in total.

Then we add together all the coefficients of relatedness for all the common ancestors:

Jessica = 0.125

Gem = 0.0625

Shade = 0.125

Total = 0.3125

What does this mean? A pairing which shows a coefficient higher than 0.125 is likely to show signs of inbreeding. This could be a genetic defect, such as a crossed beak, or it could be inbreeding depression, such as reduced hatchability. It depends too, on the proportion of 'bad' genes in the populations. But as a rule of thumb, try not to pair birds related more closely than 0.125. Olivia and Pang are destined for unrequited love, and we must look for alternatives. This is how the pedigree acts as a working tool for the breeder. Similarly, if the breeder is attempting to concentrate a particular characteristic, such as waiting on ability, he might decide to line breed, building up the genetic contribution from the ancestors which performed best.

2.3 The genetic compatibility of pairings

Whatever care you may take to ensure that your pairings are the optimum genetically, there are still some aspects outside the ken of man. Some pairs are just not very compatible genetically.

They may lay fertile eggs without problems, but then some of the eggs fade out and die during development. You try every trick in the book to hatch the eggs, but still some die. We have a female saker now age seven. She is paired to a white gyr X saker. We have hatched every single fertile egg she has ever laid. And yet another saker, unrelated, paired to another white gyr X saker, lays eggs of which 50 percent always fade out during incubation.

Genetic compatibility also affects thriftiness. Some pairings 'nick' together well. They produce a high proportion of really good progeny. Others fare less well. If this is the case, the problem can often be solved by re-matching the birds.

Harris hawks in Britain and North America often show leg defects which have been attributed to inbreeding. Actually, most of these birds are well-outcrossed and the defects are preventable by adjusting the diet of the female in the pre-egg laying period (see 2.10).

Some female gyr/peregrines lay eggs but most are not a good bet for breeding.

Gyr/sakers are prolific breeders.

2.4 Selection of breeding stock: physical assessment

Once you are satisfied that a certain bloodline is genetically suited for your plans, how do you pick out the individual bird you want? Some falconers go on weight, but as we shall see in 5.8, weight is very variable, and as a selling point can be made to say whatever you want. We sold two sister Harris hawks one year, same age and size. One recipient rang back and complained that his bird flew at only 1008 g (36 ounces) and the other rang and said his was taking hares at 1260g). Similarly, a young falcon, fresh out of the hack pen, may weigh one weight, but on arrival in the Middle East a few days later may have lost a lot of weight, then with good feeding and exercise, regain the lost weight and put on even more as the autumn wears on.

So we don't rely too much on weights!

Colour is another common criterion. Arab falconers especially often have marked preferences for different colours. Some of these can be predicted early on. For example a white gyr has very white claws at hatching whereas a good black gyr has black claws, a black beak, gray fluff, and cere. Fully feathered youngsters can also be selected on plumage and we take photos of each from above and from below with the wing and tail fanned. However, many raptors have an unremarkable juvenile plumage and you don't know what the adult plumage will be like until later, when the bird has already been sold and paired up. Also, colours are just colours; they do not tell you much about the bird as a hunter. It's amusing really that so much value is placed on colour, when in reality we found that in gyrfalcons a single nucleotide polymorphism in the *MC1R* gene is highly correlated with pigment phenotype. Just a small change on one gene makes the bird look completely different. Richard Sale and Eugene Potapov examine colour in their book *The Gyrfalcon*.

Therefore we take a number of measurements which help to define the bird as a flying machine (see 1.13-1.16). These measurements are kept on a computer database and by comparing the new bird against mean values for previous records we can make some fairly exact quantitative statements about the bird. To do that, it is essential that we all take the same measurements in the same way as

otherwise the results are not comparable. Trolling the literature for data is haphazard for this reason.

The measurements we take are designed in such a way that most of them can also be taken from museum study skins, despite potential shrinkage. This has the advantage of immediate access to large quantities of data, many more than most breeders would see in a life time. The measurements we take are shown in figures 2.4.1 and 2.4.2 and they are recorded on a data sheet as shown in 2.4.3.

WING LENGTH (A B) is the straight distance from the carpus to the tip of the longest primary on the flattened wing. WING WIDTH (A C) is the distance from the carpus to the tip of secondary 5. WING WIDTH RATIO is AC /AB X 100 percent.

To obtain the Wing area, lay the open wing across the edge of a table with the leading edge straight and trace the outline onto a sheet of paper. Label the outline with the bird's identification. This will be used for further measurements and for the calculation of TOTAL WING AREA. To calculate the total wing area we take the area of both wings, together with the area of the wing root box (the area of the body between the wings). The quickest and most accurate way to calculate the area in square centimetres from the tracing is to count the squares on squared paper. The total wing area is then used in the calculations for WING LOADING. The WINGSPAN is wing tip to wing tip in millimetres. Unfortunately, only the most modern museums store open wings.

To obtain the WING TIP FORMULA lay the closed wing tip onto some flat paper and carefully mark down all the primary tips with a pencil. Remove the wing and draw in a centre line following the wing chord. Using a set square, mark off all the marks onto the centre line. Knowing that in this example the overall wing length was 403 millimetres, place the edge of a ruler along the centre line with 403 aligned with the longest primary. Read off the measurements for the other feathers in millimetres. This is used to provide the wing tip profile.

The NOTCH or emargination is measured from the tip of the wing to the centre point of the notch.

The TARSUS is measured from the groove in the tibiotarsal joint diagonally along the tarsometatarsus bone to the groove in the joint with the centre toe. The TOES and HALLUX are measured with the digits extended straight and exclude the claws.

The TAIL LENGTH is measured from the tip of a centre (deck) feather (rectrix 1) to the point at which the feather emerges from the skin.

The TAIL STEP is the difference between the tip of the outermost tail feather (rectrix 6) and the tip of the centre tail feather (rectrix 1) on the same side. Where the two sides do not match, take the mean value.

The WEIGHT, measured in grams, is taken either by perching the bird on some scales and subtracting the weight of all equipment, or by placing the bird in a cotton bag and suspending it from the scales. The condition of the bird should be noted and an estimate made of the weight of crop contents.

At this time we also make a note if blood has been taken for storage for DNA profiling, and if any other treatment is undertaken. The easiest way to obtain DNA is from the chorioallantois inside the newly-hatched egg shell. It can be stored dry or in ethanol and sent off for DNA sexing of the chick, or can be stored indefinitely for DNA profiling. This is useful in establishing who are the genetic parents of the chick, either for a court case or because semen may have been used from more than one male.

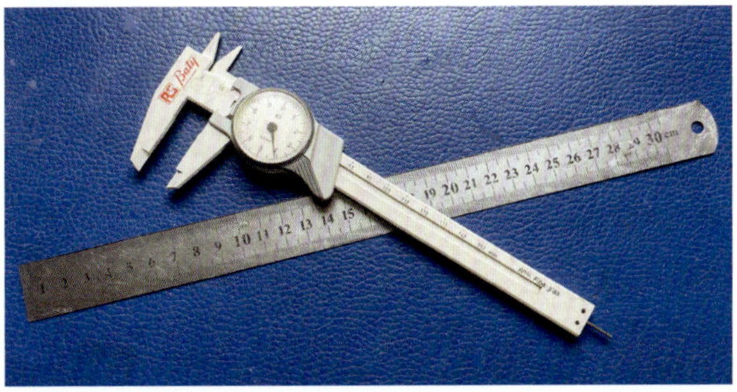

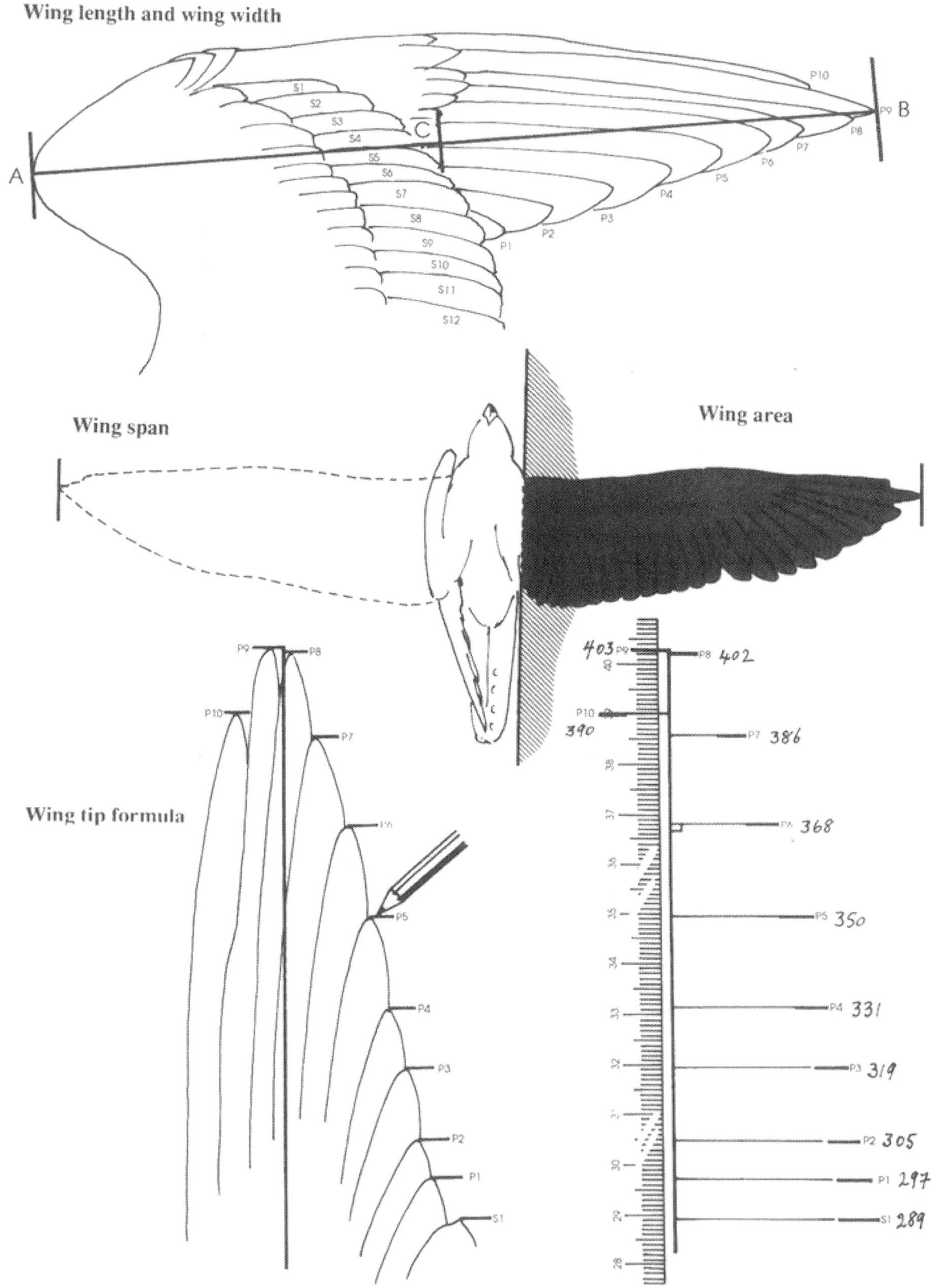

Figure 2.4.1 Measuring the wing.

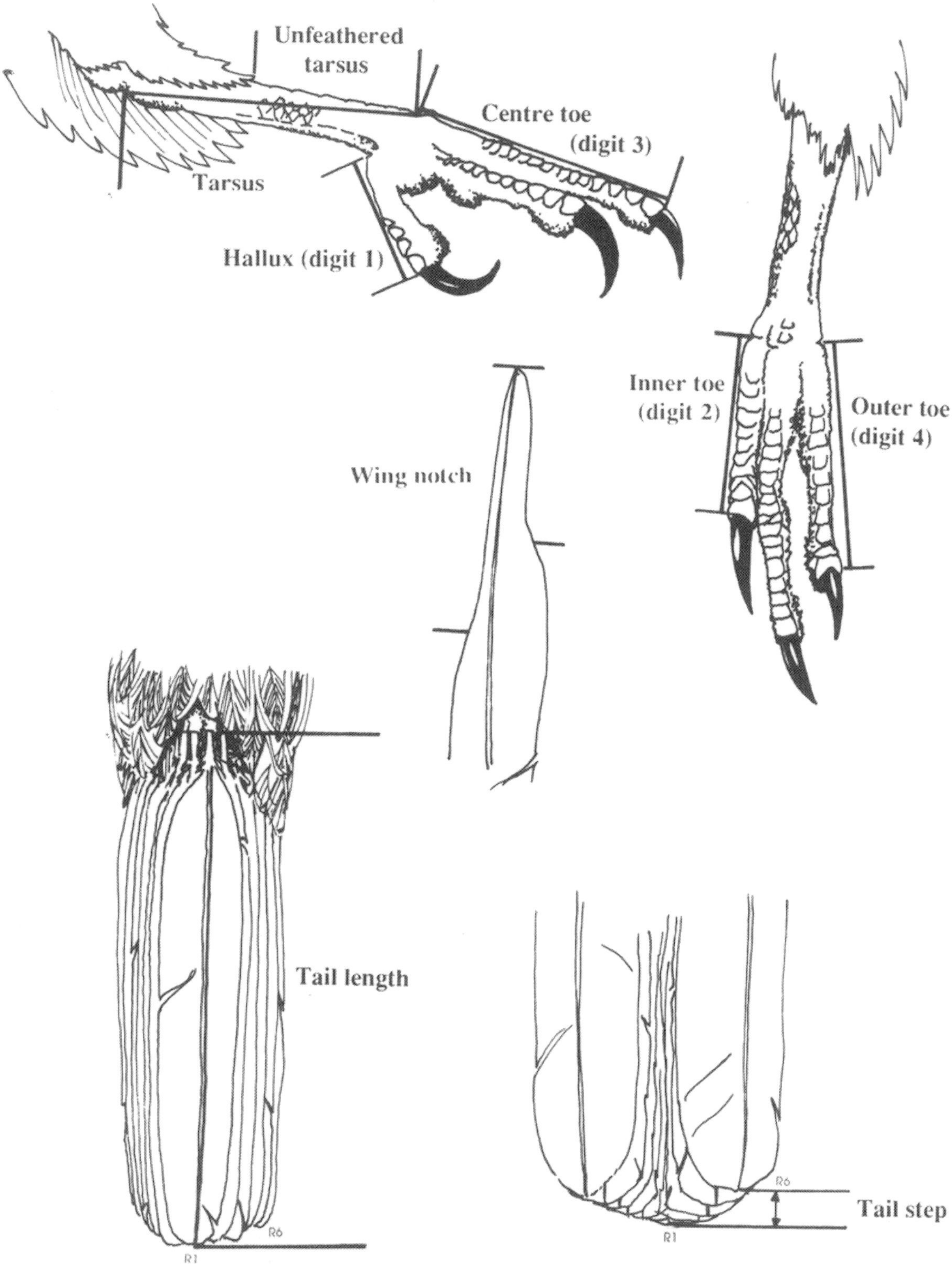

Figure 2.4.2 Measuring the tail and foot.

FALCON MORPHOMETRIC DATA SHEET

Date 1/12/22 Measured by A.N. Other

Species	Saker	Scientific Name	Falco cherrug
Sex	♀	Hatch Date	Juv ~~/AD~~
PIT Number	002-667-422	Colour Genotype	
Ring number	342-18W		

TAIL (mm)			
Tail length (mm)	225	Tail step (mm)	5

WING (mm)	P10	P9	P8	P7	P6	P5	P4	P3	P2	P1	S1
Length	376	390	385	368	347	313	287	267	239	213	225
Notch leading vane		96	70	50							
Notch trailing vane	85	67									

Wing width of S5	225	Both wings area (cm^2)	1835	Wing loading (weight/wing area)	0·65
Wing width ratio (%) (S5/longest primary x 100)	57·7	Wing span	1240	Aspect ratio (wing span/S5)	

LEG (mm)			
Hallux digit 1	33	Inner digit 2	35
Centre digit 3	56	Outer digit 2	36
		Tarsus	61

Weight (grammes)	1190	Condition Lean	Fat / Average / (Lean) / Thin

Comments:

Figure 2.4.3 Morphometric data sheet.

2.5 Performance assessment

On the back of the physical assessment form we keep records of the bird's performance (figure 2.5.1). Although we are experimenting with quantifying performance in terms of top speed in level flight, endurance, and so on, we are still working on this and so at the moment we record performance based on a subjective assessment. Because we train and hunt quite a lot of falcons every year, it is possible to judge them and to make comparisons. But you need to do a dozen or so individuals of each species before you can reach any conclusions.

We also keep records of training and hawking. All these records are kept on the bird's file and help us to assess its merits. Of course many birds are reared specifically for breeding and are not flown at all. The likely performance of these birds must be assessed by the achievements of siblings, parents, or progeny, bearing in mind that these relationships share on average 50 percent of their genes.

Norman Mackenzie with an albino Australasian harrier. Should albinos be bred from?

2.6 Mental suitability for breeding

Whether or not two birds will pair up depends on what they think of each other. Love is the same the world over! We tend to have an image of our ideal future partner, based on our experiences in childhood. It is the same for raptors. The raptor's image of its future sexual partner is formed partly by instinct and partly by its parents. If it imprinted on its natural parents (see 4.8 and 4.11) then all is plain sailing, but if it has been hand-reared for part of its development, it may be imprinted on two types of parents, hawks and humans. This puts it in conflict. A frequent outcome of this is that the bird may pair up with another of the same species but be frigid and refuse to copulate. Therefore it is important to rear future breeding stock carefully. The Westermark effect or incest taboo is discussed in 4.13.

There are two ways round this problem. One is to rear the youngsters using an imprint female. It will grow up identifying itself as a hawk but used to breeding enclosures and treating humans like passing sheep or clouds - moving but harmless and irrelevant. The other way is to rear groups of youngsters together. This is called 'social rearing' in America or 'creche rearing' in Europe.

This identification of future partner is not just about hawks and humans. It also occurs between different species and colours. If you rear a peregrine with a peregrine and then in adulthood put her in a pen with a saker, she may refuse to have anything to do with him. A few motivated birds may pair up, but many go year after year without much sign of breeding. If you want to make a hybrid by natural pairing, then it is best to rear the two species together or cross-foster them. This is the mechanism which prevents species all getting mixed up in the wild. It is not that breeding is prevented because the birds are genetically incompatible (many are genetically compatible), it is because the birds tend to pair up only with what they consider to be their own species.

This effect is not just at species level; it also occurs between different colour phase birds of the same species. If you take a baby falcon and rear her for a few days with a black gyrfalcon and then remove her and place her in the nest of a white

Sofia mothers some chicks onto an imprinted falcon for rearing.

gyrfalcon, she shows every sign of terror and retreats from the new parent. If you had left her all the time with the black gyrfalcon and then in later life tried to pair her with a white gyrfalcon, she would have a major sexual turn off. You can easily imagine the effects of different colour siblings and differently programmed partners. Some combinations will thus do well, others will always be inhibited.

Some breeders set great store by 'compatibility.' In my experience this is rather a myth. If the birds have been reared in the right way, are settled in captivity, and internally are coming into breeding condition, then they should pair up and go at least some way toward breeding. The problems arise when birds are obtained from a variety of backgrounds of unknown history. Such motley birds may be inhibited for various reasons outlined above and this is interpreted as 'incompatibility' as if it was on a personal level. The reasons lie deeper.

If the conditions are right, personal likes and dislikes may not come into it. For example, tundra peregrines migrate back to their breeding grounds in the spring. Males arrive first and wait for passing females. If one comes past, they may have a brief greeting ceremony and then copulate within just an hour or two. Also, DNA profiling shows that many birds, including raptors, indulge in extramarital sex when the opportunity arises. Similarly, if one of the parents is killed, a new partner may take up residence and nesting duties within hours. It has even be known for chicks to start life with one set of parents, for one parent to be shot and replaced, then the other, so that the chicks fledge with completely different parents!

Another aspect of mental suitability for breeding is the extent to which a raptor is stressed in captivity. Domestic raptors, properly reared, are completely at ease in breeding pens. I have a white jerkin here called Icicle whose pen covers one of my office windows. If I come in soon after dawn he chups to me for a while in greeting, then goes back to sleep, a headless white fir-cone on my windowsill.

But wild-caught raptors, especially rehabilitated ones additionally stressed by injuries, may not adjust well to captivity. They

Figure 2.6.1 The male (right) approaches with a confident 'fronting' posture. The female is not intimidated – she wants to play.

Figure 2.6.2 The male is sleek and alert, bowing and bobbing and staring with wide eyes at the female. He wants to copulate.

may appear not too bad, but breeding is the acid test and many fail to come into breeding condition. It is particularly important for these birds to have quiet, roomy, natural-type secluded pens. If such a bird is paired with a nice randy, relaxed partner, it might overcome its inhibitions and breed. Some stressed birds appear nervous and show no sign of courtship. Others are even more high strung and may kill their mate. More often though, a female kills her mate through frustration and tension in their relationship (domestic violence) than through captivity-induced stress.

When you watch your hawks in their pen, how do you know if they are happy and are getting along with each other? You must learn to read their behaviour and understand what it means. The figures 2.6.1-11 are of New Zealand falcons and are fairly typical for falcons in general. Harris hawks on the other hand have a completely different behavioural repertoire and even if you understand falcon language you have to go back to school to learn Harris language. Even within a genus there are differences, so when I go down a row of mixed imprints of different falcons, I must remember to talk in saker language, then

2.6.3. Angel threat display. Don't come any closer!

in gyr, then in peregrine, and so on. Sometimes I get in a muddle! Some play behaviour is very close to aggression and some birds can switch, cat-like, from pussyfooting to daggers drawn. It is important to understand their basic behaviour to interpret how they are getting on and what, if anything, can be done to remedy a problem.

The two-winged stretch – 'warbling'.

Scratching the face with the centre toe. Raptors scratch with the foot in front of the wing, not behind it.

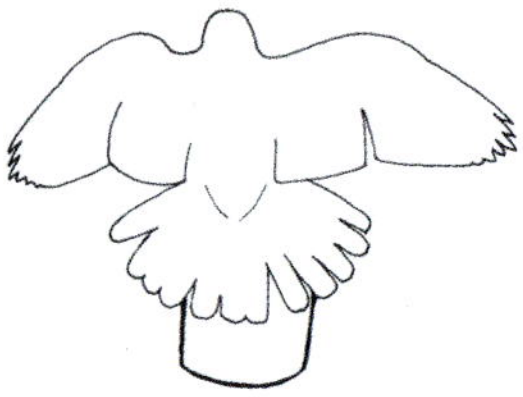

Sun bathing with back to the sun.

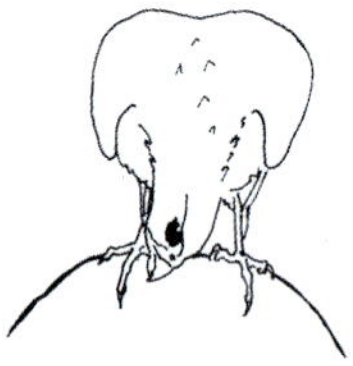

Feaking or cleaning the beak using an up and down movement of the side of the beak.

Winnowing or wing exercising.

Rousing.

The one-wing, one-leg stretch.

Mantling over food.

Figure 2.6.4.

Bat falcon *Falco rufigularis* sunbathing.
Photo Tom Cade.

A New Zealand falcon in a natural pen.

2.7 Breeding facilities for natural pairs

When I first started breeding in 1974 I had to catch my own falcons and start from scratch. I figured that the nearest I could get to Mother Nature, the happier they would be. So I built a pen 50 feet long, full of trees, with ten nest sites. You could sit in one of the two hides for hours and not see a bird. And they bred!

Then I started to get clever. The next time I had to build pens, this time in Britain, I started to eliminate all those features which I thought were superfluous. I aimed for the minimalist, cost-effective aviary. Some of these worked well too, depending on the birds. A typical minimalist aviary of this design is shown in Falcon Propagation and in figure 2.7.1. These pens work quite well for medium-sized raptors which do not require much stimulation to breed and we use them for most 'commercial' production.

Trickier species have special needs. Females of species which are strongly dimorphic, such as the accipiters and the New Zealand falcon, often kill their mates, so it is useful for these to have double pens connected by windows. These windows can be solid, or meshed, or open, or barred so that only the males can pass through. To find the spacing of the bars, we place the male in a box with temporary bars covering one end and leave him there for some hours to see if he can get out. We adjust the bars until he can just get through. We do the same for the female. Then we design the real bars at a distance halfway between the two measurements. Many pairs do better if they do not see each other all the time. Absence makes the heart grow fonder. Gyrs have a lovely clamorous greeting ceremony but get bored with each other when present all the time. They need stimulation. Nervous male accipiters are happier if there are bushes in the pen for them to escape behind.

The skylight and seclusion pen, designed by Dr Leonard Hurrell, works well for nervous raptors. But even the scattiest ones like a window; it steadies them down and is an endless source of interest for them. Of course their window need not be the same as our idea of a window; it can be just a very small hole. For example, a window 5

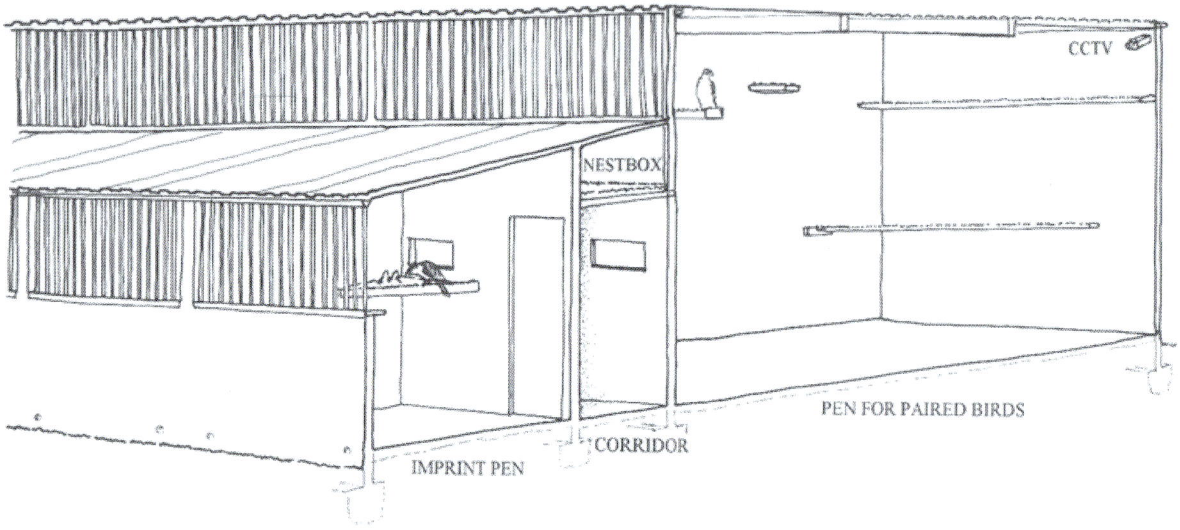

Figure 2.7.1 A section through our breeding pens with imprints along the front and natural pairs behind the central corridor.

centimetres in diameter with a perch next to it is enough for the hawk to look out comfortably and yet feel itself unseen. It can easily retreat into the peace of the pen if people come near. Having a small window with a perch wide enough for two falcons also encourages the pair to be within close proximity to each other, reinforcing the bond.

A few features need to be incorporated when building a pen. The first of these is strength. A pen should be built to last 10-20 years without major work done on it. This is how long the pair may

Get advice before building your pens. This is a failed saker breeding project in China.

Aplomados in more minimalistic pens.

A simple pen for a natural pair.

be in it and building activity will disrupt them. Lapboard and cheap timber are not suitable. They soon deteriorate and either hawks escape or predators get in; the result is expensive. We build our pens of concrete block rendered smooth. Various steel profiles on a timber frame are good too, but are not much cheaper. Flat galvanized steel is fine for interior walls but can be noisy in a storm. We have one block like this called 'The Thunderbox.' All timbers and roofs should be bolted on with steel straps and all plastic or fibreglass sheeting should be covered with mesh because plastic will shatter in time.

Poor aviaries result in lost or dead hawks, usually through taking short cuts with materials. Non-reinforced corrugated plastic roofing sheet gets brittle and easily breaks in a storm. All main structures should be reinforced with steel strapping; most aviaries are prone to wind damage. Lapboard warps and soon lets in rats. Knots in the wood drop out, allowing hawks in adjacent pens to foot each other through the holes. A lack of double doors has caused the loss of many hawks. Food chutes and access points should be properly made to prevent escapes. Food drawers are better than chutes so that uneaten food can be removed. Netting should be checked regularly to detect holes or damage. Loose string netting can tangle a hawk. Sharp, knotted, netting should never be used. Dense foliage, such as nettles, should be removed, young hawks land in it and can't get out, or chill and die of pneumonia.

The pen should be completely vermin proof, which means something substantial up to waist height. Rats may move in the autumn and we kill them with a poisoning campaign before they can get near the pens. Cats get onto the pen roofs and can keep an incubating bird off the eggs all night, resulting in mysterious failed hatches. An electric wire around the roof usually keeps them off, but persistent offenders may need sterner measures.

The pens should be built so that every part of it can be viewed from one place or another. Each bird should be seen and checked every day for signs of illness. To view the birds we use peepholes, one way glass windows, and close circuit televisions. The materials, especially the perches, floor and food trays, should be properly disinfectable with a comprehensive disinfectant such as 'Virkon' or F10 otherwise the pen will become 'sick' over the years. We mainly use concrete floors which are cleanable. They can have pea gravel or sand in heavy use areas and an astroturf mat in any places where birds habitually thump onto the ground. Deep gravel is cheaper but tends to build up bacteria and can then be a major chore to change. Larger pens can be grass but it is important to keep the vegetation under control either by blitzing it or with grazing sheep. Tall nettles restrict the available flying space and are a hazard. If a bird lands in them it may be unable to get out. No cut vegetation should be allowed to mould in a pen because of the risk of aspergillosis.

All our perches are covered with coarse astroturf, even for the merlins. Choose the bent leaf type, not the spikey sort used for door mats. Astroturf will reduce the risk of bumblefoot provided it is kept in good condition. Worn out, filthy astroturf is asking for trouble. The main perches are removable for cleaning and all are pressure washed before disinfecting. Any objects or surfaces near a perch are covered with a smooth material so that feathers will not be abraded. Harris hawks and other species which suffer from the cold have heating tape wrapped around their perch under the astroturf; this prevents frost bite and dry gangrene. Some of the perches are big and wide, giving plenty of room for two hawks. Then they can adjust their personal distances gradually, as they get better acquainted. The perches vary from very low to quite high so that the hawk can gain height by hopping up from perch to perch. This is essential for some of the big falcons; when a female is moulting and laying, she cannot fly up steeply and could otherwise be unable to regain the nest to lay. Some of the perches, usually near the nest, have a moulded concrete area. The hawks feak on this and it helps keep their beaks in trim. Food trays, baths, and so on should be sited to avoid mutes from overhead perches.

Our pens face south over a quiet valley so the birds get plenty of sun through big high windows. Each pair thinks they own the valley and remonstrate strongly if a fox comes into view 300 metres away. The gyrs flirt outrageously with passing buzzards and herons.

The nest boxes are large and roomy to accommodate large broods. The substrate is fine pea gravel for all species including stick nesters. No peat, hay, or straw is used, but modern alternatives such as various types of cat litter, pine bark or shavings can provide an alternative. Stick nesters have their nest over the pea gravel so that if inexperienced birds make a poor nest the eggs will still be properly supported. Each pair have at least two nests to allow for recycling. They almost always use the alternative site for the second clutch. There are peep holes to each nest; some of these are covered with black sticky tape and only opened or pin pricked as needed. Sitting birds are very sensitive. When we have a female which must be checked through the night for egg laying so that we can inseminate her, we install a small night light on a dimmer. It is used only for two or three nights when actually needed.

We use a variety of food hatches and trays. The main thing is that they can easily be checked and cleaned and that all feeding points are kept clear of mutes. This stops an infection cycle. Baths are refillable from an outside hose connection and are removed and sterilized as necessary.

For medium-sized raptors such as peregrines we aim for a minimum floor area of about 250 square feet (23 sq m) and a height of about 12 feet (3.6 m). A greater height than this tends to cause more problems than it solves. All windows should be as high as possible and unapproachable from the outside. Of course, if the pens will be on show to the public then it has to be possible to see the birds easily. This raises a lot of further considerations and one is then not designing a breeding pen for the birds so much as an exhibit for the people.

You may have noticed that I haven't said anything about security against thieves. The first rule with security is not to discuss it. So I won't.

Imprint pens.

2.8 Breeding facilities for imprints

Our facilities for imprinted raptors are very straightforward. Each has an individual chamber about 2.4 metres (8 ft) by 3.0 metres (10 ft) with access through a door to a central corridor or double door system. This size is large enough for the bird to get some exercise but small enough so that the handler is all the time inside the bird's individual space (figure 2.8.1). This means the bird does not have to nerve itself up to approach the handler to interact. It will not flit around the pen trying to avoid the handler because there is no space to go into.

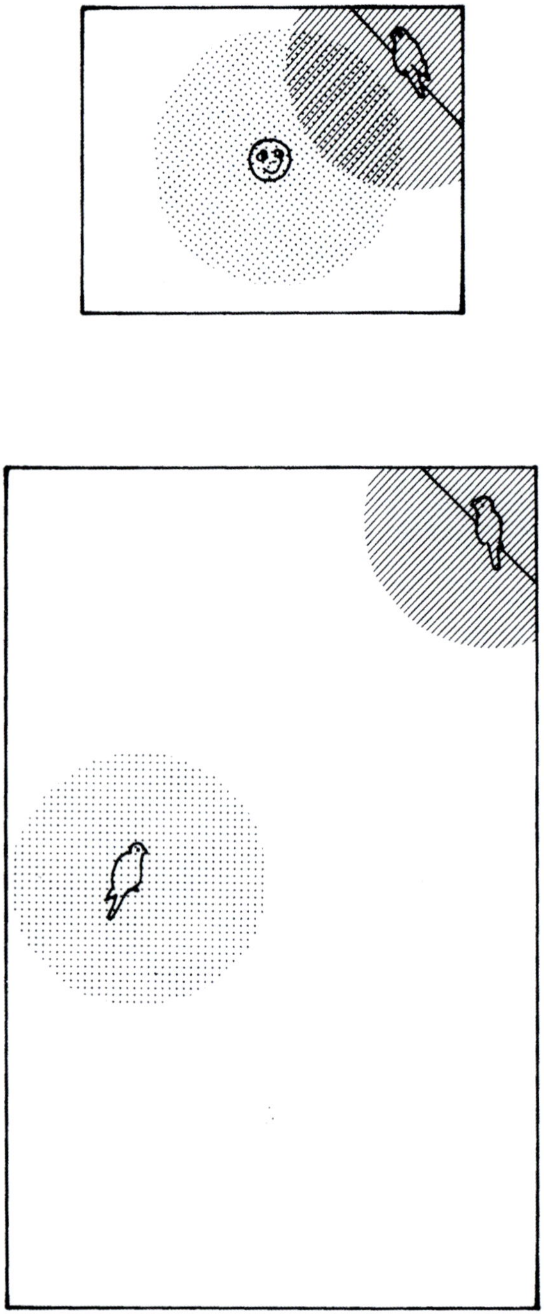

Figure 2.8.1. Top: Man and imprint. They cannot avoid being in each other's space so they stand their ground.

Below: Unmated pairs need room to avoid each other until they pair up.

There is a large, barred front window with a wide astroturfed windowsill perch at waist height overlooking an area where the birds can see things going on without being threatened or spooked. A bath and low perch are placed under the nest to keep them free from mutes. The floor is concrete with a thin layer of sand. Each pen has a corner nest at chest height with a window directly at the nest into the corridor. The nest is the central focus for imprints and it should be designed so that both bird and handler can use it together conveniently. The window glass is clear so that it is possible to interact both ways through the window. This is particularly useful with some females which can turn crabby suddenly and grab your face. Later it is useful for monitoring eggs and chicks in the nest and, if the female is sensitive, we cover the window temporarily with cardboard and make a peephole.

Imprint pens have covered roofs with one of the sheets clear. With males especially, it is important to have some control over their environment. Cold or wet days will lower their semen production and they need to be kept dry and in even temperature conditions for maximum steady production.

2.9 Incubator and broody rooms.

The first consideration for incubator and brooder rooms is hygiene. The floors, walls, ceiling and all surfaces should be sterilizable by light spraying with Virkon or F10. Nothing is kept in the rooms except what is actually needed. All the rooms and equipment are thoroughly fumigated at the start of the season after cleaning. No unnecessary staff are allowed in the rooms and child bolts are fitted to the tops of the doors to prevent access.

The air in the rooms is maintained at a steady

temperature of about 18 degrees centigrade and there are extractor fans to change the air as needed. In extreme climates it is also necessary to control the humidity in the rooms. The rooms are placed internally in the building, with no windows, to avoid fluctuations in temperature and to allow complete darkness for egg candling during the day.

The electricity circuits are fitted with anti-surge devices and with power-off alarms and there are petrol-driven and battery-powered standby generators. Any experienced breeder knows that if there is a chink in the system, fate will sooner or later find it.

There are a variety of small incubators on the market, some of which are not really good enough for raptor eggs and are designed for the small poultry breeder or for schools. Still-air incubators are gentle on the eggs, cheap, and good for hatching and early brooding, but they can be tricky for taking eggs right through, especially if the room temperature varies.

A few basic points to look for in incubators are:

a. simple design with nonabsorbent, easy to clean parts,

b. good electronic thermostat with back up,

c. electronic humidity control,

d. adjustable turning mechanism with adjustable rollers,

e. even temperature spread throughout the machine, and

f. all windows double-glazed and able to maintain heat for at least an hour during a power failure.We use Grumbach forced-air incubators run at 37.3 degrees centigrade. Some of our units are still in use after 28 years, although we replace small parts periodically.

g. For the first ten days of incubation, proper development of the embryo is critical. We use Brinsea Contaq incubators to start the eggs. These incubators have a warm membrane resting on top of the egg, mimicking the mother's brood patch.

Of course raptor eggs must be incubated in a separate room away from poultry eggs or other biohazards.

The eggs are weighed about twice a week on an electronic balance and candled using an electric candler. Commercial hand candlers are available, including some which use UV light to penetrate thick eggshells. Brinsea now make a specific candler which is very effective, also a simple LED torch with some duct tape to create a "Funnel of light" is just as effective with falcon eggs. One thing to be careful of is that the older candlers can emit heat, which if left toolong in close contact with the egg can effect the temperature of the embryo. Records are kept on a simple egg chart (figure 2.9.2).This is a basic chart for the home breeder. Commercial breeders use more sophisticated charts and software that calculates and records weight loss Digital recording is good, but paper records are often simpler and more permanent. Normally the eggs are incubated at 20 percent relative humidity, less if they are overweight and more if they are too dry.

If you live in a place with poor power supply, you can use small broody bantams. Their hygiene and management need careful attention and the flocks need to be managed so there will be sufficient proven sitters throughout the spring. An embryo that has been well incubated by a bantam is strong enough to overcome most minor problems at hatching and early rearing. For the small breeder, some bantams for incubation, together with a Polyhatch lid for hatching and early brooding (see figure 2.28.1) will give the best, most reliable results most cheaply. Broodies are particularly good for the breeder who has to be away at work during the day time and can't be continually hovering over incubators, or for field workers operating in difficult terrain. No hay, straw, or peat should be used in broody nests because they will transmit *Aspergillus* spores into the eggs and kill them. Use pea gravel and sand as a base and heat reservoir with sterilized chemical-free wood shavings, shredded paper, or cat litter for the cup. But if I had a small project, and a good power supply, I would opt for a modern, contact type incubator. They are expensive, but so are dead eggs!

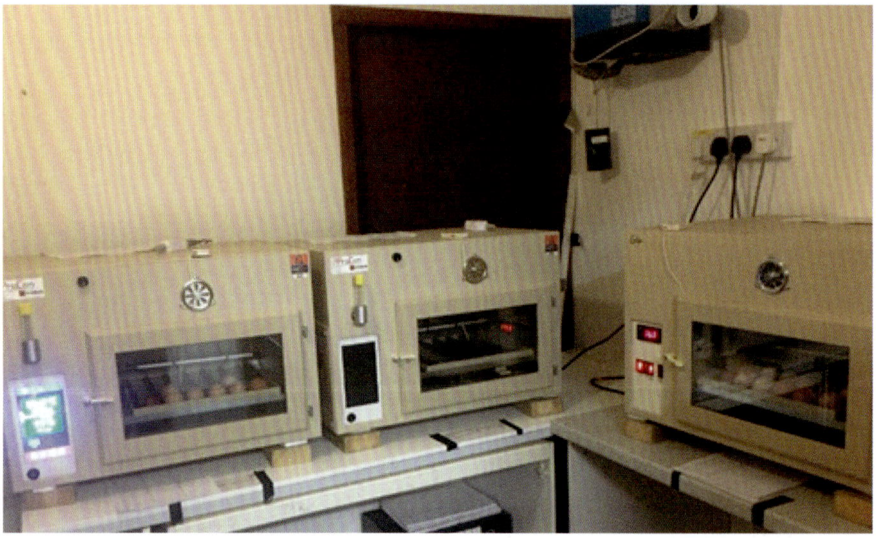

Forced-air incubators in a temperature-controlled room, with a de-humidifier.

This is a 'contact' type incubator, suitable for starting eggs.

A simple forced-air incubator for the small breeder.

2.10 Food supply, storage, vitamins, and food borne diseases

What food to feed the breeding stock is a balance between cost, palatability, nutritional quality, availability, and health risk. It also depends on the different stages of the breeding cycle.

Quail, pigeons, rats, and mice are the most expensive and most complete diets. Chickens are cheaper. Hares, rabbits, and day-old cockerels are the cheapest. We don't use large mammal meat, such as beef, because we don't keep any really large raptors. Hawks find all these diets palatable except, sometimes, lab rats and mice. Some buteos and owls eat these with relish but many of the falcons refuse to touch them. All fresh-killed food is more palatable than frozen food. We used to breed our own quail but now we contract this out. One can also purchase frozen quail. Hawks find these less exciting than twitching fresh quail, also one has no control over vitamin levels and fat content. Ex-layer quail are usually very obese and should be fed only very sparingly.

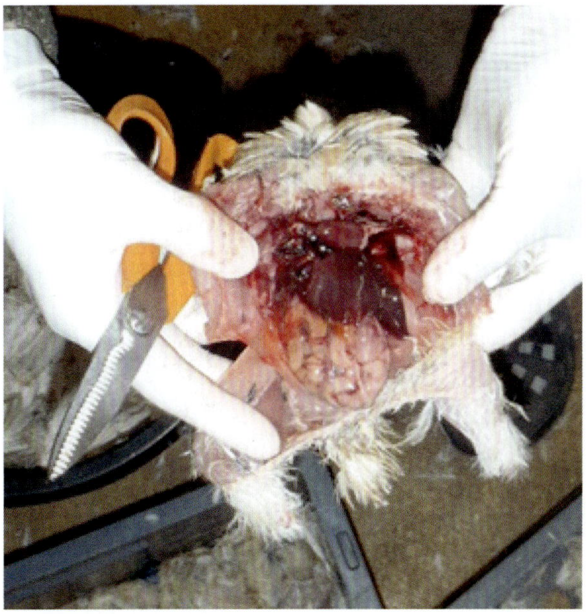

Removing beak, feet, wings and intestine from a quail, but leaving the heart and liver.

Chickens in the form of your own grown-on poults are fine, ex-layers and chicken heads less so, primarily because of health risk. If you visit some of the poultry farms the last thing you want your precious hawks to eat is something that has come from there! We have a lot of large falcons here which can be prone to obesity and a proportion of light, low fat food such as rabbit and hare is good for them. These have been hawked, netted, dogged, or rifle shot and most of the diseases they are prone to are not transmittable to birds. They are not nutritious enough for merlins and small accipiters.

Frozen day-old cockerels from hatcheries are available in many countries and are very commonly fed to hawks. They are a whole animal diet, cheap, palatable and easy to use. We use them in the non-breeding months but not as a staple for young hawks. Of course there are hawks and hawks. A diet that would rear healthy buzzards would kill an adult merlin. The yolks of day-old chicks are high in cholesterol and so many people remove them. The hawks themselves like the yolk and the yolks are good for growing strong feathers, but they can make nest ledges very messy and invite bacteria.

Pigeons and wild-source birds are a health risk. All should be frozen for at least two weeks to kill off the trichomonads but freezing will not destroy any viruses or bacteria. Shot wood pigeons are a good diet provided they have been shot with steel shot, not lead shot, and are carefully inspected before use. Hand-held metal detectors will not pick up small lead fragments. The pigeons should be in good, well-muscled condition, with a clean liver. Any diet with suspected shot or shattered bone fragments should also include plenty of casting material. Vets may tell you not to feed wild birds but we have fed 15,000 wood pigeons (not town pigeons) per year, and so have other breeders. Wood pigeons are solitary, nesting and roosting apart, whereas town pigeon often roost communally in conditions where diseases are easily transmitted. Even so, some wild peregrines live almost exclusively on them.

One can never provide a completely pathogen-free diet, nor is this desirable. Low levels of suitable potentially pathogenic organisms are needed in the diet so that the hawk can build up a resistance. Otherwise when eventually a young hawk gets to eat its own wild prey she could become ill. But pathogen levels in the diet should be at planned low levels; it is not an excuse for poor hygiene.

For a week before killing, the quail are fed a diet free of coccidiostats and antibiotics; they are also given a vitamin boost. They are not fed for 24 hours before killing. Then we pull their necks and feed fresh, or lay them out to cool off before being placed in the freezer room on open trays to freeze quickly. The next day they are packed in sealed and dated bags for storage. Before feeding, the wings and feet are removed. With frozen birds and when feeding young hawks, we remove the intestinal tract but always leave the liver, heart, skin, and feathers. In some countries, particularly the Middle East, it is common practice to skin and eviscerate the quail. Although this keeps the falcons tidy, it deprives them of most of the benefit of a whole animal diet and predisposes them to bumblefoot. Falcons need the skin, liver, and a little bird fat; it improves their skin and feather quality.

Freezing and storing the food tends to destroy the fat-soluble vitamins. Artificial and monotonous diets can also lead to nutritional deficiencies which, while not being apparent in non-breeding adult raptors, can lead to problems in breeders, especially in egg quality and hatchability and in the health of young raptors. A deficiency in the mother leads to a deficient egg which can lead to irreversible problems in the small chicks. Supplementing a chick under five days old is risky and still cannot compensate for an earlier deficiency because the internal organs and metabolism have already been based on the egg contents and because for the first 3-4 days of life the chick's nutrition is mainly from its own yolk.

Captive birds, especially in the wet British climate, tend not to receive as much sunshine as wild ones, and they need this to help make their vitamin D3. This is also one of the fat-soluble vitamins and therefore we provide extra in the form of cod liver oil. Many of the vitamins cannot be stored in the body and are needed every day. The basic supplement we use is Nekton S at 0.5 g/kg of hawk daily. To this we add 2 drops of cod liver oil per kg of hawk and make the mixture up as an emulsion in water, fresh each day. This is placed in an inverted bottle tubed to a sheep injector which gives premeasured injections of vitamins into each item of food (figure 2.10.1). The food is thawed in trays overnight and the water base to the supplements helps to counteract the effects of dehydration from freezing.

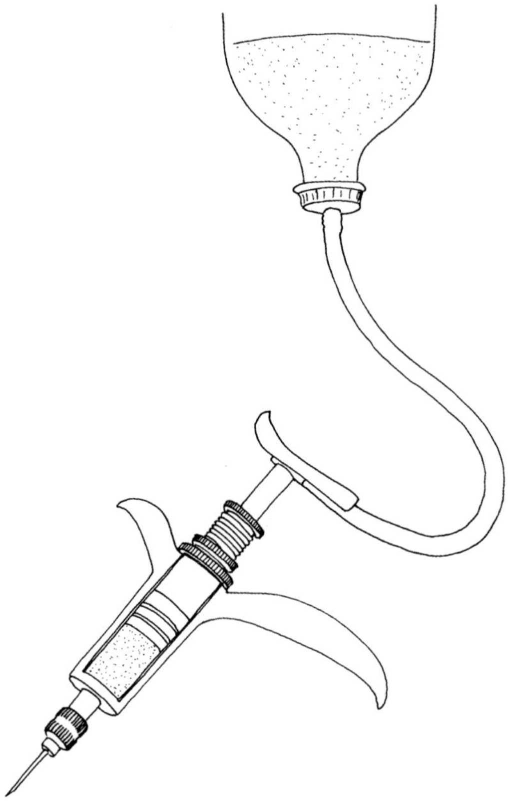

Figure 2.10.1 Injecting vitamins into food.

During the breeding season we also feed carophyll orange at 0.1 g/kg. These carotene pigments are normally available to raptors in wild bird fat and are responsible for the orange colour of the soft parts of adult raptors. An orange cere and feet are courtship signals for raptors and even if not entirely essential for health, they certainly look more natural and healthy to me.

About a month before egg laying we also feed Nekton E at 0.5 g/kg to promote courtship behaviour, semen production, and breeding especially in first timers or difficult birds. It is discontinued once the last egg has been laid or, in the case of males, when semen is no longer needed.

We have not arrived at these doses of supplements by scientific method. There have been no quantitative controlled nutritional studies of diet in breeding raptors that I know of. In addition there is a bewildering variety of different species, climates and basic diets. But for our birds, in our conditions, we have reached these decisions, using hatchability and survivability as the key criteria. This subject does not stand still and you have to make your own decisions according to what is available to you. We change from year to year, depending on what quality food is available.

Our annual regime for large falcons works out like this:

Throughout the year: Nekton S at 0.5 g/kg of hawk. Cod liver oil at 2 drops/kg of hawk.

From 1 Jan include: Carophyll orange at 0.1 g/kg of hawk.

1 March or 4 weeks before laying, add: Nekton E at 0.5 g/kg of hawk.

Feed quail or woodpigeons 3 days per week, rats 2 days per week, day old chicks 2 days per week, rabbit or hare 1 day per week (winter only).

Discontinue Nekton E and carophyll at the end of egg and semen production.

During rearing: quail, woodpigeons and rat until the young can fly, then wean in chicks and rabbit.

Rest of the year: mainly chicks and rabbit, with a starvation day once a week for big raptors.

Obviously for small raptors you need a higher quality diet than this, some even like insects. For bigger raptors and scavengers, you can use various deer species and farm animals. Just beware, especially for vultures, of sheep that have been dipped, or animals treated with any medicines, such as diclofenac. Some diets, such as day-old chicks, are likely to become less available in the future and you may need to breed your own food.

2.11 The programme in winter

Once the young birds have all gone it is time to clean out all the pens, disinfect them, make any repairs before the winter and check birds for parasites. Chemicals can cause a problem. One hawk was deloused with a proprietary insecticide and then offered a bath. It drank some of the water and died. Another person cleared an infestation of ants with an insecticide but his owl then ate some of the ants and died. We use Frontline sprayed under the wings (not above) and then keep the hawk away from a bath for a day.

Modern CCTV cameras are strong enough to be left in all year and provide 24/7 coverage of all goings on, such as sick/injured falcons or even staff taking a cheeky 5 minute break whilst pressure washing! All equipment is cleaned and repaired, spare parts ordered and fitted, and stored away properly. The birds should have just about completed moulting and are now on a low protein maintenance diet. Some of them will go through an autumn courtship or *herbstbalz* at the equinox when they are on the same day length as in the spring. This is a good time to match up new pairs. Staff take their holidays at this time and it is a convenient moment for a nervous breakdown, or a busman's holiday - a hawking trip.

As far as possible it is good to keep the birds lean and fit. They should fly down promptly when food is put in and can spend time tiring on joints of rabbit and hare. This continues until after the shortest day. From then on it is time to prepare for spring. I'm assuming that your programme is on natural daylight and daylength. Be warned: if artificial lights, such as street lights, shine into your pens it can screw up the birds' photoperiods.

2.12 The programme in spring

The falcons and their pens are cleaned, checked and settled at least two months before the first egg. Sensitive ones are done earlier, 'Regulars' are done late. The falcons have their beaks and talons coped (see figure 1.2.4) and the clumsy ones are crutched; the fluffy feathers around the vent are trimmed so that they do not impede copulation. All nest material is changed, everything is checked and CCTVs are installed and aligned using two people with walkie talkies. The food ration is gradually improved (see 2.10) and they are fed 4 - 8 times per day to stimulate courtship. We use automatic feeders to cover early and late feed times.

Figure 2.12.1 Male New Zealand falcon cliff-racing and mock-attacking the female.

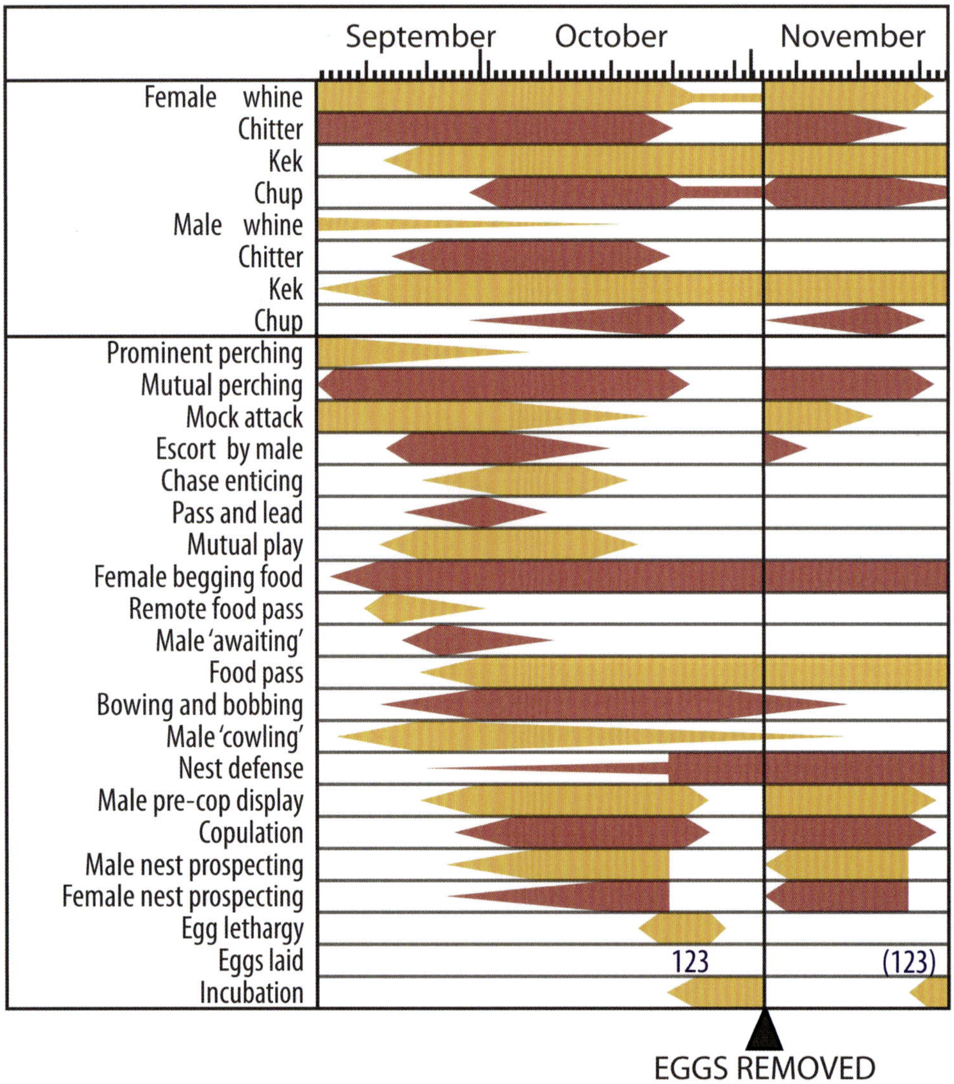

Figure 2.12.2 Courtship, incubation and rearing in captive New Zealand falcons in the southern hemisphere. The first clutch of eggs has been removed and the pair have re-cycled.

How do you know how the birds are getting on? The problem is first to properly understand the elements of courtship behaviour in the wild, and then to recognize it in a more condensed form in the aviary birds. For example, many male falcons in spring swoop at the female repeatedly and then zoom around the nest cliff (figure 2.12.1). This is known as mock-attacking and cliff-racing. When you see this in the pens you might conclude that the male is freaking out and attacking the female. So either by studying the wild birds, or by watching films, or reading books, you need to establish clearly in your own mind, what calls and behaviour you would expect of your birds, and you should be able to put it all together into a simple chart as in figure 2.12.2. Of course not all pairs follow exactly the same sequence, some get stuck early on, others who are more experienced, may skip most of it and go straight into copulation without many preliminaries. But once you are clear of the roles of each sex, and how everything is condensed in an aviary, you should be able to diagnose problems and come up with some solutions. And you will have problems!

Once the first egg arrives it is rather like white water rafting. There is no going back, you must just take each problem as it comes and try to come out the right way up.

2.13 Obtaining semen: voluntary donors and stripped semen

The best semen is given voluntarily, either by natural copulation or by copulation with another object. Natural copulation takes ten to twenty seconds, sometimes more, and at the height of courtship some pairs copulate every half an hour or so. Others may copulate only a few times each day but still show good fertility. Birds copulate at any time but most commonly in the early morning and during sunny, mild weather.

Just because the pair copulates does not mean that the eggs will be fertile. Life is not that simple. Young or early season birds often 'fire blanks' and even when semen is ejaculated a lot of it will miss the female's cloaca, and of that which does

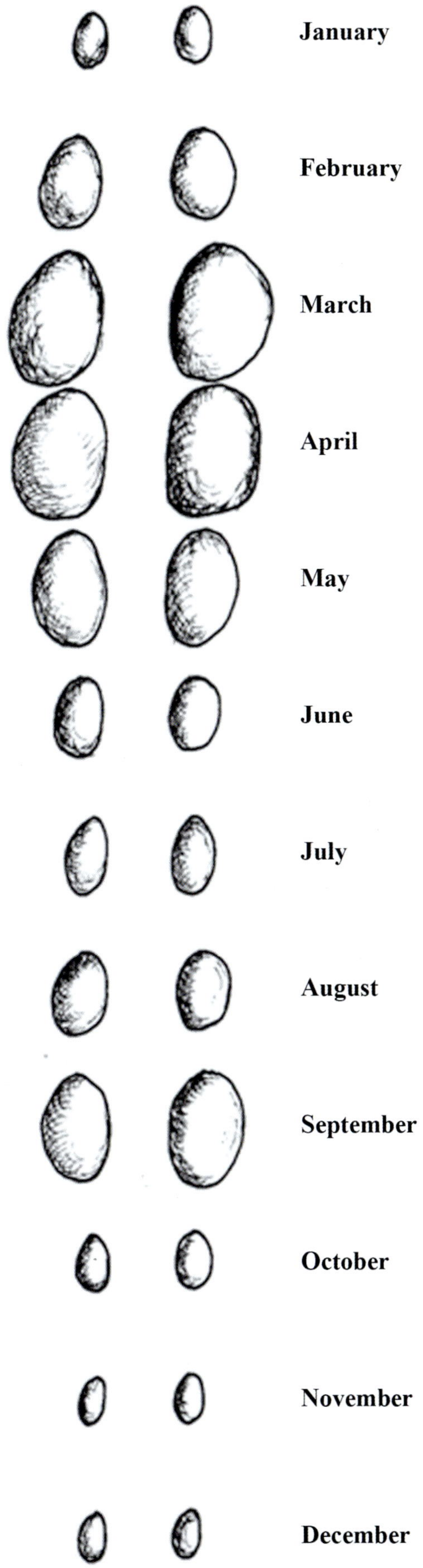

Figure 2.13.1. Monthly changes in testicular development (northern hemisphere).

reach the semi-everted cloaca, only a proportion gets drawn up into the oviduct.

Generally, it is easier for a male to produce semen than it is for a female to lay eggs. Many males have mature semen even though they have shown little or no sign of breeding condition. Many gyrs and peregrines produce semen for a short period in their second season, they do better in their third season and reach full production by their fourth season. The testes enlarge greatly in the early spring before producing mature semen (figure 2.13.1).

The other way to obtain 'voluntary' semen is to use an imprint male and condition him to copulate on a hat (a method pioneered by Les Boyd), or on your knee, glove, or a cushion. All this takes a lot of man hours of conditioning, which, being operant conditioning, is dependent very much on the bird. Some birds never become voluntary donors; many do not become sufficiently motivated until they are five or six years old. Once the bird copulates voluntarily then its mate (you) must be there in person three times a day right through the breeding season. A substitute person or your absence is liable to make him shut down either for a few days or even for the remainder of the season. So the bird is a major commitment. On the other hand, he can produce three or more good samples a day, sufficient to cover four laying females. One thus comes to rely on him. When he eventually drops dead, never at a convenient time, his absence can knock a big hole in a breeding program. Also, from so much use, he can rapidly genetically dominate a breeding program. After three or four years his daughters come on stream and then one is looking for different semen. Voluntary donors are thus a real asset but one should not be tempted to rely too heavily on one individual.

In order to train a male to copulate with a hat, we are asking him to cross quite a big behavioural gap. Copulating with hats is not part of his natural repertoire! It is possible to make the whole sequence more natural by teaching him to copulate with a stuffed or dummy female which he has been introduced to at an early age. A tape recording of female precopulatory whining can be played from a box on which the stuffed bird is perched. In other words, the biological signals are as close as possible to those programmed into his instincts. A great advantage of this method is that the stuffed female can be introduced into the male's pen via a sliding door by anybody, not just a specific person. This frees up the handler and allows the male to stay in a good settled regime. Once the male is starting to tread the female, it is possible to substitute a pad or cushion which is more convenient for collecting semen. For some reason many imprints have a foot fetish and may start to copulate with your boot.

If semen cannot be obtained voluntarily, then the male can be stripped. Anyone contemplating stripping a male should be thoroughly and three-dimensionally familiar with the anatomy of the male urinogenital tract. The best way to achieve this is by dissection of an adult male of the same species. Although stripping is a non-intrusive

Falcon copulating on a dummy.

A semen donor copulating on Tom Spink's hat.

Semen collected in the cells of a hat.

technique and is simple once one can do it, it is not something to be recommended to untrained people. A great deal can be learned by watching someone doing it, followed by practicing on chickens, but after that there is no substitute for hands-on experience and few breeders will be happy to let you use their precious birds. Stripping should involve only gentle, controlled pressure used in a full knowledge of what you are doing. Rough or ignorant handling, apart from being inhumane, could injure the bird.

A good imprint male, properly treated, will provide semen every day or every other day for six or more weeks. To do this he needs to be carefully hand-raised and made well to the hood. Preferably he should be flown to the lure in his first season. By his third season, sometimes sooner, it should be possible to go into his pen and he should court strongly. Either he can be free in his pen, with non-removable jesses and a chest high nest, or he can be tethered to a block with a car tire nest filled with pea gravel adjacent to the block. The nest is an important focus for him. He should not be fed for at least twelve hours prior to stripping otherwise the semen may be contaminated with mutes. One therefore needs to be able to predict when semen will be needed and diet the male accordingly. When you go into the pen, even though he is fat, you should be able to take him onto the fist and hood him without any fuss or alarm.

We normally take him then to a special room which is cool and very well lit. His claws should already be well trimmed to prevent self-injury. While I hold him by the jesses an assistant then holds him across the body with both hands. We then wrap a casting jacket quickly and fairly tightly round him. It is wrapped around the chest of the bird with the trailing edge in line with the tip of the allula on each wing, just in line with the point where the legs emerge from the body. The other edge loosely forms a hollow tube over his head to prevent him biting the underside of my right arm. I monitor his breathing and heart rate throughout using my right hand.

I sit on a chair holding him on my left thigh on his back so that his head is in the middle of my lap and his tail projects out across my left leg. I place my right hand lightly over his chest and hold his right leg between my little finger and ring finger and his left leg between my longest finger and my index finger. In gyrs, the left leg is held between my index finger and thumb. The legs are held at the 'hock' joint, between the tibia and the tarsus. With my two middle finger tips and thumb I draw up the vent feathers to expose the cloaca. Meanwhile I lay my left hand along the length of the tail which is pressed gently down until it is at an angle of 45 degrees to the floor. My left little finger and ring finger curl under the pygostyle, supporting the tail and creating a bed of positive light pressure from below. My left index finger and my thumb exert gentle pressure in the area between the vent and the pubic bones. This is not the only way to do it; other breeders use a variety of different methods and with a good male you don't need a casting jacket and it only takes a few moments.

The first time this is done each season I carefully 'crutch' him by trimming all the fluffy feathers for a radius of about 2 centimetres all around the vent. Then the area is damped down with cotton wool soaked in warm water. The vent is blotted dry by dabbing with a tissue, rather than by wiping which makes him sore.

Meanwhile my assistant has prepared to receive the semen. He has a warm glass plate with a series of hollows, each containing a small droplet of semen extender. Microscope slides, cover slips and a set up microscope stand warm and ready. He holds an ultipipette with a new disposable tip ready to collect the semen (figure 2.13.2).

We are now ready to obtain the semen. There is no preparatory massage, no abdominal pressure, no force of any kind. Even if mutes are present, the pressure will usually be insufficient to draw them into the cloaca. First, using the finger tips and end knuckles of my two middle fingers of my right hand, I slightly catch the surface skin above the urodeum and draw it very slightly toward the bird's head without exerting any pressure at all. This slightly opens the vent. Then, continuing this movement, I exert a little pressure with the knuckle of my right long finger over the junction between the coprodeum and the urodeum. This exerts an inward pressure on any mutes in the coprodeum and an outward pressure on any semen in the seminal papilli in the urodeum. Some semen may appear and is collected from the semi-open vent. Then I draw my left index finger and thumb along the vent side of the pubic bones back to the cloaca, keeping pressure against the underside of the pygostyle. This milks semen down the lower reaches of the ducti deferentia. There are no semen storage organs present there in raptors but even so, semen standing in the tubes of the ducti deferentia is drawn down. I then maintain the pressure on the gathered semen using the end knuckles of my right ring and long fingers. This enables me to slacken with my left hand and reposition my fingers. This time, using my left index finger and thumb, I place them either side of the vent and, without pushing downwards (which would drive the semen away) I stretch the vent outward slightly in a side to side direction. This causes the cloaca to be somewhat exposed and the semen is available for collection held in a well of the vent. Once this is collected, or if none appears, I then press downwards slightly with my left finger and thumb and pinch them in again toward the cloaca, thus driving semen toward the vent. I continue with this movement, everting the vent and exposing the lips of the urodeum. Inside these lips are the two papilli of the ducti deferentia. My finger and thumb gently push the semen up from the ducts, through the papilli and out through the lips of the urodeum where it

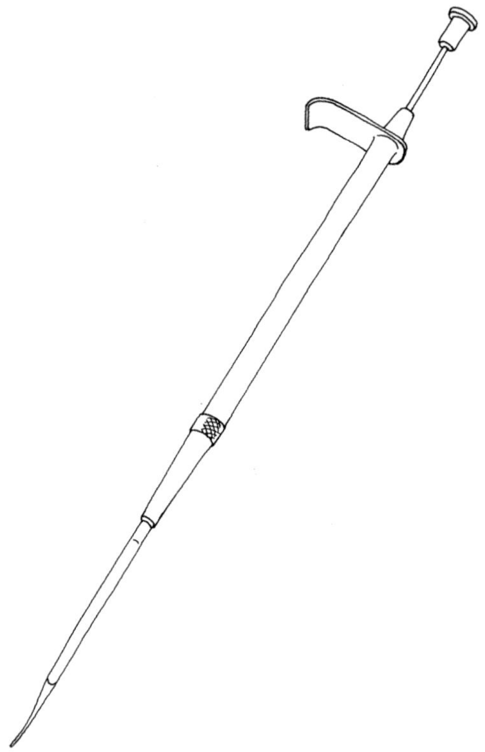

2.13.2. Ultipette semen collector with disposable tips.

wells up and is promptly collected in one or two seconds with the ultipipette. None of this pressure has been on the rectum or coprodeum.

All this procedure takes about ten or fifteen seconds. We repeat it four or five times. While my assistant unloads the collected semen I relax my hold on the bird slightly and he will usually give a little kick which exerts some abdominal pressure. When this happens, semen in the testicular end of the ducti deferentia flows down into the cloacal end which has just been emptied. You can often actually hear it move. This semen is then in a position to be stripped out. After the four or five attempts, the whole length of the two ducts have thus been drained. Sufficient semen should thus be collected for one or two inseminations. If not much semen was collected, it is because it was not there. There is no point in further handling of the bird; reasons need to be sought elsewhere. Checking that the vent is once more inverted, we unwrap the cloth and let the bird stand up on the fist. He is then returned to his pen, unhooded and a few moments are spent talking to him and giving him some quail. Males will normally immediately revert to displaying in the nest and it is important to overcome any negative effects of handling.

If necessary, the same procedure can be followed with a male which is paired up with a female. In this case it is important to catch him before he has fed in the morning. The rest of the procedure is the same. Many males which do not copulate still have good semen and can thus become fathers, if not proud ones.

In pairs which lay infertile eggs the breeder will usually inseminate immediately after the first egg, and perhaps again after the second, using semen from the male or from another male. Many pairs, once they have reared young, settle down and copulate normally in subsequent years. Others are hardy wallflowers and require intervention every year.

2.14 Handling and assessing semen quality

With semen, it is often a question of using what little one has, and being grateful for it. Also, once familiar with a certain semen donor, a quick glance under the microscope is usually sufficient to assess the semen before using it. However, anyone undertaking an insemination program should have a clear understanding of the different factors, not only because a poor sample stands little chance of fertilizing the egg, but because a contaminated sample could infect the female and cause fatal peritonitis or sterility. It has even been known for people to inseminate with mutes, thinking it was semen.

Semen consists of live unprotected cells in suspension. They must not be subjected to temperature shock, chemical insult, or changes in pH or osmolarity, otherwise they quickly die.

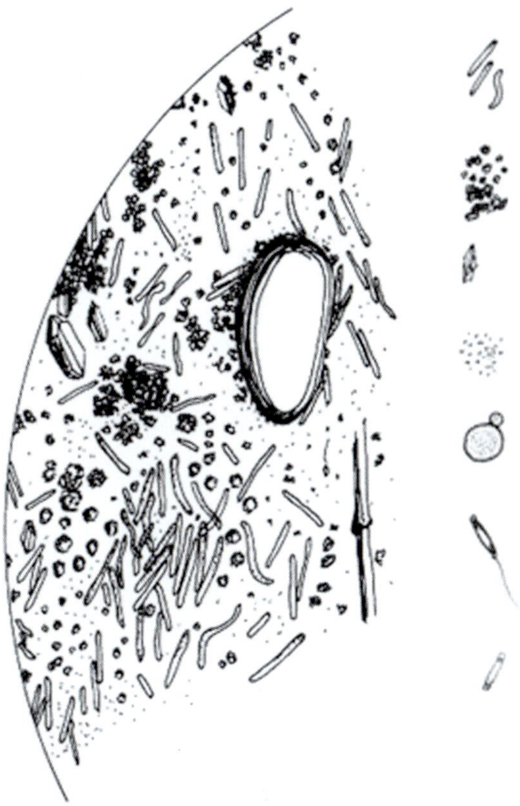

Figure 2.14.1. This is not semen. It is a faecal sample from a peregrine. At first sight the masses of seething ciliates look like spermatozoa but they have no tails, no acrosome, and are slightly flexible with a different method of moving. There is a dense background of bacteria, all floating in an acid bath of urates. There is also an air bubble and some feather detritus. The introduction of any of these contaminants into the oviduct could cause peritonitis, sterility or death.

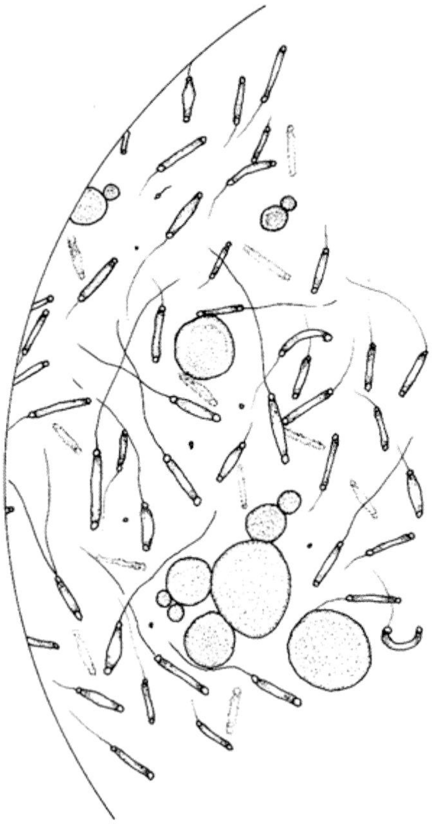

Figure 2.14.3 A usable semen sample from a gyrfalcon. There is some yeast contamination but no faecal contamination or urates. In the total field of view at 400X magnification at least 200 spermatozoa are motile and some swim across the whole field in about 60 seconds. One twin-tailed spermatozoon and two crooked ones are visible, but the rest are mature and well-formed.

So the first step in handling semen is to warm all utensils which will be in contact with it. This is best achieved by placing glass plates, microscope slides, cover slips, and extender in an incubator for fifteen minutes before use or laying them on a heat pad used for reptiles. Immediately before taking the semen we then place a small drop of extender into each small well of a hollowed glass plate using a 250 microlitre ultipipette with disposable tips. As the semen is collected it is pipetted and rinsed into one of the drops of extender. When a concentration of 1:1 has been reached we start a new well, or, if it appears that a particular sample may be contaminated or of a different quality to the rest, we place this in a separate well. The extender we use is Beltsville chicken extender; it may be possible to design a better one for falcon semen but we have found this to produce good results. Thus the semen goes straight into warm extender and is kept isolated from contaminants.

The normal procedure at this stage is to examine a small amount of semen under 400X magnification, quickly assess it and then load up the remainder and use it on a female. But to make a quick assessment you still need a broader understanding of what to look for. Additionally, there are times when it is desirable to assess more fully one or more semen samples to assess the production of a particular male and its likely effect on the various females. So let's look at semen more closely, step by step.

The first factor is quantity. As little as 10 microlitres can be used successfully but a normal useable sample will be 50 microlitres, which extends to 100 microliters. This is about one drop from a dripping tap. There is not much point in using more than this because the excess is liable to bubble out of the everted oviduct and be wasted. It is often better to use the surplus for another insemination.

The next factor is contamination. Normal semen appears slightly milky, often with a slightly yellowish tinge. Sometimes it is almost clear, sometimes straw-coloured. To the naked eye, it can often be confused with watery mutes and hence it is essential to examine all samples under the microscope before use. Sometimes a small swirl of solid white urates contaminates the sample. If the semen is valuable it may still be possible to decant off the semen from the more solid contaminant and obtain a useable sample. Contamination is dangerous for three reasons: Urates are strongly acidic; they appear as sharp sugar like crystals of uric acid. Sperm which are swimming strongly now, are in fact swimming in a lethal acid bath and will be dead in a few minutes. Certainly they won't survive long enough in the oviduct to do their work. Secondly, irrigating the oviduct with acid is at best introducing an unwanted irritant, at worst it could provoke a bad reaction resulting, for example, in a soft-shelled egg. Thirdly, examination of semen contaminated with mutes will also reveal myriads of hostile organisms, ranging from bacteria incessantly jogging about with Brownian motion, to flagellates almost as large as the sperm themselves (figure

2.14.1 and 2.14.2). Putting these into the oviduct can infect the oviduct, or they can get right up into the body cavity causing peritonitis, or they can sit in the oviduct alongside the waiting sperm cells and infect the ovum at fertilization causing early death of the embryo. Thus, contaminated semen can cause sexually transmitted infections possibly killing the egg, or the female, or rendering her temporarily or permanently sterile. The message is: if in doubt, do not use it.

Semen can show other contaminants which are less deadly. Many samples contain yeasts which appear as spherical bodies. These are usually benign and do not cause a problem. Red blood cells also are not a problem but should not be there. They indicate some tissue damage to the male which may point to a poor stripping technique. You should never see them in the semen. If they occur, the male should be rested for some days to allow the injury to heal.

Having obtained a useable quantity of clean semen, the next step is to look at sperm count. This of course depends on how much you have diluted it with extender. As a rough guide, a sample which is ready to use should have about 100 sperm cells visible in a 400X magnification field (2.14.3). There might be several times this amount, but less than about fifty may be too diluted. If you wish accurately to assess the actual sperm count of a particular male then the semen must be undiluted and a volumetric analysis made with a hemocytometer.

Perhaps more important than actual sperm count is the assessment of progressive motility. At least ten sperm should be moving in a 400X field and preferably this should be vigorous movement with individuals moving rapidly right across the field of view within about 60 seconds. This is the best measure of fertilizing capacity. A good sample will have hundreds of sperm moving in all directions. Motility is very temperature dependent. A sample stored in the fridge will appear inactive but as it warms up to body temperature, the sperm should become fully active. Also, a low volume of liquid on the slide may cramp the sperm under the cover slip and prevent them moving; it is important to have enough volume to assess motility.

If the male has not been stripped for three or four days, semen will have been waiting in the ductus deferens and some of the sperm cells will be aged or even dead. Therefore it is better to strip every second day so that the semen is fresh. If you have a sample which looks relatively inactive it is sometimes possible to differentiate between live and dead sperm cells using nigrosin stain. Dead sperm are stained pink whereas live ones remain transparent. To do this, place five small drops of stain on a slide and warm to body temperature. Add one drop of semen to the stain and mix well with a fine needle. Leave for one minute for the stain to penetrate. Taking another slide, smear the sample out across the slide in the same way as a blood smear. Dry this out quickly over a heat source and examine carefully under the microscope, preferably under oil immersion. Semen from different species take up stain variably and there seem to be no completely reliable stains at present for determining sperm survival. Obviously, if it turns out that most of the sperm are dead, then do not use the sample and review your procedure with the male.

Another aspect of the sperm to consider is whether they appear normal or are deformed. There are always some abnormal sperm cells, possibly because the male is a hybrid or is not in full production or is in poor health. Surprisingly, it is often the misshapen sperm which are most motile. Some have Y-shaped bodies, some are long and thin, and some have two tails (see page 86). Provided there are still plenty of motile normal sperm, the semen is useable because fertilizing ability is dependent only on the acrosome and the head itself. So do not worry; you will not end up with a falcon with two tails!

During all this handling you may have altered the osmolarity of your sample. If the precious drop of semen has dried up a little, then the concentration of the fluid medium will have increased. The osmolarity around the membrane of the sperm cell will tend to pull water out of the cell and desiccate it. Conversely, if the extender is too dilute for the sperm, the cells start to swell up like lemons. None of this is good for them.

Semen may look the same but behave differently. The best semen results from frequent voluntary copulation either in a natural pair

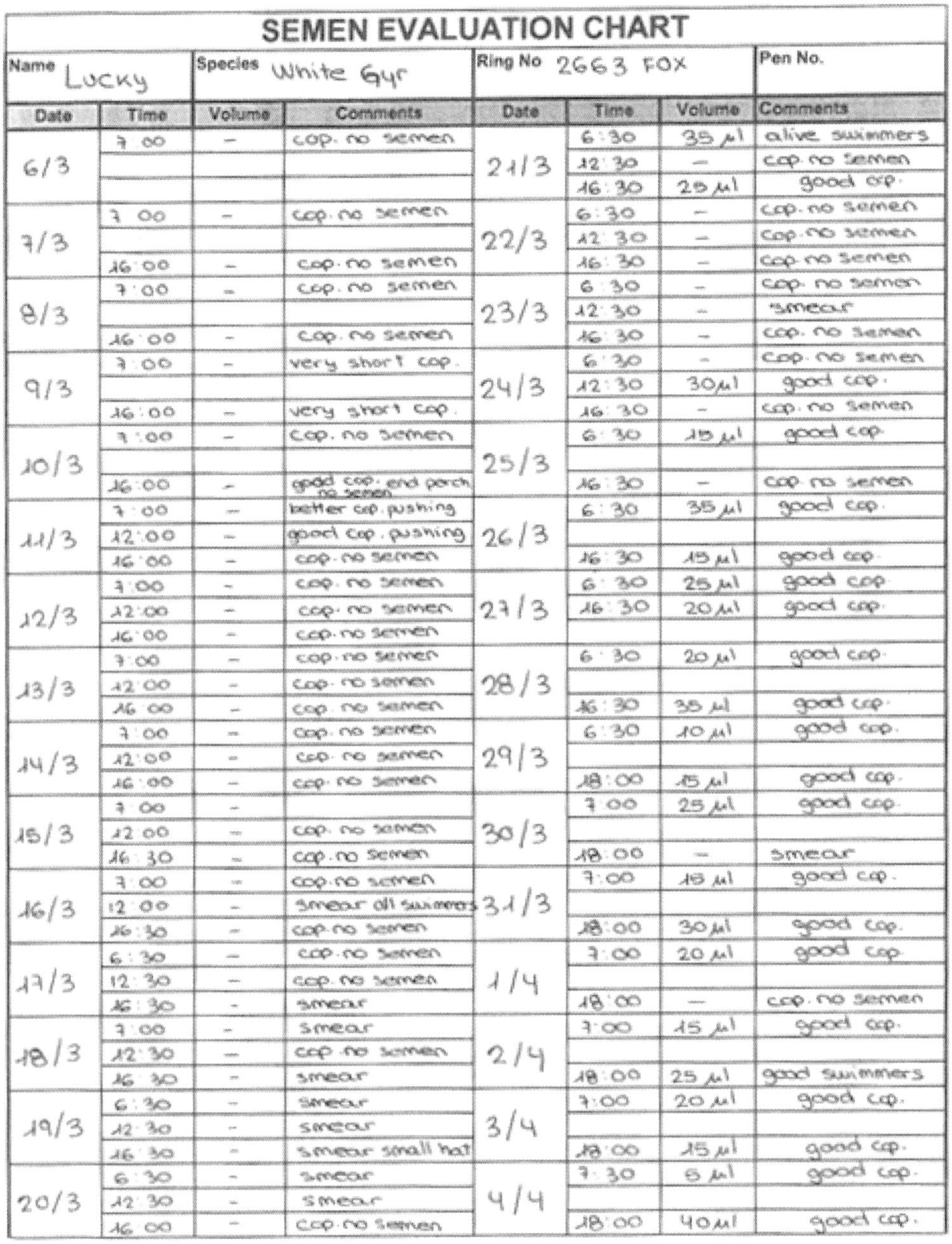

SEMEN EVALUATION CHART

Name: Lucky	Species: White Gyr			Ring No: 2663 FOX			Pen No.
Date	Time	Volume	Comments	Date	Time	Volume	Comments
6/3	7:00	–	cop. no semen	21/3	6:30	35 µl	alive swimmers
					12:30	–	cop. no semen
					16:30	25 µl	good cop.
7/3	7:00	–	cop. no semen	22/3	6:30	–	cop. no semen
					12:30	–	cop. no semen
	16:00	–	cop. no semen		16:30	–	cop. no semen
8/3	7:00	–	cop. no semen	23/3	6:30	–	cop. no semen
					12:30	–	smear
	16:00	–	cop. no semen		16:30	–	cop. no semen
9/3	7:00	–	very short cop.	24/3	6:30	–	cop. no semen
					12:30	30 µl	good cop.
	16:00	–	very short cop.		16:30	–	cop. no semen
10/3	7:00	–	cop. no semen	25/3	6:30	15 µl	good cop.
	16:00	–	good cop. end perch no semen		16:30	–	cop. no semen
11/3	7:00	–	better cop. pushing	26/3	6:30	35 µl	good cop.
	12:00	–	good cop. pushing				
	16:00	–	cop. no semen		16:30	15 µl	good cop.
12/3	7:00	–	cop. no semen	27/3	6:30	25 µl	good cop.
	12:00	–	cop. no semen		16:30	20 µl	good cop.
	16:00	–	cop. no semen				
13/3	7:00	–	cop. no semen	28/3	6:30	20 µl	good cop.
	12:00	–	cop. no semen				
	16:00	–	cop. no semen		16:30	35 µl	good cop.
14/3	7:00	–	cop. no semen	29/3	6:30	10 µl	good cop.
	12:00	–	cop. no semen				
	16:00	–	cop. no semen		18:00	15 µl	good cop.
15/3	7:00	–		30/3	7:00	25 µl	good cop.
	12:00	–	cop. no semen				
	16:30	–	cop. no semen		18:00	–	smear
16/3	7:00	–	cop. no semen	31/3	7:00	15 µl	good cop.
	12:00	–	smear all swimmers				
	16:30	–	cop. no semen		18:00	30 µl	good cop.
17/3	6:30	–	cop. no semen	1/4	7:00	20 µl	good cop.
	12:30	–	cop. no semen				
	16:30	–	smear		18:00	–	cop. no semen
18/3	7:00	–	smear	2/4	7:00	15 µl	good cop.
	12:30	–	cop. no semen				
	16:30	–	smear		18:00	25 µl	good swimmers
19/3	6:30	–	smear	3/4	7:00	20 µl	good cop.
	12:30	–	smear				
	16:30	–	smear small hat		18:00	15 µl	good cop.
20/3	6:30	–	smear	4/4	7:30	5 µl	good cop.
	12:30	–	smear				
	16:00	–	cop. no semen		18:00	40 µl	good cop.

Figure 2.14.1 Recording voluntary semen donations.

or with a voluntary imprint donor. Such semen can fertilize a falcon egg several days after insemination. There are probably several reasons for this. The semen will be fresh and of the right maturity; the sperm are ready but have not been stored in the male for too long. Second, the seminal fluid is of the optimal proportions of life supporting electrolytes. Third, it will have been ejaculated cleanly without contamination. Such semen can frequently be stored for 60-72 hours in the fridge and still have a good chance of fertilizing an egg. Stripped semen is much more variable. It may be stale, it may be immature, it may have the wrong proportions of seminal fluid, and it may be contaminated. These all affect its storage potential outside the female and within the oviduct, as well as its fertilizing capacity. Stripped semen thus can vary from as good as voluntary semen to downright dangerous. The semen quality is heavily dependent on the way the breeder manages the males and it is essential to take great care in this, otherwise the whole project will be unproductive.

2.15 Storing and transporting semen

A raptor sperm cell has a life of up to two weeks between maturation and death. It is still immature when shed from the epididymis and then matures in the upper portion of the ductus deferens. From there, it may be stored in the ductus deferens of the male, it may be stored outside the bird in a container, or it may be stored in the female. Judging from experience of various permutations, rather than from any rigorous experimentation, it appears that semen does not live long in the male. If a male is stripped right out, after one day he may have a moderate volume of lively semen with a low sperm count. If untouched for two days he should have a good volume of lively semen with a good sperm count. If untouched for three days the result is much the same, but after this the volume and sperm count stay up but the ratio of live sperm goes down. Five days after stripping he may give a sample which has plenty of volume and a high sperm count, but very little motility. One assumes therefore that sperm survives only for a few days when stored in the distal portion of the ductus. But to complicate matters, some males which have not been stripped, and have not copulated, seem to empty out the ductus and hence maintain stores of lively semen all the time. Natural pairs, of course, copulate several times a day so there is a rapid turnover of semen and the sample will be fresh and active. A carefully managed insemination program aims to obtain semen from the male while it is still young. Depending on the individual male this usually means stripping every second day, ie. not more often because the volume could be reduced, and not less often because the semen could be aged; it might appear good initially but would have just a short life left ahead of it.

Once out of the male, if not used immediately, it can be stored diluted 50:50 with extender, in a fridge at 4 degrees centigrade, rewarming again gently before use. This procedure has been used by many breeders for a long time, the aim being to reduce the metabolic rate of the semen without subjecting it to temperature shock. In the chilled state, semen can be transported to distant females by storing it in small centrifuge tubes in a container in a mix of ice and salt in a thermos flask. Each sample should be labelled with the time and date it was taken and the band number of the male. Voluntary semen stored like this for up to 72 hours can successfully fertilize eggs. Stripped semen on the other hand, depending on its history, tends not to store well out of the body and is best used fresh.

Experiments with freezing bird semen have had some success. The equipment and expertise necessary for cryogenics, together with the low resultant fertility, prevent it being a viable proposition currently, except for endangered species. However recent studies using 8 percent dimethyl sulfoxide as a cryoprotectant look promising because it does not have to be removed from the semen before inseminating. The frozen semen can be stored in liquid nitrogen for several years and fertility rates of 40-57 percent have been achieved with American kestrels. Accipitrid hawks and eagles (which of course are completely unrelated to the Falconidae) have a higher success rate with freezing semen and it is a practical method for some of these.

Once in the female, the sperm is stored in grooves or crypts in the wall of the lower oviduct and in the infundibulum. A single good shot of fresh voluntary semen can remain viable there for several days and can also fertilize two or three eggs in a row. The environment in the oviduct must therefore be more congenial to it than in the male. When using mixed semen from males of different species, the species which is most closely related to the female tends to take precedence in fertilizing the egg. Also, if there is semen still present and viable in the oviduct from a previous insemination and the female is then inseminated with fresh semen, the new semen tends to take precedence over the old. The best plan is thus to take the semen from the male early in its life and get it into the female where it will live longer and has the best chance to do some good.

2.16 When to inseminate

Having obtained the semen, the next hurdle to cross is when and how to get it into the female. The solutions to these questions are dictated by the characteristics of the female reproductive system. Normally a single egg is released from the left ovary as an exposed yolk. After ovulation it floats free and then enters the funnel-like infundibulum at the start of the oviduct. Once it proceeds down the oviduct to the magnum, the albumen and shell membranes are added and it is then impossible for a spermatozoon to fertilize it. Thus, to be successful, the spermatozoa need to be in position at the top of the oviduct ready to make contact with the yolk in its brief transitory period between ovulation and further development.

Once the egg is beginning to plump up, and particularly once the shell is on, it is not safe to evert the female. She can only be inseminated safely by this method within twelve hours of laying a previous egg. An insemination carried out within about two hours, occasionally up to four hours, of an egg being laid has a good chance of catching the next egg before the membranes are added. On the other hand, if the female is everted at this time, the newly ovulated egg may be expelled out of the infundibulum away into the abdominal cavity. There it will usually be harmlessly reabsorbed. However, 'lost' eggs are not always reabsorbed; occasionally one will remain in the abdomen and mummify. This may lead to peritonitis, which could be fatal or at least cause sterility. As another egg will not be ripe for ovulation for at least another 48 hours or so, the timing of the sequence will show that an egg is 'missing'. Thus one is gambling on the benefits of catching the next egg against the possibility of losing the egg altogether. This is a risk worth taking immediately after the first egg is laid, but not necessarily after subsequent eggs. It is also worth taking if using stripped semen with only a short life span.

On the other hand, if good voluntary semen is available, it is easier and safer to go for the window of about 3-10 hours after the egg is laid. This will miss the egg already in the system but will be held over in the oviduct and may cover the next egg to be released. We call this a 'leapfrog' insemination. Either type of insemination may cover two or even three eggs, or they may not. With good semen and steady females, I prefer to inseminate after every egg. On the other hand, if the female is a little put off by the handling, it is often better to inseminate after every second egg and hope for the best.

If we have a natural pair which has started to lay but no copulation has been seen, I prefer to obtain semen from another male and evert and inseminate the female after the first egg. Then I leave them alone, if possible, secretly pulling the first egg early on. If she has had a good shot of semen she could lay 2-3 more eggs of which maybe two might be fertile. She can then incubate these, or dummies, and then be encouraged to rear young. This may well bring the male into full working order next year.

With a natural pair which never copulate but lay and rear well, then I usually inseminate every egg or every other egg, much as for imprints.

2.17 How to inseminate

If the female will stand voluntarily for copulation, she can be inseminated several times a day by placing the semen with an inseminator onto her partially everted cloaca (figure 2.17.1).

This mimics natural copulation. Because the semen is not introduced directly into the oviduct, a lot is wasted and it is better to make many small inseminations to achieve high fertility. Therefore, one must have plenty of semen available, which in practice means a voluntary donor. The advantage of this method is that it is not physically or mentally stressful for the female. It can thus be undertaken at any point in the ovulatory cycle, even before any eggs have been laid and even

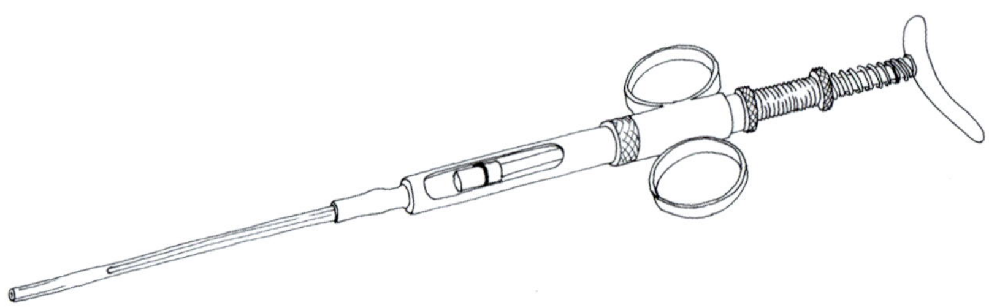

Figure 2.17.1 An artificial inseminator with a disposable semen straw.

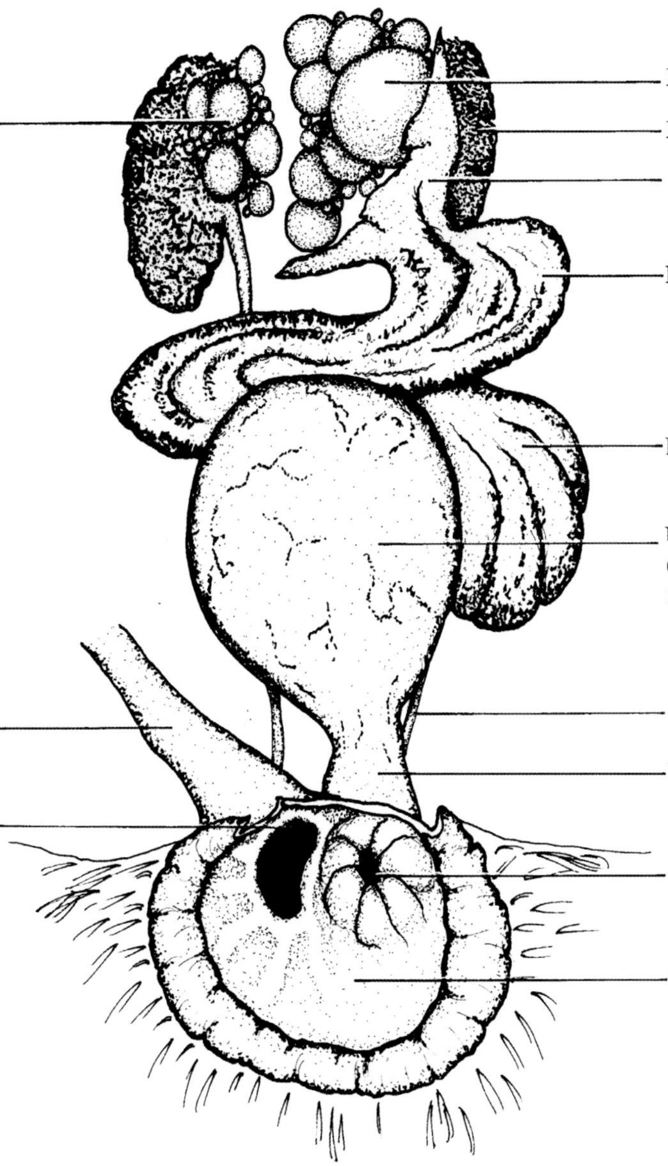

Figure 2.17.2 The reproductive tract of a female saker during laying.

when a shelled egg is almost ready for laying (figure 2.17.2). It is thus the only safe artificial way to inseminate the first egg of a series.

The more effective, but more stressful method, is to evert the lower portion of the uterus and place the semen directly into it using an inseminator. The inseminator we use is a commercial turkey or chicken inseminator which uses disposable plastic straws. It is quick, efficient, and ideal. There is virtually no risk of injury or infection to the female and there is good control of small semen samples. To inseminate the bird, I go into the pen and catch her; with an imprint this usually just entails picking her up like a broody bantam. Naturally paired females may need to be caught with a net. More often they refuse to vacate the nest and stand over their precious new laid egg glowering at me. One then has the task of collecting either the bird or the egg without damaging the egg. Usually I extend my gloved fist and the female will grab it. I then gently flip her out of the nest onto the floor, remove the egg and then net her. Alternatively, if she backs off a little, I extend my gloved fist toward the female but to one side of the nest, then quietly move my whole arm sideways over the egg. Then I slip my right hand along under my left arm to remove the egg.

An imprint female peregrine has just laid and is chupping in the nest ready for insemination.

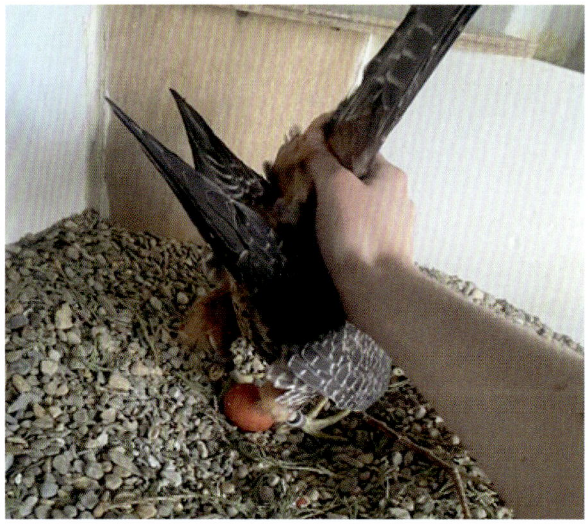

A female New Zealand falcon stands for copulation. The handler strokes her lower back.

Remember the hormones of the female at this stage are on a roller coaster ride. She has just achieved a massive mobilization of skeletal calcium to produce the eggshell in a few hours (figures 2.17.3 and 4). Now everything is suddenly in reverse and she is trying rapidly to restore her calcium levels using dietary calcium. She is in a very fragile state and should be handled with extreme care, but with speed and assurance. It is not a time for dithering. Everting a fat, sassy female gyr at 2 A.M. in your pyjamas with a torch held under your chin is a wonderful test of one's nerve and not conducive to regaining sleep.

This female New Zealand falcon has everted her cloaca ready to receive and draw up the semen.

Some females stay frozen in the copulatory position!

The female is cast in the same way as done for the male. Again, with my long finger of my right hand I ease back the skin from the abdominal side of the vent. Then with the end knuckles of my right hand I gently exert some very controlled abdominal pressure, supported by my left hand. Usually this causes the coprodeum to evacuate its mutes which flow over my left hand. Then, if the female just pushes a little herself, the oviduct everts on her left side of the urodeum, clear of the area of mutes. If any urates get onto the exposed oviduct they are removed with a dab of cottonwool. The oviduct needs to be everted for no more than two or three seconds, and only far enough out to recognize the opening and insert the straw. The inseminator straw is then winkled gently only one centimetre into the oviduct, assuming a bird the size of a peregrine. The moment the straw is in position, I relax all abdominal pressure so that the straw is now 4-5 centimetres inside the normal relaxed oviduct. The plunger is gently pressed, withdrawing the straw slowly at the same time. Then the straw is removed completely. I usually hold the female calmly for a few moments while the assistant removes any equipment, such as nets, ladders, egg boxes, etc. from the pen. Then,

Everting a female peregrine.

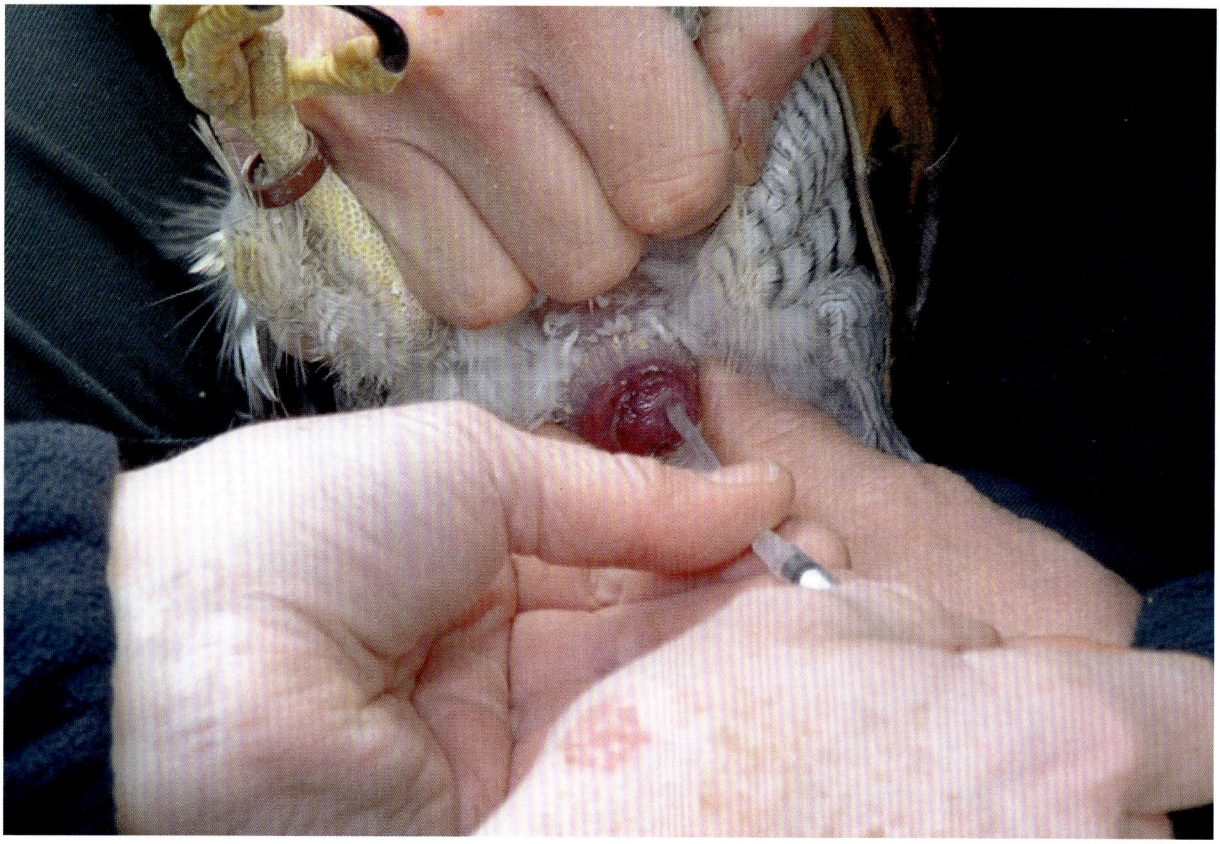

The oviduct is everted on the falcon's left side and the inseminator is gently inserted about 20mm (3/4 ").

holding her by the legs, tail base, and wing tips, I quickly unwrap the cloth and strike the hood braces. Then I place her back on the nest, remove the hood, and leave. If she is an imprint, I give her some quail and make a fuss of her. Imprints normally carry straight on with a full nest ledge display, quite unperturbed.

The main sperm storage site is near the junction of the oviduct and the cloaca. In most birds there is a secondary one in the infundibulum.

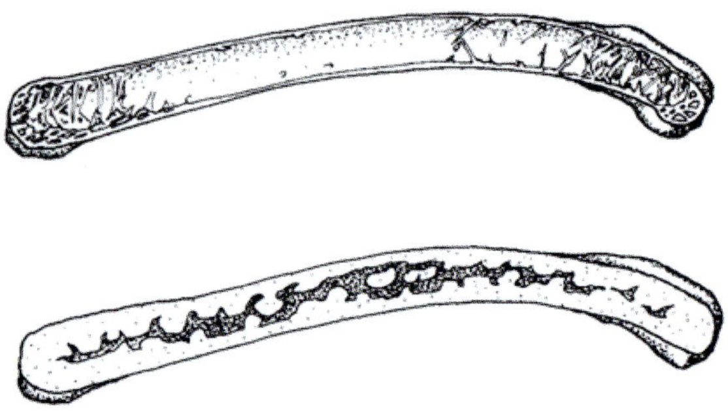

Figure 2.17.3
Top: Section through the right femur of a juvenile female northern goshawk in September, showing normal trabeculae.

Below: Section through the right femur of an adult female northern goshawk in March, showing the store of medullary bone prior to egg-laying.

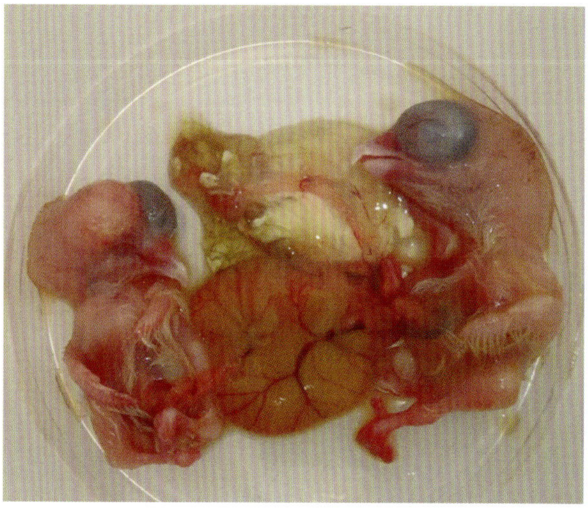

Double embryos are rare and fatal. These are New Zealand falcons.

At ovulation, sperm travel rapidly up the oviduct for fertilization. This is the only time when live sperm are found free swimming in the lumen of the oviduct. If the egg is swamped with live sperm at ovulation by an enthusiastic deep insemination, then more than one sperm may fertilize it. The egg may develop for a few days and then die early as a result of polyspermy. This can often be detected by examining tissue under the microscope. Divided cells present in a failed egg normally indicate fertility.

2.18 Variations in fertility

The fertility rate using artificial insemination is not usually as good as through natural copulation, although the results depend on many different factors. Also, if the outcross is a wide one, then there is a higher embryo mortality. Usually it is easier to fertilize the early eggs than the tail enders.

We keep records of all insemination details on a female insemination chart (figure 2.18.1). We also record the details of semen production for each stud male. These records are essential in order to learn and progress, and in decision making in subsequent years. Each bird is different and there is no blanket solution. Rather, it is a question of working out the best plan for each bird based on its individual characteristics.

It is not always easy to tell whether an egg is fertile. Just because an egg has gone rotten or addled does not mean it had been fertile; it just means bacteria have got in. When an egg has been diagnosed as not developing normally and

FEMALE INSEMINATION CHART

Name Dama	Species White 7/8 G x S		Ring No 1303 FOX		Pen no.
Egg No	**Date & time laid**	**Date, time & volume inseminated**	**Fertility (F or IF)**	**Semen quality**	**Stud male**	**Insemination type/quality**
	9/3	07:00 25 µl		S	Stone	Very Good
		12:30 25 µl		S	Stone	Very Good
		19:00 25 µl		S	Stone	Very Good
	10/3	07:00 25 µl		S	Stone	Very Good long C.
		12:00 20 µl		F	Stone	Very Good
		19:00 25 µl		S	Stone	Very Good
	11/3	07:00 20 µl		F	Stone	Very Good
		12:00 20 µl		S	Stone	Very Good long Cop
		19:00 15 µl		S	Stone	Very Good
	12/3	08:00 25 µl		S	Stone	Very Good
		12:00 25 µl		S	Stone	Very Good
		19:00 25 µl		S	Stone	Very Good
	13/3	07:30 25 µl		F	Stone	Very Good
		12:30 25 µl		S	Stone	Very Good
		19:30 30 µl		S	Stone	Very Good
	14/3	07:00 20 µl		S	Stone	Very Good
		12:00 25 µl		S	Stone	Very Good
		19:00 25 µl		S	Stone	Very Good
	15/3	07:00 20 µl		F	Stone	Very Good
		12:00 20 µl		S	Stone	Very Good
		19:00 25 µl		S	Stone	Very Good
	16/3	07:00 25 µl		S	Stone	Very Good
		12:30 25 µl		F	Stone	sucking a lot
		19:00 30 µl		S	Stone	Very Good
	17/3	08:00 35 µl		S	Stone	Very Good
		12:00 35 µl		F	Stone	Very Good
		19:00 40 µl		S	Stone	Very Good
	18/3	07:30 40 µl		S	Stone	Very Good
		12:00 40 µl		S	Stone	Very Good
		19:30 40 µl		S	Gru	Very Good
	19/3	07:00 35 µl		S	Stone	Very Good
		12:00 40 µl		F	Stone	Very Good
		19:00 40 µl		S	Stone	Very Good
	20/3	07:00 35 µl		S	Stone	Very Good
		12:30 40 µl		S	Stone	Very Good
		19:00 30 µl		F	Gru	Very Good
	21/3	07:00 40 µl		F	Stone	Very Good
		12:30 40 µl		S	Gru	Very Good
		19:30 40 µl		S	Gru	Very Good
1	22/3 seen 17:00	07:00 20 µl	IF	F	Stone	Very Good
		12:00 25 µl		S	Stone	Very Good
		19:00 35 µl		S	Stone	Very Good
	23/3	07:00 35 µl		S	Stone	Very Good
		12:00 20 µl		S	Stone	Very Good
		19:00 20 µl		F	Stone	Very Good
2	24/3 seen 17:00	07:00 30 µl	IF	S	Stone	Very Good
		12:00 30 µl		S	Stone	Very Good
		19:00 30 µl		F	Stone	Very Good
	25/3	07:30 30 µl		S	Gru	Very Good
		12:30 35 µl		S	Stone	Very Good
		19:30 30 µl		F	Stone	Very Good
	26/3	07:00 30 µl		S	Stone	Very Good long Cop
		12:00 40 µl		F	Gru	Very Good
		19:00 30 µl		F	Stone	Very Good

Figure 2.18.1. Female insemination chart.

all hope has been abandoned, then it is best to open it and examine the blastodisc carefully for cell development. If you leave the egg until after the breeding season, bacteria will get in and you will never know whether it was fertile, or at what stage it died, let alone have any idea of cause of death. If, on the other hand, you know that an egg is definitely infertile because the female had not been inseminated, it is often useful to incubate it for two or three weeks. This will tell you its weight loss characteristics which is useful information when making plans next year (see 2.26). The eggs of individual females tend to follow the same weight loss curve every year.

2.19 Laying

The onset of breeding is brought about by several factors such as temperature, rainfall, diet, and the presence of a nest site and a suitable mate. But the key factor is the photoperiod or day length. This is what tells the birds' biological clocks what time of year it is. Of course, not all birds breed in the spring. Some, such as the Eleonora's falcon, delay breeding to coincide with the autumn migration of passerines through the Mediterranean. Other species, in hot climates, time their breeding to coincide with the rainy season and to avoid the dry season.

Birds are sensitive to periods of daylight, even of low intensity. One 40 watt bulb in a large room is sufficient to register with the bird as 'daylight' and one breeder had merlins laying in January owing to extra day length provided by a nearby streetlight! Red light is the most important wavelength and is registered by the bird both through the eyes and through the top of the head where it stimulates the hypothalamus. Especially in situations where the birds do not get much direct natural sunlight, it may be worth using daylight bulbs that replicate a full natural spectrum of wavelengths of light.

Although the intensity of the light above a minimum level is not important, the duration of the light and the way it is administered is critical. Do not think that by leaving a light on all night that the hawks will come into breeding condition. British raptors normally need 14-16 hours light per day to come into breeding condition and this is best worked up to in stages over a period. It is thus possible to bring them into condition earlier or even give them an extra breeding season. With New Zealand falcons I have put them through an extra winter so that they scraped a year earlier than normal. However, playing around with photoperiods needs a thorough understanding of the subject, otherwise it can have unforeseen effects. Once birds have come into breeding condition on an increasing day length, they go through a long refractory period in which day length does not affect them. It can happen, for example, that one gives a pair extra lighting early in the season, attempting to bring them on, but they fail due to other factors, such as temperature, holding them back. Then, as real spring progresses, the birds are on a static day length and do not have the increasing day length to bring them on. They may anyway have gone into a refractory period and even perhaps started to moult. On the other hand, if one is attempting to breed birds from a different latitude, emulating the annual changes in day length in its country of origin may well swing the balance. Photoperiods are best left natural unless the breeder has a specific reason for changing them. When you are dealing with migratory species, such as calidus peregrines that breed in the high Arctic, but winter in Africa or Asia, you need first to plot where these birds normally are in the world at each month of the year, then look at the daylength of those places in those months. Some of these migratory birds enjoy a perpetual summer.

The presence of a suitable mate is also important for breeding, especially in non-migratory species. For those working with imprinted raptors for artificial insemination and who therefore must do the 'courting' themselves, the work of Barfield on ringdoves may be of interest. He found that solitary females exposed to a male for 15 minutes each day showed oviducts of 1.1 grams, for one hour a day 2.1 grams, and on continual exposure 3.6 grams, after seven days. One cannot therefore shirk one's duties as a mate!

Once the female starts laying, you have to decide what to do. If she is a first-time layer, leave well alone. Just try for a single clutch and a successful rearing. If her eggs are infertile, give her some other chicks to foster and rear. If she has bred before, you might decide to either extend her

A good clutch of saker eggs in one of our artificial nests in Mongolia.

clutch or double clutch, or both. If you remove the first egg as soon as it is laid, she should carry on laying, but there is a risk that she will behave as if the nest has been predated (which it has!) and then go and lay in another nest or even off the perch. Most raptors are semi-determinate layers apart from those species which only lay one egg. Most falcons, for example, lay about four eggs. When she lays her first egg, she leaves it unincubated. About 48 hours later, she will lay her second egg. As soon as she lays this second egg, remove the first egg so that she still has only one egg in the nest. When she lays egg #3, remove egg #2. Leave egg #3 and she will lay egg #4. If she is left for any length of time with two eggs in the nest, she will start to brood the eggs sporadically and her ovulating system will start to shut down. But there will already be two further eggs developing and she will lay egg #5 and start to incubate steadily. Her brood patch will start to develop. Finally she lays egg #6, which means that she is now incubating eggs 3,4,5 and 6, and you have eggs 1 and 2 in the incubator.

If you keep removing eggs as soon as they are laid, that is to say, pulled sequentially, the falcon may lay 6-10 eggs, occasionally up to twenty. This is very taxing for her health and often the last eggs are thin-shelled. On the other hand, if a dummy egg is added to the first so that she feels two in the nest, she may start to shut down and lay only two more eggs, making her total three. Some females are determinate layers, laying their standard quota of three or four eggs regardless of what manipulations are attempted. We had a 15 year old imprint buzzard called 'Beast that lurks'. Every year BTL laid two infertile eggs, no more, no less, before going on to foster rear up to 12 chicks on a 'conveyor belt' system. We also had an imprint buzzard which, without pulling, laid a setting of 11 eggs. She was called 'Omelette'.

Females of small species usually first lay in their second season. Some peregrines and gyrs also can lay at two but many do not lay until they are four or five. Some of course don't lay at all. Domestic raptors, thoroughly at home in breeding pens, always lay better than wild-caught birds. The first clutch is often small; once birds have reared young, they lay and recycle better.

Should a first clutch fail or be removed, many females will recycle and lay a second clutch, occasionally even a third. During the first half of incubation the balance of hormones in the body of the female is changing to prepare her for the next stages in the breeding cycle and for the moult. Only if the balance has not swung too far will she be able to ovulate a second clutch of eggs and this will take two to three weeks after the eggs are removed before she is ready to lay again, depending on species.

Females have a limited number of ova predetermined before they are themselves hatched. Some people think that therefore if you pull too many eggs early in a falcon's life she will 'run dry'. In reality, falcons may run dry in old age but not for this reason. Many can lay 12 or even 20 eggs a year for ten or more years. There are two other aspects to consider: 'saving' a female for the future by only single-clutching her will not necessarily result in more of her genes being passed on. She might die prematurely. Secondly, the more young she produces early in life, the quicker they will mature and themselves breed.

The oviduct is the original production line. The ripe ovum, on its release from the ovary, is briefly unrestrained at the mouth of the infundibulum where it is fertilized by semen which has either newly travelled up the oviduct or which has been stored in grooves in the oviduct wall awaiting ovulation.

Once fertilized, the egg travels quite quickly through the infundibulum to the magnum where the albumen or egg white is added. In chickens this takes 3-4 hours. The next region, the isthmus, takes 1 1/4 hours to add the two shell membranes. The final portion of the oviduct, the uterus or shell gland, takes about 6-7 hours to 'plump' up the egg with fluid and the female drinks frequently at this time. A further 13-14 hours are required to add the shell and pigmentation. In raptors, most of which are on a 50 hour egg-laying cycle compared to the chicken's 24 hour cycle, these stages take considerably longer and it is not known if this extra time is spent proportionately at each stage or whether the egg is delayed for longer at some point in the cycle; at least 24 hours is spent in the shell gland.

In the common buzzard, the shell weighs about 5 grams and is about 98 percent calcium carbonate, the remainder being structural protein. About 70 percent of this calcium is obtained directly from the diet and 30 percent is endosteal calcium which has been stored in the female's bones prior to egg laying. About 12 percent of the bone substance in the female's body may be mobilized in this way, and, if too many eggs are pulled from her, apart from the eggs being thin-shelled, she may show symptoms of bone weakness. No calcium is stored in the shell gland itself; it is all transported there and removed directly from the bloodstream.

Later, a considerable amount of the eggshell calcium will be reabsorbed by the developing embryo for bone formation. This makes the shell weaker at hatching time. Weight measurements of shells for shell thinning analysis in pesticide studies should take embryonic development into account.

A female common buzzard weighing 900 grams (32 oz) lays an egg weighing about 70 grams (2 1/2 oz). If she lays four eggs in seven days, this represents a considerable investment of nutrients. Particularly if eggs are being pulled from captive raptors it is essential that the egg-layers are fed on an adequate balanced diet, otherwise the hatchability of the egg suffers and the health of the female may be endangered, or she will go off the lay. Some Harris hawks lay readily on a quail diet but stop if changed to day-old chicks. Unfortunately, we know very little about the reproductive or nutritional physiology of birds of prey, which are themselves a variable group. One must rely on and extrapolate from studies of other species, especially poultry, and these may not be truly representational.

Two or three days before laying, the female goes into 'egg lethargy'. She may appear slightly ill, feathers loosely ruffled, weight on both feet and with large, slow mutes. Her appetite becomes poor and she may tend to tear at the food and then flick it about and discard it. Some, on the other hand, show little sign. Often if the weather turns cold in the spring, or if the female is unfit, she may retain the egg in the oviduct and have difficulty laying it. This can become serious very quickly. The problem can be diagnosed by feeling the shelled egg in the abdomen, but there is a grave risk in catching and handling her that you may break the egg inside her. This could rapidly lead to peritonitis and death. Neglect her and the risk is the same. To induce egg laying, we give an intra-muscular injection of oxytocin together with 5 millilitres of warm 10 percent calcium borogluconate subcutaneously. She may lay down before or after laying but stand during the actual laying, which takes only a few seconds.

2.20 Incubation

Once the egg has been laid, it can remain un-incubated for several days and yet still hatch. About 86 percent of fertile eggs should hatch under optimum conditions and storage for three or four days has little noticeable effect. After about a week, hatchability falls rapidly and after two weeks few will survive. Before incubation proper starts, the parents turn the eggs periodically and brood them lightly at night to prevent them freezing. In the day they may shelter them from the hot sun. At temperatures above 21 degrees centigrade (70°F) the egg starts to develop. Unlike chickens and ducks, which lay a lot of eggs, birds of prey lay small clutches and few eggs wait more than four or five days before they are incubated. Also once incubation starts, the parents sit on the eggs very steadily indeed and do not leave them exposed for long periods. Mean periods off the eggs for raptors in cold or temperate climates are usually less than two minutes; frequently the male, having brought food in for the female, will replace her on the eggs immediately after she comes off them. Even in tropical climates the eggs are seldom exposed for more than ten minutes. In general therefore, raptor eggs are not preadapted to tolerate erratic incubation to the same extent that game bird or waterfowl eggs can. Although eggs might survive several hours of exposure once, if they have been properly incubated otherwise (indeed I have hatched one with frost on it), frequent 1/2 -1 hour exposures weaken the embryos and reduce eventual hatchability.

Particularly if cooled rapidly to below 21 degrees centigrade, both eggs and chicks can go into suspended animation. Chilled chicks can appear completely dead with no signs of breathing and yet, if warmed up again to about 36 degrees centigrade, may revive. Accidentally chilled chicks should always be warmed up before being discarded for dead. Once I had a friend telephone about a brooder failure he'd had the night before. I told him to retrieve the dead chicks from the bin and warm them up. A few of them came back to life and my reputation as a miracle worker was well established!

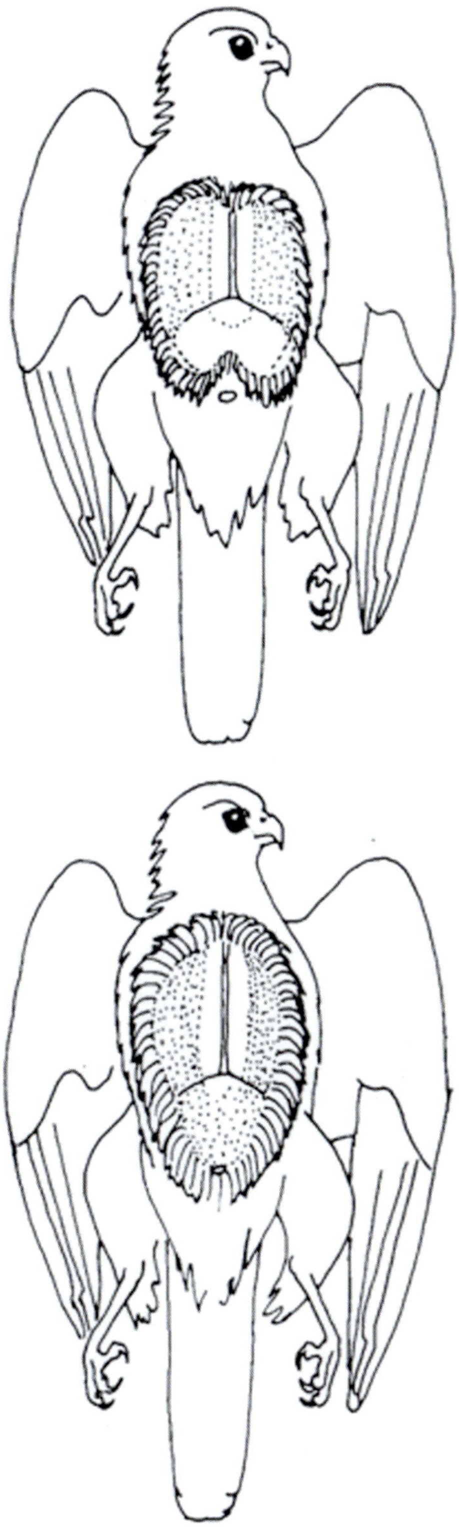

Figure 2.20.1
Top: The brood patch of a wild female Mauritius kestrel *Falco punctatus* after one week of incubating eggs.
Bottom: The brood patch of a wild female Mauritius kestrel while brooding 7-10 day old chicks.

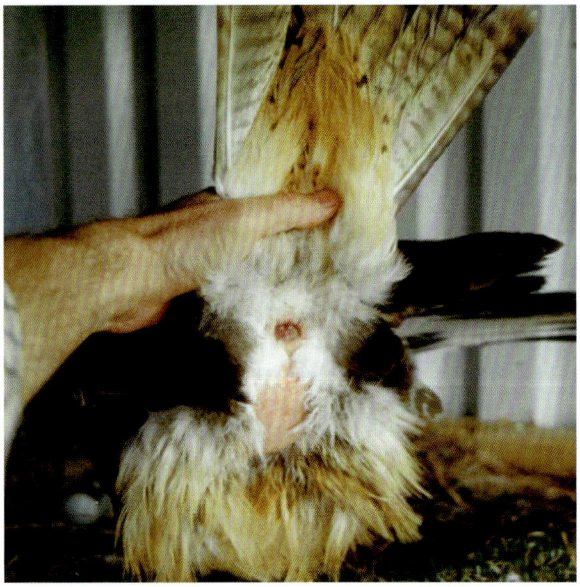

The brood patch of an incubating common buzzard.

With eggs, the problems arise when they are partially cooled to between 21-35 degrees centigrade. Development still continues but at a slower pace and so abnormal cell divisions occur. The eggs may die soon or may survive and die later in development. Embryos die quickly if their temperature rises above body temperature (41 degrees centigrade); they have no temperature regulating mechanism.

A good incubator heats evenly across the whole tray. But make a temperature map before the season starts and avoid cold spots. As embryos grow, they generate some heat of their own and you may need to make allowances for this.

A 'contact' incubator. When the lid is replaced, the membrane is blown gently back into contact with the eggs.

Once incubation begins, the eggs slowly warm up over several days by heating applied through an incubation or brood patch which has developed in the female in response to oestrogen and prolactin. The feathers on the breast and belly drop out leaving a large area of naked skin fringed by quite long feathers of the ventral tract which act as a heat seal. The skin itself becomes swollen with jelly-like fluid like a large blister and the blood supply increases. The effect is rather like laying a soft hot water bottle on top of the eggs. Full development of the brood patch is reached about one third through the incubation period. The breast feathers start to regrow again at about the same stage as the chicks are growing their own body feathers.

2.21 The structure of the egg

The structure of the egg is shown in figure 2.21.1. The shell is made primarily of calcite on an organic matrix laid down in layers but with thousands of tiny pores. These pores allow oxygen to enter the egg and water and carbon dioxide to escape. The shell and some of the pores are covered with a thin cuticle which is water repellent and hinders the entry of bacteria. Any pigmentation is laid down on the outer layer of the shell.

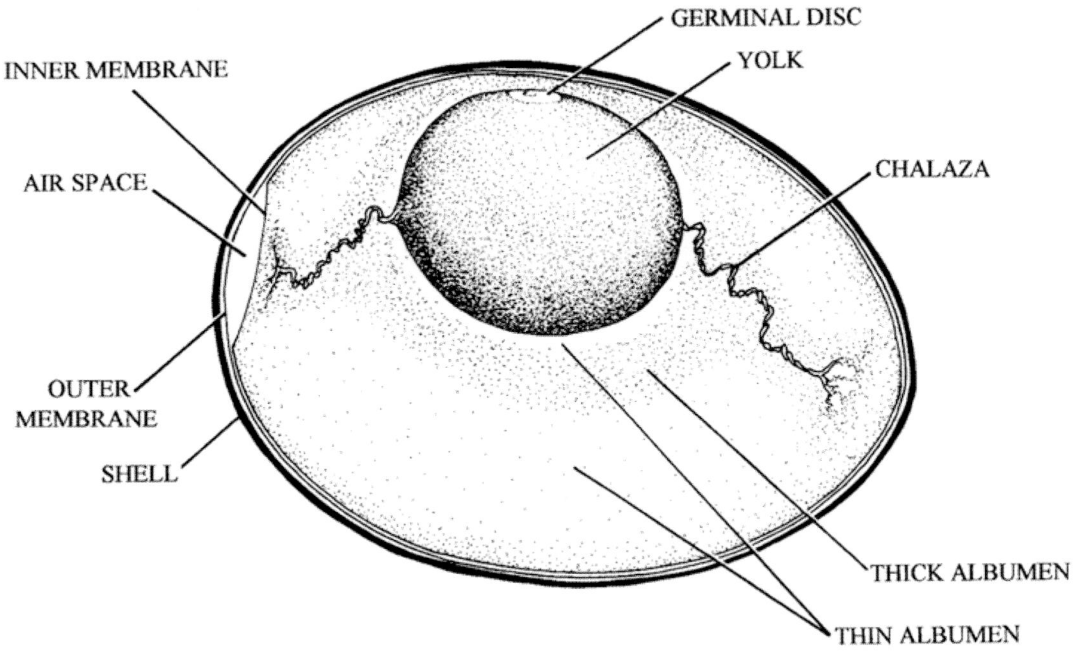

Figure 2.21.1 The structure of the egg.

Inside the shell are two fibrous membranes. One lines the shell and the other encloses the contents. The two membranes separate at the blunt end to form an air-cell. Occasionally these membranes fail to adhere and the air-cell then moves anywhere within the egg. This condition is fatal. In an electron microscope study with Professor Sally Solomon of 62 peregrine, saker and gyr eggs, we found that developing embryos started to draw down calcium from the mammillary caps of the palisade layer at about day 18, up until hatching. The embryo relies on this calcium for its developing skeleton. In another study we did, as the mother falcons get older, from about year ten in peregrines, egg and eggshell quality gradually declines resulting in decreasing hatchability and viability of chicks.

Calcium has been withdrawn from the mammillary caps inside this hatched peregrine egg.

Within the membranes is the egg albumen composed of a protein and water. The albumen forms a thick envelope surrounding the yolk, floating freely in thin albumen. The yolk itself is suspended at each end by the chalazae, which are twisted strands of protein fibres.

The yolk is contained by porous membranes and consists of lipoproteins which nourish the growing embryo. Floating on the top is the germinal disc 3-4 millimetres in diameter. If fertile, this develops into the embryo which lies on the yolk just under the upper surface of the shell, in contact with the source of heat.

In medium-sized raptors, the shell comprises 8.8-11.9 percent of the fresh weight. In falcons, the yolk makes up about 23-25 percent of the fresh weight, although individual birds sometimes produce yolks of only 16 percent. Red kites, Harris hawks and common buzzards produce yolks in the range 15.5-21 percent of fresh weight. It seems that the desert species, such as the Saker falcon and Harris hawk, have smaller yolks and more albumen than temperate species, and perhaps could tolerate greater weight loss during incubation. We need more data to reach firm conclusions on this.

2.22 The development of the embryo

As the egg progresses through the various stages of development its appearance as seen through a candler, changes as in figures 2.22.1-8.

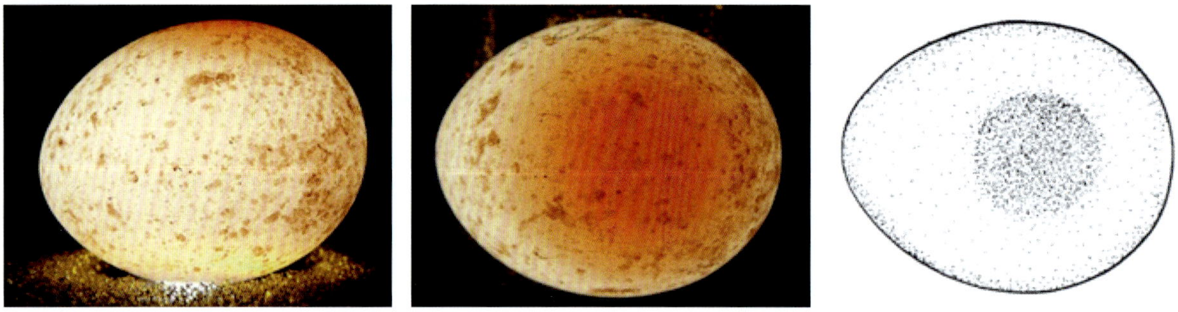

Figure 2.22.1 A fresh gyrfalcon egg from the side, and from above.

A fresh falcon egg, viewed from above, appears relatively transparent (figure 2.22.1). There is a small air cell, but it is hard to see it. The yolk is visible as a vague dark mass which floats to the upper surface. Both fertile and infertile eggs appear the same but after a week an infertile egg will remain looking like this, apart from the air cell which enlarges slightly.

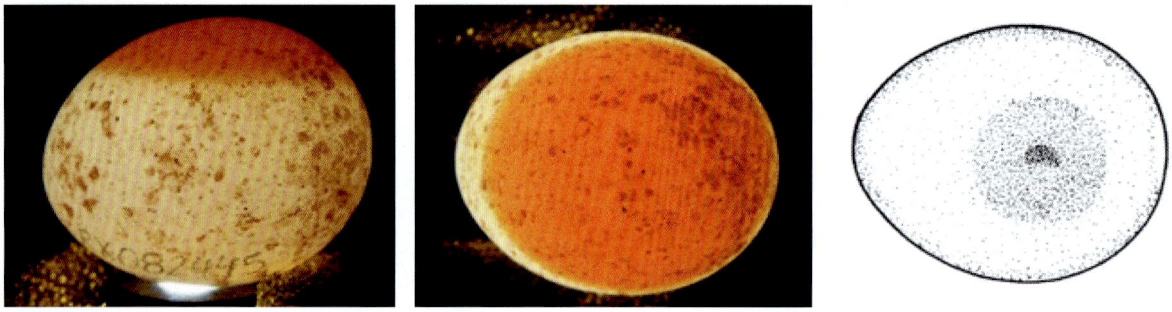

Figure 2.22.2 At five days, half shading starts and a small embryo dot is visible.

After five to seven days, in the fertile egg a small dark blob is visible on top of the yolk, which is still free floating (figure 2.22.2). A day or two later it becomes noticeably elongated and kidney-shaped with the embryo lying on its left side. In eggs with a strongly mottled shell the embryo will not be easily seen this early; instead, the most noticeable change is that the outline of the yolk becomes harder and slightly larger than before.

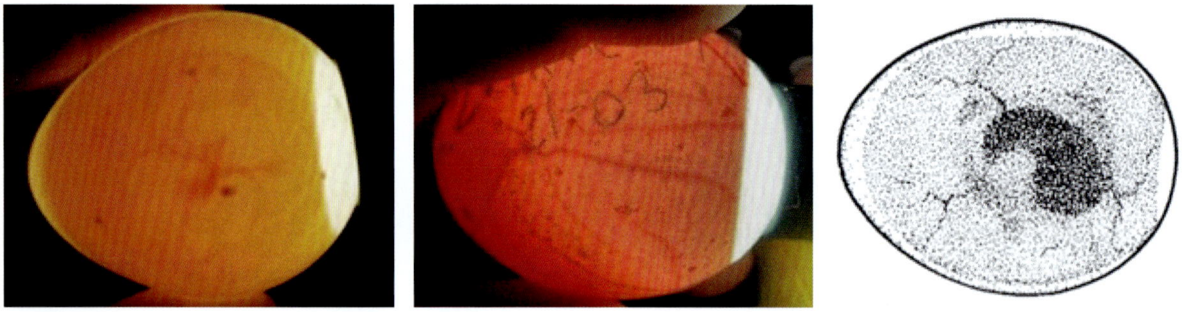

Figure 2.22.3 At eight days, seen from above, a halo of chorioallantois is developing and blood vessels radiate from the kidney-shaped embryo.

By the eighth to tenth day the yolk itself is no longer visible but the embryo has become larger and darker (figure 2.22.3). An area of shading daily enlarges around the embryo gradually spreading around the whole egg. Blood vessels may be visible radiating from the embryo.

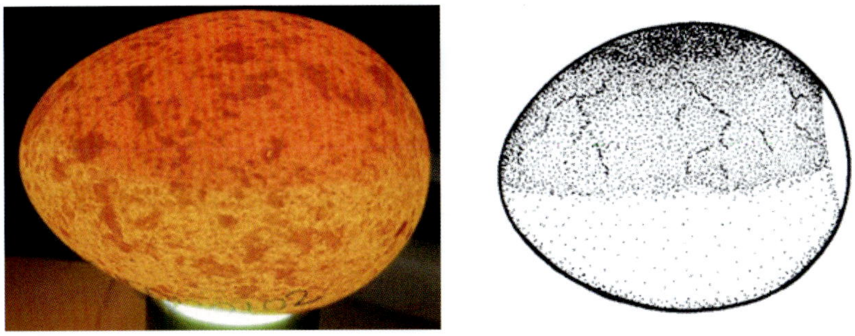

Figure 2.22.4 At eight days, seen from the side, half shading is visible.

Viewed from the side, the half shading first covers the upper half of the egg and is free floating (figure 2.22.4). Seen against the half shading, the air cell is clearly visible. The shading itself is made up of the growing chorioallantoic membrane. This is the life support system for the growing embryo. Its fine capillary system is responsible for the uptake of oxygen through the shell and the discharge of carbon dioxide.

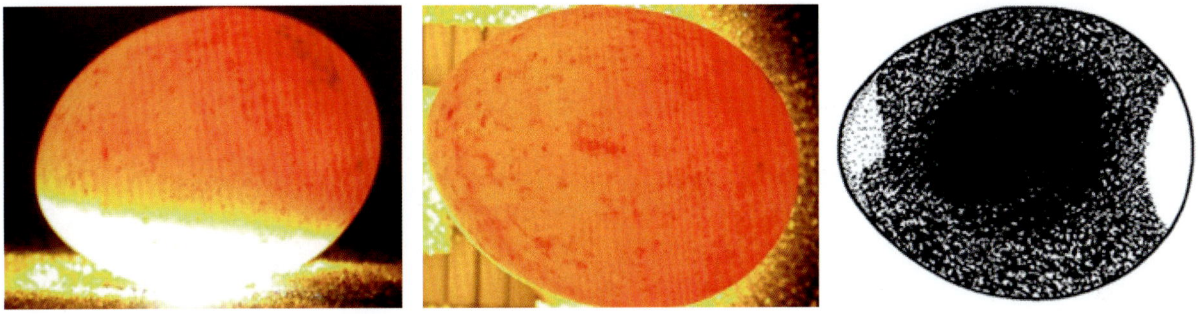

Figure 2.22.5 By mid-term at day 15, it is beginning to black out.

At about mid-term, the shading has spread around the entire egg although some clear albumen may remain visible at the pointed end (figure 2.22.5). The air cell is easily seen and has enlarged considerably. The embryo appears as a large, irregular, poorly defined dark mass which swoops around spontaneously under the shell propelled by the amnion.

Figure 2.22.6 The egg at 25 days.

Soon after mid-term the contents of the egg 'black out' and remain too dense to see anything throughout the rest of incubation (figure 2.22.6). Eggs which are overweight may still show some clear areas. The air-cell is clear and continues to enlarge slowly.

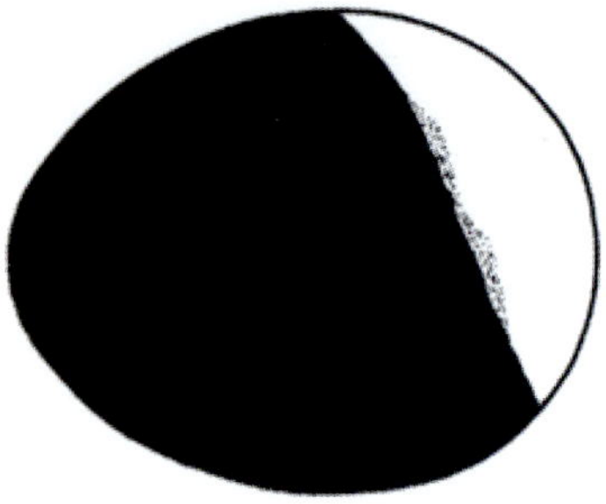

Figure 2.22.7 The egg at 28 days.

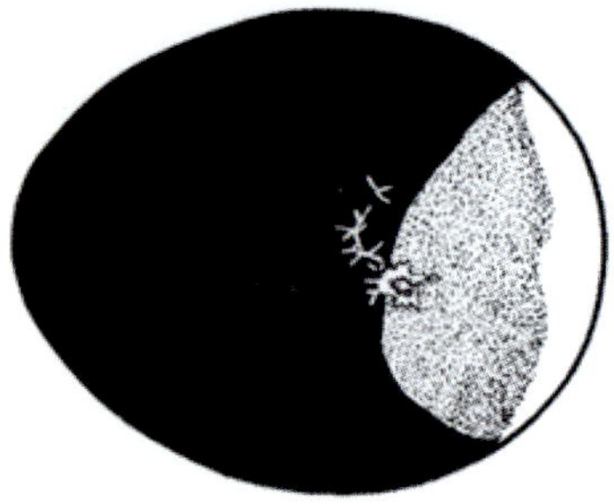

Figure 2.22.8 The egg at pip from above.

One to two days before pipping, the air-cell increases in size rapidly and draws down on the upper surface of the egg along the equilibrium line (figure 2.22.7). Seen here from the side, the asymmetry is very noticeable and careful candling may show a moving shadow when the embryo presses against the air cell membrane.

Soon after the initial pip, extensive shading can be seen in the air-cell (Figure 2.22.8). The chick has successfully split part of the air-cell membrane and now occupies some of the air-cell. This gives it additional space so that it can rotate to cut out.

2.23 The temperature of the egg

Inevitably during incubation, the top of the egg warms up most while the underside remains close to ground temperature. The heating power of the parent is not very great; Drent found, for example, that if herring gull eggs were left uncovered for 60 minutes at 11 degrees centigrade air temperature it took a further 110 minutes of incubation to bring them back up to temperature. Effectively this total time of 170 minutes is time with reduced development and the eggs will take that much longer in total to hatch. After 100 minutes uncovered, the eggs cool almost to ambient air temperature. Raptors overcome this by having only short and infrequent change over periods. In very cold temperatures the egg may be exposed for only a few seconds as the parents change over. Merlins will even occasionally incubate side by side, sharing the eggs between them.

Using the very detailed data of Drent and Baerends on gulls, together with studies of other species, and our own data (Smith and Fox 1997) on incubation attentiveness and characteristics, I have constructed temperature curves for the eggs of a typical medium sized raptor such as a peregrine, goshawk or buzzard. Obviously these vary slightly from species to species and according to climate, nest type, and ambient temperature but figure 2.23.1 shows the general trends in temperature throughout a clutch of eggs during natural incubation.

During the first half of incubation, whatever the position of the egg, the embryo floats to the top and lies just under the shell where it is maintained at a temperature close to that of the brood patch. Initially the temperature of the egg is cooler than that of the surrounding air in the nest but after a few days, owing to contact with the brood patch and then later to the heat produced by the embryo itself, it rises above the inter-egg air temperature. Once the female is sitting steadily, the embryo temperature remains within the range 34-39 degrees centigrade, the average value depending on species. The temperature of the bottom half of the egg tends to be about ten degrees lower, depending on ground temperature.

During the second half of incubation and especially during the last third, the embryo is growing and metabolizing rapidly and thus creating some heat itself. This heat is not sufficient for the embryo to maintain its own temperature, but it does contribute significantly to heat production in the nest. The still-air incubator, by simulating the temperature gradient experienced by the egg and by having a low heating capacity which warms the eggs gently, allows the growing

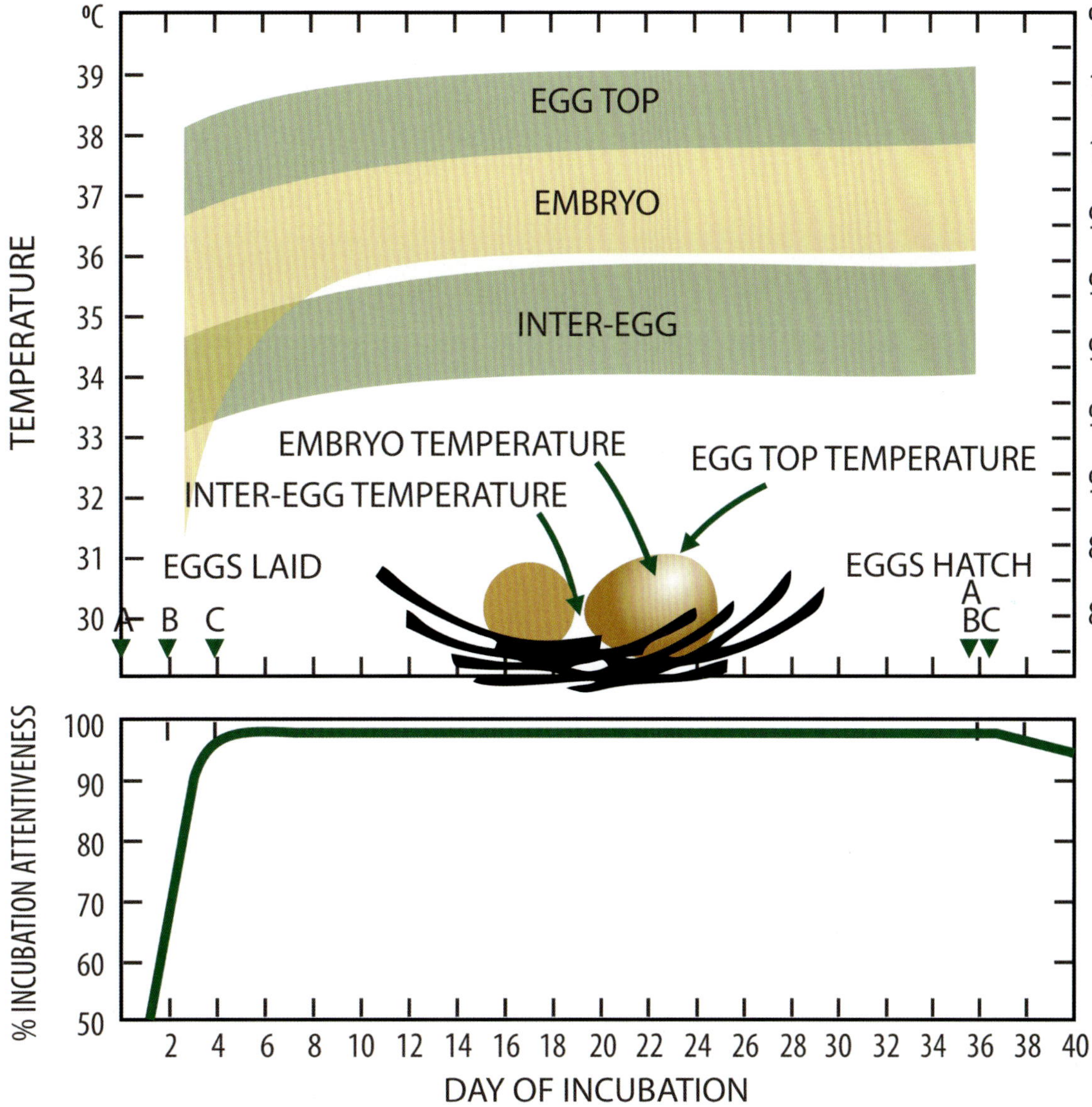

Figure 2.23.1 The temperatures of the egg at different levels during incubation.

embryo to develop its own microclimate. This mimics more closely the situation in the wild than does the forced-air incubator. But the moment you open it, you disrupt the stratified temperature.

After about the tenth day, the embryo, although still very small, is now fully formed and the remainder of the cell divisions mainly involve growth rather than differentiation. A small proportion of incorrect cell divisions are therefore less likely to prove fatal and eggs are easier to incubate artificially from this point on.

2.24 The turning of the egg

During the first half of incubation, the yolk, with the germinal disc on its upper surface, is free floating within the albumen and only loosely anchored in place by the chalazae. The high (30 percent) fat content is distributed in the yolk so that it is lighter than the surrounding albumen and always maintains the embryo uppermost. The chalazae are two light cords of protein both twisted in the same direction so that the yolk is held in place like someone twisted up in a hammock. Further rotation in one direction would tighten the twist and in the other direction unwind the twist. If the egg is rotated, the yolk gently settles back at the top, cushioned from jarring and abrasion by a layer of thick albumen.

If the egg is placed on a smooth flat surface it will be found to be evenly weighted, with no tendency for any one part of the shell to remain uppermost. The effect of the parent's beak movements at this stage is to turn the egg randomly, thus reducing premature adhesions of the extra embryonic membranes; adhesions which could flaw subsequent development and positioning. Another effect is that the yolk, on turning, comes into contact with fresh albumen and the nutrient supply contained in it. A third effect, important in species with large clutches, is to bring eggs at the edge of the nest into the centre. A difference of about 5-6 degrees centigrade exists in mallard duck nests between central and peripheral eggs but this is less extreme in raptors, most of which are able, in the females at least, to keep all the eggs in contact with the brood patch at the same time.

In a joint project with Dr Ken Smith we investigated temperature and turning rates of eggs for bantams, a captive common buzzard and a wild common buzzard. We embedded 17 mercury tilt switches into a balanced dummy egg containing a transmitter which sent signals to a receiver and data logger, which was periodically downloaded to a computer. This egg reported the angle at which it was lying to the nearest ten degrees. The results for all parents were fairly similar. The birds moved the eggs every 15-16 minutes or so. Most of the turns were in the range 5-60 degrees with occasional turns in the range of 60-180 degrees. Usually the eggs had little, bit by bit turns, rather than complete rollovers, although gradually,during the course of the day, each egg would turn over completely.

During this first half of incubation, the fine capillary network of the chorioallantoic membrane is growing. It gradually lines the entire interior of the shell. If an egg is unturned, it does not form properly. Tullett and Deeming found that if the membrane covered only 95 percent of the shell, hatchability fell to 46 percent. If it covered only 90 percent of the shell, hatchability dropped to 14 percent. Thus turning is essential until the chorioallantoic membrane is complete. Of course the embryo may not die immediately. Only later, when the embryo's metabolic demand is higher, is the defective chorioallantois unable to keep up.

By about half way through incubation the situation changes. The membrane surrounding the embryo and the shell membranes fuse and the embryo, no longer free floating, becomes fixed inside the shell. The egg is now unevenly weighted and if placed on a smooth surface, or floated in warm water, will tend to roll back to a fixed alignment. Thus the turning movements of the parent at this stage do not rotate the egg but simply free it from the friction of the neighbouring eggs for a moment so that it can roll back toward its equilibrium position. In other words the turning tends to maintain the eggs in relatively fixed orientations, not to rotate them randomly as in the first half of incubation. This is important for the attainment of the correct hatching position.

As incubation progresses further, this equilibrium position becomes more obvious and if the upper surface is marked it can be seen on candling to coincide with the furthest drawn down part of the air-cell. The first slight pip of the shell normally takes place a little to the left of this line at the edge of the air-cell and it is possible to predict and mark the exact spot where the pip will occur several days beforehand. A well-positioned pip is a good indicator that the chick itself is positioned correctly. As the pip progresses and the egg tooth is raised higher, the first actual complete hole will tend to be further to the right, on the equilibrium line itself. When the chick starts to hatch properly, the break-up will continue right-handed around the blunt end of the egg.

The parent sits assiduously but lightly on the eggs at this time and is very attentive. This reduces the chances of the egg membranes drying up and yet guards against crushing the eggs. If the hatching egg is turned hole downwards, the chick will start cheeping in distress. This call stimulates the parent to poke the eggs around until the egg rolls back upright, otherwise the chick risks suffocation.

Thus between days seventeen and the pip, depending on species, the embryo is getting itself into the right position which it must adopt in order to hatch successfully. Several malpositions can occur which may prevent the chick pipping

A common malposition. The leg is over the head.

through the shell, or rotating to cut its way out, or may cause it to pip at the pointed end. Many of these malpositions are lethal and are traceable to the ways in which the egg has been turned throughout incubation. A malpositioned chick may not be able to pip into air and thus may drown when it tries to use its lungs. Only very prompt detection in this case will save it.

2.25 The gas exchange of the egg

As incubation progresses, the embryo exchanges carbon dioxide for oxygen by diffusion through the pores of the shell into the fine capillaries of the chorioallantois which acts rather like a placenta. Gas exchange is proportional to embryonic development, slow at first but increasing rapidly during the second half of incubation. By the time the air-cell draws down, the demand on the chorioallantois reaches its peak. The embryo is metabolizing rapidly and the lungs have not yet come into use. This is when any defect is fatal. The embryo just dies.

If the eggshell is highly porous, it allows a good exchange of gases to the chorioallantois. It also allows a lot of water out and the egg may dry up too quickly. Water lost by evaporation during incubation is replaced by metabolic water during the development of the embryo. An early embryo is about 90-92 percent water, whereas an adult bird is about 60 percent water. About half the initial fat in the egg is metabolized during incubation. Any efforts to block the pores to reduce water loss through the shell also affect gas exchange with possibly lethal results. Therefore it is better to control the weight loss of a very thin-shelled egg, not by sealing the shell with wax, but by covering a proportion of it with removable cling film. This reduces water loss during the first part of incubation and helps to maintain the weight, then, later on, the cling film area can be reduced to allow for the increasing gas exchange.

The developing embryo absorbs about 6-7 percent of the calcium from the shell but this does not appreciably alter its porosity to gases. Permeability to gases does increase by about 25 percent suddenly around day seven due to the formation of the air-cell and the drying of the inner membrane. Removal of the cuticle by sanding or bleaching also increases porosity.

If the shell is too impervious, not only will the embryo lose too little water, with all the problems that that entails, but also it is likely to suffocate. Any embryos which hatch are slow to thrive.

2.26 The weight loss of the egg

During incubation the egg loses weight at a steady rate through evaporation and through metabolizing fat. Depending on species, about 16 percent of fresh egg weight is lost between setting and pipping and 4 percent between pip and hatch. Once the shell has a hole in it at pip and the chick is breathing through its lungs, water loss is quite rapid and the membranes quickly dry up if humidity is not maintained. If the chick fails to lose sufficient weight it is too cramped to turn inside the shell and may drown in its own fluids at pip. Infertile eggs tend to lose weight at about the same rate as fertile eggs. Species adapted to arid conditions have less permeable shells than damp climate species and do not lose weight as easily. Some of them compensate by having a higher proportion of albumen.

Eggs which have lost less than 10 percent by pipping are in grave danger of suffocation, drowning, or malpositioning. Eggs which lose more than 20 percent tend to be too weak and sticky. Occasionally we have hatched eggs at 8 percent and 22 percent but usually eggs fail at these limits.

There are several ways to control weight loss during artificial incubation. If an egg is losing too much weight, the first step is to raise the humidity conditions. One can also spray the egg all over every two hours with warm sterile water.

Some eggs react dramatically to this while others show no difference. The procedure effectively lowers incubation temperature and probably slows down the metabolism somewhat. It should not be used in the first half of incubation to any drastic extent.

If this is insufficient it is possible to seal part of the shell area to reduce porosity. This is a dangerous undertaking because alterations to shell porosity also affect gas exchange and it is easy to suffocate the embryo by rendering a proportion of its chorioallantois inoperative. The cling film method is safest because it is adjustable. If the desiccation really is critical I sterilize the shell over the air cell with alcohol. Then I very carefully hand drill a 1mm hole in the shell, taking care not to damage the egg contents. Then I inject about 1 milliliter of warm sterile water into the air-cell and seal the hole, either with some nail varnish, or, if I plan to repeat the performance, with a small patch of tape. The weight of the egg is recorded before and after the procedure. After about three hours, the water in the air-cell has been absorbed by the embryo. I have repeated the procedure several times without problems, but of course it must not be done too close to pipping otherwise the chick may drown.

Weighing a gyrfalcon egg.

The more common problem in a damp climate is getting sufficient weight off the egg, particularly in desert species. Eggs dry out slowly even when infertile or when in store. So if you have eggs from a mother whose eggs always have a weight loss problem, first weigh the egg as soon as it is laid. Then place it on its side in some blue self-indicating silica gel at about 12 degrees centigrade. Store it like this, turning it two or three times a day and changing the gel if it starts to go pink. When it has lost sufficient weight or after four days, whichever comes first, then set it. Some eggs refuse to lose weight with this treatment but then lose weight more rapidly once incubation temperature is reached.

If the female is stressed in some way during egg formation, perhaps some cold weather brings on a touch of egg binding, the egg may remain too long in the shell gland and receive too thick a shell. These rogue eggs can be difficult. Silica gel is not very effective in a forced-air incubator unless all the vents are completely sealed. Sanding an egg, particularly during the first ten days, risks damaging the fragile embryo through vibration or jerkiness while rubbing. The real danger comes to the embryo during the second half of incubation when it risks being suffocated as its metabolic demand increases. The easiest method to increase weight loss is to drill a small hole into the air-cell. Obviously this must be done very carefully and with a stop on the drill bit so that it does not puncture the underlying membranes. Once the desired weight loss has been achieved the hole can be covered again. It is always better to make gradual weight changes rather than abrupt ones.

We always maintain an egg chart for each egg, showing its weight loss. We also candle

These eggs have been drilled into the air-cell.

the eggs once or twice a week. This only takes one or two seconds. We do not make density calculations and we do not try to control weight by any drastic means unless the situation is serious. Weight loss is important but should not be an over-riding consideration. It is better to concentrate on giving the egg a good quality of incubation. Strong embryos overcome small weight loss variables.

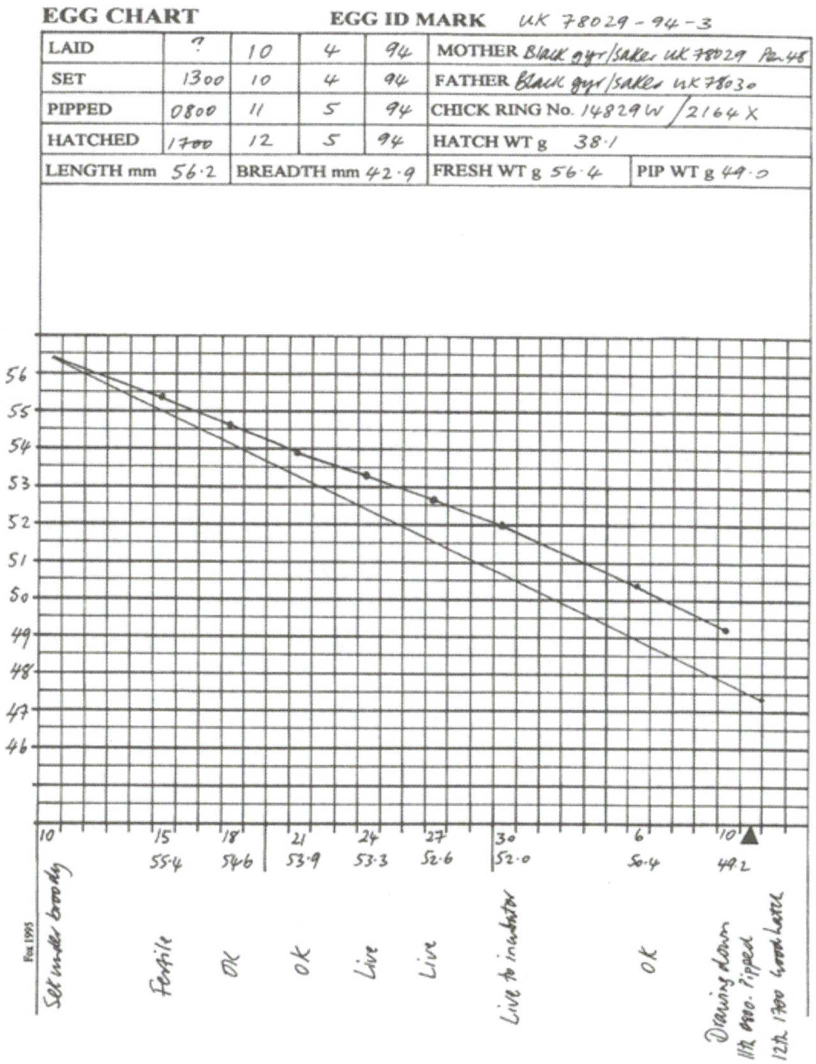

Figure 2.27 A simple, basic egg chart.

2.27 The hatch

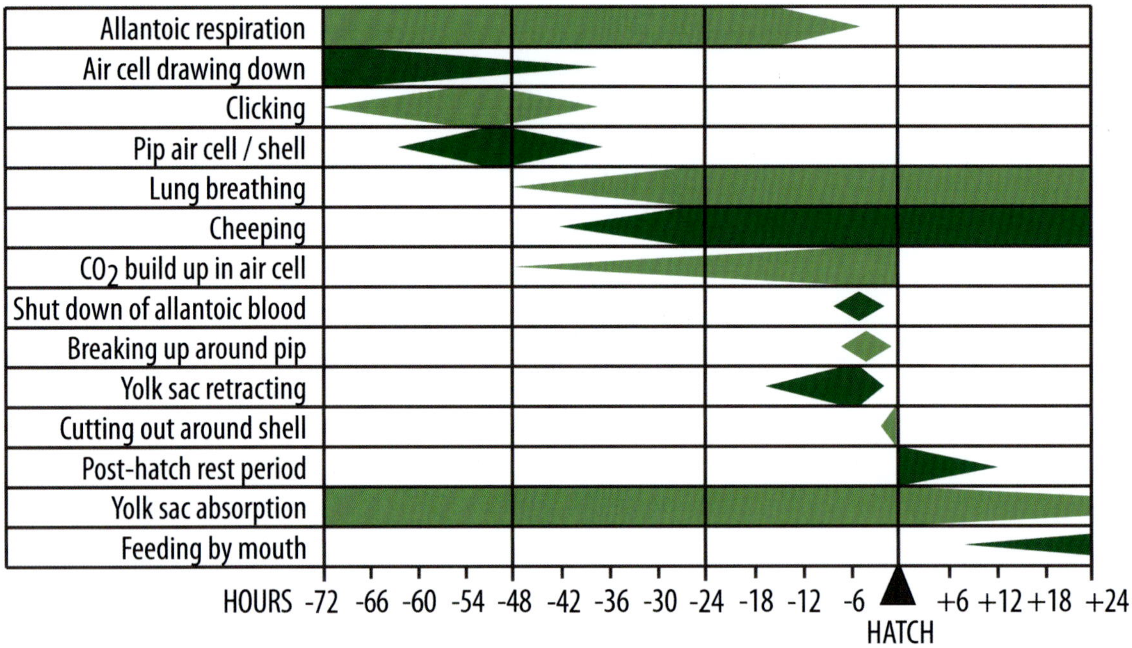

Figure 2.27.1 Events in the egg at hatching time.

About 72 hours before hatching, the egg starts an ordered sequence of events which are critical (figure 2.27.1). The chick must progress from allantoic respiration to lung breathing, retract its yolk sac and hatch and it is necessary to understand what is going on, especially if intending to help an egg with a problem.

At the time when the air cell is beginning the 'draw down' the complexis hatching muscle at the nape of the chick's neck becomes swollen with lymph and twitches spasmodically. The twitching movement extends throughout the whole body of the chick causing it to straighten briefly. These unbending movements press the chick tight against the eggshell and force the beak upwards. On the upper ridge of the beak is the small egg

Eggs drawn down and starting to pip.

Lammergeier *Gypaetus barbatus* egg pipped in the right place.

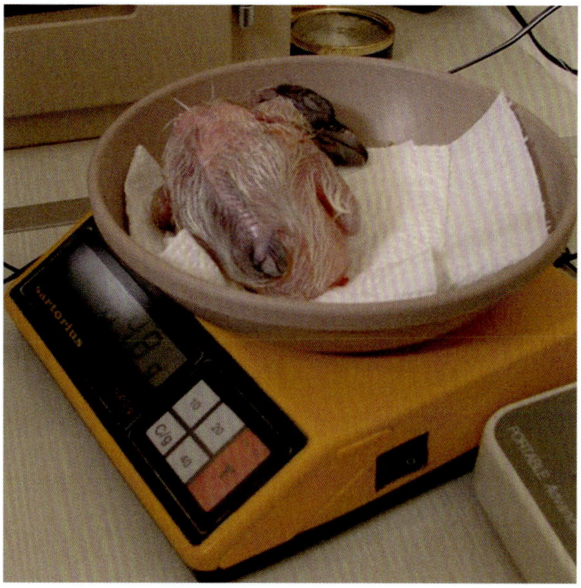

Newly hatched Lammergeier

tooth which eventually penetrates the inner shell membrane into the air cell. Once the beak is into the air cell, the chick is able to start breathing air into its lungs although it is still dependent on allantoic gas exchange. With its lungs in operation it is able to cheep and this is a sure sign that breathing has started. The beak piercing the air cell membrane may also be visible on careful candling (figure 2.27.2).

With continuous rebreathing, the air in the air cell becomes high in carbon dioxide and this stimulates the hatching muscle to further activity. With more faint tapping and pressure from the chick, the shell wall gives way and a slight lifting of a fragment of shell becomes visible. This is called 'starring'. This allows a small amount of fresh air into the air space and the chick usually becomes quiet for some hours. Working sporadically, the chick then begins to break up an area around the pip and also splits the air cell membrane much further so that, on candling, it can be seen to occupy most of the air cell and is much looser inside the shell. Gradually the chick enlarges an opening and the beak and egg tooth come into view.

The chick may rest at this point for a further twenty four hours. During this time the lungs and airsac system are becoming fully functional and gradually take over total responsibility for gas exchange. The blood circulation in the allantois slowly shuts down and the large allantoic artery and vein, which act as an umbilical cord, gradually change colour from red to white or pale blue. These vessels cross the wing and head on the right side of the pip and can often be seen through the aperture. Gentle squeezing with blunt forceps will determine if blood is still flowing. If the supply has shut down, the squeezed portion does not refill.

Once the allantois has shut down, the chick starts to rotate inside the shell. Some species, such as most falcons, have small vestigial claws on the tips of the alula, and also sometimes, the

About to cut out.

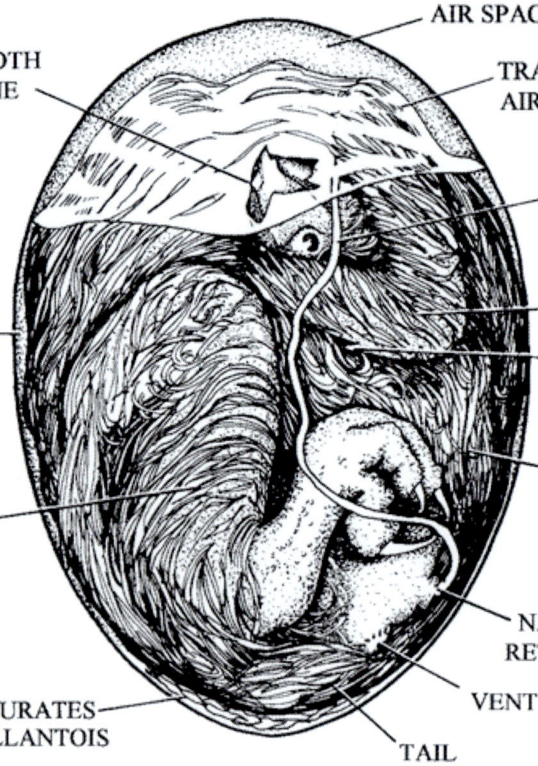

Figure 2.27.2. The position of a peregrine chick at hatching.

manus, which may help during rotation. Within about fifteen minutes, the chick circles about half to two thirds around the egg, industriously cutting out the shell as it goes. Soon it is able to push up the cap with its shoulders and kick itself free of the shell, leaving the membranes and excreted uric acid waste behind. The allantoic blood vessels quickly fall away from the navel. By one or two days after hatching, the hatching muscle has reduced in size and is no longer swollen with lymph. The total time from pip to hatch in birds of prey varies considerably from about 30-70 hours

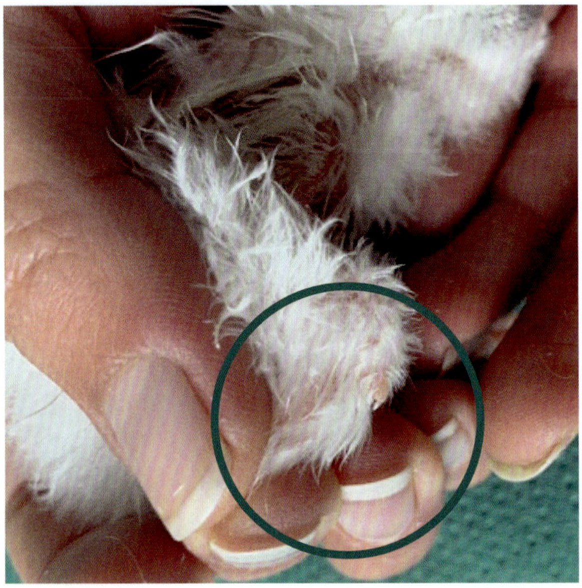

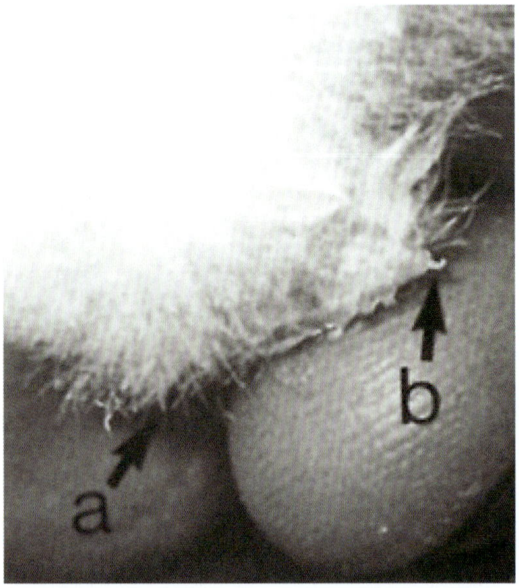

This gyrfalcon chick (left) has a tiny vestigial claw on the alula but not on the tip of the manus, whereas the New Zealand falcon has both.

but is usually complete within 50 hours.

Falcons tend to pip right at the edge of the air-cell so that they puncture the air-cell membrane and the shell more or less together. Accipitrid hawks on the other hand, more commonly puncture the air-cell membrane first and enter the air-cell. They may stay quietly in this position for 24 hours before starring the shell and then fairly quickly move on to hatch within another 24 hours or so.

Especially with opaque hawk eggshells, it can be hard to determine if the hatching egg is still alive. Everything is black, quiet and ominous. It is easy to panic at this stage, intervene, and cause irreparable damage. A quick simple test can determine the best course of action to take. First cut a piece of cling film like a handkerchief. Place the egg on it pointed end down and wrap the egg up in it so that the corners twist together to form a little wick 10 centimetres or so long. Then take a bowl of water at blood heat and let it stand until it is perfectly still. Gently lower the egg into the water using the wick. It will stay dry because of the cling film. If the embryo is alive, the egg will twitch and the slightest movement is magnified by the lever effect of the wick. This movement will show from about the second half of incubation. If the embryo is near hatching and shows no movement after one minute, it is almost certainly dead. Avitronics produce an Egg Buddy that allows you to monitor the heartbeat of the embryo in the egg and know if it is alive, but it is not 100%, so be very careful about opening an egg that you assume is dead.

In precocial birds, such as chickens and ducks, whose young leave the nest as a group together soon after hatching, it is important that they all emerge within a short space of time, even though the eggs had originally been laid over several days. They behave a bit like trade unionists; it is a question of 'one out, everybody out'. This synchronization process is achieved in two ways. First, the parents tend not to start proper incubation until the clutch nears completion and, second, the cheeping of the chicks inside the eggs stimulates adjacent slower neighbours to speed up their hatch. Birds of prey tend not to show this mechanism. Often they start incubation when the clutch is only half complete. Leaving a 'clicking' egg touching a younger hatching egg does not, from my data at least, shorten the second egg's pip to hatch interval. It is not necessarily an advantage for birds of prey to have a synchronized hatch. Indeed, in harriers and owls, the chicks may hatch over several days and the survival of the late hatchlings will depend on food availability at the time. This seemingly cruel mechanism of controlling brood size enhances the survival of the early hatchlings and maximizes ultimate productivity.

Floating an egg in warm water to see if it is alive.

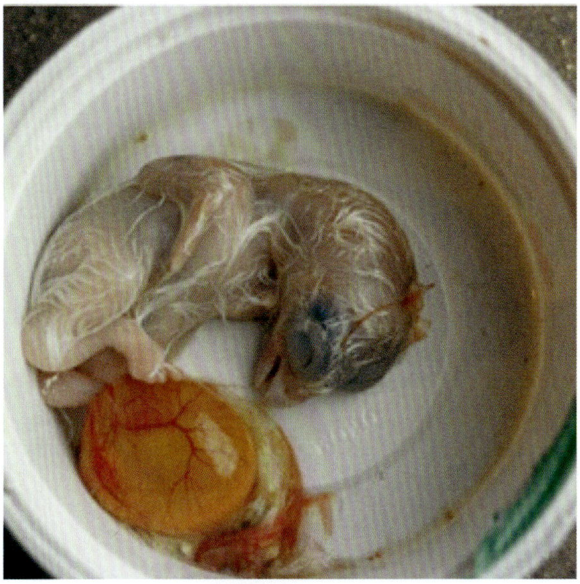

Every breeder's nightmare – an unretracted yolk sac.

2.28 Rearing

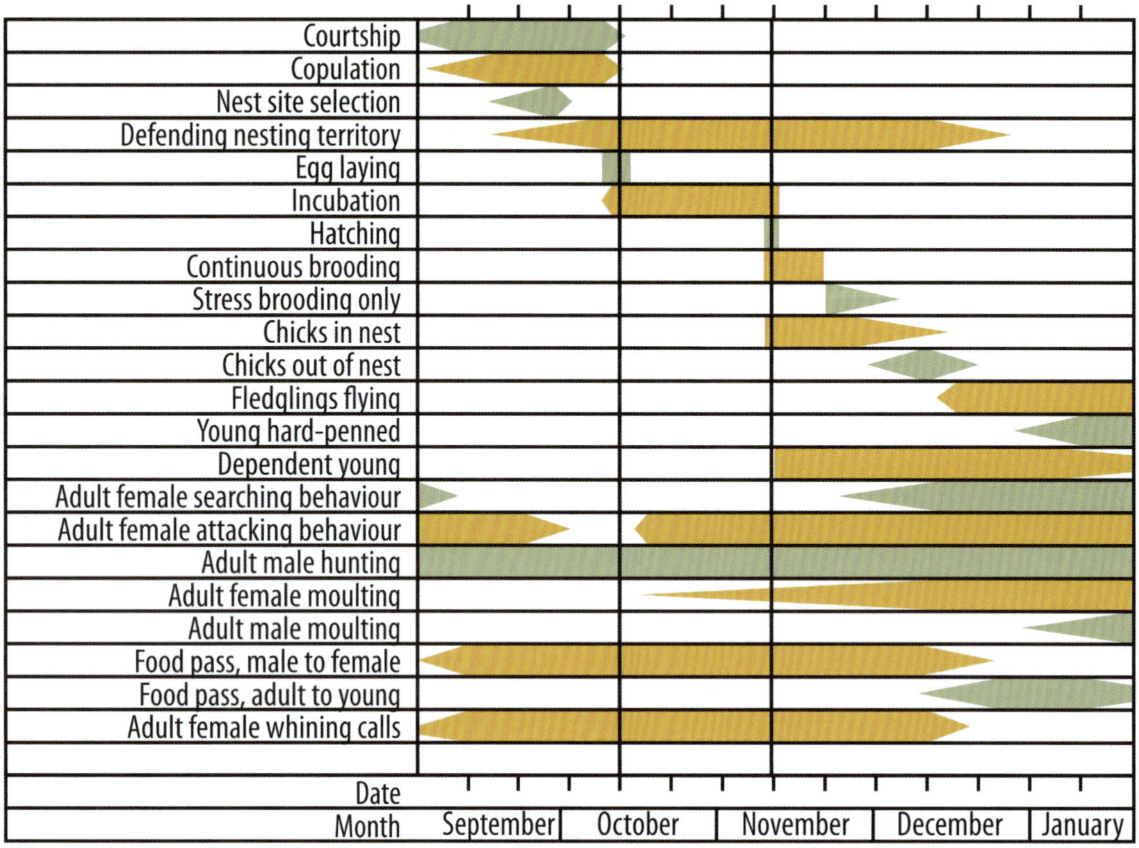

Figure 2.28.1. These are the activities around the nest of a pair of wild New Zealand falcons as the breeding season progresses. Spring in the southern hemisphere is in September.

However experienced the parents, things can easily go wrong at hatching time and chicks may disappear. Therefore, in almost all cases, we hatch the eggs ourselves in a hatcher, rear them for a few days until they are strong, and then mother them on to parents before the imprinting process is too far on.

Newly hatched chicks start to dry out and fluff up.

Once the chick has hatched, it will be tired and misshapen from being cramped inside the egg. It needs several hours of rest. During this time its natal down will dry and become fluffed out and the chick will lose some weight. As water is withdrawn from the muscle tissues over the first three days, the muscles get more tone and the chick, at first floppy and helpless, will soon be alert and able to sit up and beg. The hatching muscle will disappear after a couple of days.

Once dry, the chick is removed from the hatcher and uniquely marked with a coloured pen before being placed with other youngsters

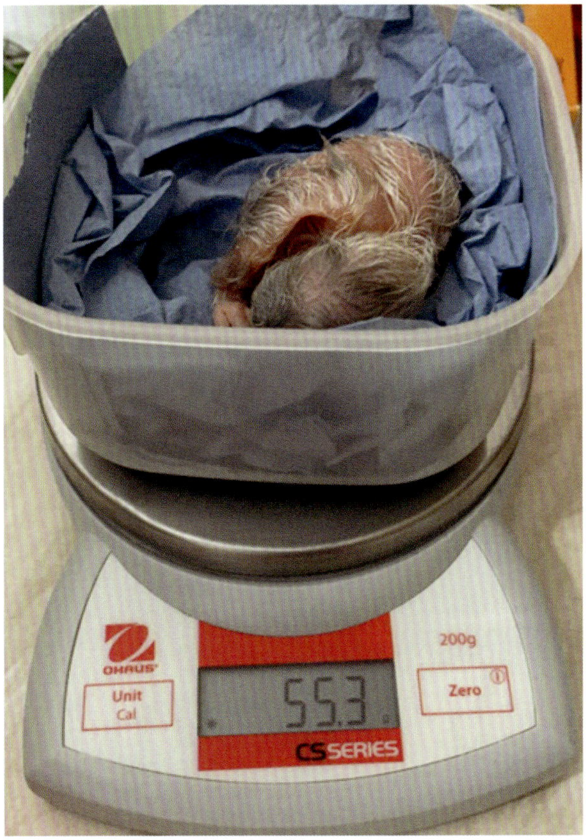

The hatch weight is the start of the growth weight chart.

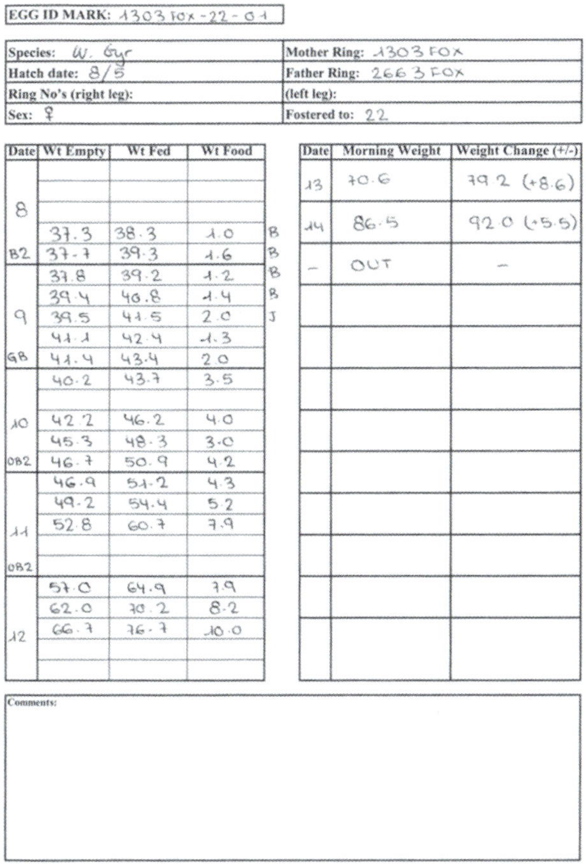

EGG ID MARK: 1303 FOX-22-01

Species: W. Gyr	Mother Ring: 1303 FOX
Hatch date: 8/5	Father Ring: 2663 FOX
Ring No's (right leg):	(left leg):
Sex: ♀	Fostered to: 22

Date	Wt Empty	Wt Fed	Wt Food	
8				
	37.3	38.3	1.0	B
B2	37.7	39.3	1.6	B
	37.8	39.2	1.2	B
	39.4	40.8	1.4	B
9	39.5	41.5	2.0	J
	41.1	42.4	1.3	
GB	41.4	43.4	2.0	
	40.2	43.7	3.5	
10	42.2	46.2	4.0	
	45.3	48.3	3.0	
OB2	46.7	50.9	4.2	
	46.9	51.2	4.3	
	49.2	54.4	5.2	
11	52.8	60.7	7.9	
OB2				
	57.0	64.9	7.9	
	62.0	70.2	8.2	
12	66.7	76.7	10.0	

Date	Morning Weight	Weight Change (+/-)
13	70.6	79.2 (+8.6)
14	86.5	92.0 (+5.5)
–	OUT	–

Comments:

Figure 2.28.2 A growth weight chart.

Once dry, the chick may open its eyes briefly and beg for food.

in a brooder. Brooders are now readily available from Grumbach, Brinsea, Masellies and other manufacturers that are very reliable and provide a stepping stone from the hatcher to the open top brooder. The chicks are in tissue-lined plastic bowls and the temperature is a few degrees less than incubation temperature. The chicks themselves are the best thermometers; if too hot they sprawl apart, if too cold they huddle. A

A commercial brooder for newly-hatched chicks.

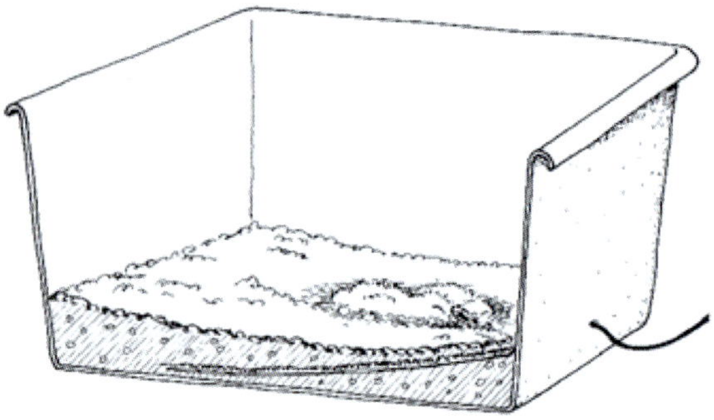

Figure 2.28.3. A simple brooder box for chicks over 3 days old, using a heating pad.

An angled, overhead heat system providing a temperature gradient.

solitary chick obviously needs a hotter brooder than a group which can snuggle together for warmth.

After two or three days in the still-air brooder the chicks are removed to a brooder box (2.28.2). This is a plastic bin box or water tank with a thin wedge-shaped layer of clean pea gravel on the floor. Over this is placed a thin flat heating element like an electric blanket. These are used for reptile cages and home brewing. It is covered with a towel to protect it and laid on the bed of pea gravel so that it is near the surface at one end and buried at the other. This creates a temperature gradient. More pea gravel is placed on top so there is a flat bed of gravel warm at one end and cool at the other. The chicks are placed on this and crawl about until they find their most comfortable position. If the room is cold or if the chicks are being trained to accept a brooding parent, we place a light hand-towel over the box secured loosely with clothes pegs. Alternatively you can place the heating plate like a little roof, at an angle. The further that the chick snuggles in under it, the warmer and darker it is. The aim is to keep the chicks as cool and fresh as possible consistent with comfort; this is how they thrive best. We do not used forced-air brooders; the chicks should be encouraged to control their own temperature and not forced into a programmed regime. The pea gravel surface encourages strong and well-aligned leg bones and is hygienic, being replaced after each batch of chicks. There are other absorbent substrates available which are also good and are biodegradable.

For the first two days or so the chicks are using up their yolk sacs and do not need much nourishment. Their feeds are mainly just gently to get the digestive tract working properly without over-loading it. They are weighed before and after each feed and records are kept. These are essential to identify a problem quickly and as a source of reference for future chicks, and are important when more than one person is doing the feeds. Small raptors, such as merlins, are fed five times a day to start with, the others four times a day. Falcon chicks sit up and beg. It is only necessary to chup once or twice to encourage them to gape and the food can be put in with blunt forceps. Accipitrid chicks, such as accipiters, buzzards, kites, harriers, and eagles, do not gape. Instead the food must be held in front of them. They will ignore it, then just as you are about to give up, they make a wild snatch and miss. But soon they get more coordinated and feeding goes smoothly. Of course if you mother baby falcons onto an accipitrid raptor the chick will gape and the mother will dangle but nothing gets fed. Also accipitrid raptor chicks will fight each other in 'Cain and Abel' battles. These must be separated and kept covered so that they sleep. An accipitrid chick will beat up a falcon chick who is completely bewildered by such behaviour and does not fight back. This is how you get falcons called 'Scarface'.

Irene and Sofia feeding chicks.

For the first two days we feed quail muscle, liver, and heart. These are stripped off the fresh-killed quails and ground up very finely in a sterilized mincer. We do about 50 quail at a time. A scoop of the resulting mix is then placed in a polythene bag on a table and smoothed out flat, about 5 millimetres thick. A ruler is pressed into it so that the result looks like a thin chocolate bar. This is labelled 'baby food', dated, and placed in the freezer. In a few minutes it has frozen solid. At feeding time, the required number of segments are broken off and thawed in a glass tub in a saucepan of warm water. A few drops of water are sprinkled on to the ration to moisten it. Specific raptor probiotics have now been developed, we use Avipro for the first feed of the day. But if the egg quality and incubation was good, the chicks will not be a problem.

On the third day we merge in 'junior food'. This is whole, skinned quail with the intestinal tract removed but heart and liver left in. A little fat is good too. This is ground up twice to make an even mix. It is frozen in the same way. Again, it is thawed and fed with some Nekton S and water. In this way, food for the raptor chicks must be prepared only once or twice a season. The food quality is better, and it is easier on the staff. Of course, only fresh-killed food can be treated thus; it is not safe to refreeze frozen food. We never feed day-old chicks to young raptors; they are a false economy.

Between day 5-7 it is time to mother chicks back on to parents. We prefer to mother them first to imprints. It is easy to check them properly three times per day and the mothers do a better job of rearing than we do. Our data consistently show that the chicks grow faster and stronger than hand-raised ones however dedicated we are. One reason may be that the imprints stuff more food into the chicks than we would ever dare. Around day 9-12 it is time to fit closed rings (see 2.29) and then the chicks can go onto natural paired parents. The chicks will be accustomed to being brooded and fed by a mother and quickly accept a different mother.

Meanwhile, the natural pairs have been kept incubating dummy eggs. These are either plaster-filled, or plastic ones or hard boiled ones. Inexperienced birds are not allowed to sit real

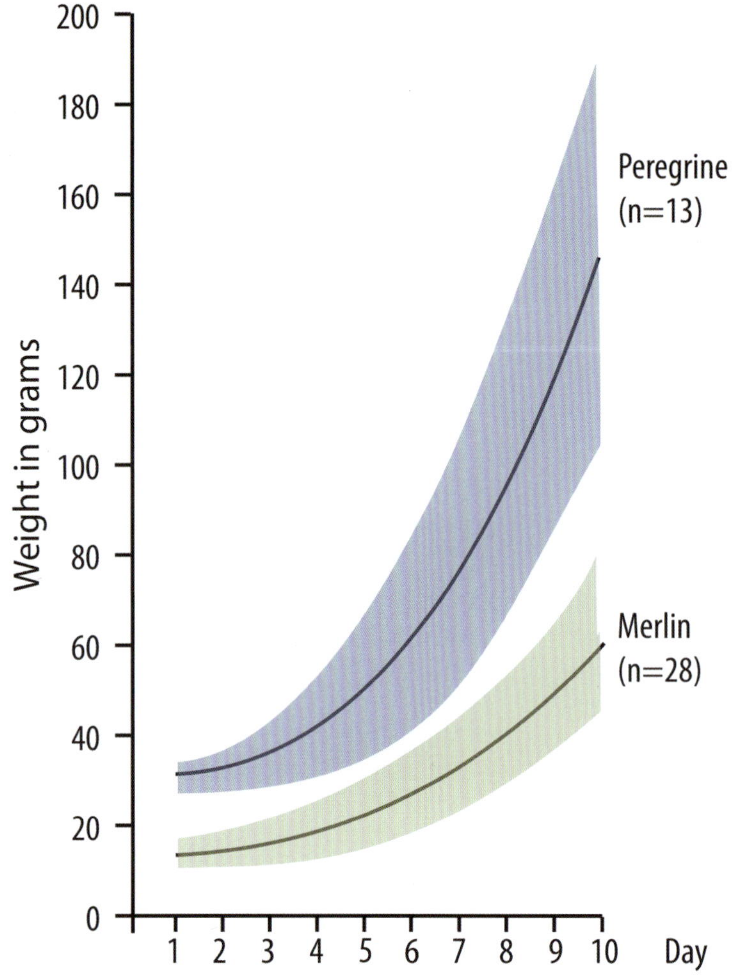

Figure 2.28.2 Growth rates of peregrine and merlin chicks showing averages and ranges of weights.

eggs for long because they often break an egg. When the time comes for mothering on a chick the parents should have been incubating for at least half the incubation period. Experienced birds will usually accept a chick without too much problem. If there is some doubt, we use the mothering-on cage (figure 2.28.4). This is a little domed cage which protects the chick if the parents attack it. The chick should then beg and trigger the mother's parental instincts. When she tries to brood the chick in the cage you know she is prepared to accept it and you can then remove the cage and leave the chick unprotected.

Figure 2.28.4 The mothering-on cage.

Inexperienced or wild caught mothers may be totally upset by the chick or put off by the cage. The problem is to make a gradual transition from inert egg to chick. When the parents sit on a pipping egg the parental bond is strengthened and a shy mother suddenly becomes a defensive tiger. It is best to try to make this happen before introducing a precious raptor chick. When both parents have incubated this they are then prepared for hatching. It is important to pay attention to the male as well as the female. The most dangerous

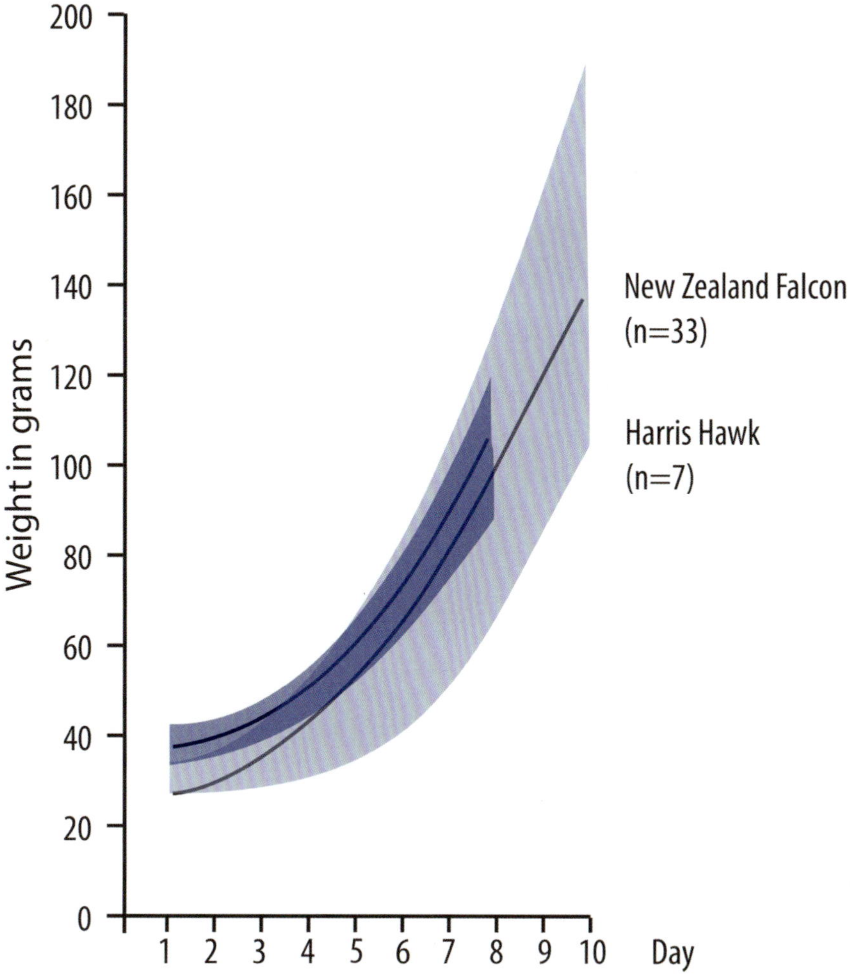

Figure 2.28.3 Growth rates of New Zealand falcons and Harris hawks showing averages and ranges of weights. Harris hawks may scream if hand-reared beyond day 7.

moment is when the mother comes back on to the nest, sees something moving and just grabs it and flies off with it. I once saw this happening, nipped in and caught the chick from the female just before it hit the ground! The way we get over this is to use the 'born again' method. The goose egg is cleaned and painted to look like the raptor's eggs, or you can make a fibreglass replica. A hole is made, removing about a third of the shell. A five-day old raptor chick is then placed in the goose egg and the egg placed in the nest with a couple of dummies, so that the chick is invisible.

The 'Born-again' egg.

The happy outcome.

The female comes back on and sits on this super stimulus egg. The egg wriggles and cheeps! The female is thrilled. She broods it and after a while the chick 'hatches'. Soon the parents are feeding it. The next day the eggs are removed and some more chicks added and there is seldom a problem.

It is worth going to all this risk and effort. If the parents rear a brood, it reinforces their pair bonding and pen bonding. This means they will breed better next year. Non-copulators may copulate, single clutch pairs can be recycled. You have established a good breeding pair. The youngsters too will grow up being imprinted on parents and will thus be suitable for falconry and future breeding.

Again, what happens in the aviary is a condensed version of events in the wild (figure 2.28.1). You should be familiar with the normal sequence of events through the season.

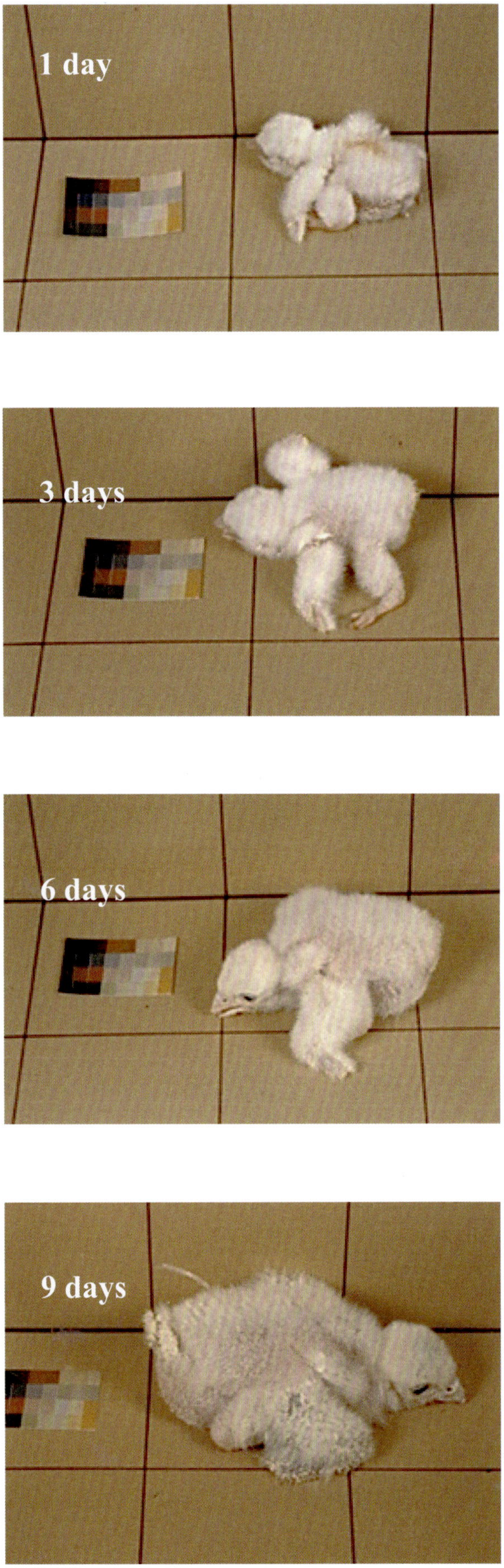

The development of a saker chick

12 days

15 days

18 days

21 days

24 days

27 days

2.29 Methods of identification

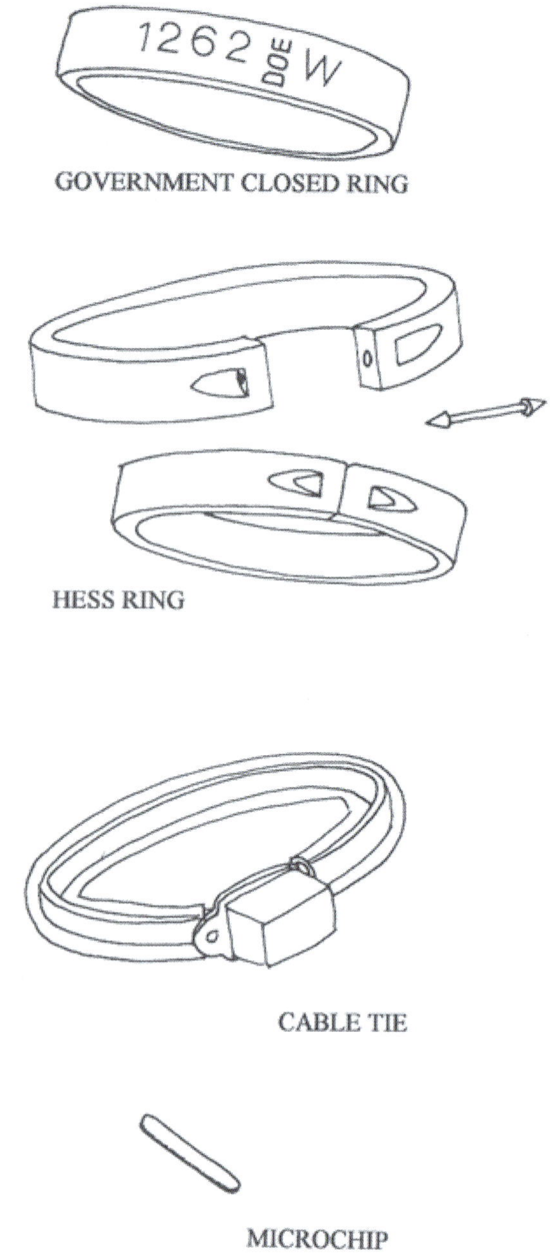

Figure 2.29.1 Identification rings.

When the chick is about nine days old it is time to fit a closed identification ring on its leg (figure 2.29.1). In America, Britain and parts of Europe this is a metal ring stamped with a serial number. Government rings also carry a government mark and in Britain at least are recorded on the central database. The ring may also carry the year of hatch and the breeder's name.

Closed rings come in various sizes from P and S for merlins to V for male peregrines, W for female peregrines and X for female gyrs. Of course, at nine days it is not easy to sex a chick and therefore it is best to put on the female ring above the male ring. They can be held loosely in place for a couple of days using Elastoplast, until the foot has grown enough to stop the ring coming off (figure 2.29.2).

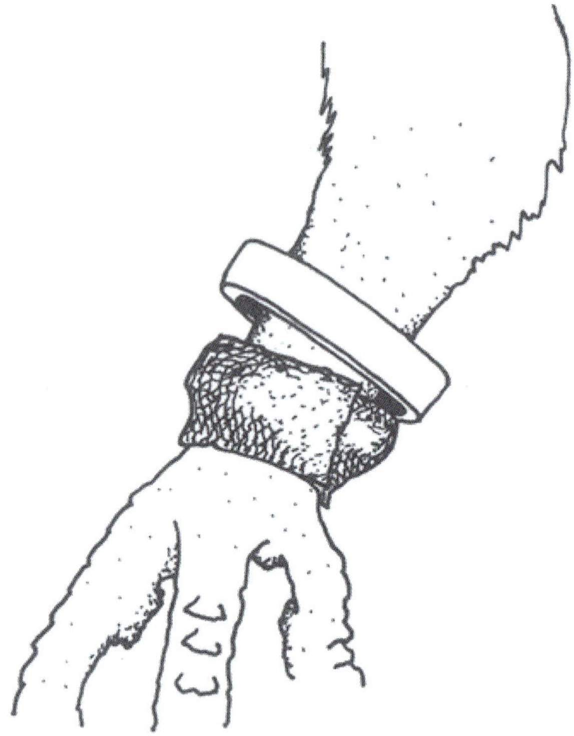

Figure 2.29.2 The female ring is placed above the male ring, which is temporarily held in place with loose Elastoplast.

These close rings are as near as you'll get to showing that the bird is aviary-bred. It will be sold as 'ABCR' (Aviary Bred Closed Ringed). Once the chick is older it can be ringed for identification purposes with another type of ring. A popular open ring in Europe is the 'Hess' ring (2.29.1). This is put on and closed, then a small steel pin is fitted through a hole linking it up. The metal part of the join is then crushed, and the barbs on the pin prevent removal without damaging the ring.

In America and Britain, the government cable tie ring has been frequently used but is not popular. Hawks easily remove it and thieves easily fiddle it. It has mainly been phased out but is still used for rehabbed birds.

Inserting a microchip under the skin.
Photo Gombobaatar.

A modern way of marking is the PIT tag or microchip, about 8 millimetres long by 2 millimetres wide. It can be implanted under the skin or into muscle and will give a coded number readout when scanned from 5-10 centimetres away by a scanner. It is a very effective system of marking but does have some drawbacks. One is that it cannot be seen by eye and so there is no immediate visual evidence supporting the bird's legality. Another is that it is only as good as the central database supporting it. These databases, and the systems themselves, are still in competition and duplication at the moment. Thus, if your bird is stolen, there is not yet a very good chance that you will get to hear of it being found. Also of course, this method does not authenticate the bird's previous history in any way; it doesn't prove captive breeding or parentage. It is not unknown for a microchip to be removed from a dead bird and inserted into another one. We have used it in wild sakers in our field projects in Kazakhstan, Pakistan, and Mongolia. The young sakers disperse and are trapped on migration, sold in Pakistan and find their way to the Middle East. There our biologists and vets in the Falcon Hospitals find them with scanners and report them to our computer in Abu Dhabi. Later when these birds are released again we can detect them back on the nesting grounds. In this way we can determine the harvest and sustainability of sakers in Arab falconry.

Rings with built-in microchips are used on racing falcons. These carry the data of the owner, allowing for quick processing at the starting gate. They are also picked up by the beam at the finish line.

DNA sexing is widely available now and results normally come back in about five working days. We send blood from the chorioallantois inside the newly hatched eggshell. This will determine the sex of the chick with about 95% accuracy. We also store blood from all the hatched chicks, some going back 40 years.

2.30 Systems of rearing the older chick

Once the ring stays on and the chick is being reared by parents, there is a short respite in which to plan ahead. If the chick is to be hacked, either for falconry or for release, she can stay with parents until about a week before she is due to fly. Then she is transferred to the hack box. If she is not scheduled for hacking or dual imprinting, it is best to leave her with the parents for at least one month after her first flight. The first two weeks are spent completing feather growth. After that, although fully feathered, the youngster's joints are still hardening and are very fragile, especially in long-legged or violent species. Also, in the wild, the youngster would still be dependent on her parents at this age. If you take her in hand then, she may start to react to you as a parent. This may include screaming and looking toward you for food, rather than getting out and catching her own. This period should be extended in species with a long dependency period, such as Harris hawks. If possible they should be left with their parents, or in sibling groups untouched by man, until at least two months after first flight. This reduces the risk of screaming and makes the bird easier to enter. Further delay tends to be counter-productive: the bird needs the flying experience and the exercise to develop her physical performance fully and learns a great deal at this age.

If the bird is intended for natural breeding, it can either be left with its parents until removed for pairing, or it can be reared away from its parents. Some parent-reared birds, particularly gyrfalcons, can be a little nervous and as a result, very slow to come into breeding condition. If reared in sibling groups, they tend to be steadier, but creche or socially-reared raptors tend to be noisy for falconry. An alternative which is perhaps the ideal, is to use a tame imprint parent to rear the birds. This leaves them properly imprinted but also well-adjusted to life around humans.

Handling imprints and dual imprints.

A hack pen in Abu Dhabi.

2.31 Finding the right homes for the young stock

This is always the worst part about breeding raptors. Having spent so much trouble producing them, it is important that they go to good homes and it is always distressing to hear that one has been badly treated. It is almost impossible to gauge the abilities of a prospective purchaser. Many falconers of twenty or thirty years experience didn't learn properly in the first place, have developed plenty of bad habits since, and have closed their minds to any further progress. Some beginners, on the other hand, although short on practical experience, will bend over backwards to do the best for their birds.

Many falconers who do not breed hawks have no real idea of the work and uncertainty involved. They ring up wanting birds as if you are a grocery store and do not appreciate that there is a lot of attrition between a new-laid egg and a finished, hacked female falcon. Almost without exception, if there is a problem, falconers blame it on the breeder rather than on themselves. If it goes well, then the falconer claims the glory! Most experienced breeders keep a blacklist of problem people.

The Middle East market is particularly difficult and fickle. It is essential that breeders fully understand what type of bird is wanted and how to prepare it so that it is suitable for Arab falconry. Therefore it is important that the birds fit their method, which has evolved originally for the passage falcons.

To the person planning on buying a young raptor, examine yourself carefully. Do you have all the necessary factors to make a success of the bird? Have all your birds turned out reliable, successful hunters? If they didn't, in what way did you fail? Have you sought help? Are you likely to improve this time? Now that raptors are more easily available, it is even more important to realize the responsibility that the trainer has for each bird. If your bird was a healthy youngster when you got it, and now it has a problem or is a poor hunter, it is almost 100 percent certain that it is YOUR fault. Face it, analyze it, and do something about it.

2.32 Maintaining breeding records

There are so many small, but vital, items of information in breeding that it is impossible to progress on any scale without good records. They are essential as a guide for the future, a basis for learning and of course for legal reasons. Unused copies of our data sheets shown in this part of the book are provided at the back. You are welcome to copy them.

In addition to our data sheets and databases, and to legal and veterinary documents, we keep a Day book. On the front of this is a schematic plan of all the pens with the occupants labelled to age, sex, species, and ring number. The book itself is then divided up with two or three pages for each pen. Any events happening in any pen are recorded each day as they happen. A new book is started each year, but the previous year's book is always close at hand for quick reference on laying dates and so on.

The other overall document is the Stocklist. This again lists all the birds broken down by age and sex (aren't we all?), species, ring number, both parents, and owner. The secondary stocklist covers all birds out on loan. The stocklists are dated and up-dated periodically, being in frequent use for CITES applications and so on.

However small your operation, careful records are the key to making progress. Software is available now to monitor your eggs, chicks and whole breeding programme. They can make life easier but at the same time you can easily lose data when you change computers or systems. We have paper records for some of species going back 15 generations to 1974. It may sound old-fashioned, but often paper records and photos seem to last longer than digital records.

2.33 Further Reading

Anderson Brown, A. F. 1979. *The Incubation Book*. Saiga Publishing Co. ISBN 0 904558 70 3.

Baerends, G. P., R. H. Drent, A. J. Hogan-Warburg, G. J. Blokzijl, A. G. van der Molen, O. C. de Vos, H. J. Baarda, M. J. Mudde, J. P. Kruijt, and R. Vodegel.1982. The Herring Gull and Its Egg: Part II. The Responsiveness to Egg-Features.Behaviour 82 (1/4) :1–416.

Budianski, S. 1997. *The Covenant of the Wild: Why animals chose domestication*. Phoenix. ISBN 1-85799-979-7.

Cade T.J. 1988. The breeding of peregrines and other falcons in captivity: an historical summary. In T.J. Cade, J.H. Enderson, C.G. Thelander & C.M. White (eds) *Peregrine Falcon populations, Their management and Recovery.* The Peregrine Fund, Boise, ID, USA.

Calle P.P., E.S. Dierenfeld and M.E. Robert M.E. 1989. Serum alpha-tocopherol in raptors fed vitamin E supplemented diets. J. Zoo and Wildlife Medicine 20(1):62-67.

Carter, T. C. and B. M. Freeman, (eds.). 1969. *The Fertility and Hatchability of the Hen's Egg*. Oliver & Boyd, Edinburgh.

Chargé, R., C. Teplitsky, G. Sorci and M. Low. 2014. Can sexual selection theory inform genetic management of captive populations? A review. Evolutionary Applications 7: 1120–1133.

Cooke, F., and P. A. Buckley. 1987. *Avian genetics: A population and ecological approach*. Academic Press. ISBN 0 12 187570 9.

Cunningham, F. J., P. E. Lake and D. Hewitt, (eds.). 1984. *Reproductive Biology of Poultry*. British Poultry Science. Longmans Ltd.

Deeming, D. C. and M. W. J. Ferguson, (eds.). 1991. *Egg Incubation: Its effect on embryonic development in birds and reptiles*. Cambridge University Press. ISBN 0 521 39071 0.

Deeming, D.C. (Ed). 2002. *Avian Incubation: Behaviour, Environment and Evolution*. Oxford Ornithology Series. ISBN 0-19-850810-7.

Dierenfeld E.S., C.E. Sandfort and W.C. Satterfield.1989. Influence of diet on plasma vitamin E in captive peregrine falcons. J. Wildl. Manage 53(1):160-164.

Drent, R. 1973. The Natural History of Incubation. In D. S. Farner and J. R. King (eds.). *Avian Biology.* vol. 5. Academic Press, New York.

Eastham, C.P., M.K. Nicholls and N.C. Fox. 2000. Post-mortem shrinkage in large falcons. Newsletter of the Middle East Falcon Research Group 17: 1-4. (http://www.falcons.co.uk/mefrg/Falco/17/Eastham.htm (version 10/11/01))

Fisher, H.I. 1940. The Occurrence of Vestigial Claws on the Wings of Birds. The American Midland Naturalist 23(1):234-243.

Fox, N. C. 1989. Managing a gene bank: The New Zealand Falcon account. In C. Imboden (Ed) The Re-introduction of endangered species. The Wildfowl Trust, Slimbridge.

Fox, N.C. and B.I.M. Fox. 1993. The problems in establishing a domestic, self-sustaining international population of New Zealand Falcons *Falco novaeseelandiae*. pp. 194-200 *In* M.K. Nicholls and R. Clarke [eds.]. Biology and conservation of small falcons: Proceedings of the 1991 Hawk and Owl Trust Conference. The Hawk and Owl Trust, London, England.

Fox, N.C., Paterson, M. & C.P. Eastham. 1996. Hatching and Rearing Red Kites (*Milvus milvus*) to supplement wild populations in Wales and England. Abstracts of 2nd International Conference on Raptors. Urbino, Italy 2-5 October 1996. RRF, University of Urbino. 118.

Forbes, N.A. and C.J. Flint. 2000. Practical Raptor Nutrition. Honeybrook Animal Foods.

Freeman, B. M. and M. A. Vince. 1974. *Development of the Avian Embryo*. Chapman and Hall, London. ISBN 0 412 11520 4.

Gee, G. F., C. A. Morrell, J. C. Franson and O. H. Pattee. 1993. Cryopreservation of American Kestrel Semen with Dimethylsulfoxide. Raptor Research 27 (1): 21 25.

Grant, G. S. 1982. 'Avian incubation: egg temperature, nest humidity, and behavioural thermoregulation in a hot environment.' Ornithological Monographs no. 30. The American Ornithologists' Union, Washington, DC.

Haak, B. A. 1992. *The Hunting Falcon*. Hancock House, Surrey, BC. ISBN 0 88839 292 3.

Halgrain, M. *et al.* 2021. Eggshell decalcification and skeletal mineralization during chicken embryonic development: defining candidate genes in the chorioallantoic membrane. Poultry Science 101(10):101622.

Harvey, R. 1993. *Practical Incubation*. Hancock House, Surrey, BC. ISBN 0 88839 310 5.

Johnson, A. D. and C. W. Foley (eds.) 1974. *The Oviduct and its Functions*. Academic Press.

Moore, H. D. M., W. V. Holt, and G. M. Mace. 1992. *Biotechnology and the conservation of genetic diversity.* Oxford Science Publications. ISBN 0 19 854030 2.

North, M. O. and D. D. Bell. 1990. *Commercial chicken production manual*. AVI. Van Nostrand Reinhold. ISBN 0 442 31882 2.

O'Connor, R. J. 1985. *The Growth and Development of Birds*. John Wiley and Sons. ISBN 0 471 90345 0.

Parry Jones, J. 1991. *Falconry: Care, captive breeding and conservation*. David and Charles. ISBN 0 7153 8914 9.

Slack, J. M. W. 1991. *From Egg to Embryo*. Cambridge University Press.

Smith, K.W., N. C. Fox and C. P. Eastham. 1997. Measurements of incubation temperature and egg-turning rates for a still-air incubator, a bantam *Gallus domesticus*, a captive Common Buzzard *Buteo Buteo*, and a wild Common Buzzard. Incubation and Fertility Research Group.

Solomon, S.E. 1997. *Egg and eggshell quality*. Manson Publishing. Iowa State University Press. ISBN 0-8138-2827-9/

Schwartz, A., J. D. Weaver, N. R. Scott and T. J. Cade. 1977. Measuring the temperature of eggs during incubation under captive falcons. Journal of Wildlife Management 41 (1):12-17.

Thear, K. 1987. *Keeping quail: a guide to domestic and commercial management*. Broad Leys Publishing Co. ISBN 0 906137 15 2.

Thoroughbred bloodlines. https://www.bloodlines.net/TB/PedigreeMatters/Fams/EquineGeneticGenealogy.htm

Tullet, S. G. (ed). 1990. *Avian Incubation*. Poultry Science Symposium 22. Butterworth Heinemann, London. ISBN 0 7506 1002 6.

Van Vleck, L. D. 1983. *Notes on the theory and application of selection principles for the genetic improvement of animals*. Department of Animal Science, Cornell University, Ithaca, NY.

Van Vleck, L.D., E.J. Pollack and E.A.B. Oltenacu. 1987. *Genetics for the Animal Sciences*. W.H.Freeman, New York.ISBN : 0716718006.

Watson, P. F. 1978. *Artificial Breeding of Non Domestic Animals*. Symposia of the Zoological Society of London 43. Academic Press.

Weaver, J. D. and T. J. Cade. 1983. *Falcon Propagation: a manual on captive breeding*. The Peregrine Fund Inc., Ithaca, NY, U.S.A.

Woodard, A. P. Vohra, and V. Denton. 1993. *Commercial and Ornamental Gamebird Breeders Handbook*. Hancock House, Surrey, BC. ISBN 0 88839 311 3.

Zhan, X.J., A. Dixon, N.C.Fox and M.W.Bruford. 2012. Missense SNP of the *MC1R* gene is associated with plumage variation in the Gyrfalcon (*Falco rusticolus*). Animal Genetics 43(4):460-462.

C3 Equipment and Facilities

Some of this section is supported by a film in the Bird of Prey Management Series:
Basic Training

3.1 Obtaining good equipment

When raptors are being managed using falconry techniques, either for falconry itself, or displays, or rehabilitation, then specialist equipment is needed. Now that so much falconry equipment is available commercially there is little need to make it yourself. But if you are going to buy your equipment you do at least need to know how to recognize good workmanship and good design. Unfortunately, although some furniture for sale is excellent, some is absolute rubbish. If you wish to make your own equipment Beebe and Webster, Glasier, Upton, Ford and Kimsey and Hodge provide useful descriptions in their books.

There is very little margin for error in managing captive raptors. All the small details must be attended to carefully or a hawk could be quickly injured or dead. Destroy all faulty or old equipment so that you won't be tempted to use it again, and if you hear yourself saying words like 'that will do for now', go back to square one.

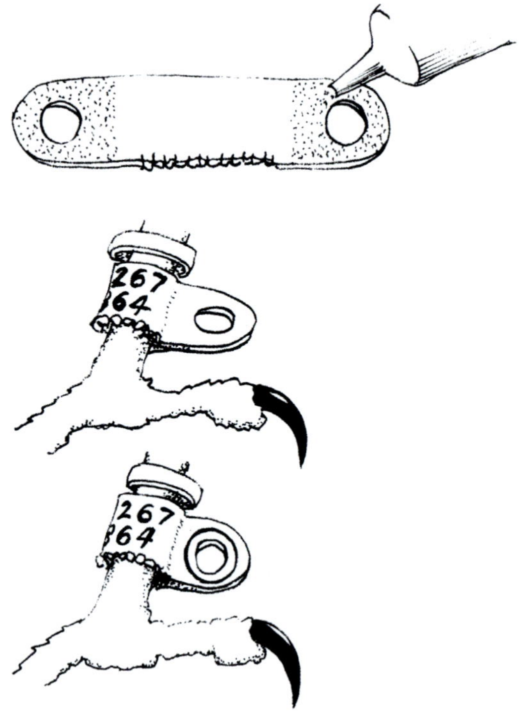

Figure 3.2.1 Fitting anklets.

3.2 Anklets and jesses

The Aylmeri anklet is the best way to attach jesses. Traditional jesses are now obsolete. In America they are illegal, and rightly so, because of the risk of the hawk getting caught up on the slit in the strap. There are various ways of putting on the anklets but this one is very simple:

a. Cut a strap of light, tough, pre-stretched leather of the required width. If the leather is thick, chamfer both edges with a sharp scalpel and in the centre, make about 15 small cuts so that the edge curls outward and will not chafe (figure 3.2.1). Do not put in cuts near the eyelet area or you will weaken it. We use a white, chrome-tanned leather. Our telephone number is printed on the front of one anklet and the hawk's name on the other. Do not oil the anklets at this stage otherwise the glue will not stick.

b. Place the anklet around any object, such as a pen, about the same diameter as your hawk's leg and mark where the ends meet each other. Place an eyelet onto the tab and press it down to mark the position for the hole. Punch a hole through both tabs and round them off, cutting just outside the rim of the eyelet.

c. Put rubber contact glue, such as Evostick, on the inside of each tab, around the eyelet holes.

d. Place the anklet around the hawk's leg and bond the glue tightly with a pair of pliers.

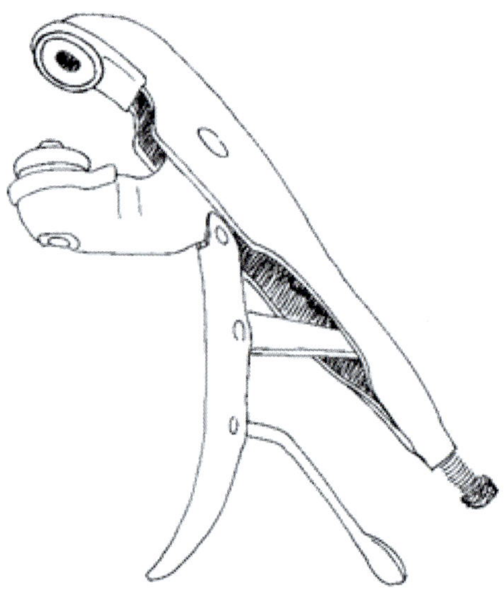

Figure 3.2.2 Mole grips for fitting eyelets.

e. Place the eyelet in position from the outside of the leg with the eyelet washer on the inside. Crimp the eyelet with an eyelet punch. If the eyelet is too large for the pliers-type hand-punch weld or fit the eyelet punch parts onto a big pair of pliers or mole grips to make a sturdy hand-held eyelet punch. Avoid using the anvil stamp tool for crimping, or the type that come in two halves and screw together.

f. Apply a light coat of leather oil such as Neatsfoot oil, Saturol, or Mars oil. Oil penetrates better than grease, and does not hold sand and grit so much. The complete operation can be done in two or three minutes with the hawk hooded on the fist.

If the hawk is a young one, this is also a good time to snip off any unwanted closed rings. Rings should stay above the anklet, not below it.

Anklets can cause several problems. Over-greasing may cause sand to stick inside, which can abrade the hawk's legs, as can prolonged bating. In small accipiters a common problem is an awkward sore at the back of the leg.

There are three ways to reduce these problems. Man the hawk so that she seldom bates, use a shock absorber in the leash, or manage the hawk in such a way that she is never tethered but kept in a flight pen when not in use. In hot climates a few ventilation holes in the anklet may help.

Mews jesses are easily constructed, always in pairs, from pre-stretched leather (figure 3.2.3). Take care not to weaken the leather when making the slits. There is no need for a rip-stop hole punched at the inner end of the slit. There is no magic leather for jesses; the tannage is as important as the species. Badly tanned kangaroo is just as bad as sheepskin. If in doubt, make a cut in the leather and pull to see how easily it tears. There is no point in using a stainless steel swivel

Tight jesses, swollen legs and overgrown blunt claws on this white goshawk. The Falconer should be ashamed of himself.

that could tow a car, if the jesses couldn't pull a chicken off its nest. Recently I recovered a lanner somebody had lost and I could tear the jess slits with one hand. No wonder she had escaped!

Mews jesses should just be long enough to go down through the palm of the hand and out between the fingers, exposing the swivel. Any longer than this and the bird is in danger of straddling the block or of getting her jesses wrapped around her upper tarsus. As these jesses are not used when hawking there is no need to make them any longer. For a peregrine, the jesses are about 15 centimetres (6 inches) from the button to the far end of the slit. If the hawk bites at her jesses an application of anti-nail-biting fluid, mustard, or curry powder often solves the problem.

Figure 3.2.3 A flying jess and a mews jess.

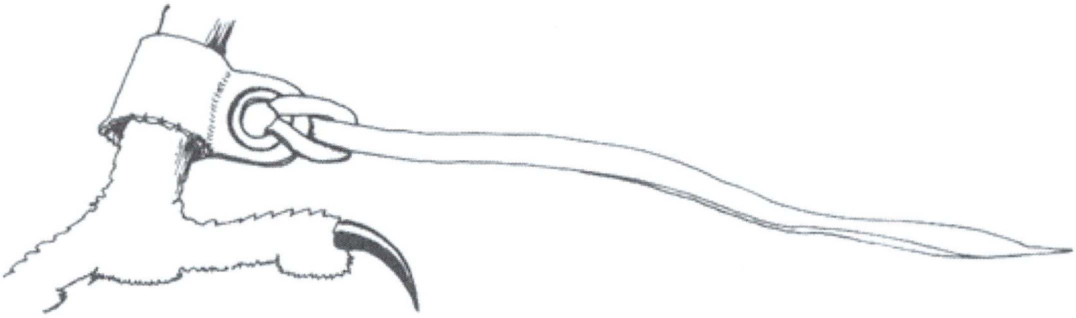

Figure 3.2.4 A 'non-removable' flying jess.

Never fly a hawk with slitted jesses. Sooner or later she will get hung up. If you have seen, as I have done, somebody's hawk caught by one leg, hanging from a live electricity crossbar, or another in the top of an unclimbable tree, or another thrashing amongst the strands of a barbed-wire fence, you will swear never to use such things in the field. Nor should you be tempted to commit even worse a sin: turning the mews jesses round so that they are held to the anklets by the slits and the button end is left dangling, waiting to get caught in the fork of a branch. I saw someone still doing this at the races in Britain in 2021. There have even be examples of really stupid falconers flying their hawks with the swivel still on the jesses!

Normally it is best to fly falcons with no jesses at all. But sometimes, particularly if accipiter hawks are being flown, more control is needed and field jesses are one solution. They can be like lightweight slitless mews jesses (figure 3.2.3), or, if the hawk is in the habit of promptly removing button jesses, try using slit jesses of the type shown in figure 3.2.4.

There are various braided or bullet type jesses available, or you can make your own. We have tried all sorts of designs but braids tend to go stiff and also do not hold nicely into the eyelet. Leather ones, properly fitted, stay sitting in the eyelets and you can make half a dozen pairs while drinking a cup of tea.

Arab jesses are very light and are designed only to be used when people are in attendance. Falcons constantly escape from them but they withstand the harsh climate better than leather.

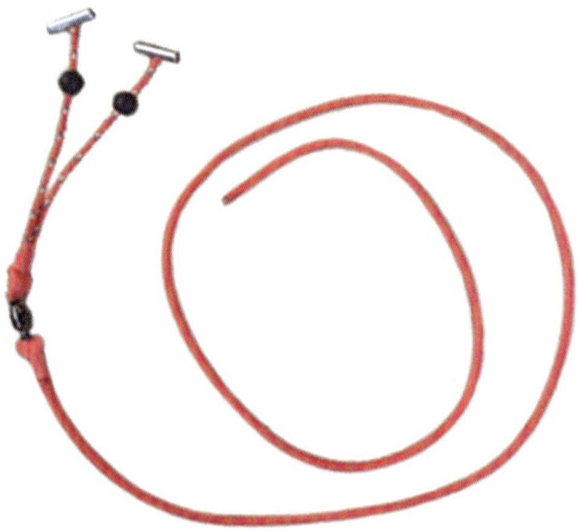

Bullet jesses with swivel and leash attached.

3.3 Identification

All domestic-bred raptors carry a ring. This ring bears a number which identifies the hawk and corresponds to any documents she has, such as CITES papers or her passport. The number is supposed to be unique but, without a central recording system, breeders can put on any numbers they wish. The problem is that, unlike the old varvels, these rings do not help you get your hawk back. You really need to have your contact details on your bird somewhere, and a good plan is to use a ring with your telephone number on it, or use a service such as the Independent Bird Register (www.independentbirdregister.co.uk). Rings should always be fitted above the anklets, not below.

In the old days, vervels were used to identify owners, either through their shields of arms, or by name. Some were flat rings on the ends of the jesses, acting like swivels. Others were smaller rings on the strap of the jess. In 1603, Sir Thomas Sherley described a better form of vervel, still used by some to this day, fitting around the anklet. I have a vervel found near Winchester, close to where I was born. It is engraved *Sir Benjamin Tichborne,* who was knighted by Elizabeth I in 1601. He would have known Simon Latham, and Latham's mentor, Henry Sadler of nearby Everleigh in Wiltshire. Imagine how many hawks were lost in those days, their skeletons eventually rotting away, leaving only their vervels to be found centuries later by metal detectorists. What a story these rings could tell!

3.4 Swivels

Provided that a good quality swivel is used, this item should cause no problems at all. The difficulties start with cheap swivels. Remember that although an item of equipment may be temporary, the resulting accident could be permanent. The swivel should be welded stainless steel or at least phosphor bronze or brass, brazed, or silver-soldered. The joints should be well-fitted and the shank head threaded and welded on. The shank itself should be visible to check for wear. Of the 13 types of swivel or clip shown in figure 3.4.1 and 2, only A-C are safe for tethering birds. They are widely available commercially. Avoid the Pakistani cast swivels, you can pull some of them apart with your bare hands.

The Sampo range of game fishing swivels are strong and work well. They should be replaced after a year or so as wear cannot be detected. Their problems start when fine sand gets into the bearings. The jesses are attached to the fixed end. The cheap imitations of Sampo swivels do not rotate so easily and need proper welded steel rings put on them. Key rings should never be used as they have very little strength. Try doing a load test on one and you will be shocked. The old Mollen type swivel (J) was unsafe but the modern steel or titanium ones of this pattern are fine.

The quickest and most trouble-free way of attaching the swivel to the jesses is to thread both jesses through together. Avoid any kind of spring swivel or clip, especially type E. The jess has only

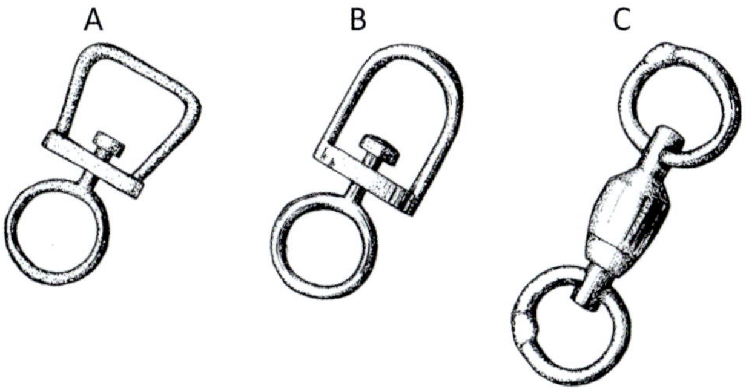

Figure 3.4.1 Swivels A-C are safe.

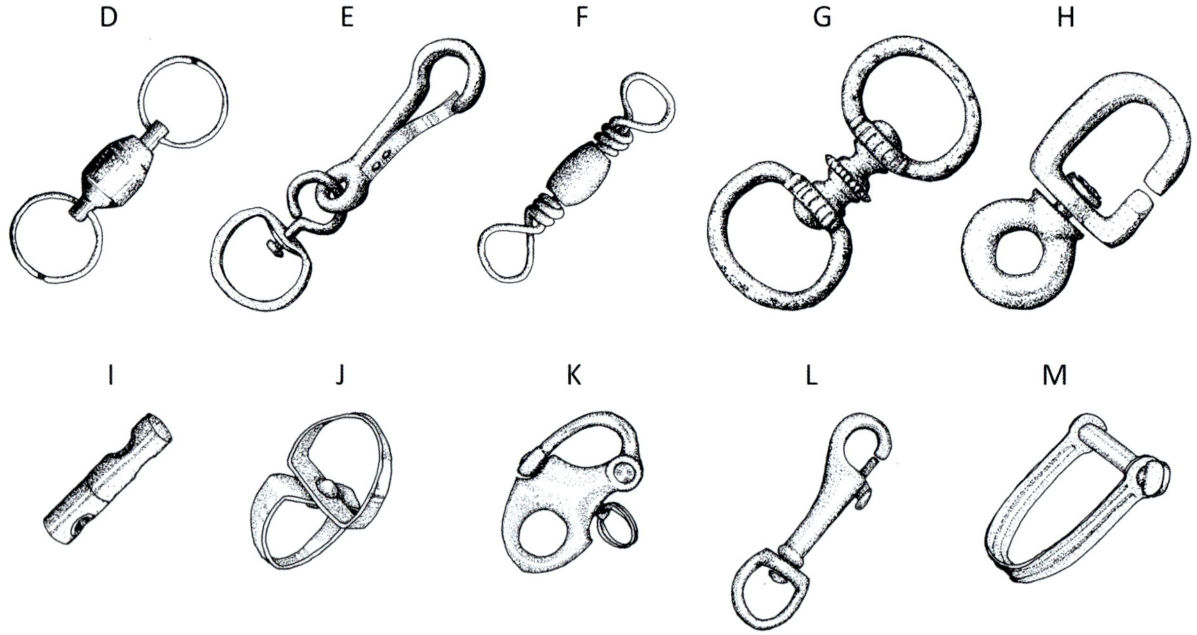

Figure 3.4.2 Swivels D-M are unsafe.

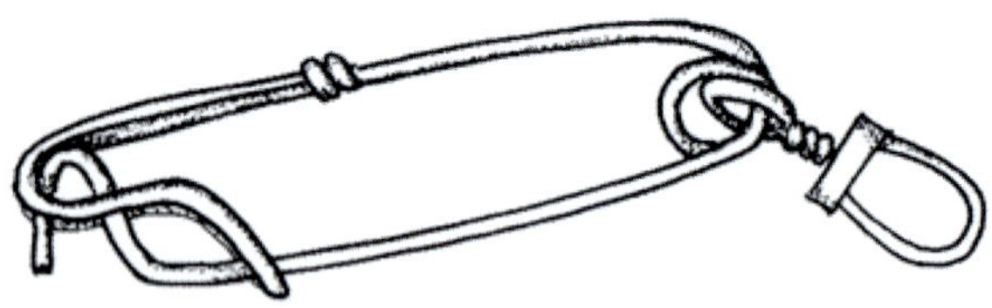

Figure 3.4.3 A safe spring clip.

to cross the outside of the spring for them to fail. This cannot happen with the clip shown in 3.4.3

Within a day of writing this I heard of someone who has just lost his peregrine. She freed herself from a spring-clip on a screen perch, flew out of the mews door, which had been left open, and was found later hanging dead in a tree. The falconer has now attached his goshawk to the same spring-clip! Some people never learn.

Clips and fastenings are common hazards and frequently cause death or injury. Avoid any clip which has a leaf or coiled spring. They all can fail. Americans are very fond of clips, but they can cause loss or injury just as easily as knots. Also avoid any clip which includes a key-ring. Key-rings are very good for keeping keys. Recently a friend tethered his peregrine/saker using a key-ring. She got free and killed his female Finnish goshawk tied nearby. Even on the lure, key-rings are a hazard. They can wrench apart allowing your falcon to fly off with the lure. Cable-ties have caused several injuries when hawks have got beaks or claws caught under them. Poor cable-ties will also fail unexpectedly, allowing transmitters to come off.

3.5 Leashes

Leather leashes are not safe and have been the cause of many deaths and losses. A good terylene leash can be easily made for a trivial sum and will not break without showing wear. Take some round woven terylene cord, preferably of the closely woven soft variety. Cut it to the length required and pull out the central core leaving a flat hollow tube which will hold a good knot. Tie a neat overhand knot in one end and slide on a leather or plastic washer. Heat seal both ends and cover the knot in a thin coat of epoxy resin so that it cannot loosen. A few minutes work will last for years.

If you wish to incorporate a shock absorber into the leash cut a length of inner tube from a car tire, or bungee, one third of the length of the leash and tapered so that it slides up inside the hollow tube and through the washer into the knot (figure 3.5.1). Leave the lower two thirds of the leash for tying the falconer's knot. Having tried many types of clip-on and parallel shock absorbers, I have found nothing to beat the inserted absorber. Young accipiters are especially prone to leg injury if no absorber is used because the terylene has little or no elasticity. The inserted absorber is very convenient and discreet, it's basically a kind of bungee. Nobody will even notice that you are using it - except your hawk.

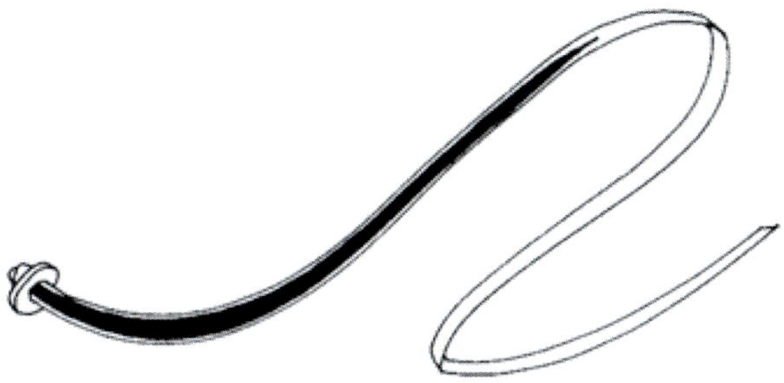

Figure 3.5.1 Traditional leash with rubber insert.

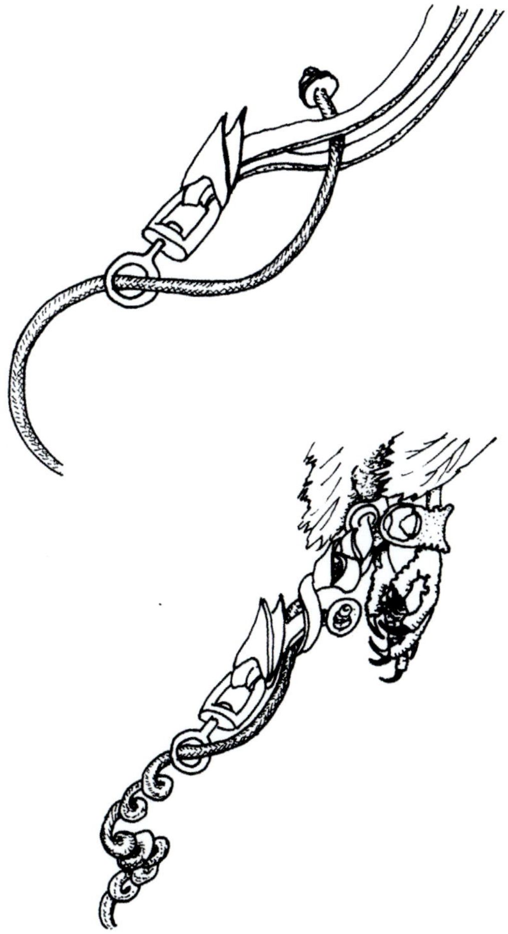

Figure 3.5.2 Tangling caused by the traditional leash.

A decorative line in leashes.
Photo Ben Crane.

This poor owl has not even been tied with a falconers knot, and the perch is a flower pot filled with smooth cement.

The traditional leash has two design faults which have caused injuries to many hawks and, in some cases, death. One problem is that the button rides up between the jesses, the hawk turns round a few times and is soon completely trapped, its legs abraded, sprained, or even broken (figure 3.5.2).

The second problem is the falconer's knot which can be undone or can work loose through the activities of the hawk. I have seen a New Zealand falcon completely undo a double falconer's knot in 20 minutes (fortunately indoors). If the hawk escapes with her leash trailing, you can count yourself lucky if she doesn't kill any of the other hawks out weathering before she flies off and hangs herself. This is an unnecessary disaster.

To overcome this I have, since 1979, used the loop leash with completely satisfactory results. It is made of woven terylene with the core removed. A 90 cm length of cord is folded back inside itself and threaded through this tube leaving a 10 cm (4") loop which ends in a small tab to aid threading. The other end has a resin button. For big falcons we use loop leashes 45 cm (18") long with a blue button, and some 53cm (21") ones with a green button. If you are no good at threading terylene with a fid or pulling needle, simply stitch the cord along its double length. It only takes a few minutes.

The perch ring has a small steel loop with an internal diameter of about 15 millimetres (5/8 inches). The leash is threaded through this and is left there with the perch when the hawk is taken away. When you approach with the hawk, first thread the leash up through the hawk's swivel, then double the leash back down through the perch ring. Slip the leash button through the loop, pull and the loop will slide up and lock on the swivel. All this takes just a moment and the leash cannot come undone or jam the swivel. In addition, there is now an extra swivel point at the perch end of the leash so that tangles are eliminated with no extra weight, clips, leash extenders or other devices. A similar but shorter leash is used in the mews (figure 3.5.5). When on the fist, the hawk is attached with a cord on the glove.

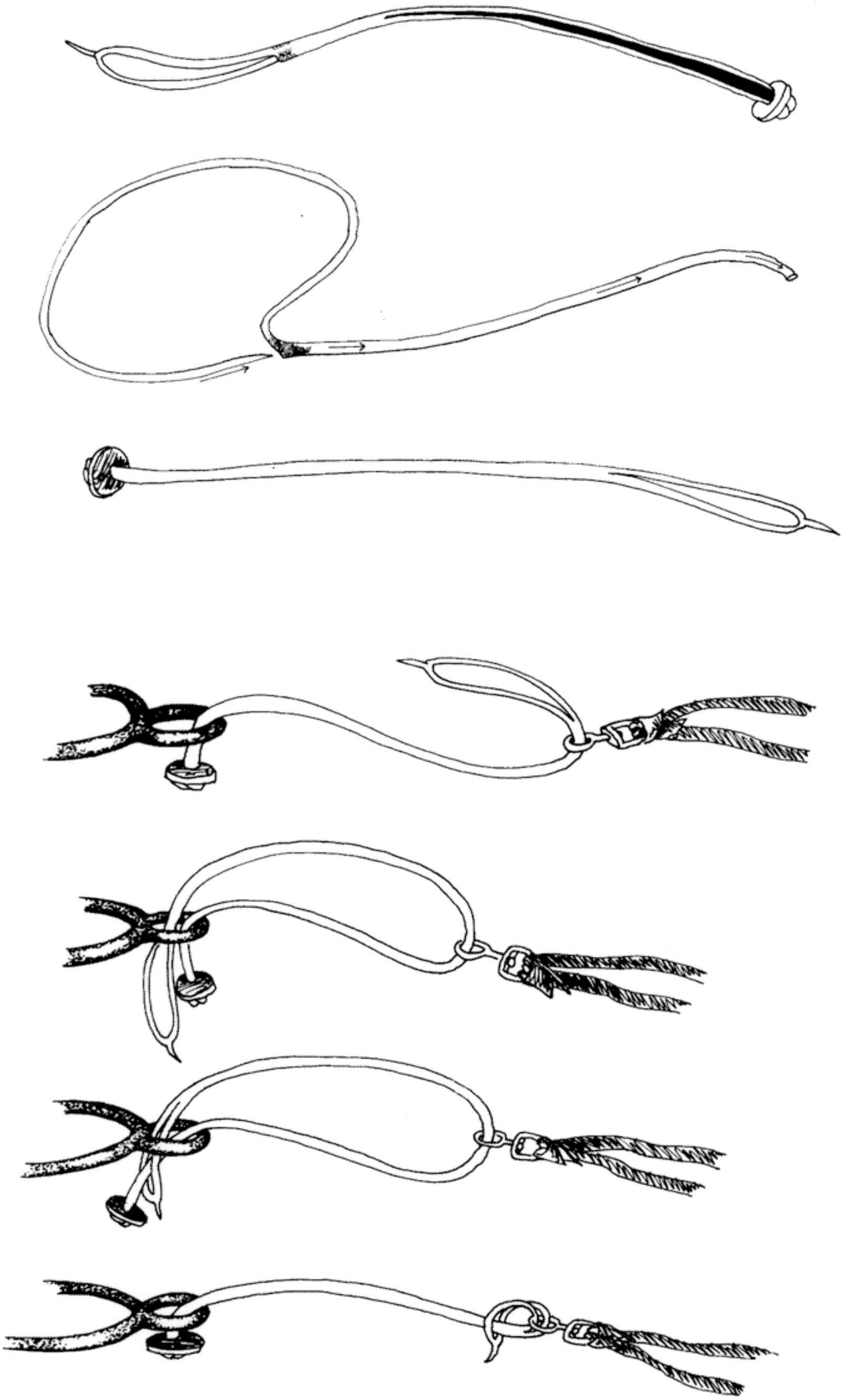

Figure 3.5.4 Making and using a loop leash.

Figure 3.5.5 A short loop leash on an indoor perch.

Another common fault is to fail to tie up a hawk while she is hooded. Hawks may bate while hooded and can easily fly off. Never leave a hooded hawk unsecured or able to reach a neighbour. They can easily crab and quickly cause a bad injury. Hungry hawks hooded on a cadge easily bite out blindly, and often the first thing they find is the carpal joint of an adjacent falcon. One bite on this joint can cripple a bird for life. If a hawk bates with its hood on, it will fly high up into the sky until exhausted. In that state it is an ideal decoy for a passing eagle. Also hooded hawks can bate unexpectedly, the result being a broken wing, shut in a car door.

The Arab jess and leash system.

The Arab leash system was never intended for tethering unattended. Falcons easily escape from these leashes and if you attend an indoor falconry event in the Middle East you will usually see a falcon or two flying around amongst the ceiling supports. Outdoors, the consequences are more serious and the bird is unlikely to survive for long with a trailing leash. If you ever have to leave your hawk unattended for any length of time (which we all do), then your tethering system must be fail-safe. An enclosed weathering is mandatory in some countries.

3.6 Bells

The size of the bell and the way it is mounted is more important than its weight. For example, medium-size bells for a male peregrine weigh 3.8 grams (Lahore type), 5.5 grams (Asborno type), and 7.3 grams (typical British handmade brass bells). A large size Lahore bell weighs about 7.0 grams. The weight differences between them are not even equivalent to a good mute! It follows then that we are looking for bells which are small for the sound that they give; transmitters are the ultimate solution and we'll look at them later.

Neck bells are popular for large accipiters in some cultures, such as Pakistan and parts of Asia. They are often used together with a fine leather or braided neck strap called a *jangoli* or *halsband.* This extends to below the hawk's feet and is held tightly by the falconer so that he can launch his hawk, like bowling a cricket ball. Smaller accipiters can also be thrown this way, but they are small enough to be grasped in one hand for throwing, and this is more controlled. Hawks can get themselves tangled up with jangoli so it is best only to use them when the bird is in constant supervision day and night. This fits in with some cultures but not for the person with a nine to five job.

A passage goshawk with jangoli in Pakistan. *Photo Farooq Khan.*

A passage sparrowhawk with jangoli in Turkey.

There is no doubt a neck bell works really well and a hawk in cover cannot pluck prey without ringing her bell. For goshawks I have used a rubber band 7mm wide and 80mm long when laid flat. When the bell is attached the remaining loop is 60 mm and can be stretched over the hawk's head while she is hooded. It needs replacing every three weeks or so.

The last place to put a bell is on the hawk's leg. It may be convenient for you, but look at it from the hawk's point of view. As was shown in section 1.18, the raptor foot is a very highly evolved instrument finely tuned by the rigors of survival through hunting success. Already we are obliged to use some sort of anklet and a closed registration ring. These can be designed to be reasonably neat but it is sheer folly to clutter a bird up with a bell or a transmitter on its leg. Can you imagine trying to play a game of football with

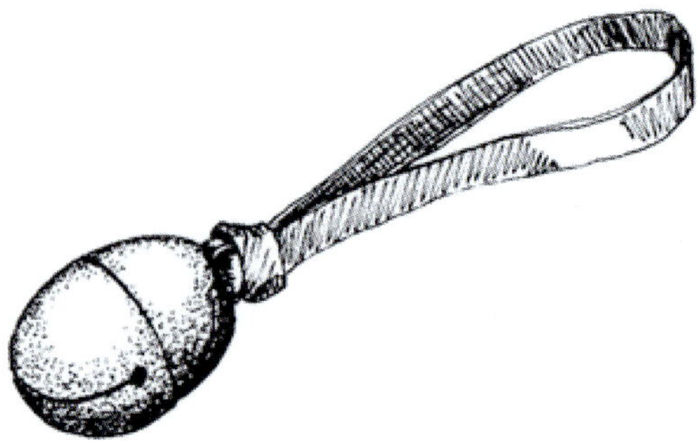

Figure 3.6.1 A neck bell on a wide rubber band for a goshawk.

a football strapped to each ankle? The problem is that falconers are so used to seeing cluttered up legs and so rarely have a chance to see the nimbleness of wild hawks that they take their hawk's poor performance for granted. Leg bells can be dangerous too. I have twice seen goshawks trapped in forked branches by their leg bells.

We used to use leg bells during hacking because the tail is not hard-penned then. If you have to use a leg-bell, look out for the following: first make sure that the bewit is double looped so that no bell metal comes into contact with the hawk's legs where it can cause sores. Second, make sure that the metal strap on the bell has been flattened and will hold the bell up at right angles to the tarsus. Lahore bells are made with rounded loops which require altering. Third, use pre-stretched leather for the bewit and put it on with sufficient tension to hold the bell clear of the leg without restricting blood flow. Slack bewits and cumbersome bells are extremely irritating for a hawk. Plastic cable ties should not be used directly on the tarsus because of the risk of overtightening and because they can be hard and abrasive. The alternative of attaching the bell with a cable tie to part of the anklet is also a clumsy arrangement; better to get the bell off the leg altogether.

This shikra *Accipiter badius* has little chance of catching anything with this lot on its legs.

Nowadays transmitters are replacing bells, but for accipiters down on a kill in woodland, bells still beat transmitters and GPS. Radiotracking accurately and quickly in woodland is tricky, and GPS may lose signal under the canopy. And the hawk may not be far away; she could be tantalisingly close, crouched over her kill in a thicket. You know she is there somewhere, but where? But she cannot fill her crop without ringing a neck or tail bell and giving herself away, as she can with a transmitter.

The best site for a bell is on the tail. It is out of the hawk's way, it rings well and is held clear of debris, snow, etc. However, many falconers do not use it because they cannot put one on securely. The problem is to attach the mount securely and to ensure that the tail feathers do not get damaged or fall out.

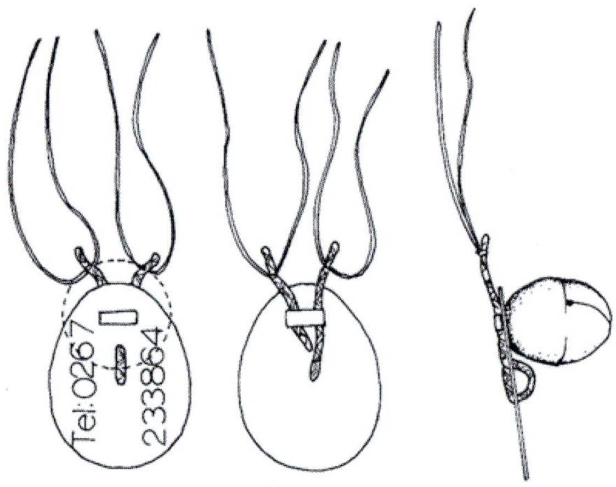

Figure 3.6.2 The 'plectrum' tail mount, from above, from below, and from the side.

You don't need to make your own tail-mounts anymore, 3D printed ones are available commercially and attach to two tail feathers. You can either attach a bell to it, or a transmitter, or both. Fit a length of terylene creance cord to the mount and heat seal the ends of the two tags to stop them unravelling. Attach a rot-proof terylene thread to each by passing it through on a needle and knotting it. This is to prevent it sliding off the tag. Cast the hawk and expose the base of the tail. Some people give the hawk tranquilizers but if the situation is that traumatic it is probably better to take the tranquilizers yourself! It is easier if you mask off the rest of the tail from the feathers you are working on by sliding in a big white sticky label and pressing it down lightly so that it stays in position if the hawk moves.

The mount will be attached as near to the body as possible; there is seldom any need for trimming the feathers but you may need to clear some fluff away. Being so close to the preen gland the shafts of the tailfeathers are often oily and repel glue, so you may need to roughen them a little at the attachment sites. The two tags are bound by the threads to the two feathers each side of the central deck feathers. This distributes the weight of the mount over four feathers, not just two. This reduces the risk of the follicles being damaged and of the feathers being dropped prematurely, or pulled out. Place some rubber-based glue such as Evostick (not epoxy or superglue which will eventually crack and flake away) onto the tags, the threads, and the feather shafts. Pass the threads around each shaft three times, ensuring that the glue is all round the shaft, then tie off tightly with a reef (square) knot. Put a dab of glue on the knot and trim off the surplus ends. Do the same for the other tag and for very active hawks repeat the bindings. A sprinkling of talc stops the feathers sticking to the wet glue. Leave the hawk hooded quietly until the glue is dry.

Do not actually pierce the feather shaft with a needle because of the risk of weakening it. With small hawks, such as merlins, take care not to pinch the shaft when tightening the thread. If necessary, the shaft can be reinforced with a sleeve made from 1 centimetre of shaft from a slightly larger feather, well glued of course.

Such a tail-mount will last all season and from the hawk's point of view is the least uncomfortable site for a bell. It irritates far less than leg bells and the transmitter is safer there because the hawk cannot get a good pull at it. Should she so wish a falcon can destroy a leg-mounted transmitter in a few moments.

Nowadays telemetry has virtually outmoded bells as a means of finding the hawk but a tail bell is still useful for a goshawk or any hawk flown in woodland. It also helps to protect the hawk from shooters when out of sight and it is useful at home to know what the hawk is doing by the sound of her bell. We don't use bells at all on the falcons. At the moment we use a GPS tag and a conventional radio tag, but technology is improving every year.

Fitting a tail mounted cockerel feather onto Mr Li Shi's goshawk in Yunnan, China.

3.7 The tail sheath and taping

Tail sheaths are useful when training an accipiter or any bird with a vulnerable tail. It is a light tube which wraps around or slides on to the tail, preventing the tail opening and thus protecting it. This is preferable to taping because it can easily be removed for flying and then be replaced for traveling. Although hawks occasionally break feathers in the field, for example by falling into a tree when holding struggling prey, most feathers are broken through mismanagement at home. The general attrition of bad handling causes the feathers to weaken and eventually to break.

There are various designs of tail sheath including wrap-around ones with velcro attachments. A really violent hawk may need a sheath reinforced with piano wire or nylon bristles from a yard brush. The following design,

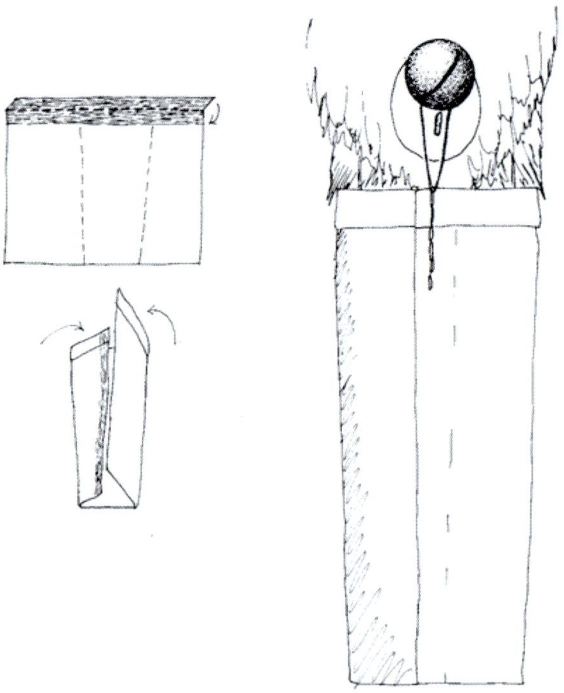

Figure 3.7.1. A plastic removable tail guard.

if adjusted for the size of the hawk, has stood the test of time and is easily made:

a) Take a square of old X-ray sheet (figure 3.7.1). With sandpaper, roughen a 25 mm strip along one edge, apply a flexible contact rubber adhesive such as Evostick and allow to dry. When dry, warm the glued area to activate the glue, fold over, and stick down. This makes a strong rounded top rim to the sheath. If you wish you can flare the top 20 mm so that it doesn't dig into the sides of the tail.

b) Measure the width of the hawk's tail at the tail mount and at the tip and mark out this tapered shape down the centre of the polythene. The two sides will now fold over inward with a 15 mm overlap for gluing. Mark and cut these accordingly, roughen and glue. Note that the top rim folds outward leaving a smooth surface inside the sheath.

c) Using the finest 1 mm braided terylene, sew about six stitches down the centre leaving a loop 4 centimetres long which is used to loop over the tail-bell to hold the sheath up. Glue the knots. Alternatively, some designs Velcro the tail-guard to the tail-mount.

d) Trim off the end of the sheath so that when it is in position it is 5 mm longer than the tail. For a goshawk such a sheath should weigh no more than 5 grams.

During air travel, and sometimes for fresh-

-caught hacked birds, we tape a simple tail guard on, together with wing guards for the outer five primaries. A narrow band of masking tape holds it in place and is easily removable.

3.8 Gloves

There are various gloves available, from buckskin falconry gloves to goatskin rodeo gloves. Mostly it is a matter of personal taste and depends on the type of hawk being carried. I make my gloves with the first two fingers together. Gloves should be looked after properly and kept clean, otherwise they are a major source of bacterial infection. Cuffs or mangalas are used in parts of the Middle East and eastern China, but for most purposes, gloves are better.

A cuff or mangala used in eastern China, a culture linked to the Middle east through the old trade routes.
Photo Alan Gates.

The glove should have some means of attaching the hawk to it. Wrapping or draping the leash around the fingers may be traditional but it is just not good enough. It is so easy for a sudden bate to pull it out and in cold weather, or on rough ground, the falconer can easily stumble and let go. I have seen somebody let go of a sparrowhawk, leash and all, and watched her disappear into the sky to certain death. Each summer when we ran courses at least one person each week, sometimes more, let go of the hawk by mistake, only to realize with a sigh of relief that he had been made to tie the hawk to the glove.

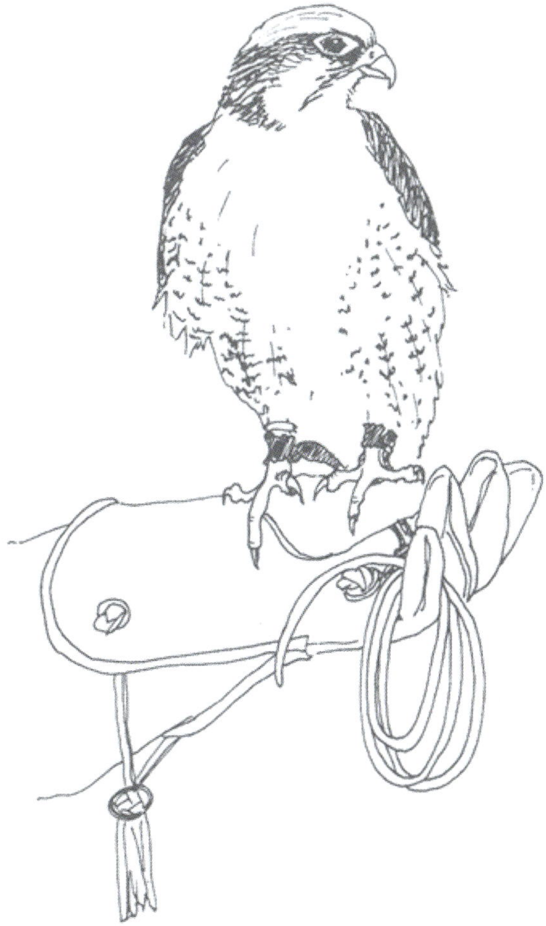

Figure 3.8.1 The traditional way of securing a falcon to the glove. If you trip, you don't just lose your balance, you lose your hawk.

When using a traditional leash, the simple solution is to tie it with a falconer's knot to a D-ring or the tassel loop which should be a proper continuous loop of braided terylene not some light or ornamental strap. The system I use with the loop leash is similar to that described by Jack Mavrogordato in *A Falcon in the Field.* Below the wrist of the glove is a brass eyelet to which is attached two lengths of braided terylene (figure 3.8.2). One 35 cm (14") length ends in an 8 cm (3") loop; this is the lock cord. Note that the lock cord is stitched double all the way up to the glove with no binding or knot to impede it sliding smoothly through the anklet. The loop itself is pinched at the end to give a little threading tab about 15mm long. The other 50 cm (20") cord remains single; this is the slip cord. We used to

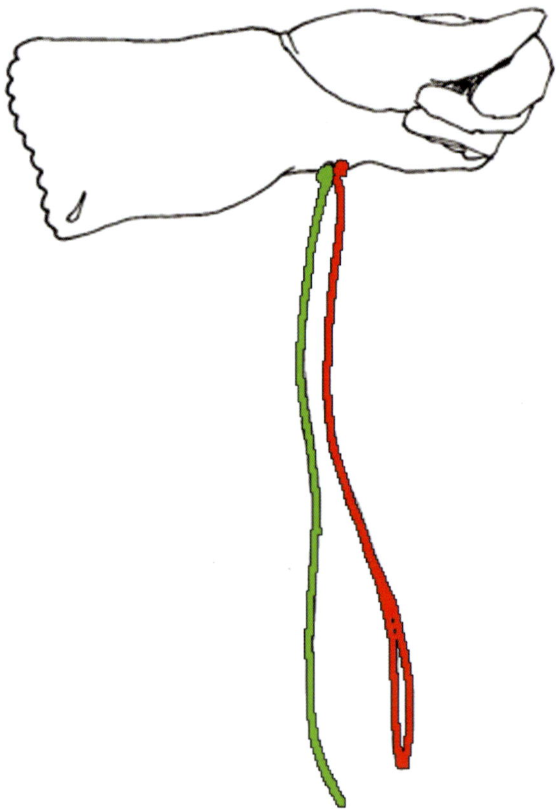

Figure 3.8.2 The slip cord (green) and the lock cord (red).

use two lock cords but if you slip using a looped lock cord there is a risk that it will catch on a toe when sliding out of the anklet. This is why the slip cord has no loop.

On picking up the hawk, the bird is first called onto the fist and the slip cord is tied to the swivel with a falconer's knot. The loop leash is then undone and left on the perch. At no stage is the hawk unsecured. When ready to fly, first check that both transmitters are attached and working, then undo the slip cord from the swivel and remove the swivel. Then remove one mews jess and thread the looped cord through the anklet and back, locking it by threading your ring finger through the loop. Then remove the second jess and put the slip cord through the anklet, looping it tightly round your fingers. So now you still have secure hold of her by both legs and can walk or ride with her without letting her go by mistake (figure 3.8.3). If the slip is not imminent, tie the slip cord with a falconer's knot in the interim. When ready to slip, remove the lock cord from your ring finger and from the anklet leaving her tied by just the slip cord. Open the braces of the hood. When ready, remove the hood and slip out the slip cord leaving the hawk free and unfettered without a loop catching on a toe as the hawk flies off the fist. In this way we can get a quick clean slip every time, and yet be secure until needed. For pursuit flights and hunt racing a clean slip is critical. The falcon herself should go like a bullet the moment the hood clears her beak.

Figure 3.8.3 Slip and lock cords holding the falcon securely.

3.9 Hoods

Raptors live in a visual world and, by temporarily blind-folding them with a hood, they can be shielded from visual stimuli which would otherwise stress them. Instead of being frightened or bating wildly and risking injury, the hawk stands quietly and many will often go to sleep. Thus hoods have an important role in managing raptors in captivity, not just for falconry but for veterinary and rehabilitation work as well. If you have raptors on the premises, you should have well-fitting hoods to hand for each species and sex; they are important for their welfare.Vets and rehabilitators can unwittingly stress a bird by failing to hood her adequately. Some vets have treated hawks without using a hood and caused the bird to become neurotically afraid of humans. Others have crammed on hoods which are too small, causing eye infections.

There are plenty of books on how to make hoods and hoods are easily available commercially. The definitive book is '*Hoods, hooding and hood-making*' by James West Nelson. His book will tell you all you need to know and the Archives of Falconry in America has a nice virtual exhibition of hoods on its website.

A good hood should meet several criteria: the hawk should not be able to get it off. It should be easy to put on and take off. The hawk should be able to feed and cast through it safely. It should be light in weight, sit comfortably on the hawk's head without causing condensation, irritation or trailing braces which annoy the hawk. It should be light-tight. It should not cause sores around the gape or pinch the nape feathers. It should be easy to make and to duplicate. It should be tough, durable, weather-proof, and easily cleaned or renovated. It should be elegant. Hoods which constrict the mouth opening can choke a hawk to death if it needs to cast or vomit. Traditional Dutch and Bahreini hoods with small beak openings can be unsafe.

We make all our hawks of all species to the hood. Hooding makes training easier because it enables you to 'switch off' the hawk when necessary, protecting her from undesirable sights which not only might upset her but also might be linked in the hawk's mind, to her handler. Some hawks are more difficult to hood than others and this is something which requires skill. There are a lot of little tricks and wrinkles according to circumstances, and

This hood is too soft. The leather has collapsed in and rubbed against the eye.

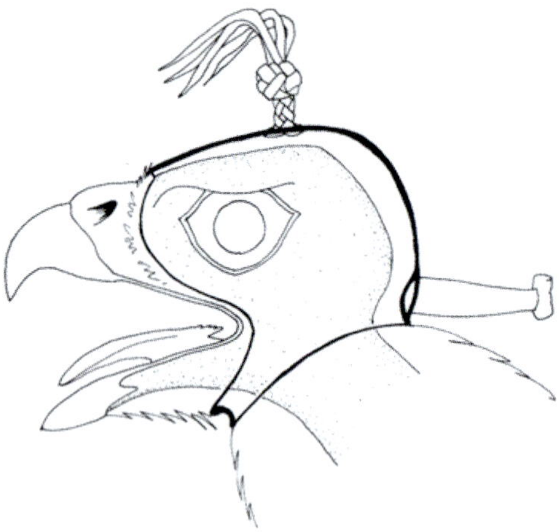

Figure 3.9.1. How the hood fits on the head.

A spring hood.

not many short cuts. Hooding is discussed in section 5.12, but Jim Nelson's book goes into all details.

A well-fitting hood should not cross the rictus, the angle of the mouth, because it will either chafe or let in light, owing to the way the mouth opens (see figure 1.2.3). At the back it should fit snugly under the occiput (figure 3.9.1) so that it cannot come off. It should not extend any further than this otherwise it will impede neck movements and press on the crop. During early training we usually use a hood that is too large. It slips on and off easily and the hawk can see through it like blinkers on a harness horse. If you have plenty of hoods, you can cut 1 cm holes so that the hawk has limited vision forwards. Often they start to feed on the fist this way without being distracted. The medieval practice of 'seeling' the eyes has no place in modern raptor keeping. If you, or your hawk, struggle with hooding then it is best to abandon the hood rather than create a counter-productive nightmare.

For me, hoods are unfinished business. Current designs have two major design issues. First of all, over the eyes, we expect the leather to be firm, yet around the beak, soft. Other than soaking the inside of the eyepatches with resin, this is asking too much of the same piece of leather. And when leather gets wet, it loses its shape. Where we hawk, we get wet! But the

The old traditional pattern of Japanese hood!

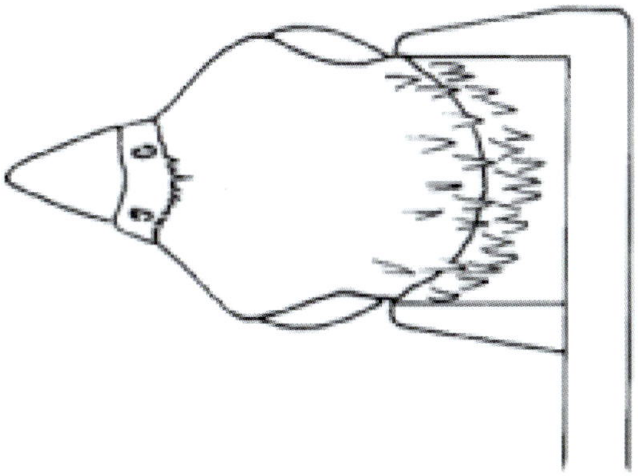

Figure 3.9.3 To measure a hawk for a hood, take the width of the head behind the eyes, in millimetres. Measurements are provided in Appendix 1.

major problem is braces. When the hood is on but the braces are open, the hawk can easily toss it off. This happens when hooding a wary hawk, or when about to slip a trained one.

In most situations, hooding and unhooding can be done at your leisure. But when you are slipping a hawk at prey 'out of the hood', the hood must be removed quickly and cleanly at the critical moment. You cannot easily carry a hawk with the hood braces open for long or she may flick it off. Ferreira (p57) recommended brace-less hoods for this purpose, and the Asian eagle hunters use brace-less hoods for the same reason. If the hawk flicks an open hood off when you are on a horse and about to slip, this can be very discombobulating.

A Mongolian Kazakh berkutchi using a braceless hood. Notice the Eagle Owl feathers. *Photo Alan Gates.*

I've made a lot of leather hoods, bridles and saddles; so much so that I no longer have fingerprints left. This is embarrassing at Passport Control. But there are a lot of other materials available now, such as resins and plastics, some of which can be 3D printed. The problem is that hoods have three fits: one when the hood is open to put on, one when the hood is on and closed, and one when it is on but the hawk needs to open her beak, perhaps to cast. If you use a material that is flexible, it will not maintain sufficient shape. But if you use a material that is rigid, it will not flex. So I've been designing spring hoods for many years, struggling to solve this conundrum. Finally, I think I have succeeded; I have hoods that are comfortable for the hawk, fit nicely and look OK. I'm not yet fully satisfied with my efforts, so I tinker on… But hey – falconry is not about the destination, it's about the journey!

Prototype spring hoods in resin, thermoplastic or 3D printed.

Prototype side-opening spring hoods

Prototype back-opening spring hood.

3.10 Telemetry

There have been huge advances in telemetry in the past forty years. No doubt there will be even more to come and for detailed information follow the websites and instructional videos of the major manufacturers. But despite the technological advances, the basic tenets hold true. Reliability is key. Get a system that works in the cold, in the heat, in the rain, despite impacts, that will stay on the bird without hindering her, and lasts sufficient days for you to recover her. Many systems don't work unless they are line of sight, or receiving a mobile signal, or not under a canopy of trees, or not submerged in water. Find all this out and practise using it when not on the bird. The system

Tracking from a Hercules.

may be sophisticated but if you don't know what you are doing, don't expect it to work. Don't do as we once did, fly all over Baluchistan trying to get a signal from a lost falcon, only to find later that the falconer had put the battery in upside down and then not checked the signal before slipping…

Following on from that, make yourself thoroughly familiar with the gear under field conditions, especially around hills, houses and pylons. Carry it with you out hawking. Don't leave it at home or in the car or, most frustrating of all, distribute it in bits for your companions to carry. When you need it, you need it. The first few minutes of loss are critical, as many have found to their cost.

Check your gear before leaving home.

Co-ax cable easily breaks undetected, so keep a spare. The antenna should be firmly attached to the transmitter and fully electrically insulated. Always check your transmitter signal and tuning immediately prior to flying. There are various ways of attaching the transmitter. Some still use bewits or cable-ties, but most tags nowadays use spring clips. These are not infallible. Check carefully that you have fitted the springs into the tube properly. Tags with magnet switches are better because you don't need to take the tag off each time. But before you finish for the night, just check nothing in the mews is still transmitting.

Transmitter design is changing constantly. Everybody wants more range. In open desert, this is useful, but in Britain a transmitter operating further than about 10-15 kilometres is usually wasted output because you cannot often get a line of sight signal that far away; there is usually something in the way. Better to trade the extra power for longer battery life. If your transmitter has a life of five days and you have already used it for three, and you then lose your hawk, you have only 48 hours to get her back. Of course, the battery could last longer. Or not...

The yaggi is a piece of equipment which is easy to hate. There are more user-friendly designs with arms made of flexible tape for use in woodland, or which fold down completely to a 30 centimetre pack which fits in the hawking bag. Car aerials are useful for the more desperate situations, together with large-scale maps. You can use a magnet-mounted omni-whip, or a rotating elevated five element yaggi which will give you a bearing while you are driving.

Some of the more traditional falconers regard telemetry as 'last-resort' equipment, to be used at most only once or twice a season. This is a perfectly valid approach, but I don't subscribe to it. Having been in the habit of using telemetry for long periods during research on wild falcons and on hack falcons one comes to regard telemetry as an extra sense, particularly useful in my case because I'm deaf in one ear and cannot tell the direction of sounds. One comes to look at a landscape from the viewpoint of radio waves, very much in the same way that flying teaches you to 'see' air. I flew falcons for fourteen years prior to telemetry and being used to the discipline which that demanded, I appreciate all the more the possibilities and opportunities in hawking which have been opened by telemetry. A great deal of the hawking which I enjoy now would be folly to attempt without telemetry. Most of the really spectacular hawking is to be had in hilly country with powerful air currents, the sky's answer to white water. To hawk such country requires control, trust and reliable telemetry. Paradoxically, the person who is using telemetry on a daily basis and constantly checking and becoming familiar with the equipment, tends to be less likely to lose a bird than the person who only uses it in a crisis.

We have young people here every year who learn basic radio-tracking. They get

Bulky tags are great for tracking but not for the falcon carrying them.
Photo Stuart Grimwood.

into difficulties for a number of reasons, for example they don't properly understand what all the different knobs on the receiver do, such as gain, volume, headphone socket, diminutor, and different frequency settings. They don't understand about peaks and nulls, about bounce and interference. Most of all, they fail to think carefully and deduce as much as they can from the signal. They don't know what a falcon soaring up and down just over a horizon sounds like, what signal to expect from a hawk down in a ditch or gully, what a faulty battery connection does to the signal, or a faulty antenna connection, how to sort out several signals all on close frequencies, and so on. All these things only come with experience, and one can either wait until it happens for real, or you can experiment first without risking your bird. The technology is really only as good as the person operating it.

Many hawks will quickly damage or remove a leg-mounted tag. There is no point in using a

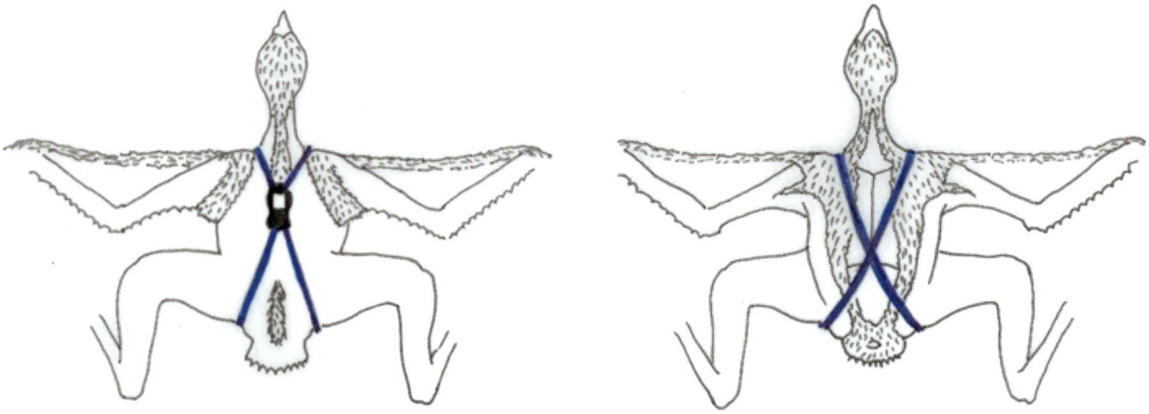

Figure 3.10.1 The position of the Teflon ribbons for the back pack.Left: The back. Right: The chest.

Drs Eugene Potapov and Andrew Dixon training Russian and Mongolian field biologists using a dead falcon.

transmitter with a battery life of two weeks if the hawk removes it ten minutes after you lose her. If you have to use a dummy transmitter to condition your hawk to accept it, then stop and ask yourself if she really likes what you are doing. Better to mount it on her back or tail where it will not annoy or hamper her so much. The harness configuration whereby the abdominal straps go behind the legs is gaining popularity and is more forgiving when the hawk puts on more pectoral muscle. Ricardo Velardo has a good YouTube video showing this with a Marshal tag. https://www.youtube.com/watch?v=PzQll8xbG78. All this equipment must be fitted carefully otherwise it will impede the bird.

Young Kazakh sakers with solar powered satellite tags.

Electronics are wonderful, but in thick woodland or broken terrain, they can be a frustrating nightmare. If you are hawking pheasants in woods you may find your hawk more quickly if she is also carrying a bell. A single tag may be sufficient for an accipiter, but for falcons I would always use two tags: a GPS tag on the back and a VHS tag on the tail. You can be sure that, for one reason or another, there will be occasions when a tag does not work. But you are welcome to attend the school of hard knocks.

The new GPS-based tags are now proving their worth and

can be linked to your phone, without the need for cumbersome gear. They show you exactly where your hawk is, not just her direction. This means that you can reach her quickly if she is in a dangerous place, or feeding up on a kill. It also enables you to record her flight track, speed and altitude, which is useful during training. Try not to get too dependent on it. I have young falconers now who, the moment the flight goes out of sight, start groping for their phones and pocket links. While they are busy with their heads down, pounding their buttons, they have not noticed a crow dipping in the distance, marking the kill site…

There is a whole debate to be had about the merits or otherwise of using technology in falconry. With increasing urbanisation, fencing and so on, anything that can swing the balance in favour of the falconer is welcomed. But the use of cars has resulted in over-hunting and any Houbara that has the temerity to land in the Arabian peninsula now faces cars full of falconers arriving within hours. Falconry is *the art of hunting wild prey in its natural state by means of trained hawks,* and this is a definition worth hanging onto. That means *wild prey*, not farmed or bagged prey. And *natural state* means we must conserve habitats and hunt in a natural way. What chance have wild animals against our barrage of technology? Cars, walkie-talkies, transmitters and so are all stack the balance on our side. Try to regard them as evils to be minimised. Use them to reach the hunting ground or to recover hawks safely or minimise time wasted through miscommunication. But otherwise, try to keep the natural balance between predator and prey. Falconry is not about bragging rights.

3.11 Lures.

The different uses of lures are covered in 7.4 and obviously the construction of each lure depends on its function. The basic idea is that the lure mimics the prey so therefore the most powerful lure is actually a live prey on a line, or a dead one. The use of bagged game is banned in most countries for good welfare reasons, but a dead quail or pigeon will be very attractive for falcons and half a rabbit on a line will bring in goshawks and Harris hawks. You can even wind up wild

Now what to do?

ones that way. Always carry a creance with you so that you can quickly attach a previously killed prey or lure to a long line. Sooner or later your hawk will kill something on an unsafe barn roof, or on a cliff, or on a quaking bog or lake. The creance enables you to throw the lure as near as you can, then walk around to the opposite side, drag the lure to the hawk who should grab it, as well as her prey, then by twitching and dragging, you can pull her in close enough for you to reach her.

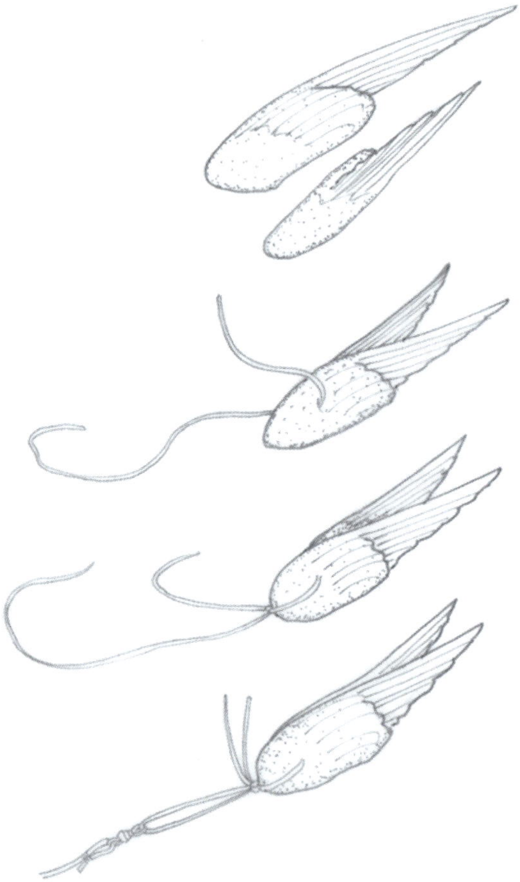

Figure 3.11.1 Constructing a winged lure.

A simple recall lure.

A rotating lure for pole-luring.

Everyone out hawking should carry a simple recall lure of some sort. In medieval times this was carried in a loop over your shoulder, but the problem then is that the lure is visible. This may not matter for a peregrine, but for an accipiter or an intelligent falcon like a New Zealand falcon, they will grab it wherever they see it. The best is to have a basic lure made from a couple of wings, with a swivel, 3m (10 ft) of line and a lure stick. In wet climates dried wings soon become rotten. Instead one can use rubber inner tube, shredded to look like feathers. But this can be quite heavy so when using dried wings we first thoroughly soak the bony parts of the wing in latex that has been coloured with water-based acrylic paint to match the wing colour. This is worked in well, leaving only the primaries and secondaries exposed. Once the latex is thoroughly dried, you have effectively encapsulated the wing. Then, select two right wings or two left wings and place them together so that when swung they will rotate like a propeller and flutter attractively. Taking about 50 cm (18") of creance cord, thread one end onto a large needle and push it through both wings, leaving one end short and the other long. Tie them very tightly together at the wing butts. Then leave a loop on the longer piece and bring the tail back and tie it tightly next to the first knot. Dunk the wings again into latex to seal the needle holes and lock the knots. You now have two tie strings at the head end of the lure, visible which ever way the lure lands, and you will have a strong but light swivel that will rotate freely under tension, away from the lure and unobstructed by knots. Such a crow or duck wing lure weighs about 60g (2 oz) and a teal (*Anas carolinensis*) or snipe (*Gallinago gallinago*) wing lure about 35g. This makes them suitable for being swung by hand or from a pole.

The poles we use are made from carbon-fibre telescopic fishing poles. Fully extended, they are about 2.5m (9ft) long and about 12mm (1/2") at the tip. Select a sampo swivel that sockets into the tip nicely. Glue it in place with epoxy, with the rotating end of the swivel outermost. The inside can have a twist of wire or heavy thread attached to the inner ring and embedded in epoxy inside the tube to make sure it cannot come off. The outside of the tip has a turn of glass fibre tape set in epoxy to reinforce the area. A 6mm (1/4") bungee is next firmly attached either directly to the sampo or via a clip. (By the way, don't use a simple loop on the pole tip instead of the embedded sampo – it won't work…) At the far end of the bungee attach a sampo swivel on a loop, tied off securely and if possible covered with heat shrink tubing. The loop of the lure now loops onto the swivel and I will leave you to figure out how to get it on.

Such a lure packs away quickly in the car, and out hawking, the pole can be used as a walking stick. Put the lure itself in your bag and then you can walk across the grouse moor, but still have an effective lure to hand at a moment's notice. Because the lure can be swung in a much bigger arc, the pole lure will recall a falcon from about five times as far as will a hand lure.

These lures can also be used for accipiters and buteos provided that you let them have it first time. But the commonest lure for them is the

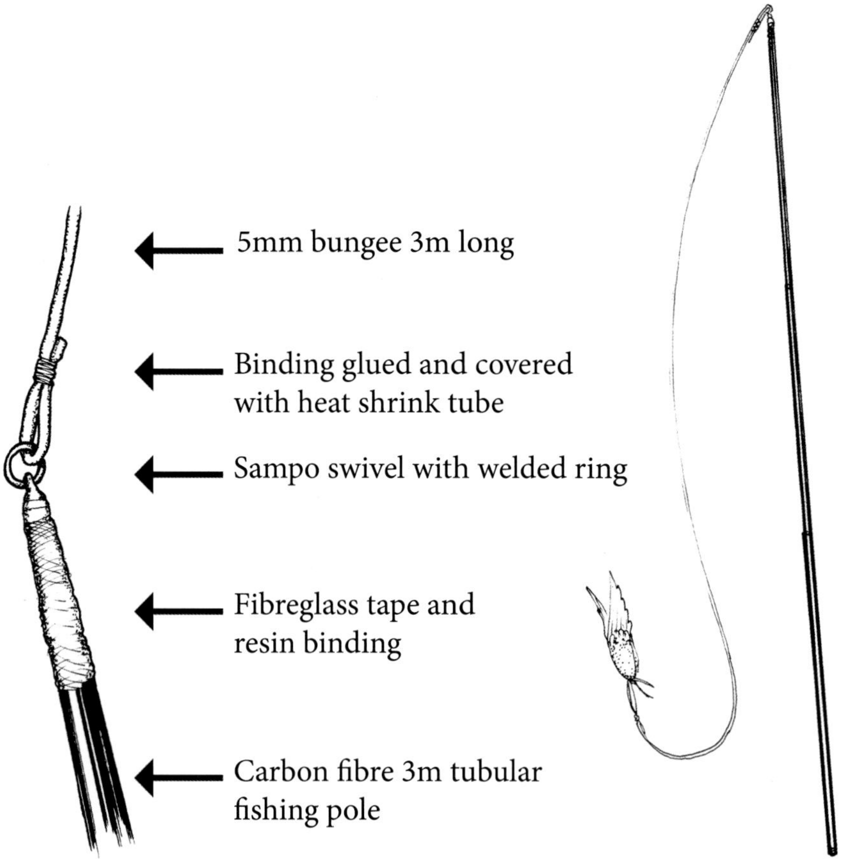

3.11.2 A pole lure.

'dummy bunny' which is a dog retrieval dummy covered in rabbit skin or fake fur. By attaching one of these to a motorised line you can create chases for them, the same as in terrier racing, with a zigzag course. Lure-towing rigs are available commercially.

Thrown lures are used when you want the falcon to catch the lure going away from you. They can have a short cord and toggle like a dog dummy. For young falcons at tame hack, some of them love you to hit tennis balls out for them, but then they leave them lying all over the place and you need a dog to retrieve them.

The ultimate lure is the robotic prey resembling a chosen prey species. I will cover these in a separate section.

Badly-designed lures containing lead shot, or metal, kill or maim falcons when a stoop is misjudged. Splintered bone ends on dried wings can cause foot injuries. Poor lure technique can cause the line to tear the propatagial membrane. Unhygienic lures can cause foot or gut infections. Single-sided lures land with the meat underneath out of sight. You can get away with these small errors for years. But one day they catch up on you.

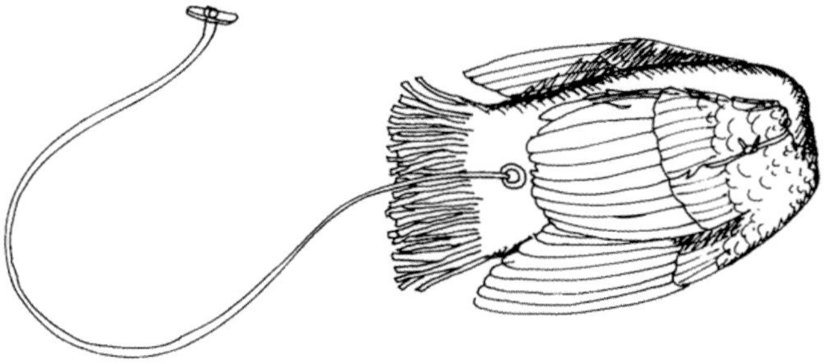

Figure 3.11.4 A throwing lure.

3.12 Creance

This is a light hunting creance. Use a heavier line for a bigger bird and never attach it to anything solid. Thread it through the arm of a coat lying on the ground, or anything that will just drag rather than jerk. People use all sorts of fishing spools and so on to wind up creances, but actually it is very simple. First, the creance line should be bound and glued to the stick so that it cannot come off. Assume that you have your hawk on your gloved left fist, maybe feeding up after a lesson. Hold the creance stick between your fingers and thumb of your right hand, with the line feeding through the fingers of your gloved left hand. This controls the tension you need.

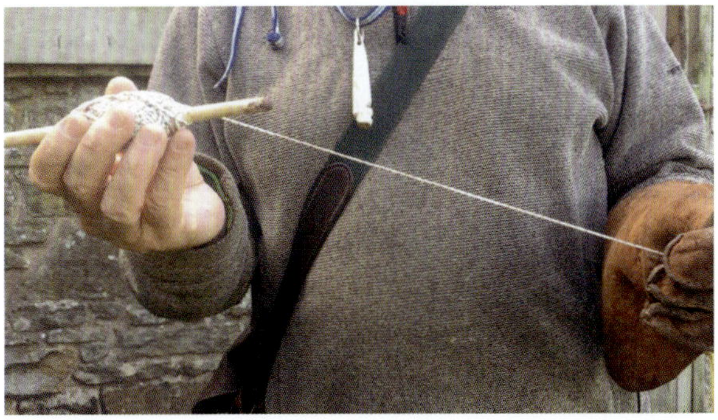

Then hook the bottom end of the stick away from you and back over the line, pulling the line towards you and rotating your grip so that each loop lies further on than the one before. Next push the top end of the stick away from you and back under the line.

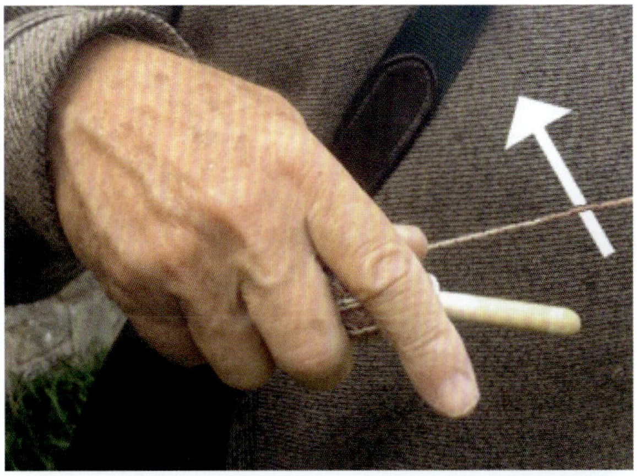

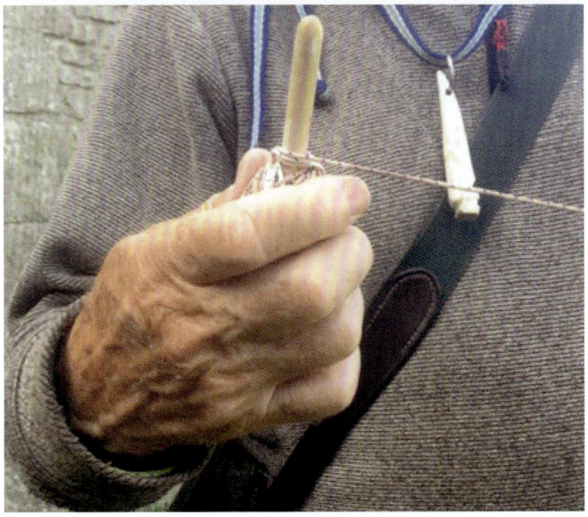

Now you are back where you started. Each move gathers up some line and you just keep going until it is all rolled up. The lure line is wound up in the same way.

When we are training several hawks, the easiest way is to have a running line creance. Stretch a strong braided line between two posts, 100m or so apart, aligned along the training field, and preferably the usual wind direction. Make it tight so that it is off the ground and keep the grass mown all along it. Have a bit of foam at each end of the line as a buffer. Some foam pipe lagging cable-tied on works well. Have a strong steel ring about 4-5cm (2 inches) running on the line with a 15m (50 ft) creance securely tied to it. This enables the bird to fly up and down without much drag on the ground and without swinging off into nearby trees etc. If you think your hawk may shear off, you can make the side creance shorter, only a metre or so, which forces her to go where you want. This is the Arabic *gulss*. But a longer side line allows the hawk to fly more freely, taking her own flight line approach, so that she hardly realises that she is tethered.

Creances can cause injuries. Hawks suffer sprained joints or a broken tibiotarsus when pulled up too abruptly. Hawks get tangled up in trees and strangle themselves or cut themselves with the thin line running over their legs or wings. Hawks escape with the swivel on when the creance has been tied to the swivel, not to the jesses. There are even people still using unsecured dragged creances. One peregrine flew off with a 50 metre creance, never to be seen again in this life!

Another type of creance is the *Zai Wo Zi*, a traditional Chinese lightweight creance used for throwing Sparrowhawks to hunt small birds. I have hunted this way in Beijing and in Hohhot in Chinese Inner Mongolia. The fine braided line is first wound in a figure-eight fashion around a wooden plug. This is then pushed into a small wicker basket strapped over a cotton glove on the falconer's right wrist. The plug is gently removed, leaving the line nestled inside with the loose end emerging from the centre, to be attached to the jesses. When the hawk is thrown, the line flows out freely, about 20 metres (70 ft) long and not attached to the wrist. Most of the hawks are tiny passage Japanese Sparrowhawks *Accipiter gularis* which man down quickly and are released at the end of the season. They don't seem to mind being thrown and the line makes it easier to pick them off the kill if they think of carrying.

The wooden plug used for winding the creance onto.
Photo Alan Gates.

Fitting the cord into the basket.
Photo Alan Gates.

A merlin ready to throw.
Photo Alan Gates.

Our mobile team in Beijing, off to hunt sparrows.

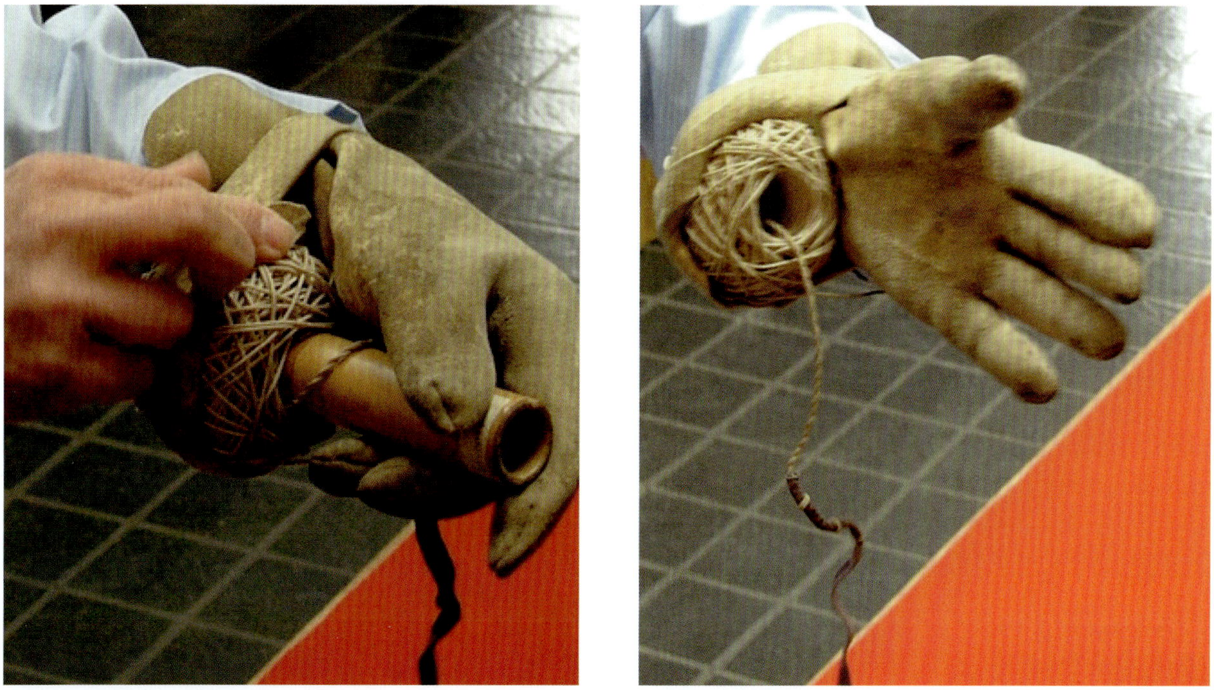

The Japanese version of the *Zai Wo Zi*.

3.13 Field equipment

It is a good idea to get together all the field gear you need in one bag or coat so that when you pick it up, you know it's all there. Alternatively stick a checklist on the dashboard of your car. What gear you take with you out hawking depends on what you are doing. Some people scarcely leave their car, so bits and pieces are kept there. But most falconers must be self-contained and take with them what they need. Some use bags, some use waistcoats or fishing jackets, some use all sorts of pouches. I prefer a bag which has a plain flap covering it, with nothing that can tangle in vegetation. It has no buttons which might snag a lure-line and is high and secure so that I can run fast. Some 'traditional' bags are a liability in this respect. It contains a light 50 metre hunting creance, a minimal recall lure, a pocket for a pick-up piece such as a rabbit leg, and a bigger pocket for a couple of dead crows. Outside I have a hawking knife suitable for killing or cutting, and a magnet for switching on tags. When I slip a hawk, the hood, jesses and swivel go into my coat pocket and my other pocket has a phone and nowadays, a pocket link to the GPS. On my saddle I keep a walkie talkie, a pair of pliers and a receiver with yaggi. Don't leave your yaggi behind! Many a falconer has spent a night out because he left his yaggi back in the car. Spare batteries and maps are kept in the car. Each member of the Field carries a hood, glove and lure in their bags. These lures are simple recall lures with a 2.5 metre cord wrapped round them.

I always carry a lightweight pair of field-glasses from the day the first hawk is on a creance. It amazes me how many people manage without them. Often they don't. During exercise and when crow-hawking by car I take a good pair of Zeiss glasses which are invaluable for assessing the falcons' fitness and for spotting crows.

Some people take a whistle for calling their hawk. A good horn or bone whistle is handy for the dogs or for the hawks. Metal ones stick to your lips in winter. A light crook stick is good for beating and hooks over my left arm when I need my right hand free. Sometimes I loft the hawk over the pointing dog so that she has a good start. When I am training her to do this, I put my hawking glove on the handle.

3.14 Perches

During intensive handling, such as during training, it is often necessary to tether a raptor on a perch. This should be regarded as a necessary evil; as soon as it is no longer necessary, the bird should be free-lofted. It will stay healthier and fitter untethered.

Many accidents arise from tethering. Tethering raptors is not a good idea generally, unless there is a specific reason for doing so. It should never be done as a convenience for the falconer. Think of it in the same way as tethering a dog or a horse; you would not tie them up for more than short periods, and certainly not day after day. Obviously you need to tether during training and flying, and while travelling. But tethering all through the moult is not good for the hawk; better to build her a flight pen and make her comfortable. Some people oppose the keeping of raptors in captivity because of tethering, but, just as you would not let a child or a horse run around in a busy street, there are times when animals need to be kept responsibly under control.

Perches are the source of an unending catalogue of disasters. Birds either escape from them, or get injured by them or develop foot complaints from them. A well-designed and made perch costs little more than a bad one so it makes sense to use the best equipment from the start.

Many tragedies have occurred because perches have been faulty. Bad welding, perches not secured properly so that the bird escapes or attacks an adjacent bird, birds tied to unbarbed staples hammered into wood but which come out, perches light enough for the bird to drag and kill another, these are common bad mistakes. Then there are poorly-designed perches. Some of these designs are still recommended in current books. These perches allow the bird to get tangled up, causing feather damage, sprained joints, or broken limbs. Some, such as the screen perch and the pole perch, cause death overnight on a regular and frequent basis. Rope bound around perches comes slack, causing claws to get caught and wrenched off. Other perches kill by slow attrition, causing bumblefoot to the victim. These include smooth, flat-topped blocks and filthy perches. If you have any of this faulty equipment, go out,

Figure 3.14.1 An unsecured perch allows one hawk to reach and kill another.

Figure 3.14.3 When jesses straddle the block.

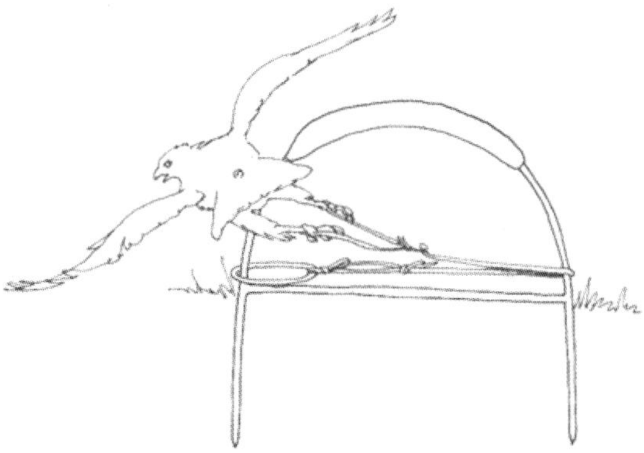

Figure 3.14.2 A badly-designed bowperch.

Figure 3.14.4 When the leash wraps around the block.

smash it up, and bin it. Don't give it to young friends. Don't wait for the accident to happen. Get rid of it. Get rid of it in your mind as well. Don't hang on to old 'traditions.' This gear is dangerous. Give your birds the best available, in the end it is the cheapest.

In 1974, I sent a falcon to an acquaintance who placed her on a brand-new lathe-turned block 'just like in the books'. I advised him to change the polished flat top for something more domed and rough. Human nature being what it is, he didn't want to ruin his smart new block, so he left it. Six weeks later I had a phone call: the bird couldn't stand properly, could she be sent down to me for treatment? The flexor tendons to the hind toes had collapsed. I hacked her free for three months and the condition stabilized, but there was no cure. Foolishly perhaps, I sent her back, and she was dead in a few more months. Her brother, which we kept, lived to be 16 and we now have his great-great-great-great-great-great-great-great-great-great-great-great-great-great grand children.

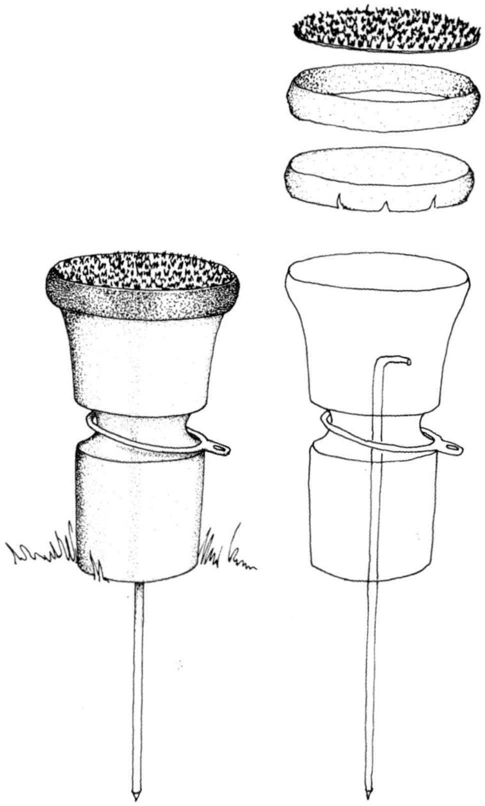

Figure 3.14.5 A falcon block made of concrete, nylon, resin or wood.

Traditionally, falcons are kept on cylindrical blocks which supposedly equate to a rock, while accipiters and buteos are given a narrower perch, similar to a branch. There is no real anatomical or ecological basis for this; if an accipiter prefers a block then use one. Have you asked your bird what she prefers?

I was a teenager in the sixties and rarely wore shoes. I still often go barefoot in the summer months when I can. I can really sympathize with a falcon on a block, and appreciate how comfortable or uncomfortable different surfaces are. We use astroturf on all the perches, for over 300 raptors, most of them large falcons, and we seldom have a bird develop bumblefoot. Some of these birds have had bilateral surgery for bumblefoot in the Middle East before being retired for breeding and they are fine. Bumblefoot is 99% preventable, and it's all to do with perches, diet, and hygiene.

The best descriptions of perches and their construction are given by Kimsey and Hodge. We opt for the very simplest designs with the minimum of moving parts. When the climate is variable, wood tends to split and iron rusts, so it is worth using more permanent materials. Quality plastic and stainless steel perches are now widely available online.

When placed out to weather on a day perch the hawk is very vulnerable to the weather. Every year in hot climates, hawks are killed because a careless owner has left them in the sun, or where they can get into the sun, or where the sun comes round even though the hawk was in shadow to start with. A sequel to this, common in the Middle East, is to place the hooded falcon on her wakr, close to the car in the narrow shadow. Someone then backs the car over her or opens the door on her. All hawks need proper protection from extremes of weather.

The opposite is to leave the hawk tethered near the ground at night in cold weather. The hawk then gets frosted and develops dry gangrene in her toes or manus. Alternatively she is left in a damp place with mouldy material nearby in the shape of rotting orchard fruit, hay, straw, lawn clippings, mushroom manure, shredded wood bark, or old sacks. Never use a barn or stable which has had hay in it, or take a hawk into a room which smells musty. The result can be aspergillosis.

Figure 3.14.6 A good bowperch for an accipiter.

A compromise to this is the trolley system in which a cable is stretched tightly low to the ground between two perches several metres apart. One perch is in a sheltered porch and the other is exposed to the sun. So the raptor has a choice throughout the day and also can get a little bit of flying.

Eagles, buzzards, harriers, goshawks, owls, crows, magpies, polecats, mink, ferrets, cats, foxes, dogs, badgers, raccoons, and snakes have all killed hawks out weathering. A black eagle was killed and swallowed at night by a python, a goshawk killed by a genet cat. Cattle may break in over the fence. A trained hawk flying nearby may attack a tethered one, and humans may injure, frighten, or steal a hawk. Two people I know, who leave their hawks out all night in open porches, have both lost four in one night to foxes and yet both still leave their (remaining) hawks out. Many a hawk placed under a shady tree has been killed by a falling branch. Often, when on hawking trips, hawks are put out to weather on unprotected lawns; in this case take some electric sheep netting with a small power pack, to keep off strange dogs and unwelcome visitors. Designs for safe weathering areas are given by Kimsey and Hodge and they are essential for the person who has to leave his hawk unattended while tethered. In many countries enclosed weathering areas are compulsory.

The worst predator of hawks is the ferret. Ferrets are notorious escapologists and if they get out of their traveling box in the car, or out of their hutch at night, they will quickly kill even a large goshawk. If ferrets are kept at the same establishment as hawks, there should be a fail-safe system of keeping them apart.

During the day, depending on the weather, the hawk should be offered a bath and drinking water. But there is no need to be too fanatical about this. If you offer your hawk a bath in the morning and she doesn't use it within half an hour or so, remove it. Otherwise it will just be left there day and night, collecting mutes and making the grass underneath go yellow. Most baths are too small or too shallow. Wild hawks often bath in water out of their depth and many is the time I have trotted downstream as my goshawk has had a thorough ten-minute bath floating down a river. Young hawks will try to bathe anywhere, including warm fresh cow pats! The bath water should come halfway up her chest. Shy or stressed hawks will not bath except in natural type conditions in privacy and it is quite easy for a rehabilitating hawk to become dehydrated standing right next to a pan of water. The desert falcons and kestrels often like a dust bath and some fine dry sand can be offered in a tray. I have seldom seen a bath used in the Middle East. Spraying a hawk with water cools her down and quietens her, but is not a substitute for a good bath. Many hawks learn to drink from a cup or bowl, which is useful out hawking on a hot day. A handful of pea gravel as rangle (see 6.5) is often appreciated by tethered raptors.

3.15 Night quarters and temporary quarters

While raptors are being kept tethered, they need somewhere absolutely safe as night-quarters. In moderate climates they can stay out in secure sheltered weatherings, but in cold or wet weather they must be brought in otherwise they could suffer illnesses caused by damp or frost. In the wild, most raptors, except harriers and some owls, roost high off the ground in an airy place sheltered from the prevailing wind and from ground predators.

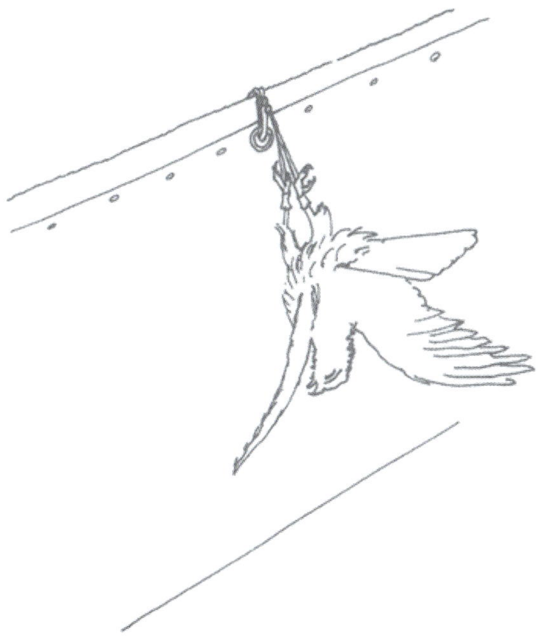

Figure 3.15.1 The screen perch in use.

Screen perches such as these are used during the races in the Gulf.

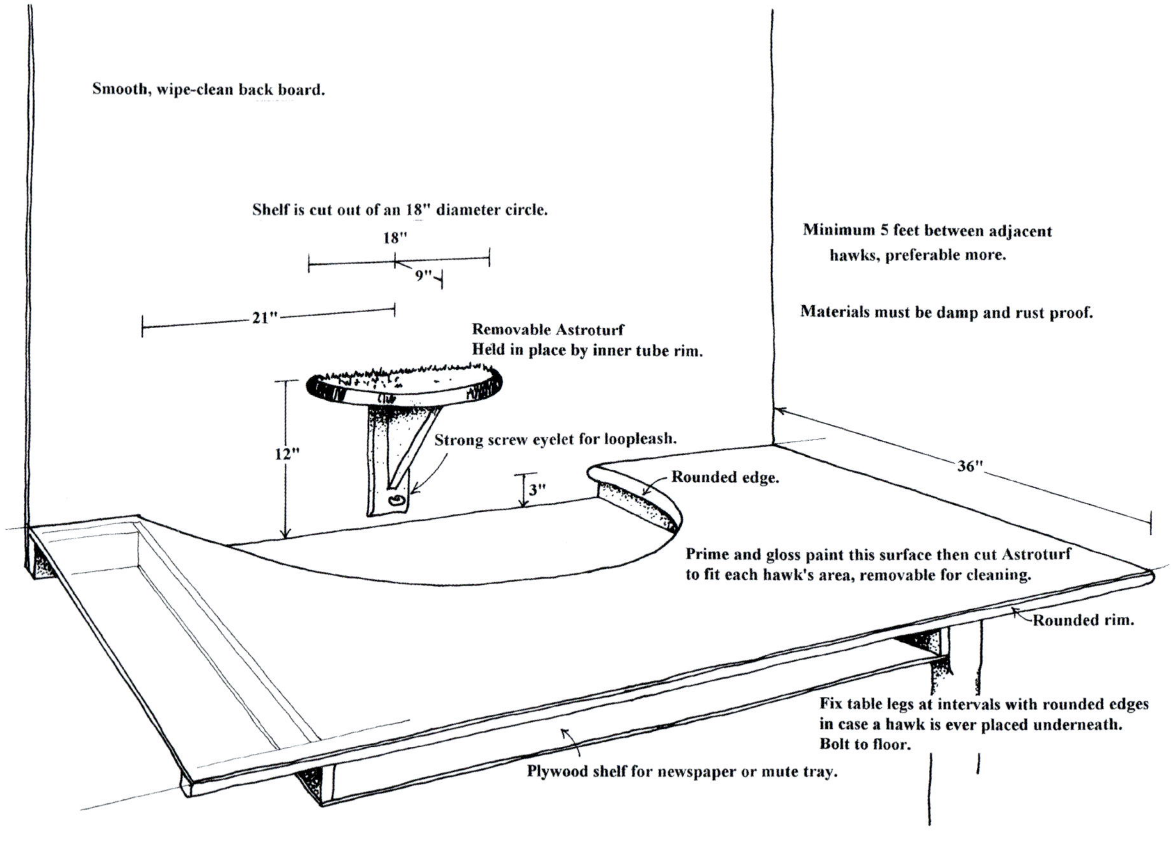

Figure 3.15.2 A simple shelf perch design.

3.15.3 A simple shelf perch.

The screen perch (figure 3.15.1) has killed many hawks, including my old hawk-eagle 'Gorgon' who was killed within two days when I lent him to a friend with a screen perch. The best alternative for falcons is the individual shelf perch (figure 3.15.2 and 3). It is important that all the measurements are exactly right. The measurements in 3.15.2 are for any species of falcon on a loop leash 16 inches long and jesses 6 inches long. For the first few days of training, while the falcon is still jumpy, she is kept hooded at night.

Accipiters can also be placed on the shelf perch but tend to be decorative with their mutes when facing the wall. For them we use a long perch (figure 3.15. 4). The perch is covered in astroturf on a batten wrapped in plastic so that it is all sterilizable. A removable polythene sheet runs down the back wall and across the shelf. Newspaper or corrugated cardboard soaks up the mutes and is changed daily. Some accipiters take to this readily, but others are hopeless and break tail feathers. For sensitive hawks the best plan is a totally darkened mews, possibly a roomy, airy box. They may feel more secure like that, and settle better.

Our mews for 20 falcons in training.

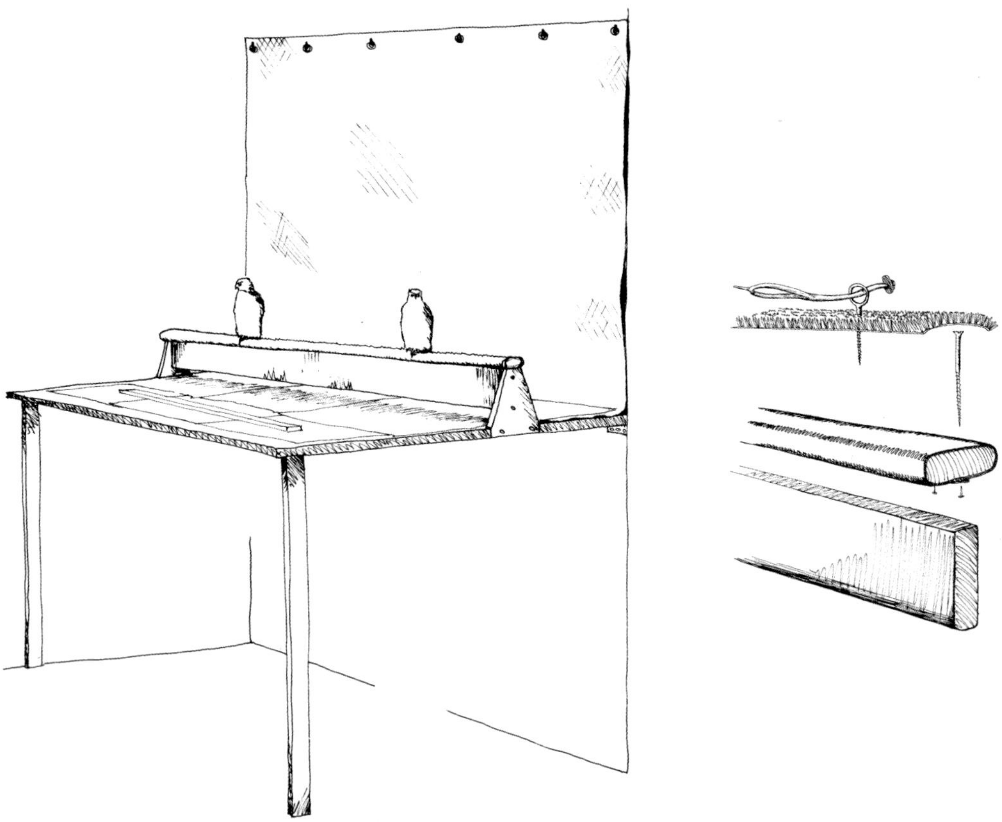

Figure 3.15.4 A safe alternative to the screen perch with (inset) construction detail.

Figure 3.15.5 A temporary night perch.

If your plan is to free loft your bird, a pen similar to the imprint pens shown in 2.8 is a good start. Remember you need a double door system. Obviously it needs to be airy and hygienic but you may also wish to design it with lights so that you can control the photoperiod during the moult.

Both the shelf perch and the long perch can be designed to fold down against the wall. A garage can thus accommodate several birds during the flying season but be returned to normal use for the rest of the year. When away from home, it is often necessary to use various sheds as temporary night quarters. Years ago grouse hawking at Forsinard, our mews was the waiting room of the railway station. But making-do is when accidents happen. The commonest is that one hawk kills another.

Figure 3.15.5 shows a temporary arrangement which is fairly safe, provided that no predator can enter the room. Two squared blocks of wood are nailed from below onto the underside of a pole. This stops it rolling and raises it off the ground so that newspaper can be slid underneath. The bark and any soft wood is blazed off with an axe so that a staple can be hammered in securely. The best staples are the large barbed ones. The staple attaches a Sampo swivel to which the hawk is attached by a short loop leash. If there is any doubt about the staple, tie the Sampo with a cord right around the perch.

It is never a wise move to tether a hungry accipiter in the same room as other hawks. She will be bating to kill them and they will be uneasy. Remember your Shakespeare:

Neither the king, nor he that loves him best, the proudest he that holds up Lancaster, dares stir a wing, if Warwick shake his bells.

3.16 Travelling arrangements

During trips, accipiters are best kept in boxes during the night. In Britain it is illegal to keep a hawk in a box too small for it to spread its wings except when transporting it. Car hawk boxes are a very sensible arrangement and the hawks love them. Most will bate to get into their box and often will rouse and tuck a foot up before you have finished securing them. These traveling boxes should be well-ventilated and cleaned out daily. Hawks can easily over-heat or dehydrate in the back of a car, even when the driver is pleasantly cool. Make sure you have a fail-safe system so that you do not have an awkward moment when the door is open but the hawk is not secured. She could fly out suddenly with her swivel on.

For travelling a team of falcons in the Hawking Van we have plastic perches with dividers and a clear panel so that the driver can see all the falcons at all times, as well as see through the rear window. Each falcon is clipped in with an extensible cord to the clip. Eight falcons can be safely carried that way, with a dog or two below. There are many excellent designs available for

Mrs Arthur Newall, circa 1887.

travelling hawks in cars. They are convenient and safe. There is only one thing worse than a warm goshawk mute down your neck while driving – one down your spouse's neck!

Hawks on traditional cadges in cars should always be physically separated and attended in case they crab. Boxes or partitions are a better solution and also discourage interference from the public. Boxes should be kept clean and dry and free from mould or ammonia fumes.

One person went grouse-hawking on his own and forgot where he left his cadge on the moor. He spent the rest of the day frantically searching for it. Hawks should never be tethered in the open when other hawks are being flown; this is common at public displays and meetings. Sooner or later a free hawk will bind to a tethered one and small accipiters and falcons rarely survive this.

Traveling boxes are especially useful in that hawks can be left unattended in them for several hours provided the weather is cool, and that inquisitive people can't see in. Traveling is a source of accidents and kills many hawks. When traveling by car, hawks should not be placed in the boot (trunk) where they can succumb to petrol or exhaust fumes at much lower levels than those tolerated by dogs or humans. Nor should a mews or aviary be sited next to a garage where fumes can reach the hawk.

If falcons are travelled hooded on long-perches, they must be attended at all times. A variety of designs are given in Kimsey and Hodge which is worth obtaining before you start. Traveling boxes should always be darkened with no mesh or sharp objects inside. If it is warm, each box should have fan-assisted airflow. Many welfare or rehabilitation organizations still transport injured wild raptors in weld mesh cat cages, which is at best disgraceful. The approved airline (IATA) design for raptor traveling boxes is not much better, with drinking and food troughs and various access holes. Boxes should be strong enough to keep hawks in and tea-breaking airport staff out. Consult the latest IATA regulations for box specifications. Musty hawk boxes are a major source of Aspergillus infection.

The clip on the bungee stretches to allow you to attach the hawk. Once the hawk is on the perch the white toggle clips into the slot to keep the clip firm at the front of the perch. The round bar supports the hawk's tail so that she doesn't rub it on the astroturf.

Wooden and plastic air-shipping boxes. Wooden boxes are for single use and can be re-cycled.

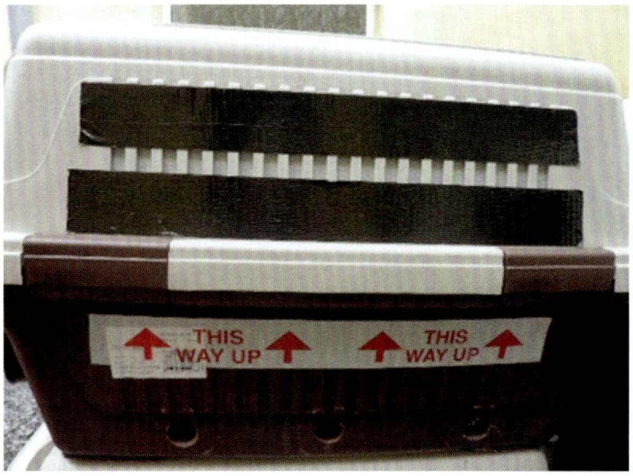

A plastic hawk shipping box. The upper vents have been half-taped on each side so that air can still flow but the hawk cannot see out. Extra vents have been drilled lower down. The mesh door has been blanked off with plastic. A temperature data logger records what temperature the hawk has experienced on the journey and a 'Shockwatch' records any rough handling.

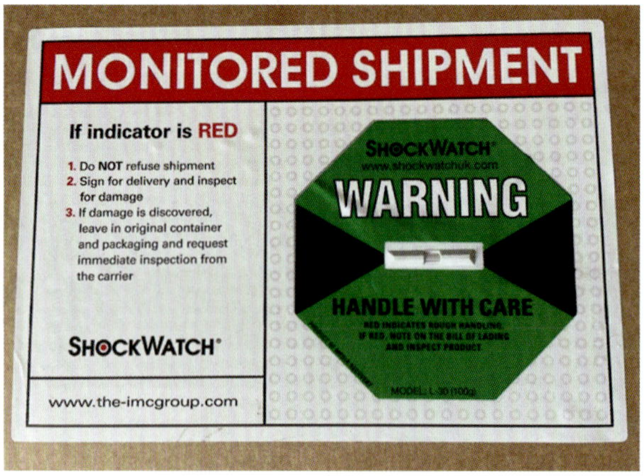

One of the handiest items of equipment to have is a large plastic portable kennel, designed for shipping dogs by air. With a mat of astroturf on the floor, small ones can be used for air-freighting raptors or moving injured or sick raptors and they can be used as overnight observation boxes for sick raptors needing quiet and warmth. They take down into two halves for cleaning and storage. The size that will take an Alsatian dog make good hospital cages for birds undergoing prolonged treatment and needing to be kept quiet, for example, when mending a broken leg.

A hospital box.

3.17 Casting jackets

Many years ago we used to wrap a towel or fine cloth around a hawk to cast it. On one occasion a hawk reached out blindly and bit the young lady holding it on her nipple, and wouldn't let go. What is a polite Englishman to do? Once the situation had been eye-wateringly resolved, I designed a casting jacket and these are now commercially available.

They have elastic and Velcro straps so that the jacket can be quickly secured and the hawk held not too tightly, but unable to wriggle out because the top strap sits above her shoulders. A large loose collar allows the hawk to breathe but protects the handler from biting. These casting jackets should only be used for a few minutes while fitting equipment or imping etc because the bird is completely unable to thermoregulate and, especially if she struggles a bit, she can rapidly over-heat. Most birds in a jacket feel the constraint and lie relatively quietly, whereas if you hold them with your bare hands they feel as if a predator is grabbing them. Then they struggle and the handler reacts by clutching the hawk

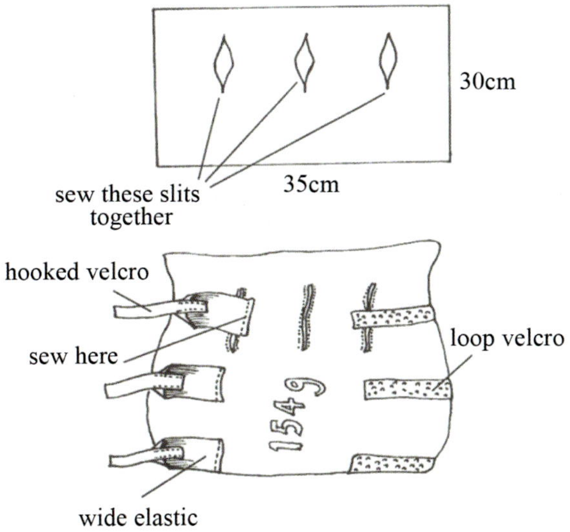

Figure 3.17.1 Casting jacket made of soft material suitable for 1 kg raptor.

more tightly. This can lead to unintentional injuries.

If you are working on a wing, fit the jacket so that the wing protrudes between the straps but the body and other wing is restrained. Feet can be taped up but it is not usually worth it for just a short space of time. The main danger is of a foot snatching blindly and puncturing itself or the other foot. If the hawk has just been lifted off someone else's gloved fist, the glove can be taken off with the feet are still clenching it, and she can keep hold of it throughout the operation.

There are other kinds of restraints or brails that hold the wings but allow the bird to stand up normally. The simplest design is a soft leather tape with two slits in it for the wing butts. The two ends are tied carefully. Check out Ken Riddle's *The Art of Falconry from Arabia Westward* for information on the cloth *guba, guddi* or *abba*, and Jim Nelson's *Hoods, Hooding and Hoodmaking* for modern alternatives.The Pakistanis use the guddi a lot for early manning of their passage goshawks. The Chinese and Japanese have their own designs of brails. The guba is cooler for the falcon, but takes longer to put on. Socks or sleeves are not suitable for travelling. In one case in Pakistan falcons were transported in sleeves for three days. Unable to mute properly, they got kidney failure and died. This is very common when falcons are smuggled and concealed in luggage.

When you first catch a falcon, such as this haggard saker in Mongolia, gently hold the wings, tail and legs together without putting pressure on her body. Lay her along your arm to take her weight, then tuck the chinstrap of the hood under her beak and slip the hood over her head. This will keep stress or injury to a minimum.

Falcons being smuggled in Pakistan.

Releasing a smuggled falcon from a guba. Its eyes had also been stitched.

A Chinese guddi for goshawks.

3.18 Daily records, routine, and hygiene

While a hawk is tethered, it is normally either in training or being flown daily. A record chart is important to monitor progress from day to day and from year to year. I sometimes refer back 15 or more years. An example of a simple basic chart is given in figure 3.18.1 and a blank is in the appendix. Charts are invaluable when you are flying several hawks at once, or when more than one person is flying a bird, or when teaching a young falconer. On hawking trips, I keep a hawking diary with names of visitors and descriptions of flights, but this is all icing on the cake. Your daily chart is a working tool.

You should also have a daily and weekly routine to keep everything clean and tidy. Saturday or Sunday morning is a good time to have the weekly thorough clean out, removing all soiled material, disinfecting perches, checking for faulty equipment and generally setting everything fair. Young handlers of any animals should learn this basic discipline right from the start, before they actually take responsibility for a bird themselves. As you sit and read this, would you be happy to show me round your birds right now, or would you need to do a quick clean up before I came?

Some years ago I was at a nephew's wedding in Ireland and, having a couple of hours to spare, had a little drive around the country. At a very pleasant estate by a lough, I noticed a sign saying 'Falconry Centre' so I drove in to have a look. A nice young lady came out to greet me, took one look and disappeared again. Soon a whole gaggle of staff arrived. Consternation all round! Apparently the owners of the centre (which was very tidy) had been using the first edition of this book to train their staff and used to quote the above paragraph to them. "Imagine what Nick Fox would say if he came here and found this mess!" I hastened to assure them it was purely a fortuitous visit and that all that was needed to make things perfect was a cup of tea…

Routine cleaning will help prevent disease and keep birds healthy into old age. As part of your equipment you need proper poultry grade disinfectants, hoses, scrubbing brushes, possibly a pressure washer, fumigation materials, and a basic first aid box for both raptors and humans. You also need proper facilities for storing and preparing food. There is no veterinary section to this book because the emphasis is strongly on prevention, rather than cure. Your job is to prevent illness and check the birds carefully so that, when something does look worrying, you can contact a specialist vet promptly and with good back-up information.

So far in the book we have looked at the bird itself, breeding a youngster, and assembling the equipment you need before you can start any hands-on training. We still need to look at how the bird's mind works before we are ready to take a raptor on the fist.

3.19 Further Reading

Beebe, F. L. and H. M. Webster. 1994. *North American Falconry and Hunting Hawks* 7th edition. Denver, CO.

Beebe, F. L. 1992. *The Compleat Falconer*. Hancock House, Surrey, BC. ISBN 0 88839 253 2.

British Falconers' Club: Journals 1937 - present.

Dixon, A., D. Ragyov, G.Purev-Ochir, M.L. Rahman, N.Batbayar, M.W. Bruford, and X. Zhan. 2016. Evidence for deleterious effects of harness-mounted satellite transmitters on Saker Falcons *Falco cherrug*. Bird Study 63(1):96-106.

Ford, E. 1992. *Falconry, Art and Practice*. Blandford, London. ISBN 0 7137 2248 7.

Glasier, P. 1978. *Falconry and Hawking.* Batsford. ISBN 0 7134 0232 6.

Kenward, R.E., R.H. Pfeffer, M.A. Al-Bowardi, N.C. Fox, K.E. Riddle, E.A. Bragin, A. Levin, S.S. Walls and K.H. Hodder. 2001. Setting harness sizes and other marking techniques for a falcon with strong sexual dimorphism. Journal of Field Ornithology 72: 244-257.

Kenward, R.E. 2001. *A Manual for Wildlife Radio Tagging*. Academic Press.ISBN 0-12-404242-2.

Kimsey, B. A. and J. Hodge. 1992. *Falconry Equipment*. Kimsey/Hodge Publications, Houston, TX.

Lewis, M.J. and I. Richardson. 2019 *Inscribed Vervels: A corpus and discussion of late medieval and Renaissance hawking rings found in Britain*. ISBN: 9781407316789

McElroy, H. 1977. *Desert Hawking II*. Privately printed.

Morel, P and D. Horobin. 2018. *Hawk's Vervels*. Blurb books ISBN 9781389040245.

Musée International del la Chasse – Gien. 1994. *La chasse au vol au fil des temps*. Privately printed.

North American Falconer's Association: Journal 1961-present.

Nelson, J.W. 2016. *Hoods, Hooding and Hood-making*. Western Sporting. ISBN 978-1-888357-18-9.

Parry Jones, J. 1994. *Training Birds of Prey*. David and Charles, Devon. ISBN 0 7153 0142 X.

Peels-Mollen, J. 2020. *Karel Mollen: the Last Falconer*. Coch-y-Bonddu Books.

Rayner, J.M.V. 1988 Form and function in avian flight. In Johnston, R.F. (ED) Current Ornithology volume 5. Plenum Press.

Riddle, K. 2009. *The Art of Falconry from Arabia Westward.* Motivate Publishing. ISBN 978-9948-01-308-2.

Sherley T. 1603. *A Short Discourse of Hawking to the Field.* Reprinted by the Archives of American Falconry 2004. ISBN 0-9619839-4-9.

Upton, R. 1991. *Falconry, Principles and Practice*. A. & C. Black. ISBN 0 7136 3262 3.

Videler, J.J., G.Vossebelt, M. Gnodde and A. Groenewegen. 1988. Indoor flight experiments with trained kestrels 1. Flight strategies in still air with and without added weight. Journal of Experimental Biology 134:173-183.

C4 Development and Behaviour

This chapter is supported by a film in the Bird of Prey Management series:
Behaviour and Learning.

A female calidus peregrine with satellite tag.
Photo Andrew Dixon

4.1 Physical development

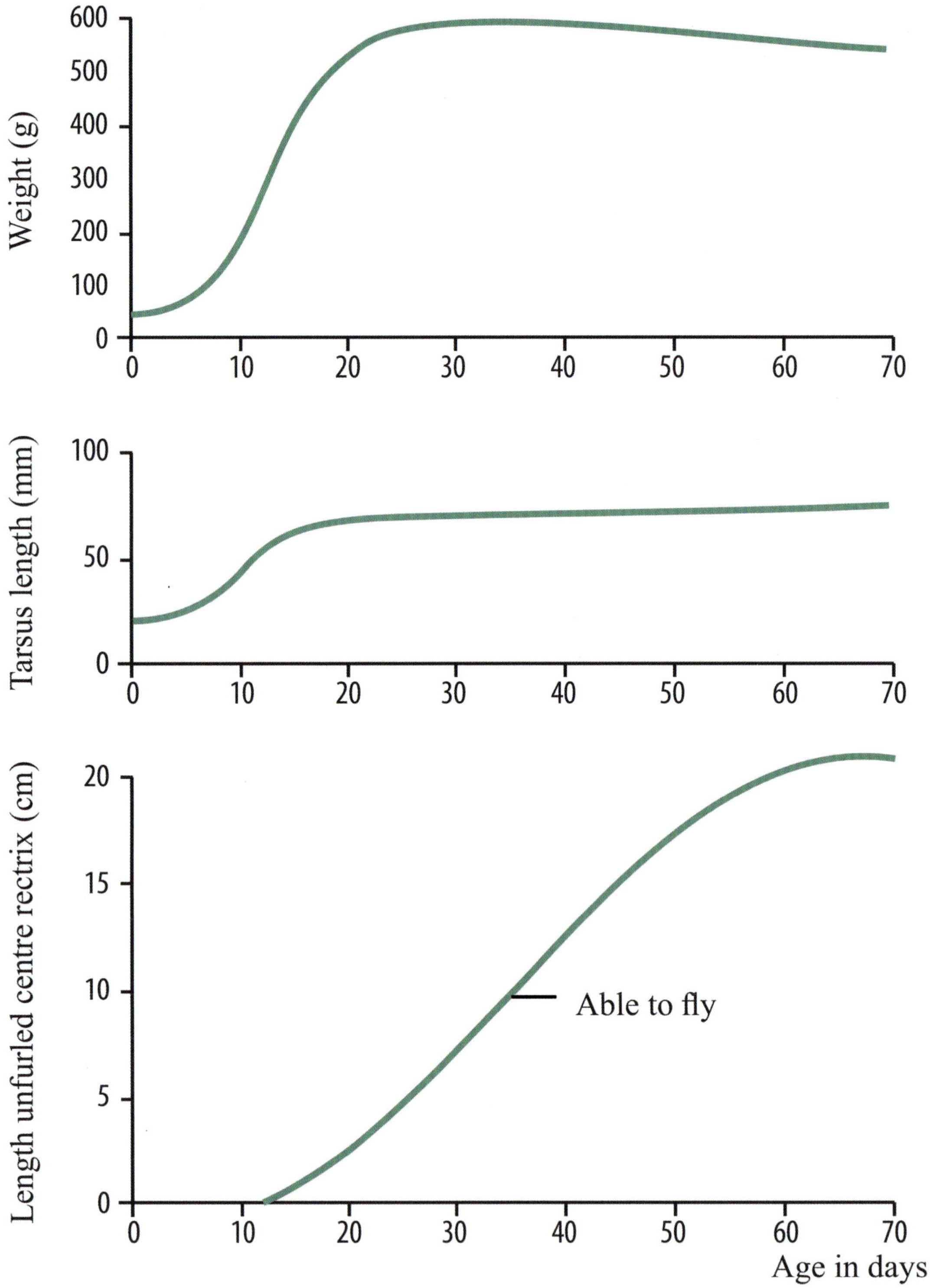

Figure 4.1.1 Physical development in New Zealand falcons.

Physical growth follows two stages. First the chicks energy is devoted to bodily growth and then, as she nears full size, her energy is directed into growing her feathers. Newly-hatched raptor chicks are covered with a fine coat of natal down, dense in species from cold climates and sparser in the hot desert species. The occipital muscles at the back of the head, which are used for pipping the shell of the egg, are still swollen and the chick barely has strength to lift her head. Often the eyes are closed. The chick responds to the soft calls of her mother by raising her head briefly and gaping for food.

During the first week the chick rapidly gains weight. Although the yolk sac is used up after about three or four days, the abdomen grows tremendously to cope with the food, intake and the chick is virtually a living gut, a food processing machine.

After 7-8 days for small hawks, and 12-14 days in large hawks and falcons, the second, woollier down coat grows through the natal down and enables the chick to control her own temperature and survive comfortably for short spells without brooding. This frees the mother for hunting and she is able to assist the male in catering for their almost insatiable appetite. The parents spend most of their time obtaining sufficient food. During this period the long bones of the body are growing rapidly, as is the whole skeleton. By about 18 days for small hawks, and 22-26 days for larger species, the chick has reached full adult weight. Any deficiency in the quantity or quality of the diet during this growth stage will result in a slowing or stunting of growth and may also result in porosis, rickets, or other deficiencies, some of which are lethal or irreversible.

As the long bones begin to harden, the bird begins to look more hawk-like and may be able to shuffle about or take a few faltering steps. Now her efforts are put into feather growth and the primaries and tail feathers are sprouting well. In a few more days the chick is a harlequin of feathers and down and then, quite soon, she appears fully feathered. Any dietary deficiency at this stage will cause stress marks on the growing plumage. Although the chick may appear fully feathered, the flight feathers are still short and heavy with the sheaths still in blood. Much time is spent in preening, playing, and wing exercising with bouts of snoozing in between.

At this time of bone and feather growth, the chick's senses and brain are also developing. She can see and hear well and recognizes her parents. This is the most sensitive period for the various aspects of imprinting (see section 4.7). How she is treated both physically and mentally during these brief weeks will profoundly affect her for the rest of her life.

By four weeks old, small hawks are starting to fly. Larger species follow at 5-6 weeks. But it is some further time before they are hard-penned. Feathers which appear to have lost their sheaths may still have soft shafts inside the feather sockets and it is best to keep the youngsters well fed for a week or so beyond this point. The skeleton too is still calcifying and the joints firming up. The heads of the long bones are still not properly fused to the shafts. Newly-fledged hawks should be handled very carefully to avoid excessive strains or violence. Young accipiters, if tethered, may not only sprain a limb but also may damage their leg scales which are still soft at this time.

Newly-hatched saker chicks in Mongolia.
Photo Andrew Dixon.

Saker chicks at 7 days old.
Photo Shane McPherson.

Saker chicks at 14 days old.
Photo Andrew Dixon.

Saker chicks at 21 days old.
Photo Dimitar Ragyov.

Saker chicks at 28 days old.
Photo Andrew Dixon.

Saker chicks at 35 days old.
Photo Andrew Dixon.

Saker chicks at 42 days old.
Photo Andrew Dixon.

Modern methods of domestic breeding and rearing chicks have resulted in a high success rate in the physical side of development. Most of the problems which arise are on the mental side and therefore most of this part will be devoted to behaviour and its origins. It is much easier to diagnose a physical problem than a mental one. Before we can go on and train a hawk, we have first to understand the bird's natural behaviour and the factors influencing it.

Figure 4.1.2 Newly-fledged youngsters chase their parent for food.

4.2 The components of behaviour

While the bird is growing physically, developing its *phenotype*, (the shape and form we recognise as a species), its *genotype* is also programming its mental development which collectively we call behaviour. The behaviour of a bird of prey is the product of three components and their interactions. The first component is innate or genetically inherited and as such is virtually unalterable. It is its suite of instincts which collectively make up its *ethotype*. The second is imprinted on the bird in early life and is largely unalterable thereafter, but the third, learned behaviour, can be modified during the life of the bird. It's a bit like a computer: the hard drive (ethotype) is what it is, according to the specifications that you bought from the shop. You can load it up with software as you wish (various ways of imprinting) and then you can enter data (training and experience). For those rearing or handling captive raptors, who may wish to rear a bird suitable for release to the wild, or one that is optimal for falconry or for breeding, it is essential to understand what can, and what cannot, be done.

These are the components:

Inherited behaviour patterns (ethotype)

1. Specific instincts
2. Individual temperament
3. Biological urges and drives
4. Memory

Imprinted behaviour

1. Imprinting on parents
2. Imprinting on siblings
3. The fear response
4. Imprinting on future sexual partner.
5. Imprinting on environment on nest type, site, and habitat

Learned behaviour

Involuntary learning

1. Habituation
2. Classical conditioning
3. Operant conditioning.
4. Association
5. Trauma learning

Voluntary learning

1. Insight learning
2. Experience

4.3 Inherited behaviour

Every raptor is hatched with its own individual biological program, a legacy from its forebears who themselves, generation after generation, have been among the elitist percentage which has survived long enough to breed. Their program has three separate but interacting components: specific instincts, individual temperaments, and biological drives. Each bird too has a different capacity for storing information in its memory.

This genetic or inherited component has the potential to make the bird behave in certain ways, if and when the appropriate circumstances arise during the bird's lifetime. In captivity, these circumstances may not happen and so the bird may never express that particular behaviour, or, alternatively, it may express it in a different or displaced way which may be puzzling to the handler who does not understand the origin of the behaviour.

For example, the Peregrine, *Falco peregrinus*, is a packet of certain genes, its *genotype*. Using gene sequencing techniques we have been able to specify the entire genome for the species. A multinational team of us (Zhan *et al.)* published this in 2013.

So now we have a complete list of all the genes in a peregrine, that make it look like a peregrine (its phenotype) and behave like a peregrine (its ethotype). This opens the door to a whole gamut of studies, but looking into the peregrine ethotype, we could start to analyse a 'migration gene' connected to long-term memory. Our field team, headed up by Dr Andrew Dixon, had already spent several years fitting satellite tags to arctic *calidus* peregrines. We knew their migration patterns and we had their DNA. We knew that the young peregrines didn't follow their parents south, they journeyed alone, so the only way they could know how to navigate was genetically. They had an instinct to take the routes they followed and they remembered for subsequent journeys.

They left Siberia mainly in September, travelled 2,280-11,002 km in about 27 days and arrived at their wintering areas mainly in October. Individuals that we tracked for more than one year used the same routes each time during migration, used the same wintering sites and

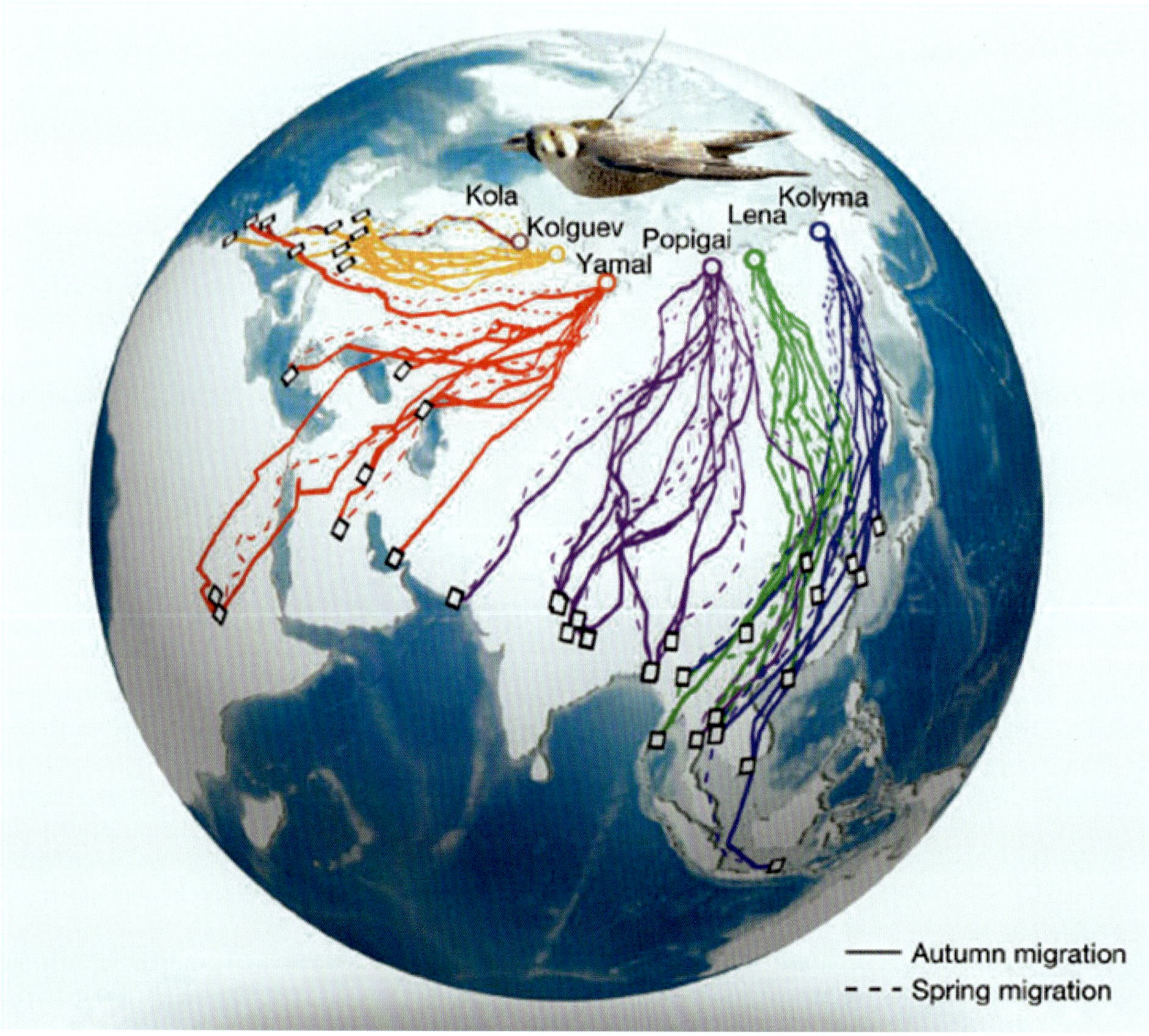

Five migration routes for 56 peregrine falcons we tracked by satellite.
Image credit: Gu et al., doi: 10.1038/s41586-021-03265-0.

nested on average within 5.37 km of where they were raised. We even found peregrines still wearing Arab jesses nesting in the Arctic.

We found that the birds used five migration routes across Eurasia, probably established between the last Ice Age 22,000 years ago and the middle-Holocene 6,000 years ago. Using whole genome sequencing we found a gene, ADCY8, which is known to be involved in long-term memory in other animals, associated with differences in migratory distance. ADCY8 had a variant at high frequency in long-distance (eastern) migrant populations of peregrines, indicating this variant is being preferentially selected because it may increase powers of long-term memory thought to be essential for long-distance migration. (Z. Gu *et al.* Climate-driven flyway changes and memory-based long-distance migration. *Nature*, published online March 3, 2021; doi: 10.1038/s41586-021-03265-0).

4.4 Specific instincts

These are instincts which are characteristic of the species collectively making up the 'ethotype'. The buzzard, for example, quite apart from differences in structure and physiology, is quite different from a goshawk. Its temperament matches its anatomy and is geared toward catching small, slow-moving prey. Similarly, although the peregrine and the goshawk are both rapacious in that they are exclusively predators of vertebrates, their temperament and their approaches to the problems of hunting are quite different. Their courtship and nesting behaviours differ too.

In general, those species which attack active quarry, such as birds, over short distances, tend to live in or near woodland and to have very fast reactions and an impulsive nature. To our eyes they seem to over-react to stimuli and appear fearful. Their nervous system systems are so finely tuned that, as a group, they are the most prone to fits and nervous disorders. It seems as if the reaction from stimulus to response is so fast that it cuts out the conscious part of the brain which therefore gets small chance to identify and make a decision about the stimulus. Thus the goshawk may see a small movement and impulsively launch an almost immediate attack. Only when the bird is actually in flight does its brain process the signal, identify it as a piece of uninteresting thistledown, and abort the attack. As well as the accipiters, the forest-dependent hawk-eagles and the *Micrastur* forest-falcons show similar behaviour. In a wooded habitat, events happen suddenly and the successful raptor is the one which can react in time to exploit the fleeting chance before the prey can reach the safety of nearby cover.

Species which attack active quarry over long distances live in the open and their attacks are often premeditated, tactical, and unspontaneous. These species, such as the falcons, tend to have more equable temperaments. Of the large falcons, the peregrine is probably the most specialized predator and tends to have a fairly consistent and uninquiring temperament. This can make them reliable performers in the field. The merlin, the gyrfalcon and the New Zealand falcon, even when several years old, will often spend an hour at a time playing like cats with a twig or a prey item, catching it again and again. It is as if their mental energy is overflowing and needs an outlet. This can be fine if you like that kind of thing, but frustrating when you are trying to hunt.

More of bird behaviour is instinctive than appears at first sight. In many situations, especially those which occur only once or twice in a bird's life and for which it is important that it is done correctly, it is more advantageous for the bird to do them instinctively than to have to learn them. For example, courtship behaviour, copulation, nest-building, incubation and rearing of young are all ethotypic instinctive behaviours. There is a small learning component - parents do learn to improve their feeding of small chicks for example - but by and large, first time breeders have almost as good a chance of producing young as experienced birds have.

It is not always advantageous for the whole program to be instinctive. Some elements may need to be learned, and learned rapidly. This is where imprinting (see 4.7) comes in. Birds learn to identify their own parents quickly and certainly through imprinting and then their instinctive behaviour patterns (food-begging and so on) can take over and make the bird act appropriately. Otherwise the chick could waste its energy begging at a tussock of grass or a passing seagull.

Some aspects of a bird's life rely entirely on instinct. It cannot be learned because the bird has no chance to learn it; it must get it right first time, as in the migrating arctic peregrines we studied.

Other aspects of a bird's life cannot be covered by instinct because they are much too variable. It appears for example that the various searching behaviour and attack behaviour patterns described in chapter 8 are basically instinctive. They appear in a variety of genera in recognizable ethotypic forms and when they are used by a young hawk for the first time they occur as complete sequences, such as the glide attack (8.13) or the ground-skimming indirect flying attack (8.11). Although the young hawk is equipped, according to her species, with her own repertoire of instinctive attack patterns, instinct cannot properly prepare her for the wide variety of prey she is likely to encounter, the prey's evasion tactics and the problems imposed by terrain. She must learn these by experience and use whichever attack styles are appropriate. She also learns of course, to improve her flying skills, her footing ability and her basic attack strategies. These are based on instinct but are improved by learning. The more specialized the hawk species is, the more its hunting behaviour is governed by instinct. The more of a generalist or opportunist a species is, the more its behaviour is shaped through learning, and this usually requires a longer dependency period.

It takes time for a hawk to gain this experience and develop hunting skills. Often two thirds die before succeeding. Only when a hawk has gained sufficient hunting skills to support not only itself, but also its family (an extra hunting load of perhaps 4-6 times), is it fit to breed. There is no point in a hawk becoming sexually mature before its hunting skills are adequate. Large falcons take two or more years to become sexually mature and they are also slow to peak in their hunting skills. Small hawks, such as merlins, kestrels, and sparrowhawks, not only become sexually mature sooner, but also become proficient at hunting more quickly. This may be because for them the direct flying attack (see 8.10) and, in the case of the merlin and sparrowhawks, the tail-chase (see 8.12), are the most important attacks. These are the basic instinctive attack strategies. The more advanced strategies, such as the glide attack and indirect flying attack require detailed experience of the prey's reactions to be developed and this requires a maturation process.

This underlines the evolutionary balance between stupidity and intelligence. A hawk which can live by ethotypic instinctive hunting behaviour is ready to breed before one with a long learning period. The latter may well die before breeding at all. The smaller hawks tend to have a shorter generation interval, breeding earlier and dying earlier. The strategists have a high juvenile mortality but are often long-lived.

Skilled adult strategists can exist in areas where there are perhaps only one tenth of the prey numbers which an unskilled juvenile would need. Most of the attacks launched by experienced strategists are at opportunities which the naive youngster doesn't even recognize as opportunities. It is like the country boy coming to the big city looking for work. There are opportunities all around him but he doesn't realize where to look.

Nesting is timed so that at the period of peak food demand, when there are big chicks in the nest, there is also a maximum food supply. This timing is most obvious in species such as the Eleonora's falcon which depends on migrant passerines. But it is equally important that this supply of naive young prey continues long enough for the young raptors to develop their hunting skills, otherwise they will miss the boat and many will die.

Thus breeding, which has all the major components stereotyped or easily learned through imprinting, is mainly instinctive, whereas hunting behaviour, although basically structured through instinct, has a large learning element so that the hawk adapts to the wide variety of contingencies it has to face to catch prey. The differences in emphasis of these two aspects of a hawk's life - breeding and hunting - are reflected in the failure rate; first time breeders are only marginally less successful than experienced ones whereas inexperienced hunters have a much higher risk of starving than have adults. This selection pressure through hunting success is extremely finely tuned both phenotypically and ethotypically. The peregrine and the lanner, although close, have slightly different hunting methods, and this is reflected in slight differences in their morphology. Have a look at the toe lengths of the two species.

4.5 Individual temperaments

Individual birds of the same species have different temperaments, just as humans do. Some are placid, some are excitable, some are easy to teach, some are withdrawn. Except in cases involving changes in body chemistry, brain damage or genome editing,there is little scope for altering temperament; this is beyond the frontiers of science at the moment.

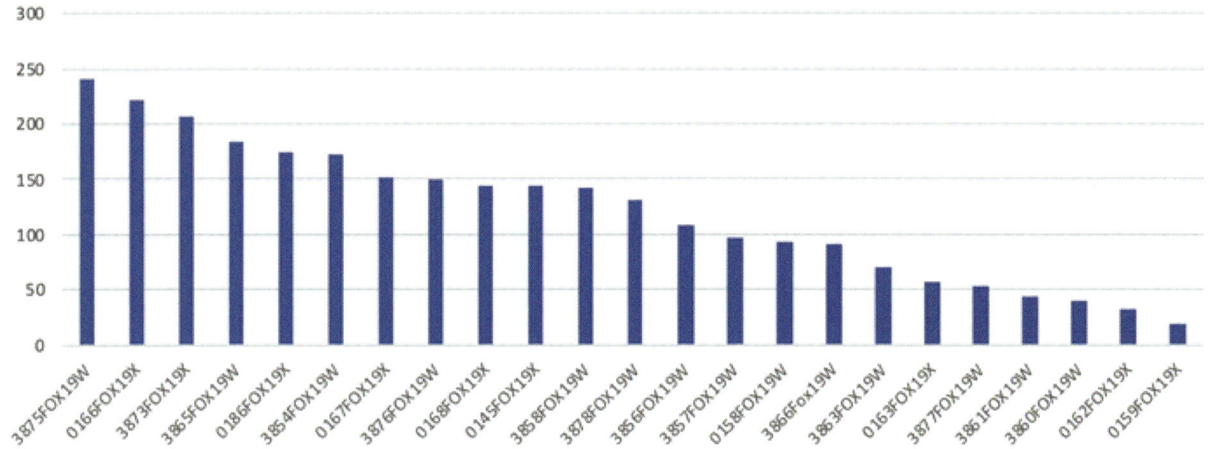

How can we assess temperament, particularly if trying to create a good hunting falcon? A falcon that is a poor hunter is no use; she cannot adapt and become a vegetarian. She is an obligate carnivore. But to find out if a falcon is a good hunter you need to train her and take her out hunting, and, like race horses, that entails a lot of work and wastage. I wondered: could we devise a personality test for young falcons? So I asked one of our staff, Sofia Hurtado Munoz, to set up a monitoring system. When the young falcons are learning to fly we put them in groups of up to 30 birds in free flight pens 25 metres in diameter. They are fed remotely by food falling through a chute onto a food tray so they never see humans and we have one way glass viewing windows. Sofia fitted RFID microchips to the leg of each falcon and a chip reader along the landing perch at the feeder. A datalogger recorded when each falcon visited. Would their food visits be random, or would some falcons be more forward than others, consistently? Sofia created a simple scoring protocol. The results confirmed what we already suspected: maybe a quarter were always in first, half were fairly random, and about a quarter were always reticent, holding back. One male never came in at all – he must have fed at night when the logger was switched off!

This was just a pilot study and the results are marginal statistically. We did three cohorts of falcons. There are several variables, such as species, that need controlling for. Some species, such as the hardy desert falcons, like sakers or prairie falcons, are renowned for their slow metabolism compared to the more voracious peregrines. But, all other things being equal, if you were choosing a male from this bunch, which would you pick?

Of course Sofia was assessing 'dominance' in terms of the first falcons consistently to go for the food. Equally she could have put a choice of perches in the pen and measured which birds dominated the most 'desirable' ones. We don't know if position scoring in our study translates to subsequent hunting performance. To get a useful

sample size is very difficult because there are many confounding variables, such as the skill of the trainer. But maybe one day we will be able to take this study through to its logical conclusion.

4.6 Biological urges and drives

Raptors are not static machines. Changes are going on in their body chemistry as they grow up and as the seasons go by. For example, young hawks chase their parents for food. The parent, rather than live quarry, is the stimulus for them to chase. Quite suddenly, perhaps triggered by a small loss in body weight, or by the parent releasing live prey, the youngsters' programs switch to prey seeking.

Once fully grown, other stages in the biological program occur. Some species are migratory. Abruptly one autumn day the otherwise trained and responsive hawk may fly away, completely over-riding all you have taught her.

Later in life raptors come into breeding condition. The two sexes diverging into different roles, females tending not to hunt but to beg her partner for food, males hunting in excess of their own requirements and tempting the female with food and by courtship displays (see 2.12). Males have to fit in with the breeding cycle and can be triggered by appropriate stimuli. For example, Noel Hyde and Debbie Stewart recorded a wild adult male New Zealand falcon who came down to a pen of Muscovy ducklings and tried to feed them daily for several weeks. He brought in plucked prey to them and, when they didn't eat his offerings, he cached them nearby. Other raptors have 'adopted' chicks of similar raptors or even Herring Gulls *Larus argentatus*.

Certain species, particularly falcons, have a strong drive to store food. One female New Zealand Falcon of mine caught five small birds in a row, caching each one before chasing another. Sometimes this urge to chase and then hide the food seems totally to overcome the desire to eat, and the two activities, hunting and eating, become entirely dissociated. This happens too in house cats which hunt assiduously and bring home countless dead prey but which then refuse to eat them and eat only from their food bowl.

Thus these urges and drives, triggered by environmental stimuli such as increasing spring day- length, together with internal maturation and mental processes, unleash instinctive programs so that the birds relate properly and attempt to breed in a way that is appropriate and adapted to their situation

4.7 Memory

Generally the memory of raptors is good and once the bird has been trained very little is forgotten. Fortunately their sensory world is similar to ours and this is reflected in their memories. Their main senses cover sights (in colour), then sounds, then to a lesser extent taste, touch, and, very little, scent (see 1.4). As in most humans, their comprehension and recall is better for sights than for sounds. Even after more than a year of absence, hawks recognize previous owners and recall the significance of the lure and hood. As predators, their visual acuity and recognition are much better than ours and it takes an inexperienced falconer some years to realize just how much a hawk recognizes. Usually of course the bird does not demonstrate this recognition and it is only occasionally that one gets an insight into it. For example, my old goshawk sitting silently on the lawn on her bow perch, surrounded by my students, used to call to me immediately when I showed just part of my head at the laboratory window. Similarly, hawks recognize places where in the past they have caught prey and will sometimes divert in order to check out the place.

Memory should not be confused with intelligence. You can be very bright, but have a poor memory, or be quite stupid and have a good one. Don't forget that!

4.8 Imprinted behaviour

I define imprinting as: *a distinct genetically programmed learning mechanism in which there is a permanent development, during a specific sensitive period, of an innate imprinted behaviour pattern towards specific objects which thereafter become important elicitors of that behaviour pattern.*

Rather a wordy definition, I know, but it is important to be clear what we are talking about.

Scientists have struggled with imprinting. It is a drawn out process requiring study of an animal often for several years, and to do this, and get meaningful sample sizes, is almost impossible for research budgets and university departments. But the biggest problem is that early researchers, such as Oskar Heinroth and Konrad Lorenz, first documented it in precocial species and got very hung up on the concept of 'sensitive' or 'critical' periods. These are dramatic in precocial species such as chickens, ducks and sheep and were over-emphasised by early workers. The essential thing to appreciate is that it is the <u>permanence</u> of this imprinting that is important. Altricial species, such as raptors, dogs and humans, also imprint just as much as precocial ones do, but their sensitive periods last longer. In humans imprinting continues until the child is about seven years old.

The other aspect to understand that imprinting has several strands. While the duckling imprints on and follows its mother, this is only one element, albeit the most obvious one. Imprinting on your siblings, home and place of birth come into it too. Even, in humans, the ability to pick up languages is best in the under sevens, so is religion. The Jesuit St Ignatius Loyola wrote: 'Give me the child till he is seven years old and I will show you the man'.

Imprinting is a form of learning linked closely to the ethotype and assimilated easily, without the need for repetition and reinforcement and so on. It is uncritical. The duckling can imprint almost as easily onto a human foot as it can to its real mother. And once it is fixed it is very hard to undo.

So untangling imprinting in raptors is extremely difficult and I have struggled with it for almost 60 years. Of the 5,000 or so falcons we have bred here, about 4,600 have been fully naturally imprinted, although many have been conventionally trained once full grown. The other 400 or so have been imprinted in many diverse ways, some solely on humans, and some dual imprinted. Most of these have been kept right through to sexual maturity at 2-6 years old. It is impossible to do all this in a totally controlled manner. About 300 staff were involved over the years, and facilities changed too. But gradually, bit by bit, trends emerged and I am summarising them for you in this book. I could never publish this in a peer-reviewed scientific paper, but just because something is not easily studied using scientific methods does not mean it is unimportant; far from it. Science can be a very clumsy tool. And studying something is just the start; application is the acid test and I will examine applied aspects of imprinting in chapter 5.

In precocial birds, such as chickens and ducks, which leave the nest soon after hatching, imprinting takes place during a critical sensitive period - usually between 13 and 16 hours of age. This rapid identification of mother, siblings, and danger is of obvious survival value to precocial chicks, so that they all stick together when they leave the nest.

Altricial birds, on the other hand, stay in the nest for some time while they develop physically. Most raptor chicks keep their eyes shut for the first day or two and their imprinting is much more protracted-weeks rather than days or hours-and therefore are regarded as having susceptible sensitive periods rather than critical periods for imprinting.

These longer sensitive periods in altricial birds mean there is much more of a chance for things to go wrong and, as this is also the age at which raptors are taken for falconry, various forms of malimprinting are common.

There are five main parts to imprinting in raptors and these overlap chronologically:

1. Imprinting on parents.
2. Imprinting on siblings.
3. Development of fear response.
4. Imprinting on future sexual partner.
5. Imprinting on nest type, site, and habitat.

The susceptible periods for the five aspects of imprinting in birds of prey can be shown diagrammatically. It should be emphasized that these vary from species to species and under different conditions and the exact plotting of them, if feasible, has yet to be clarified.

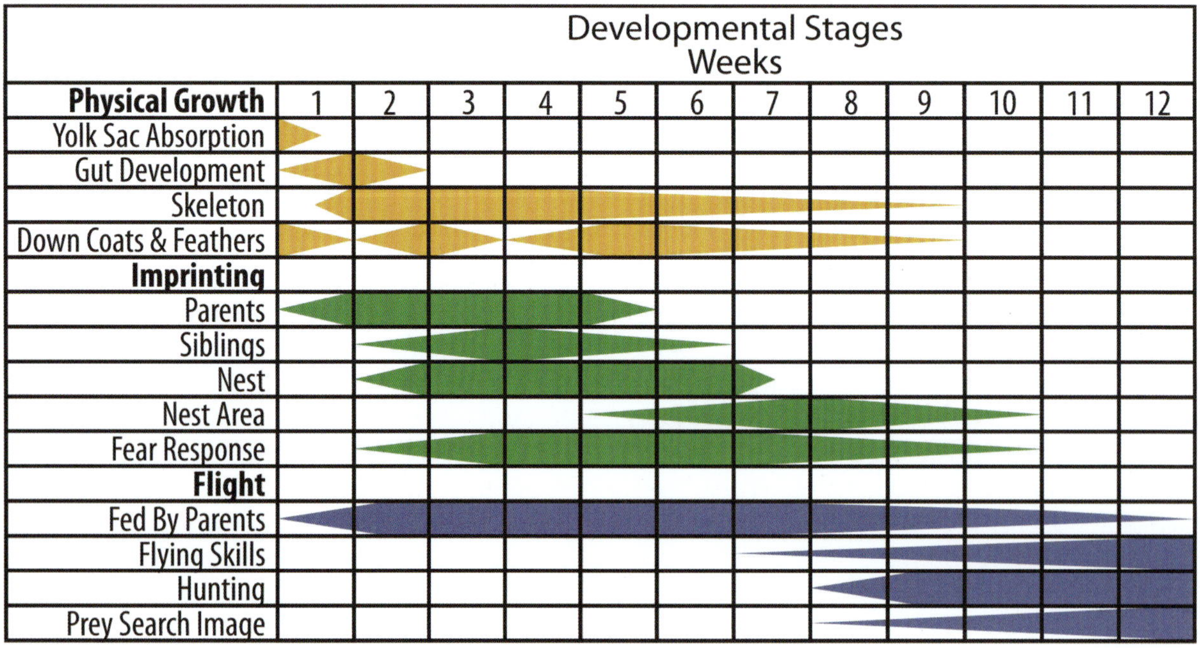

Figure 4.8.1 Developmental stages of a young peregrine.

Contrast this with a precocial species, such as a sheep:

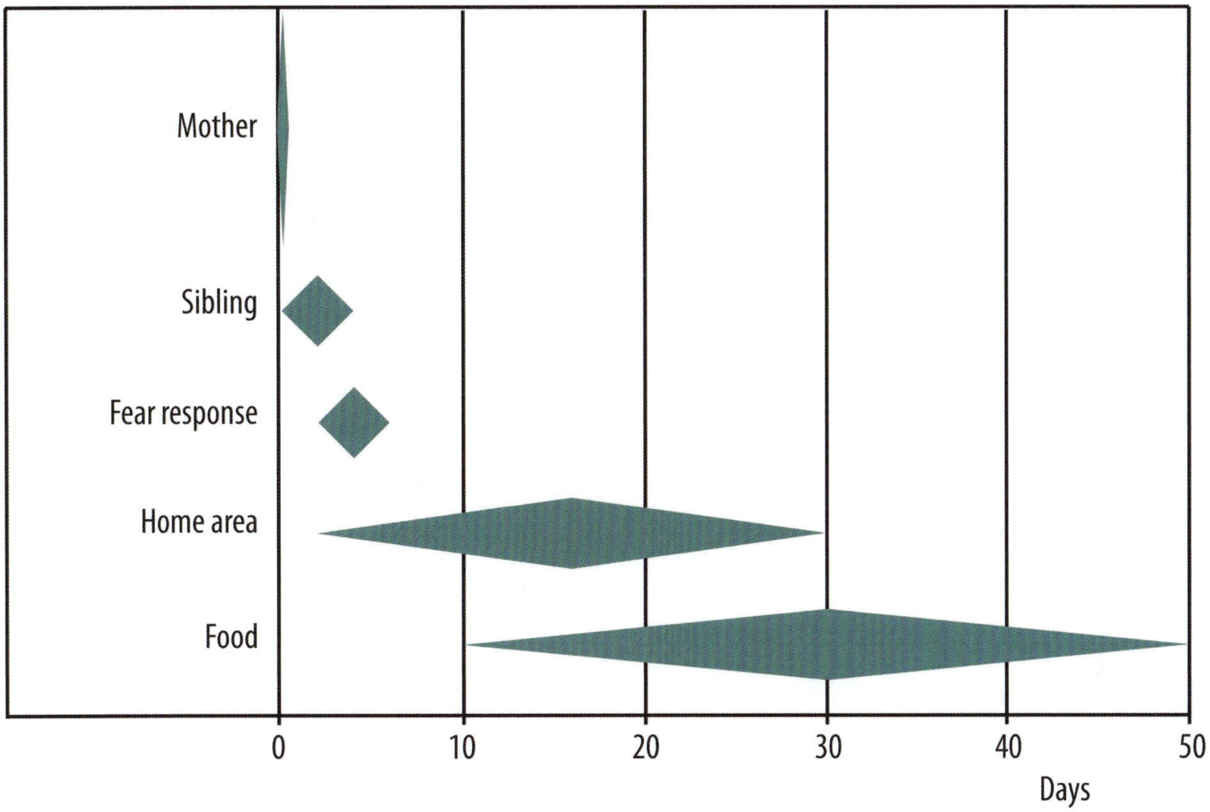

Figure 4.8.2 Imprinting in sheep.

4.9 Imprinting on parents

Newly hatched raptors, being poor sighted, at first respond to the call of the mother; in falcons this is the *chup* or *eechip* call. The chick responds by gaping for food. By the time the second down coat has replaced the natal down, the chick is much more aware of its surroundings. This is a very rapid period in its mental development. At this stage it begins to imprint on the parents as the provider of food; food begging is directed specifically at the parent and not at any random shadowy movement. This imprinting continues during the whole period in the nest. Once the parent has been identified, the chick develops fear of 'nonparent' objects and if it is switched from natural to human parents or vice-versa will show initial fear of the new, strange parent. But, provided that the new parent persists, the chick soon settles down and continues its imprinting period, this time on the new parent. Peregrines start to imprint on their parents from about the fifth day.

Although chicks will imprint on objects quite dissimilar to their natural parents - humans for example, they imprint more strongly and more rapidly on objects which closely approximate their natural parents. In other words there is a genetically programmed parent image, albeit a hazy one, built into its ethotype. The rather nasty experiments of Harry Harlow on monkeys demonstrated how certain elements, such as cuddliness, warmth, movement, can make an object more or less attractive for imprinting.

How does the hawk develop her own self-image? If the hawk is reared by a human, she may identify that human as a parent, and later, a human as a sexual partner. But does the hawk think that she also looks like a human, or does she think that she looks like a hawk? If she can't understand a mirror, how does she know what she, herself, looks like? I'm still working on this little problem. For example, some years ago I fostered some New Zealand falcons to a pair of gyrs. Then

I trained one of the young females. She had been used to bossing her mother around in the breeding pen and was certainly not intimidated by a bird of gyrfalcon size. Out hawking (at 550 grams) she would happily attack any common buzzard that came near, even though they weighed around 900 grams. (This is not for falconers with poor nerves or slow legs.) As far as I could see, she believed that she was at least as big as the buzzards. When she got older, she bred a lot of nice youngsters, paired naturally with a male New Zealand falcon. But of course she had imprinted on them as siblings. Did she think she was a gyr, or a New Zealand falcon? Or did she think she was a gyr and that a New Zealand falcon male was an appropriate mate?

Interestingly we had a young white male gyr in a hack pen recently, together with three black ones. When we added a new, slightly younger, white one, the original white bird kekked at it, never having seen a white bird before. Had he thought he was black too?

Although this subject is complicated by ethological problems of observation and interpretation, it is important because of applications to management of rehabilitated birds (see 11.4) and of hybrids (see 11.6).

4.10 Imprinting on siblings

As the chick becomes aware of its mother, so too does it realize the presence of its siblings.

These brothers and sisters are a mixed blessing. At first they are a source of warmth, but as the chick becomes older they represent competition for food and then, later, positive dangers as robbers of food. The chick responds by mantling over and defending the food, or by carrying it away, but at the same time the siblings are reasonably inhibited against injuring one another. Some of the accipitrid hawks, such as buzzards, kites, and some eagles, are prone to Cain and Abel battles in which the oldest chick attacks a younger one. To some extent this behaviour is genetically programmed, but is also dependent on food. During fieldwork on Welsh red kites, video cameras revealed this going on at most nests, often with fatal results. We also noticed it when hand-rearing baby kites. However, when we put these kites under an old common buzzard as a foster mother, the behaviour stopped within hours and never recurred. This buzzard was such a good mother that even with seven kite chicks she would keep them constantly fed. Whenever one so much as opened one eye, she was there with food in her beak, waiting to feed it. Also, during the early days at least, she would sit firmly on the chicks. The only time she raised the lid was to feed them, so the chicks had no daylight time in which to attack each other.

4.11 Development of the fear response

Once the chick has imprinted on its parents and siblings, then the fear response develops. The imprinted objects are now familiar and any new objects are henceforward experienced as unfamiliar and to be feared. Chicks first show fear of an intruder to the nest defense just as they grow their second down coat, before feathers begin to show. They respond by crouching down and huddling, or by sitting up and kekking or screaming. Females tend to be sullen and crouch, whereas males tend to be more active and vociferous. Sometimes when checking nests of wild New Zealand Falcons in the Southern Alps I would be attacked by nestlings scarcely more than ten days old and hardly able to move out of the nest bowl. They would still sally out valiantly and try to grab some part of my anatomy!

4.12 Imprinting on future sexual partner

The way in which raptors imprint on sexual partners is complex and poorly understood. It seems to start after the initiation of the fear response, and is independent of the presence or absence of siblings. Hand-rearing chicks together with their siblings certainly does not fully prevent them imprinting on humans as sexual partners. The chick seems to identify primarily with the food source (normally the parent) as the future sexual partner, but a hand-reared bird, reared in a group, if isolated from humans, may pair with another of the same species. Raptors that have been foster-reared by other, related species both in the wild and in captivity appear to be able to pick the right partner when the time comes, but ecological requirements and courtship differences are probably important in preventing hybrid pairs from forming in the wild. Many of the experiments on re-introductions by cross-fostering, are inconclusive because of the presence of synchronous dual imprinting which we will discuss in 4.15.

Some birds, such as cuckoos, are not reared by their natural parents and so their future sexual partner must be selected through instinct and not through imprinting, although the foster species they select for egg laying may be determined by imprinting. In raptors it seems that both factors contribute.

In some species, and this sometimes happens with captive raptors, a male and a female reared together often fail to form a sexual pair bond with each other. This is because they maintain their sibling relationship. If placed with other members of the opposite sex, or if separated for a couple of months during their first winter, they usually pair up normally.

4.13 Environmental imprinting

Once the active participants (parents, siblings, and intruders) have been identified by the growing chick, the static factors make an impression. At first these are the immediate surroundings-the type of nest, whether a stick nest, rock cavity or on the ground. Then, once the chick can fly, it imprints on larger aspects of the nest-the tree, cliff, or man-made structure-and the surrounding habitat. There is a strong correlation between the type of nest in which chicks grow up and the type they will subsequently choose as adults. It is even possible for two populations of a species to exist in one area in a state of virtual genetic isolation; the ones which use stick nests seldom breeding with others which use cliff nests.

As well as this site, or limbic, imprinting which is well documented in raptors, there are examples of continuity of attempted use of nest sites over many years which, owing to parents and young being killed every year, must have been chosen through factors other than imprinting. In these cases the aerodynamic properties of the nest site and area (see 1.16) and other geophysical factors may be the lure for continuity. Thus, although a falcon may cruise a cliff looking for a ledge, because that's its idea of 'home', the decision on exactly which ledge to use may be because a little upcurrent of air just there makes take offs and landings much easier than on a neighbouring ledge which although a better shape, is in a bad down-draft.

4.14 Practical aspects of imprinting

As imprinting in raptors is so drawn out, rather than compressed into a tightly scheduled sensitive period as it is in ducks or sheep, and as there are at least five different aspects which overlap, so there can be various combinations of ways in which hawks become imprinted, depending on their treatment during development. The two common extremes are shown in figure 4.14.1. Naturally imprinted, wild-caught birds are usually fairly suitable for falconry. Their main disadvantages are their well-developed fear response (which makes accipiters particularly difficult to use), and their nest site imprinting which can make aviaries appear so alien that the bird is too stressed to come into full breeding condition. Their hunting ability is usually good, although sometimes they range too independently of the falconer to be of much practical use.

The opposite extreme is the raptor that has been openly hand-reared by man on its own. It incessantly screams at its human 'parent' for food, it treats humans as siblings and mantles over its food, it is aggressive on kills and may even fly at and attack the handler. Because falconry demands that the hawk comes to the falconer for food, and that the falconer must rob the hawk of its kill, the falconer acts as both parent and sibling and reinforces this jealous behaviour, unless he works his handling system out very carefully. The natural process of independence from the parent does not take place and the raptor's behaviour is arrested at a juvenile dependent stage. Such birds, if continually taken hunting without a proper chance at prey, may vent their frustrations on the escorting 'parent' (the falconer) or other providers of quarry, such as the dogs. If it kills, it may leave the prey to drive off any competitive 'siblings' (the falconer and his dogs).

	Wild-caught bird	Bird reared openly by human from day one.	Domestic falconry bird
Parental imprinting	Begs for food from own species.	Begs for food from humans, rejects own species.	Begs from own species, doesn't scream at humans, ignores them.
Sibling imprinting	Shows competitive aggression/ jealousy, mantling, etc. with own species.	Treats humans as siblings. Instinctive predator response on kill, to own species.	No aggression or jealousy over food toward humans.
Fear response	Fearful of other predators, particularly man.	No fear of humans, especially of handler. May fear own species.	No fear of humans, own species, dogs, etc.
Mate imprinting	When mature, female begs from male of own species. Male courts female.	Female screams at human "husband" for food in spring. Male courts human "wife."	Responds sexually to own species only.
Nest site imprinting	Responsive to natural nest sites.	Prefers artificial nest type in which it was reared.	Prefers artificial nest and is relaxed in aviaries.

4.14.1 Different ways of imprinting.

Such imprints can go into a kind of trance, standing on the fist all bristled up but refusing to actually eat the meat, maybe exploding if you put your hand up. The remedy for this is not to feed lumps of meat at all. Always pre-cut the food into small mouthfuls and offer it in a bowl. Maybe she recalls her earlier rearing, but for whatever reason she will normally settle down and eat straight away. Such birds also go into a trance on the kill or on the lure. The best remedy then is to walk away out of sight for a while. Once no humans are around they normally start feeding.

There are several antidotes to these problems, all based on minimizing the falconer-parent-sibling image and maximizing independent game-catching attitudes and these are examined in section 5.3.

Changing the falconer often produces an immediate change, especially in accipiters which seem to identify individuals more specifically than falcons do. But the new relationship that the hawk builds with the falconer is usually based on the old imprinted images of falconer-parent and before too long it is screaming again, this time at its new 'parent'. It needs to mature through to developmental adulthood and break away from the relationship with the parent figure.

The Westermark effect, which essentially is an incest taboo inhibiting siblings pairing and mating with one another, also works with parents. This does not require the siblings or parents to be genetically related to the individual. The mechanism works through the process of rearing, so it is just as powerful with foster-siblings and foster-parents. The trick is, while we may want the chick to imprint on humans, we don't want it to form too close a bond with a particular individual human foster parent. That doesn't matter if you only want to continue the relationship as a hunting partnership, in fact it makes the bond even stronger, but if the foster parent wants to go on and turn the relationship into a pair bond for breeding then the Westermark effect could inhibit things. The obvious solution to this is to get someone else to do the rearing, or even several people, so that the chick is not imprinted onto any one individual as a 'parent' figure.

When human-imprinted raptors come into breeding condition the females tend to refuse to hunt but beg to the falconer 'husband', while males make courtship overtures to the falconer 'wife'. A wide variety of unnatural objects can stimulate them into copulation. These birds are well-adjusted to artificial environments and readily come into breeding condition in captivity. But although they are adjusted to humans, they show varying unnatural responses to their own species and may kill each other if placed together in pens.

In recent years, imprinting techniques on humans have become more refined, either to prevent the bird from pairing naturally so that, if a hybrid is lost it would not pair up with a wild bird, or to produce a bird for breeding, either as a semen donor or egg layer, which will be used for artificial insemination. Such imprints are hand-reared from day one in isolation. If the bird is intended for hunting, as soon as it can feed itself, hand-rearing is stopped and instead the bird is fed finely chopped whole animal diet in a feed bowl. Food is never brought openly to the bird. Either the youngster is hooded and then the food bowl placed beside it, or the chick is itself placed next to the bowl. Hooding practice continues several times a day from the start of the second down coat until the bird is at least hard-penned. Throughout the youngster's entire growth period it is kept constantly in sight of humans going about their daily business. Once it can walk, it can be outside for part of the day. It can be flown at tame hack (see 5.4), although this is not necessary for birds intended solely for breeding. Such birds progress naturally and quickly to hunting and should not scream. But if the method is done half-heartedly, it is disastrous and also these birds must not have their weight reduced much at all in their first season. For this reasonsuch falcons are better suited to waiting-on flights at game rather than pursuit flights requiring more weight control.

The imprinting of accipiters has now been developed into a fine art, explained by Mike McDermott and others, although actually little has changed since the detailed descriptions by people like Diogo Ferreira, four hundred years ago. Nowadays many austringers prefer to fly imprint accipiters rather than spooky parent-reared ones.

In between these two extremes of natural and unnatural imprinting is a wide variety of possible combinations of malimprinting depending on the treatment the bird receives at each stage of its development. The label 'imprint' or 'malimprint' is a very general term which doesn't mean a lot unless more detail of the condition is given. Unless the developmental history of the bird is known, it takes quite a lot of time and skill to identify in what ways a bird's imprinting is unnatural and what the objects of the imprinting are.

Many falconer advocate keeping the young raptors together in a pen without seeing humans until the birds are fully fledged. They have then been horrified that the birds have started screaming. This is because the chicks have imprinted on one another as siblings but have had no parent object on which to imprint. In these circumstances of deprivation it seems the susceptible period is drawn out until after the birds have fledged. Thus they may imprint on the falconer, being the first object with which they have a parent relationship, and of course respond by screaming for food. This bond is normally not as strong as the natural parent imprinting, but it is permanent and is reinforced by subsequent calling to the fist in falconry training. If the birds are almost fully fledged (branchers) when first put into the pen they will already be imprinted on their natural mother and, provided this is not overlaid by human interference, they will remain naturally imprinted. Some falconers like a young hawk to come out of the rearing pen 'good and wild.' The wildness is an indication that the fear response has not been overwhelmed by human interference and thus that other aspects of imprinting have not been interfered with. But the fear response itself is not a desirable thing in a falconry or potentially breeding bird and it can be selectively removed during development without affecting other aspects of imprinting.

The third column in figure 4.14.1 shows the domestic falconry/breeding bird. While being a fearless falconry bird it has no imprinted ties with the falconer and is suitable for normal pairing in breeding pens. Rather than attempt to avoid malimprinting by isolation, the developing chick can be offered a definite, recognizable parent object on which to imprint. There is then no problem with late stages of imprinting. Probably the best way to do this is to have the chick raised by a human-imprinted female of its own species. This has two distinct advantages; first it means that the chick becomes naturally imprinted on its own species: it identifies its adopted mother as food provider and will not scream at the falconer. Second, because the adopting parent is not afraid of humans, it is possible for the rearing process to be done where there is plenty of human activity in the immediate area. The adult female does not show fear reactions to humans and in the absence of these signals the chick grows up not fearing man, or any other objects which are present, such as dogs or machines. This of course is provided that the chick starts before its fear response has developed. It is important that the humans show absolutely no interest in the chick, or in the food supplying process, otherwise the chick may imprint on them either as food providers or as siblings. Instead the chick must identify humans as being harmless parts of the scenery, similar to trees, bushes or clouds. Once a number of chicks have been raised by foster mothers, they themselves usually breed satisfactorily in aviaries. If these pens are exposed to human comings and goings (but not interference or feeding) the parents will rear the next generation with no fear of man, in other words your birds will then be truly domesticated. This of course is provided that the chick starts before its fear response has developed.

Skylight and seclusion aviaries which are good for nervous breeders such as accipiters because they prevent the birds seeing things outside which might disturb them, are bad for rearing youngsters. Ideally the pen should have a window, high up. This window need not be large; it is solely for the bird to see out, not for you to see in. Thus one window per bird with a perch below it, the window being the size of the palm of your hand, is sufficient. Whenever it is quiet, even the nervous birds will be up at the window watching out, retreating to the rear of the pen when disturbed. If necessary the windows can be blocked during the early breeding season. But once the youngsters start to feather, it is important that they can see the big outside world. Otherwise, when they are drawn out for training the whole experience is too traumatic and stressful, and the bird may be 'highly strung' to the point of being impossible to handle.

Another way of obtaining natural imprinting but no fear response, is inferior to the first and uses a replica parent, usually a hand dummy. The first method is followed as closely as possible but the role of the mother is played using a hand dummy resembling the natural parent. It has the advantage of not requiring a possibly erratic foster mother but it is harder to prevent the chick associating humans with the food supplying process. It must be done scrupulously carefully, otherwise the chick will realize the whole thing is a cunning plot. I regard it as a last resort.

A third method of rearing is 'crèche' rearing, in which groups of chicks are reared together by human foster-parents. This produces very steady birds which are suitable for breeding in natural pairs. But some species, especially Harris hawks, tend to get too familiar and if cut down a little for falconry, quickly degenerate into nasty screamers.

If you are buying a bird, this very important stage in its development will be out of your control. One breeder's definition of 'parent-reared' may actually entail hand-rearing until 10-14 days old. This is usually done to improve the chick's chances of survival and to wait until the closed ring stays on. But falcons start to recognize and imprint on their feeder from day five and, if put out at day ten to natural parents, these early few days are often still recognizable in the subsequent behaviour of the youngster, even after a wild hack. If you value your bird, you should inquire closely into exactly how it was reared, and you should also inquire as to the integrity of the breeder beforehand.

4.15 De-imprinting

Imprinting, as I have defined it, is irreversible. However, examples do occur in which it appears to have been reversed. Examples spring to mind of passage hawks, sometimes even wild-caught adults, which have become 'screamers'. Others, of birds imprinted on man that have gone on to breed naturally in the wild or in captivity. In the naturally imprinted wild birds that appear malimprinted this is caused by conditioned learning (see section 4.17) and is often associated with injury; the injured bird is even more dependent on the handler and may revert to juvenile begging behaviour. Being conditioned rather than imprinted, this is easily reversed.

In examples of malimprints subsequently breeding naturally it can be seen from the imprinting chart that many varieties of malimprinting can occur, including dual imprints. Dual imprints can occur when both imprint objects are present at the same time (synchronous), or when one is present after another (asynchronous), as when a bird is taken half-developed from the nest (and

These chicks are growing up imprinted on the falcon parent but without fear of humans.

thus 'half-imprinted' naturally) and completes the remainder of the imprinting period being hand-raised. When this happens the bird usually appears fully imprinted on humans (depending of course on the proportions and extent of exposure), but on reaching sexual maturity, if presented with only a natural mate, may breed successfully, or at least show some natural pairing behaviour. Synchronous dual-imprints on the other hand, those which had both natural and artificial parent/mate objects present at the same time during the entire imprinting period, may well breed naturally and be additionally stimulated by the artificial 'mate'.

Asynchronous dual imprinting is usually at the root of most problems and misunderstandings in domestic raptors.

Accipiters, unlike falcons, might have a reversible fear response. Certainly some accipiters which are aggressively tame malimprints go quite spooky when moulting without human contact. They seem to become very tame again when handling recommences. Human imprinted falcons and buteos, however fat and unattended, usually stay relatively fearless (although they may become highly territorial!). The way in which fear, imprinting, physical condition, mental conditioning, and stress are interlinked in accipiters needs more study; we are hardly beginning to understand it. Perhaps the fear response in accipiters is less strongly cross-linked with the other imprinted behaviours than it is in falcons; the two genera are not closely related.

Only birds that are completely naturally imprinted in all five aspects are potentially suitable for release to the wild. If you take on the responsibility of rehabilitating an orphaned raptor and fail to rear it so that it is naturally imprinted, then do not release it to the wild. You are not doing anybody any favours by releasing such a bird and you must examine yourself and see what can be done to prevent further failures (see 11.4)

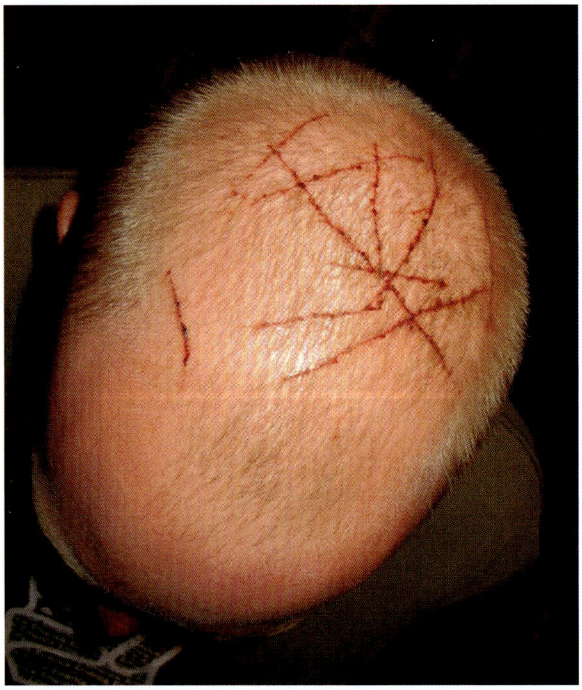

If you go near a New Zealand falcon nest, the parents will have you. The chicks in the nest pick up on their alarm calls and come to regard you as an intruder.
Photo Shane MacPherson.

4.16 Learned behaviour

This section is a hard one to write because it brings together the fields of ethology, animal psychology and empirical training methods. Each has its own followers with their own terminology and a skepticism of other approaches. Each has a contribution to make to the total problem and yet each on its own is too narrow or too shallow an outlook. I will give just a simplified resumé of what is an extremely complex, interrelated, partially understood subject which is not really susceptible to scientific investigation.

To me, there are two facets to learning: involuntary and voluntary learning. Involuntary learning creates an involuntary response or lack of response through the autonomic, nonconscious part of the nervous system. The trained behaviour is prompt, automatic, and unthinking. It includes habituation, classical conditioning or reinforcement, operant conditioning, association, and trauma learning.

Voluntary learning produces conscious, voluntary responses which are not automatic and not necessarily immediate. It includes insight learning and experience. Intelligence may be regarded, in a simple way, as the ability to learn; the product of intelligence and experience being wisdom.

In chapter 5 we will get down to the practical nitty gritty of training and motivating raptors. Here in chapter 4 we are looking at the theory and principles behind behaviour and learning. Once these principles are understood, not only do the solutions to practical problems become fairly obvious, the whole process of training can become effective and streamlined, taking a lot less effort than before.

4.17 Habituation

Habituation entails exposing the hawk to a stimulus so often that it ceases to respond. It 'gets used to things. This was traditional 'manning.' The hawk was simply tied to somebody's fist day and night until it was too tired, physically and mentally, to react. And so the wild accipiter sat quietly on the fist despite all the nearby stimuli which formerly alarmed it.

There is one big problem with habituation: like conditioning, it wears off. This means that the habituator's work is never done. The Elizabethan austringer used to man his goshawk in the morning until she had roused three times, before flying her. Most of the hard work in training a wild-caught accipiter is habituation. Basically the hawk is exposed to unpleasant stimuli such as people, machinery, hoods, and so on, and this is repeated frequently until the hawk gives no response. Because there is no response there is no reward.

One of the main reasons for weathering hawks is 'maintenance manning' or habituation, but careful positioning of pen windows will often keep a hawk sufficiently tame without the risks and disadvantages attendant on conventional weathering.

A good use of habituation is in teaching a goshawk to ignore white ferrets. After a few days of seeing the ferret in its hutch, the hawk pays no further attention. Similar habituation can be used to overcome 'hang-ups.' My old goshawk used to have a neurotic dislike of plastic containers such as buckets, hawk baths, and ferret boxes. Leaving these in her mews over the moult, especially hanging on elastic so that they moved and bobbed in the breeze, totally cured her. Often when a neurosis is originated by trauma learning (see section 4.18), habituation seems to make a long-lasting, almost permanent cure.

Here is an example of unintentional misuse of habituation: a friend telephoned to ask why his falcon persistently ignored the lure. He had introduced the lure to the falcon by garnishing it with food and throwing it down beside the falcon on her block. This was what his books recommended. The falcon had jumped down onto the lure and eaten the food, checked the lure for any other food and hopped back onto her block. The falconer left the lure there all day while he was at work. Next day, and thereafter, the falcon ignored the lure and would only come to the fist. The falconer had successfully habituated her to ignore the lure. The remedy was to start again with a new lure of different appearance and to pick the falcon up with a tidbit on the fist as soon as she had eaten the lure food.

In general, habituation is more effective and more persistent in young birds than in old ones. Young hawks are still developing their habits. Old hawks, like dogs, horse, and humans, often get too set in their ways to do much with. If habituation is stopped, such as at the end of the season when the hawk is put down to moult, the process is achieved more easily the next time. Each season the hawk needs a little less manning to reclaim her.

Habituation to one type of stimulus often deadens the response to other similar stimuli too. For example, a goshawk which refuses white ferrets may refuse white rabbits as well. On the other hand goshawks and Harris hawks can be very discriminating and may work amicably with one dog and yet be totally stressed in the presence of another, even one of the same breed.

To speed and accentuate the effects of habituation it is often possible to introduce some positive reinforcement, for example, by placing food in the hood for a hood-shy hawk, or feeding the hawk while carrying during manning. Great care must be taken that the hawk does not bate too much from the fist or it will begin to associate the fist with an unpleasant experience and thus train itself to become even more wild. This is where the sensitivity of the falconer is paramount; it is not a question of patience.

4.18 Conditioned learning

Conditioned reflexes work through the nonconscious autonomic part of the nervous system and the response is rapid and automatic. An example is the way the sound of water trickling onto tin evokes a very rapid and embarrassing response! The response starts less than a second after the stimulus and requires no thought at all. Frequently it is only one's own response that draws one's conscious attention to the original stimulus.

In the same way, when you see a person or animal yawn and even before they've finished yawning you find yourself yawning. Try it out next time you are in the train or on a bus after a long day. Contagious yawning like this is a reflex at root, but strongest between emotionally close individuals.

Classical or Pavlovian conditioning follows this order: stimulus or cue (eg whistle), response (eg flying to the fist), reward (eg eating food), all occurring rapidly one after the other. This is positive conditioning or reinforcement.

Cues can be very small indeed. Given that raptors have better sight and hearing than us, they can notice cues even beyond our perception. Cues can be very brief and very slight and if you find the hawk is doing something puzzling, look carefully at the situation; she may be reacting to a cue you haven't noticed. I remember once bringing my female goshawk into the living room one evening in early spring to dry out after a day's rabbiting. I perched her on the back of a chair in an alcove. She preened and settled down and then did something odd: she glanced at a piece of electrical flex dangling against the wall. I thought "She wants that wire. She must think it's a branch. Maybe she wants to nest". So I went out and got some twigs and dumped them on the chair. Sure enough, down she jumped and spent a happy evening building herself a nest in the chair! Similarly, if you watch Steve Martin's films, you can see how his parrots react promptly to small cues which we normally wouldn't even notice. During falconry we use lots of cues, such as a whistle or call, hand-signals and swinging the lure. We actually give out a lot more cues than we really intend, and often hawks react to these without us realizing it. Horses and dogs, being social animals, are even more perceptive and empathetic; it is not easy to hide your feelings from them.

Usually this is no problem, the hawk is getting to know our personal ways of doing things. When it becomes inappropriate, it is called superstitious behaviour. We have a male Harris here at the moment. He has only been trained a couple of weeks, by someone using a dark brown glove. We took him out hawking in the rushes and found that he would come to anyone who held up a dark brown glove, but not to anyone with a light-coloured glove. So we will have to sort out that little bit of unintended superstitious behaviour!

Once the animal has noticed the cue and done the required behaviour, then the act can be reinforced. As Karen Pryor or Steve Martin would say: Positive reinforcement is anything that occurs in conjunction with an act that tends to increase the likelihood that the act will occur again. Positive reinforcement is the best way to train raptors. The reward is usually food or quarry but there are many other, more subtle rewards. Once you have worked out what motivates the hawk in any given situation, what it really wants, then you can give an appropriate reward.

Negative reinforcement is something the

For young falcons going out to the Gulf, in can be quite a shock to see people in white flowing robes. So we give them some positive conditioning, associating the Arab falconer with food while they are growing up in the hack pen. The food chute is converted into a falconer, so they learn quickly to come to him for food.

subject wants to avoid. It occurs in conjunction with an act and tends to decrease the likelihood that the act will occur again. For example, if you wave a hood around in a hawk's face and then proceed to hood her in a rough and ready manner, she will soon associate just the sight of the hood with the forthcoming nasty experience and will bate as soon as she sees the hood. Notice that the action and the reinforcement occur close together so that the animal is able, by her actions or inactions, to change the reinforcement. Raptors are quite capable of learning by negative reinforcement (see trauma learning) and by more subtle negative conditioning such as when a falcon starts to refuse a prey she finds unpalatable. But the falconer must take infinite care that the negative experience is not linked to him. Otherwise it will undo the tenuous bond he has worked so hard to build up. Also, with negative reinforcement she will only do enough to make the negative reinforcement stop, whereas with positive reinforcement she will make bigger and bigger efforts in order to get more reward.

Operant conditioning is when the behaviour comes first, without a stimulus, and is then reinforced. It is a tricky part of training because you are dependent on the animal doing the behaviour before you can reinforce it. She may do so accidently or you may have contrived the situation as best you can so that the behaviour is likely to happen. This is where natural bred-in instincts are so important. For example, when a peregrine waits on just a little, you try to give positive reinforcement by serving quarry under her. Day by day she waits on higher, longer, and steadier. The trainer should always be looking for ways to motivate his hawk to do things so that they can be positively reinforced. Often this means trying to break down the behaviour into a lot of small intermediate steps, rather than trying to get there in one big jump.

An example of operant negative reinforcement is when a dog jumps on to the best sofa and is given a prompt slap. After a few performances the undesired behaviour ceases. It needed the dog to do the undesired behaviour first before you could train it not to do it.

Whatever the nature of the reinforcement-positive, negative, real, or implied, it must be given as close to the behaviour as possible, otherwise it is not linked with the behaviour. So if you beat a dog for running off only when it gets back, the dog links the punishment to coming back. Thus this is punishment, rather than negative reinforcement. Intelligence can in some measure be taken as the extent to which a person or animal can stretch these links between the cause and effect, in time, in space, and in abstraction. In general, the shorter the link the more effective the lesson. Contrast this with a game of chess in which you are trying to think through a whole series of potential consequences to the reward or punishment of a move.

While you are training or shaping the behaviour you want, it is best to give a positive reward every time the act occurs. But from then on, if the behaviour is not reinforced, it may fade away. If you don't reward the hawk for coming to you she will, after a while, become slow to come and eventually stop coming at all. On the other hand, if you reward her every time, she will start to take the reward for granted and become complacent. It is better from then on, to reinforce using a variable schedule which in practice means rewarding about every fifth time. But the schedule of reward should be random, not regular. In other words in ten calls you might give two or three rewards in total, scattered randomly among the ten calls. You might reward twice in a row and then nothing for seven or eight times. You are in control of the reward system, not the hawk.

This is like what happens when the hawk chases prey. You might think that this would be more discouraging than if you rewarded the hawk every time, but this is not so. Imagine a light switch. Every time you switch it the light comes on. Then one day nothing happens. You try it a couple of times more. Still nothing. So you give up. Try the same thing with a flashlight. You switch it on, nothing happens. Not an unusual occurrence with the average flashlight! So you try several times more and eventually start checking the connections, the batteries, the bulb, and so on, convinced that you'll get it going.

I was out with a friend hawking some time ago and I noticed that he always offered a tidbit when he called the hawk. If there was no tidbit, the hawk wouldn't come. He had trained latency into the procedure. So I said "Why don't you call her without food?" So he tried it and the hawk refused. I thought to myself "That's odd, if it had been one of my hawks she would have come. How would I have made her come? I would just sort of willed her into coming." But, on analyzing it more carefully, I realized that I have always trained using a variable schedule (probably because I can't be bothered to mess around with all those tidbits!). Similarly, once the hawk is trained, I don't put food on the lure at all. Food on the lure is a health hazard, it encourages possessiveness, and there is always the risk of the hawk detaching it.

The most effective training is when the reward is as close as possible to the act, preferably only milliseconds. But it may not be physically possible to achieve this and any delay risks disassociating in the hawk's mind the link between act and reward. This gap can be bridged with a token reward, such as a word of praise, a pat or a smile. This works best with social animals but it can also be used successfully for birds, and perhaps this is one area where falconers are lacking. Steve Martin says the word 'Good' to his parrots, as a bridge before rewarding them. We use a bridge in the form of an ungarnished lure. We whistle and swing the lure (cue), the hawk takes the lure (bridge) we come up with the pick-up piece rewarding her for coming and positively reinforcing our own approach. This also gets around the problem of the hawk associating our approach with stealing her prey (negative reinforcement).

As well as training the hawk to come, and reinforcing the behaviour, there is a further aspect to consider: timing. When you train her, she must come immediately. She is perfectly capable of this, after all, if the cue was a rabbit or other prey, she'd be after it in a flash. The way to do this is, right from the start, never to call her for more than three seconds. If she hasn't come by then, put your fist or lure away and do nothing. Don't call her again. Turn away from her, close the incident, go to a different position. Don't walk toward her (who's training who?). If you want to get nearer, then go in at a tangent. Make her realize that she had her chance and she missed it. The mistake was yours because you lacked judgement. The experienced trainer can read his hawk and, knowing her history, can predict her behaviour. So he takes care always to be successful in what he asks. So the training goes quickly, rather like somebody running across a river on a stepping stones. He's marked his route, checked his stones, and goes confidently and quickly from rock to rock. The less experienced trainer gets halfway across and either falls in or finds that the gap between rocks is too wide and must spend time figuring out another route.

Figure 4.18.1 When you call your young buzzard, do you read her mind correctly?

Remember: never call in vain. This is training for a pattern of refusal. Every refused call is training in the wrong direction. You would do better to have stayed at home drinking a cup of tea.

Thus although a hawk may be 'trained,' there is a world of difference between one that will only come for food, is sulky or moody, and has established patterns of refusal, and one that comes when called immediately. A hawk which has been trained properly, first time and at her own speed, and kept up to it, will be better than one on which hours and weeks of somewhat confused, repetitive training has been lavished by an inexperienced person. It is important to be able to 'read' a hawk so that you are able to anticipate. Thus when a falcon is flying you 'know' what she is 'thinking' from her little signs and act accordingly so that the lesson goes smoothly without a hitch. Being half a second late in realizing something is all that it takes to ruin a lesson, or even to lose a bird. As you get better at communicating with the hawk, she will seem more intelligent. Some people are better at reading these little signs than others; curiously, a high proportion of our falconers are mildly dyslexic, they seem better with the animals than with the written word. The problem is that although these people are 'natural' handlers, they often cannot verbalize or analyse what they are doing.

Punishment is the point at which the training of raptors, excepting perhaps some eagles, differs from that of social animals such as dogs and horses. Punishment happens after the action, is

Figure 4.18.2 Did she come to you, or did you come to her?

not directly associated with the action, and the animal has no control; it cannot stop doing the action to avoid the punishment, because it already stopped doing it some time earlier. Punishment is inflicted personally by the trainer. Punishment does not work on raptors because most falconry birds are non-social. They do not belong to a structured society and it is only in a structured society that agreed morals, a moral conscience, and dominance (the means of applying morals) can exist.

A dog, for example, is a social animal with all the biologically inherited programs for social behaviour. We say he has a 'willingness to please'. that is to say he is willing to have his behaviour modified by more dominant individuals in his own society. We can thus introduce to him the concept of 'sin'. If we decide that stealing from the rubbish bin is a 'sin' for our dog, then the first time he does it, we grab him by the scruff of the neck, give him a shake, stare him in the face and growl 'No!' (negative operant reinforcement). We link the world 'No!' by association to our actively disapproving dominating shake. Quite soon just the bridging word 'No!' is enough negative reinforcement to keep the dog away from the bin. Because the dog is a social animal he is prepared to accept our commands without rejecting us, the leaders of his pack. As the dog grows up he considers, as we do, that stealing from the bin is a sin and he can be trusted to leave it alone. If occasionally a scrap of food proves too tempting for him, he shows every sign of guilt when caught. These concepts of 'sin' and guilt' are signs of morality which is a mechanism extending beyond our own human species and pre-dating it.

I define it as:

'Morality is behaviour that promotes genetic survival through social cohesion. It requires four elements:

1. *A social group.*
2. *Shared ethical values.*
3. *A judge.*
4. *A system of enforcement.*

Morality is biological in origin and has evolved in higher vertebrates. Its roots lie in genetic, imprinted and in learned'.

The hawk, quite simply does not have much of a conscience because she does not have much of a society beyond her family group in the breeding season. She cannot therefore be given a concept of sin and so punishment to this end is fruitless. Not only this, because she does not accept our dominance, any punishment we give her just alienates her. She links the punishment with us rather than with the hypothetical sin or undesired behaviour. We must take care therefore that our treatment of her is consistently good and positive in order to maintain the bond, which in a hawk not imprinted on man, is flimsy. A dog will very rarely abandon its owner however cruel he may be, but a hawk, unless imprinted that way, has no social bond with you and will soon show her disapproval in a very practical way. Of course there are some hawks that have degrees of sociality, such as Harris hawks living in family groups, with their own dominance hierarchies. But others, such as accipiters, are extremely solitary and do not have much compunction against killing one another.

This wild freedom from a social conscience is one aspect of raptors that has fascinated man over centuries. Perhaps it is a way for the falconer to escape by proxy from the bonds that hold him to his position in life, secure but at times, constrained.

Being territorial, raptors do have signals of dominance and submission, but these are usually brief before fight or flight. Also, as man is so different in size and movement, it is difficult for him to mimic these signals. Nevertheless the falconer should know them, if only to avoid them. Flying or perching in a prominent way over a hawk is a sign of dominance. So if you stand towering over a hawk which is on a low perch or on a kill, she feels threatened. A direct stare held for more than a second is a threat. Entering the personal territory of an unbonded bird who does not regard you as a mate, parent, or sibling is threatening; this distance varies from about 10 metres to about 1/2 metre, depending on many factors. Obviously to train a hawk you have to get this close and it is a sign that the bird is beginning to accept you as less of a threat once she stands on your fist eyeing things other than you.

Because raptors can't really be trained by punishment, then further whole areas of training are not possible. For example, when earlier you trained the dog not to jump onto the sofa, first you used operant negative reinforcement. This developed his conscience and reinforced your position as pack leader. From then on, if you saw the dog so much as looking at the sofa, you could growl 'No!' and the dog would know that you disapproved of what he was contemplating. Then we go into the bedroom. This time there is no need to start from scratch and let the dog jump onto the bed and so on. You just look at the bed, the dog understands what you mean, you growl 'No!' and the dog has learnt without more being said. He has extended his area of guilt. Soon you have a dog which 'understands your every word.' He is in marked contrast to a street stray which owes you no allegiance and doesn't understand you.

4.19 Trauma learning

This is learning which needs only one lesson or experience and which occurs under traumatic or stress conditions. Usually the lesson is a negative one, and it can often be the start of a neurosis or phobia. For some reason, when an animal is in severe stress or crisis state, input stimuli are perceived very clearly and, as it were, flash to the brain so violently that they fuse together regardless of whether or not there is a logical or meaningful connection. The whole conglomeration of a split second's input becomes deeply etched on the memory. Thus humans may remember every inconsequential detail of the moment before a car crash, right down to the tune on the car radio. Just hearing this tune subsequently may be sufficient to induce an automatic and unpleasant reaction. Trauma learning achieves in one lesson what would have taken many repeats by classical conditioning.

Trauma learning takes place both intentionally and unintentionally in falconry. It can be used intentionally when teaching a young dog to keep away from hawks. The puppy is allowed to sniff at a big hawk such as a redtail and the hawk promptly gives him an experience he never forgets and never needs to learn again.

Unintentional trauma learning can take place when a hawk is subjected to a violent but necessary treatment such as being cast to have her equipment changed. Many falconers hood the hawk or get someone else to do the casting to avoid themselves being associated in the hawk's mind with the experience. Or the hawk may have a bad experience with quarry, or with the vet. For the wild-caught hawk, her first experience of man may be traumatic, as she is trapped and handled. Unless care is taken at this stage, she can associate this trauma learning, not just with mankind in general, but with you in particular. As a result of this, many wild-caught hawks can be very sensitive when approached on the kill.

Undesirable behaviour picked up in association with trauma can be very hard to eradicate. Habituation linked with positive reinforcement are the best methods to use.

4.20 Insight learning

Insight learning assumes that, at some level, the bird has knowledge or understanding of a situation, through her consciousness. Conditioning, on the other hand, bypasses the conscious mind. It is easy enough to teach the hawk the significance of the lure as a possible food source. You show the hawk the garnished lure and after one or two meals the hawk has learned the lesson. Then, in the field, the process which is going on at some level in the hawk's brain is something like this: 'I feel hungry. That lure means food. I will fly down and feed on it.'

Under ideal conditions flying a hawk which has been taught rather than trained (ie conditioned), works reasonably well. The problems come when other signals also reach the hawk's consciousness and over-ride the decision. For example: 'I feel hungry. That lure means food. When it stops raining I might fly down and feed on it. It won't fly away.' Result: one very soggy frustrated falconer.

The falconer who has conditioned his bird to respond immediately has much better control over the hawk because the bird's response is automatic. But it takes skill to develop conditioned reflexes in a bird when negative reinforcement cannot be used and when she is free to move rapidly in three dimensions. Attempting to condition the bird when it is not motivated, or giving sloppy, unclear, or protracted signals, can ruin the conditioning of the bird and one is then left at the mercy of the bird's conscious decisions.

One aspect of insight learning is that it is relatively permanent. Once a hawk has learned the significance of something, she seldom forgets it. Conditioned responses wear off but knowledge is permanent. You can't unknow something. Once innocence is lost, it cannot be recovered. Once a hawk knows what the lure means, she will remember even years later. She may not do anything about it, but she will know all the same. If you inadvertently let your hawk into a secret, for example if she discovers that farmyard hens are hot dinners, you cannot then whisper in her ear and say 'Please forget that little secret.' All you can do is keep her well away from hens and perhaps try some negative conditioning. But she will always know about hens; she has learned by insight and her innocence has been lost.

Insight learning tends not to be used so much by modern falconers, which is a shame because it is very quick and effective. The best way is the natural way, that is to use another hawk to 'show' the youngster what to do. Normally she learns a lot by insight learning, watching her parents and siblings and joining in. The single falconer's bird on the other hand often has to learn every single thing on her own, without any example to follow. This can lead to problems with entering or with going up to wait on. If the bird is imprinted in such a way that she can safely be flown free with a make-hawk, then this is a great advantage. The hawk can learn from example. Care must be taken that the bird learns just enough to appreciate the situation without becoming dependent on the make-hawk. The practical way to achieve this is to have two hawks: one is experienced and the other is the youngster. Feed the youngster and place her unhooded on a perch where she can see everything around her. Then fly the old bird, doing the procedure that you want the young one to see and learn. For example we might be flying robotic prey. The youngster sees the old bird chase and take the prey and be rewarded on it. Having a crop of food already, the youngster is not strongly motivated to join in, but has gained an insight. The next day, unfed, the youngster is shown the roprey and her lessons begin with her already 'knowing' that the prey can be chased and caught. Her instinct to chase, and the motivation of hunger, should trigger the which, once the sequence has been completed, can be rewarded and reinforced.

4.21 Experience

Behaviour learned from experience is basically a mixture of the types of learning already discussed, but occurring in an uncontrolled fashion. Imitation can be a short cut to experience. Also by exposing the bird to a range of stimuli, such as different species of prey, experience teaches the hawk to discriminate. Hawks learn that ducks start off slowly, then speed up and keep going, whereas pheasants start off fast, then slow down and stop. By experiencing these differences, she learns to adapt accordingly.

The developmental age of the bird affects her ability to learn. As the imprinting process is completed, so the age of learning blooms. When the hawk can fly, she is at her most impressionable and remains so for the first year of life. The old saying 'You can't teach an old dog new tricks' has some foundation in truth. It has been shown experimentally with chaffinches, for example, that if they fail to learn their song repertoire through being isolated during the first few months of life, then they never pick it up.

Exactly how much this affects birds of prey is not known, but individual raptors do show various marked preferences, such as becoming wedded to certain favourite quarry species. Also hawks which have been kept in aviaries for their first twelve months of life are markedly harder to prepare for hunting wild quarry, not just physically but mentally also. Whereas the young hawk may learn something after just one exposure, the older bird may need several repeats to pick it up. Conversely, some old hawks which have been trapped in the wild never fully adjust to being in captivity.

All the points mentioned about hawks learning things apply equally to falconers learning falconry things. Sometimes the hawk trains the falconer faster than the falconer trains the hawk.

4.22 Mental maturation and the orientation toward quarry

Field research on wild peregrines by Dr Steve Sherrod and by others has clarified the stages through which young falcons go during their development.

These are:

a. Newly-hatched chick is offered food by parent.

b. Downy chick begs food from parent.

c. Newly-fledged chick chase parent for food. When satiated, it plays and chase objects resembling prey, but does not catch live prey.

d. Chick physically attacks parents for food.

e. Parent drops dead prey for chick who chase it and catches it. Chick's search image reorientates from the parent to the prey.

f. Parent drops live prey which reinforces connection of pursuit with food.

g. Chick has increasing difficulty in obtaining food from its parents and is left unattended for longer periods. It becomes more successful in catching its own live prey and begins to stray away from the nest area, self-hunting.

h. Chick is completely weaned to independence.

During captive management and falconry, we interfere with or alter this progression according to our own needs. For example, most falconry birds are now domestically bred and many falconers wish to use their birds for breeding during the spring and summer when the hawking season is over and the birds are moulting. This usually entails having the chick reared by natural parents in the breeding pen so that it develops with natural imprinting and is accustomed to being in an aviary. The chick is then usually removed around stages d) or e) for falconry training.

A mother New Zealand falcon arrives with prey. The youngster chases her, she drops it and the youngster catches it.
Photos Kurien Koshy Yohannan.

In the wild, the aim of the parent falcon is to make the chick become a successful hunter, independent of her parents, ie stage h). In falconry, the aim of the falconer is to make the falcon a successful hunter of wild prey but to remain still partially dependent on or manageable by the falconer. The problem with the wild-caught falcon is that she tends to be too independent spatially and is insufficiently orientated toward the falconer. She easily wanders off self-hunting, looking outward for opportunities instead of inward toward the falconer for game. Although she may have several drawbacks in her prey preferences, her strong point is that she is a superb hunter, better than most falconers could make her in several seasons' flying. It is the falconer's task to make the wild-caught falcon's life more pleasant and productive than she experienced in the wild, so that she wants to stay with him as a hunting partner, and does not see him as a potential predator, such as a coyote. Another factor to consider is that it is now virtually impossible to obtain such a bird in Britain and here at least the discussion is now purely academic. The would-be falconer is left with his young domestic falcon endeavouring to bring her to perfection in a hostile world.

4.23 Further Reading

Adams-Hunt, M.M., and L.F. Jacobs. 2007. Cognition for foraging. In *Foraging: Behaviour and Ecology*. D.W. Stephens, J.S. Brown and R.C. Ydenberg (Eds) pp105-138. University of Chicago Press.

Anderson, S.R. 2004. *Doctor Doolittle's Delusion: Animals and the Uniqueness of Human Language*. Yale University Press.

Bates, L.A., and R.W. Byrne. 2007. Creative or created: Using anecdotes to investigate animal cognition. Methods 42: 12-21.

Dawkins, M. S. 1985. *Unravelling animal behaviour*. Longman.

De Waal, Frans. 2016. *Are we smart enough to know how smart animals are?* Granta Publications. ISBN 978-1-78378-306-9.

Ferreira, D. F. 1616. *Arte de Caça de Altaneria*. English translation by Anthony Jack 1996. ISBN 0952690616.

Godfrey-Smith, P. 2016. *Other Minds*. Harper Collins. ISBN 978-0-00-822629-9.

Grant, P.R. and B.R. Grant. 1997. Hybridisation, sexual imprinting and mate choice. American Naturalist 149:1-28.

Halliday, T. R. and P. J. B. Slater. 1983. *Animal Behaviour*. Blackwells Scientific Publications, Oxford.

Hammond, N and B. Pearson. 1988. *Birds of Prey Behaviour Guide*. Hamlyn. ISBN o-540-01277-7.

Hess, E. H. 1973. *Imprinting*. D. Van Nostrand Co. ISBN 0 442 23393 0.

Heyes, C.M. 1994. Social learning in animals: categories and mechanisms. Biological Review 69:207-231.

Horn, G. 1981. Neural mechanisms of learning: an analysis of imprinting in the domestic chick. Proceedings of the Royal Society of London B213:101-137.

Immelmann, K. 1972. Sexual and other long term aspects of imprinting in birds and other species. Adv. Study 4:147-174.

Irwin, D.E. and T. Price. 1999. Sexual imprinting, learning and speciation. Heredity 82:347-354.

Kaplan, G. 2016. *Bird Minds*. CSIRO Publishing. ISBN 9781486300181.

Krebs, J. R. and N. B. Davies. *Behavioural ecology*. Blackwell Scientific Publications, Oxford.

Lorenz, K. 1965. *Evolution and modification of behavior*. University of Chicago Press.

Marzluff, J., and T. Angel. 2013. *Gifts of the Crow*. Atria Paperback. ISBN 978-1-4391-9874-2.

McFarland, D. 1981. *The Oxford Companion to Animal Behaviour*. Oxford University Press. ISBN 0 19 86610 7.

Rowlands, M. 2012. *Can Animals be Moral*? Oxford University Press. ISBN 978-0-19-984200-1.

Skutch, A. F. 1976. *Parent birds and their young*. University of Texas Press. ISBN 0 292 76424 3.

Slagsvold, T and B.T. Hansen, 2001. Sexual imprinting and the origin of obligate brood parasitism in birds. American Naturalist 158:355-367.

Slagsvold, T, B.T. Hansen, L.E. Johannessen and J.T. Lifjeld. 2002. Mate choice and imprinting in birds studied by cross-fostering in the wild. Proceedings of the Royal Society of London 269:1449-1455.

Tinbergen, N. 1951. *The Study of Instinct*. Oxford University Press.

Weiskrantz, L. (ed.) 1988. *Thought without Language*. Clarendon Press, Oxford.

C5 Basic Training

This chapter is supported by a film in the Bird of Prey Management Series:
Basic Training.

Just spending time together.
Photo Helen Nakielny.

5.1 Routes to training

Training consists of two interwoven strands: training the mind and training the body. In this chapter we'll look at how to apply some of the mental elements covered in the last chapter into real-life training techniques. During the period 1968-2022 I have kept detailed daily records of training about 330 birds of prey, a total of about 700,000 training and hunting days. Some of these were young falcons, destined for new owners, but many were young falcons being 'proven' before we retained them as breeding stock, and some were retained for hunting up to 15 seasons. About 300 novice falconers assisted in training these birds and so together we discovered all the various mistakes one can make. Our procedures had to work, to minimise stress and risk for all concerned, and be bomb-proof. At the same time we have been developing protocols for preparing young raptors for breeding and eventually domesticating them. If you only train a few birds it is easy to become prejudiced from your sample size. You convince yourself that prairie falcons are 'difficult' or gyrfalcons take two years to mature. The reality of course is that any wild young gyr that took two years to learn to hunt properly would be long dead. The arctic is a hard task master for gyrs and peregrines and the window of summer is short. It is our management that is inadequate.

All the old books were based on training wild-caught raptors; the procedures from various cultures are extremely well documented and I don't need to repeat them all here. The raw material in a wild-caught bird is one who already knows how to fly and hunt, but needs taming for management by humans. The captive-bred bird, on the other hand. does not know how to fly and hunt, and, if not handled by the time she is hard-penned, also needs taming and training, so you have the worst of both worlds. This is the kind of bird we send to clients in the Middle East, whose preferred training methods are based on those traditionally used for passage hawks. It is a lot of work, and stressful on the birds, leaving them vulnerable to diseases such as aspergillosis.

But it doesn't need to be like this. Through breeding in captivity we have the huge advantage of having access to the bird from the day she hatches. We can control how we manage her throughout her entire developmental period. We can exploit these sensitive periods so that, by the time the bird is hard-penned, most of her training has already been completed, and in a far less stressful and more welfare-friendly manner. It's like learning a language; it is easy and painless when you are a toddler but a slow process when you are a teenager.

How we handle them when young depends on what we intend for them when they are sexually mature. Do we want them to be imprinted for artificial insemination? Do we hope to use them as natural breeders in pens? Or are they destined to be introduced to free-living in the wild in a restoration programme? All of them need to learn to fly and hunt in some shape or form, and for falconers, hunting may be their main goal and priority. For future breeders, flying prey of some sort in their first season gets them fit and strong, which will prolong their active lives, it ensures that only good hunters get to breed and, if after a few years, a bird is not needed for breeding, she can still be flown in falconry.

So essentially we have main two routes to training:

1 Being naturally imprinted and not handled until at least 60 days old.

2 Being fully imprinted on humans, or dual imprinted first on real parents then on humans.

5.2 Developmental stages

As far as physical development goes, all raptors are the same. On hatching they rely on nutrients from the yolk sac while they develop muscle tone and the gut starts to function. All growth is centred on the gut to enable it to maximise food intake. As the gut develops the whole chick rapidly grows in size, too quickly for the skeleton to keep up, so the chick at this stage is virtually a big floppy bag of belly. The skeleton quickly catches up, with the long bones extending and calcifying. The fine natal down is now too sparse to cover the expanding skin area so a new, woolly, second coat of down grows, helping the chick to keep warm and control her own temperature. In a peregrine, by week 4 the first feathers start to sprout and the chick should be able to take her first staggering steps standing up. During weeks 4 and 5 the body feathers are growing and the tips of the tail and primaries are opening. The chick is at the harlequin stage and starting to pull food for herself. By the end of week 5 the tail is half down and there is not much down visible, although most of it is still present underneath the feathers. The skeleton is now full-sized but not fully calcified. By the end of week 6, after a lot of wing pumping exercise, the chick may attempt her first flight and leave the nest for rocks or branches nearby. About 2-3 weeks later all the primaries and tail feathers will be hard-penned, allowing the youngster to fly competently. The skeleton too has hardened, although the epiphyses, or caps, on the long bones of the legs are still fusing.

While all this physical growth is going on, the brain is going through some very critical phases of mental maturation. To start with the chick is almost blind but quickly she becomes aware of her surroundings, particularly her mother, the source of heat and food. During her first 4-5 weeks she identifies her mother in detail: her size, colour, voice, everything about her. Later in life the chick will still be able to identify her mother as an individual, and also if she has been reared by a black falcon, she will choose a mate that is black, and vice versa if a white falcon, regardless of her own colour. This means that if a black chick is reared by a white mother, she will look for a white mate. But the white mate may have been reared by white parents and spurn the advances of the black chick, so any relationship is dysfunctional. Thus the chick has imprinted on her parents, both as individuals, and as species. In future she will look to breed with birds of the same species, but not her own parents, because of the Westermark effect or incest taboo. If during these critical weeks the parent is changed, the chick will grow up as a dual imprint. In peregrines the critical period for parental imprinting is the first five weeks. After this time, the youngster will forever consider herself to be a peregrine, regardless of any future parent figures. Similarly, if the chick has siblings, she will imprint on them as individuals and later as species. She cannot imprint on them as species until they have sufficient feathers to be recognisable. Obviously these imprinting effects are key to understanding how to rear the chicks, depending on our plans for them.

Around weeks 3-6 the chick is well aware of her surroundings and can realise what type of nest she is in. It may be an old stick nest in a tree, or just a scrape on a rocky ledge. She will tend to choose a similar place when her own time comes. Generation after generation may have a tradition of just using one type of nest, while other families conservatively use another type. While this aspect is imprinted, the broader elements of nest building and so on are instinctive and ethotypic. They have to be. An accipiter was not alive to see her parents build the nest she hatched in. So she has no prior experience of nest-building when it comes to build her own first nest. Instinct steps in and not only does the bird make a nest, usually it is recognisable to species level. Just looking at a nest is usually enough to identify what species made it. So this is part of its characteristic ethotype.

Nest imprinting extends beyond the structure of the nest to the surrounding area during weeks 5-10. Later in life the youngster will usually come back to the same area to breed if there is a vacancy. Our *calidus* peregrines in Siberia nested on average 5.3km from where they hatched, despite having travelled thousands of kilometres away during the winter.

These imprinting effects have critical periods when they take place. They also require some sort of entity to imprint on to. But there are other processes going on as well that are more generalised. The fear response has a long critical period that extends virtually until fledging. During this period the chick accepts harmless stimuli and comes to fear harmful ones. As raptors are strongly visual, sights are what they fear most. They accept their parents and siblings, and waving branches and scudding clouds. But a stranger, such as a snake or mustelid, appears instinctively harmful, and this may be reinforced by the alarm calls of its parents. Thus the chick grows up accepting some things, but fearing others. If, as humans, we wish to be identified as harmless, then we need to be seen frequently during these formative weeks, without ill effects. The default position is that humans, even if never encountered during rearing, are to be feared. For species that have lived alongside humans, this is a sensible default position. Many of their ancestors have been killed by humans. It is a sad

A *calidus* peregrine nesting in Lena.
Photo Andrew Dixon.

reflection on us really. If you go to remote islands where wildlife evolved without humans, they are naturally unafraid. We might say they are 'tame'. But they are not tame, they are just unafraid. They will still object to you intruding into their personal space. Male New Zealand falcon chicks ten days old and barely able to walk, will kek and try to attack you (ie defend themselves) if you come to the nest. But outside the breeding season the falcons will come and land within a few metres of you.

By taking a chick from her parents at about five weeks old, she has already imprinted on her natural parents, but her fear of humans can still be allayed, so that she loses her fear of anything you introduce her to, including yourself. This fearlessness, achieved during the critical developmental period, will continue for the rest of her life. So the fear response has a critical period, and an imprinted permanent effect.

While all this has been going on, the chick has been fed by her parents and has imprinted on them. Once she starts to fly, the parents will tempt her into flying by dropping prey for her in the air. So the chick has to physically learn to coordinate her flying, with wings not yet fully developed. She needs this period to develop these skills, and if confined so that she cannot fly for a month or two, she struggles later in life to learn to fly. This may seem obvious, but some methods of handling entail keeping young raptors tethered or confined long after the period when they would naturally fly. Giving them the opportunity either by hacking or some other free flight system, is important.

As soon as the youngster has some proficiency in flying, and as her feathers harden off, the instinct to chase and hunt kicks in. If denied this opportunity to learn to hunt the youngster becomes less and less motivated to hunt and more and more dependent on the falconer to provide food or a lure. We say she is 'lure-bound'. Additionally, some species, such as peregrines, quickly develop prey preferences and search for those prey species while ignoring others. The search image may be very specific and we say the raptor is 'wedded' to that prey or group of prey. Some peregrines specialise on pigeons. Some goshawks get wedded to squirrels, or corvids. Some hunt birds but not mammals. This may be annoying for the falconer, or it may be an advantage. The last thing I want my crow falcons to look at is a pheasant; I want them just to have a search image for black crows. These developmental periods and traits are not as fixed and as well marked as some elements of imprinting, but they occur nonetheless and we ignore them at our peril.

5.3 Training full imprints.

To produce raptors for breeding by artificial insemination, we fully imprint them on humans. Accipiters too, even for hunting, are often better off being fully imprinted. This is what we do for a peregrine or similar:

Day 1-7. Hand-rearing in a brooder.

Day 7-35. For full imprints, we continue with hand-rearing.We use cardboard or plastic boxes with plenty of old towels that can be taken out and washed properly. The towels keep the chick warm and prevent legs splaying. When moving the chick, we never grab her bodily, especially not with legs dangling. This is very scary for her. Instead we use a board, the size of a bread board, covered in astroturf. We gently persuade her to shuffle onto it and then pick the board up.

Full imprints are never allowed to see other falcons, only humans. Therefore we use a box with a single window facing one way. That way we can have a row of imprints all seeing humans, but not one another. It is best to start hooding imprints as early as possible. Once they are fully feathered they do not like being introduced to a hood.

Imprints, having little fear or respect for humans, can be dangerous. Imprinted Red-tails and Harrises, if managed wrongly, may attack the falconer or children. Harrises which have been fed on fluffy, yellow day-old chicks may attack yellow pom-poms on children's hats. Imprints with a record of attacking people should not be flown loose, particularly not near children. Learn how to catch and handle hawks without either party being clawed, and if you do get clawed, soak the injured part in hot water with disinfectant as soon as possible. This often prevents an infection developing. Gentlemen should also take precautions when relieving themselves near a hungry imprinted hawk; one person's pride and

Dual imprints on a carrying board.

A simple unidirectional imprint box.

Young imprints can be very playful. They enjoy toys.

A big, easy hood slipped gently on and off accustoms the young imprint to hooding.

joy is another's pick-up piece, and I believe the record for this type of injury is eight stitches.

5.4 Training dual imprints.

To produce a good hunting bird suitable for natural breeding later in life, the best system we use for our own birds is dual imprinting. Dual imprints are fostered back and parent-reared by the same species and colour of bird as herself. This ensures her parental imprinting is normal. The parents may be a natural pair, or better still an imprint who will allow you to come and say hello to the chick without sounding an alarm call.

The key to understanding the dual imprinting method is to look at what is happening around day 35 (for a large falcon). The youngster is about 7-10 days off being able to fly. By now she has completed her bodily growth and grown most of her body feathers. She can just stand up and her wing and tail feathers are about half grown. She has imprinted as a 'falcon' but her fear response

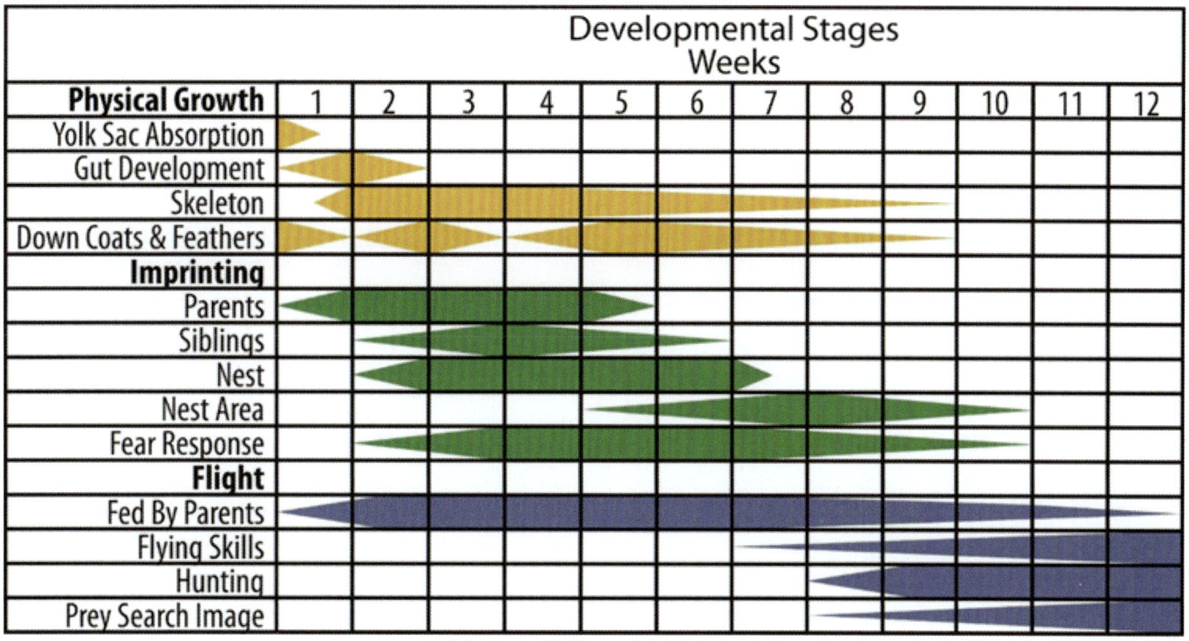

Figure 5.4.1. Developmental stages for a peregrine.

is still developing and during this critical period we can reverse it so that she becomes naturally tame. Being unable to fly, she does not react by attempting to bate, so we can stroke her and handle her, over-coming her fear, without her panicking. When we take her from the parents we fit her with anklets and place her in a circular box made of clear plastic. She can see everything going on but cannot injure her feathers and her mutes are contained. We place her in whatever room has busy people in. On her first day she will be frightened and make kekking alarm calls or hissing. We approach now and then and offer titbits of meat which she will at first reject in a trance. She can be outdoors in the shade, driven around in cars, moved from room to room and so on. Being driven in a bumpy car teaches her balance and she will travel better later on. Within a day she will stop alarming and within a day or two more will be showing a calm interest in what is going on. She can be hooded from time to time and for short periods, without pulling the braces too tight. She still has fluff on her head so an extra large hood is easier for her. After four or five days she will be relaxed without you having had to do any confrontational manning. No bating, no tethering. Her imprinted fear response is being overlaid at this stage; she will become tame and this tameness will hold, without top-up manning, for the rest of her life. She will be used to humans generally, not just one individual. She knows dogs, traffic, children, anything that might have alarmed her later in life. She is eating as much as she wants, which is not a lot at this stage. The metabolic demand on her is small and the weather is usually warm.

Day 40-45. We continue with the same treatment, but now her food is given in feathery chunks that she has to pull herself. In other words, we are making her to the 'lure'. The food can be tied to the proper lure. By now she is getting restless and wants to get out of the box. So we gently lift her out, hooded, on the fist, slip on jesses into her anklets, and a swivel, and take her outside. This is a tricky age. She is starting to fly yet you cannot really tether her. You can continue and tame hack her, or you can put her in a flight pen. But there is a danger that you will not maintain contact with her during this critical time. A simple alternative is to attach her swivel to a leash, which is attached to a short length of chain, which in turn is attached to a drag weight. This is the same principle as a boat at anchor. The links of the chain lift, progressively adding drag, without a jerk.

Fifty years ago I was tame hacking three brothers in New Zealand as dual imprints. They would fly away high into the sky but when they got tired they would often come down to a grass pile by the lawn. It was so soft and warm! Soon

A youngster investigates the lure, while a younger one watches.

Gentle hood practice.

they would be lying down, fast asleep. So now, instead of a grass heap full of aspergillus spores, we can take a pillow or cushion and drape a rug or old towel over it. The youngster can use it as a perch, or lie down, but she is not in danger of being tangled or hitting a sharp object. She can

Weathering in the shade.

wander around on foot and get used to tethering but without bates, jerks or risk of injury to her calcifying leg bones. She can have her first bath and do her wing exercising. She is hooded off and on. At night she is left in her round box unhooded, in the dark with mesh over the top or her leash attached through a hole near the base so that she cannot climb out.

If there are several of the same age they can be kept in sight of each other but kept physically separated unless supervised. This encourages them to retain their identity, imprinted as falcons not as humans.

Day 45-50. Depending on species and gender she is now starting to fledge. She is physically capable of short flights. This is a difficult age for handling because feathers are still in the blood and her skeleton is still not fully calcified. She knows the lure and will run across the lawn for it, more out of play than out of hunger. We also introduce the roprey model and she eats her food off the neck. Having travelled in a car and got use to motion, she can be carried gingerly on the fist bare-headed, avoiding any bates. At any signs of restlessness she can be lowered to the ground or hooded.

She is more co-ordinated now, preening and scratching her head. She will also be very tame, rousing and preening your ear lobe or hair for you. She is soaking up learning like a sponge. Dogs, tractors, children, passing birds, horses, all are examined with curiosity rather than fear. She is ready now to have her back-pack fitted. This can be done without casting her. It will be quite loose on her, to allow for her filling out with muscle, and she will soon preen it in until it disappears. At this age Lynn Oliphant lets the youngsters fly free at tame hack. We cannot easily do this because the youngsters interfere with breeding pairs in aviaries. We are able to put them in groups in large hack pens and continue with their lessons there for a few days while they learn 'circuits and bumps' - take offs, landings and turns. This enables them to socialise together, but do not allow a potential future pair of birds to be reared together because of the Westermark effect.

If you cannot hack at home, or use large flight pens, you may need to drive out of town

Getting used to cars, dogs and travelling.

each day and provide a 'chaperoned' hack from your car. This is described nicely by Bolinches Gómez-Torres in Patrick Morel's *The Art of Falconry*. This requires a lot of commitment and you still have to hold the bird somehow during the remaining 20 hours or so of the day. If you cannot tame hack in some way you may be able to put her on a running creance so that she can flutter around the lawn without being jerked and without getting herself into trouble. It is a tricky few days when you want her to have maximum freedom but equally protect her from danger or injury and keep her socialised.

Day 50-60. As she becomes hard-penned she will have graduated to the creance for lessons, coming a few metres to the lure which is now attached to the pole to make more interesting movements for her. Her food is becoming more rationed and the main meal is now being fed on an old roprey representing her future quarry species. The engine is revved to create a sound association. Soon, on hearing the sound, she anticipates the roprey and a meal. Then the roprey is suspended from a 1.5m (5 ft) cord hanging from a pole. We call this the 'conker crow' after the English game of conkers. Thus she quickly learns to take both lure and roprey in the air, all while still on a ten metre creance. We put up a 100m line stretched between two posts across a

At this stage don't hold on to her if she jumps off, just let her go.

paddock, with a running ring and a 10m creance. One or two calls on this give her experience of flying a little distance.

Entering to roprey.

Once she is hard-penned we can control her food intake without risk to growing feathers. We can still work mainly on appetite without much weight reduction. She has a progress chart and is weighed before flying. She has graduated to a shelf perch at night and a block on the weathering lawn during the day. She can have her tail-mount fitted and start flying loose. Using a slow gentle sweep of the pole lure, we encourage her to approach with a little speed and then turn in the air without landing on the ground. Usually in the same lesson she does two or three passes to the ungarnished lure. Her rewards should be small, but she should be given plenty of time to enjoy them. Peace can be reward in itself. The lesson is finished with a single call to the conker crow, with her meal being taken on the head. Some falcons are taking flying roprey before they are off their creance.

After a day or two on the creance to give her a chance to become familiar with the flying ground and generally confident with flying properly, we fly her free. Without the chance either to do circuits in a pen or outside, she may panic, not knowing how to stop or turn. Preferably she

Seeled? No, asleep. These falcons grow up perfectly stress-free. Two months later this falcon came third in the Flat Race at Vowley and third in the Novice Hunt Race, in September.

learns all this before her feathers are full-grown and before she is capable of fast flight. From there we build up to about 30 stoops to the pole lure, preferably in a good wind. She will also be given calls of 500-1000m long-luring to the pole lure to ensure she is good to come in from a distance.

Once she has taken a roprey flying slowly past her, her instinct to chase and hunt will kick in. In only one or two sessions she will start chasing it all around the sky. Each flight is made more difficult, concentrating on her turns, footing and fitness. After another two weeks she should be chasing very determinedly up to about 150m altitude and in strong winds. She knows what her prey looks like and she is ready for entering at real prey. Thus a chick hatched at the beginning of May is ready for entering by the end of July, having had more flying and successful hunting experience exclusively at her chosen quarry than a hack hawk. Some people maintain that the young hawk at this stage should be given live bagged prey to kill. This is not so. Yes, she needs to start 'killing', but in her mind, a hunt after roprey counts exactly the same as real prey; as far as she is concerned, she has already had a few dozen chases, all of them ending in a 'kill', but none of which could be taken for granted because they became progressively more challenging. Nothing succeeds like success in building confidence. By the end of September her flying should be indistinguishable from a passage bird's, because we will have systematically taught her all the escape moves of her prey, and how to overcome them. I cover these procedures in chapter 9. She should also be silent and good mannered in all aspects.

I have used dual imprinting now for many decades, initially without thinking about the theory behind it. After all, when eyasses were taken from wild nests in times gone by, many were effectively dual imprints. But timing is everything. Start too early and you will end up with a screaming 'tame-hearted eyas' so detested by Shakespeare. Too late and the fear response will not reverse before the hawk is able to fly. Lynn Oliphant calls this the 'Four Week Window' method and combines it with tame hacking. We avoid the tame hacking, and the use of bagged game, by substituting roprey and giving a more controlled, targeted, course of learning to fly and hunt. But the point is, as Lynn explains, to exploit that critical sensitive period when the hawk can learn so much, so quickly, so painlessly, and so permanently. And our young hawk will still be on the same developmental time scale as her wild counter-part, ready to become a strong independent hunter while the prey are still young.

5.5 Training the 'wild' raptor.

The traditional way of training entails leaving the raptor alone until she is fully hard-penned. Only then is she taken in hand and her training parallels that of a wild-caught bird. This means that conventional training doesn't even begin until the bird is about 60 days old and we have to start from square one. This is more stressful for both the hawk and the falconer, and more time consuming. With the dual imprinting method the hawk at 60 days is already flying loose to the lure, without having had any confrontational training or bates. She is about a month ahead already.

For some falconers the taming of a 'wild' raptor is a mastery to be achieved and marvelled at, a sense of dominance over the brute creation, a battle of wills. For me, some of these traditional methods of taming and training raptors involve too much stress and raise welfare issues. Now we know better ways to train, they are increasingly redundant. For other people, the relationship or bond between man and bird are key to their whole falconry thing. While an imprint raptor does have some sort of a relationship with a human, identifying the human as a parent, or perhaps as a potential mate, the late trained raptor has no imprinted relationship and, being a solitary predator, does not include the falconer within her familial relationship. Yet she has no other types of relationships, other perhaps than her instinct to shadow large mammals in the hope that they will flush prey. She is not a social animal with a hierarchy and a flock or kinship group. So the falconer is trying to construct a relationship which, for the raptor, is an artificial one. The falconer uses habituation and reinforcement techniques, and of course the raptor recognises the falconer, she has acute recognition skills and is not stupid. But the bond she experiences, such as it is, is based at this level. The falconer may project all sorts of emotional bonding on to the bird and believe it is reciprocated, but the reality is more prosaic. Shakespeare, in Othello, contrasted the emotional tie of the falconer with the lack of a bond from the old falcon:

If I do prove her haggard, though her jesses were my dear heart-strings, I'd whistle her off and let her down the wind, to prey at fortune.

Paradoxically, this means that a whole group of us can hunt with the same bird using the same signals. She will respond to any lure regardless of who is flourishing it. Yet a dog, a social animal, has his own idea of who is in his own bonded social group and may ignore a stranger's instructions, unless they mimic those of his owner, such as borrowing his whistle.

The traditional program goes something like this for the domestic-bred raptor:

Day 1-5 Hatched in incubator and hand reared.

Day 5-10 Reared by imprint falcon until ring stays on.

Day 10-40 Reared in big pen by a natural pair of falcons.

Day 40-60 In the hack pen.

Day 60 Measured, fitted with anklets, jesses, harness and tail-mount, taken up hooded.

Day 60-65 Hooded, or secluded weathering, while weight is slowly reduced.

Day 65-79 Taken in hand, trained to fist, lure and roprey, jumped to fist for exercise and flown loose.

Day 79-90 Stooped to lure, catching easy rocrow in the air and jumped to fist for exercise. Long-lured over 500 metres and introduced to dragged dead crow.

Day 91 Chasing rocrow up to 200 metres altitude and footing well, ready for entering and hunting real crows. Flown at rocrows on non-hunting days, or rested.

Passage hawks are taken any time after they are hard-penned. Domestic falcons destined for the Middle East are shipped at about two months old and treated in the same way as wild-trapped birds. These are the traditional methods of Arab falconry which are based on passage birds because very few falcons nest in Arabia and there is no tradition of training eyasses.

Our procedure starts on the day the egg hatches and the falcon goes through a very careful process of natural imprinting, learning, developing physical skills and mentally maturing. The daily food association with man is only introduced at day 60; by this time the falcon has already graduated through several major stages in her development. We used to wild-hack or tame-hack the young falcons but abandoned this now that we have roprey. Hacking risks losing the falcons, or them being injured or killed, and they can develop bad habits such as wandering, lack of focus, and chasing the wrong prey species. I will go into this

Young falcons in a hack pen. The pennants are to discourage the birds from flying too close to the netting.

in more detail in advanced training.

All training programmes must consider two aspects side by side. First is the actual mental training using the systems already discussed. Second is the physical management of the bird, making sure that she has a good appetite to be responsive to your training efforts and developing the physical condition of the bird so that she is fit and strong.

The training of a wild-caught accipiter differs in some ways from that of a young domestic falcon and therefore it would be pointless to give a potted recipe for training. Different birds must be brought on at different rates according to their responses. The ages I have given are guidelines corresponding to mental and physical developmental stages. These vary between species, genders and individuals. The skilled falconer will bring on a bird in such a way that he doesn't make her stale with needless repetition and yet doesn't venture out onto the thin ice of risk. The fastest bird I ever trained was a hacked New Zealand falcon called *Sally*. From feeding her on the fist for the first time to stooping loose to the lure took twenty minutes. Marvellous. On the other hand, I had a haggard Changeable Hawk-eagle called *Gorgon* who never flew loose, despite all my efforts. He was a very, very slow, deeply suspicious bird. He seldom bated, he just looked at me with an unchanging, impenetrable trance-like stare.

Although there are these differences, the falconer should have a fairly shrewd idea of his own schedule. If his hawking is planned to start on a certain date, he should work forward from that date and plan out his program with certain targets to aim for along the way. I examine this in detail in 10.2. The prey has its own season and there are critical periods for entering young raptors. In nature, the fledging of the raptor chicks is timed to provide the best opportunity for them to learn to catch the abundant naïve prey. A few weeks too late and the situation is very different; the young raptor will have 'missed the boat'. A wild one will die. For many of the game species, these seasons are enshrined in law, and usually developed for gun hunters. For example,

the hunting season for pheasants in UK starts on 1st October. So if the falconer is developing a pheasant hawk, what should he do with her during August and September when he cannot legally hunt pheasants? The season for Red Grouse (*Lagopus lagopus scotica*), on the other hand starts on 12th August. (Actually to fit in with the summer recess for Victorian parliamentarians rather than the biology of the species. There are still cheepers around in August.) Can the falconer have his young falcon fit and waiting on properly by mid-August? For merlins, the timing is even more critical. The young merlins are usually ready towards the end of July, at the right time to enter at young skylarks or moulting adults. The traditional season for lark-hawking therefore was August. By the end of August the larks are clean moulted and getting strong. A fit merlin can still catch them but a new youngster would quickly become disheartened if entered this late. Yet the legal season for limited licensed hawking is now set at 1st September, effectively destroying any chance for a young captive-bred merlin to become an independent hunter, particularly if she has been flown to the lure for all of August and become lure-bound. To a lesser extent, the same applies to prey such as rabbits. A January rabbit, or pheasant, is a very different animal to an August one. Many falconers, unconscious of nature's urgency, let their training drift on through summer, and before they know it, October has arrived, the nights are drawing in, the clocks are changing and preventing flying after work, the weather is getting rougher and the easy game has gone. His unentered young hawk now has the odds stacked against her.

Similar situations pertain in other countries. There may be a gap between the time that your raptor is ready for entering, and the legal hunting season for your intended quarry. For example, in USA, the seasons for upland game such as partridges or grouse may not start until September, and for ducks, even later. Falconers in some areas have negotiated an extension to the gun hunting seasons on the basis of a 'let lie' clause, but in other areas falconers use bagged game, such as pigeons, instead of hunting. Thus, we must balance the developmental stages of our young raptor, with the legal hunting dates. A gun hunter does not need to teach his gun to shoot, but a falconer must bring his young hawk on in stages. Hawks are not machines.

Each bird goes through a pre-planned program of training and, allowing for bad weather and individual differences, is ready by an approximate date. I tend to be very jealous of the bird's intellectual virginity. It is a lot easier to take a young bird and train her right first time, straight through. Her training goes bang, bang, bang, - finished. On the other hand, if someone has messed her up in some way, made her hood-shy perhaps, then everything is on stop while this is sorted out and it might take more time and effort to rectify than the rest of the entire training.

It is important that the falconer works out a simple program for his individual bird and takes the training through in a positive, progressive manner. Young hawks are tremendously receptive and to dither with a half-trained bird is to invite difficulties. There is a whole myriad of problems waiting to beset birds which are not taken through training at their natural rate: screaming, aggression, carrying, 'moodiness', difficulties in entering at quarry. The experienced falconer must learn to anticipate and side-step all of these. While it is unwise to rush your hawk, at the same time it is not good to over-repeat lessons.

As a breeder, I send out many young hawks each year to their new owners, and inevitably, some of them don't turn out well. This is almost always due to the falconer who does not have the ability to do the bird justice and who is not prepared to take responsibility for it. Instead, he blames it on the bird. On the other hand, the birds which stay behind with us turn out well, because we put extra attention and work into any developing problems, in order to overcome them.

Therefore, assuming you are soon to take delivery of a new bird, work out two schedules for yourself: the first schedule is your calendar, showing target dates through to catching quarry regularly. Your second is a list of attributes you wish your hawk to have. At present you are at

Goshawks in Pakistan.
Photo Ben Crane.

point A: early summer with an untrained, unfit, hawk. You must get to point B: early autumn with a trained, fit bird. Maybe you are getting a Harris hawk, then here are some possible targets:

- The hawk should be in top class health and feather condition.
- She should hood well, and have a well-fitting hood.
- She should be good mannered on the fist and never try to foot you.
- She should come immediately 200 metres to the fist without reward.
- She should be fit and capable of 150 high jumps in one session.
- She should be experienced in wind and hilly terrain and should wait on in suitable areas.
- She should work well with your dog and the dog with her.
- She should work well from the lofting pole.
- She should follow you well and choose suitable high perches for overlooking the terrain.
- She should take rabbits well and also have experience with feathered quarries.
- She should not scream.
- She should work with other Harris hawks to which she has been properly introduced.
- She should be pleasant to handle on the kill and throughout the day.
- She should travel well in the car.

During the course of this section I will be looking at some of the aspects you will need to cover when going with your hawk from point A to point B. For some, such as accipiters, the best route to follow might be full imprinting, as described by Mike McDermott. For some falcons, dual imprinting, as described by Lynn Oliphant, creates perfection. For Harris Hawks, which are naturally tame and sociable, full imprinting or even dual imprinting, can be a disaster, and often the best plan is to keep the bird in contact with parents during her formative period. And for other falconers, who are used to passage birds, the traditional route suits them best, with birds untouched until after fledging.

5.6 The wild hack

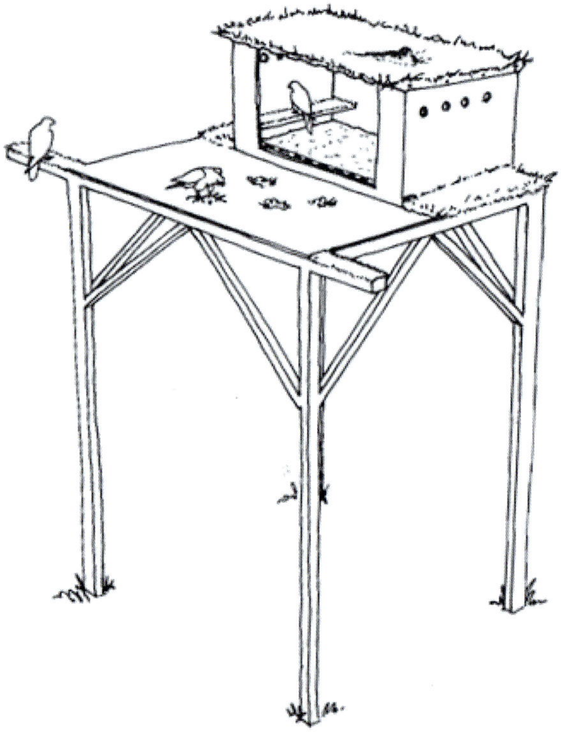

Figure 5.6.1 Tower hack box for young falcons.

Much has been written about how to hack birds of prey, systems of hack management and their pros and cons. Stevens (1955) and Sherrod (1981) cover the main points quite adequately. My own experiences have been limited to buzzards, kestrels, merlins, sakers, peregrines, gyrs, various hybrids, and New Zealand falcons. I have hacked the latter extensively both as juveniles and as adults and have even had them nesting at hack. We used to wild hack 20-30 sakers, gyrs, peregrines or hybrid falcons in Wales and in Northumberland every year. In many countries it is now illegal to wild hack non-indigenous raptors because of the risk of them escaping and breeding in the wild.

Our old hack box in Northumberland.

A hard-penned youngster is not safe to put out at hack immediately because she can already fly and might become lost. Also, if hacked from her original home, it may well interfere with other breeding pairs on the premises or on the weathering lawn. Hacking out high-spirited falcons intended for serious falconry is a different matter again from having a few cuddly kestrels or buzzards lurking around the yard. Big falcons can potentially cover considerable distances, paying particular attention to such magnets as pigeon lofts. In most of Britain they are thus exposed to unacceptable danger. Apart from human persecution and shooting, the falcons must run the gauntlet of busy traffic and a landscape festooned with wires. The hack site must be chosen so that there are none of these hazards, no gamekeepers, no rivers, lakes, or water-troughs in which the hawk might drown and the minimum of barbed wire fences. Standing crops are taboo because if a young hawk lands in one she can't relaunch herself, the same can apply to dense forestry plantations. Inquisitive, stampeding cattle are to be avoided. In short, you need to be able to control the land use over the immediate 100 hectares (200 acres) or so around the hack box. Some farm livestock and foxes can be excluded temporarily with electric sheep netting. But be warned: a young falcon can easily fly 200 km from the hack site in a day.

To get them to fly well, the box should be on a slope with plenty of hill-lift in all the main wind directions. This will draw them like a magnet and will encourage them to get into the air. It is hours in the air in strong wind which really does the good.

There can be no doubt that, if the falcon survives, she is improved physically by hacking. Her wind and muscles are better, she foots better and, most of all, she flies better. She begins to resemble the passage falcons, now no longer available to most European falconers. But what does it do to the falcon mentally?

A naturally imprinted falcon, even though captive bred, has a fear response of man and does not regard the falconer as a parent figure. Flying wild at hack (in contrast to tame hack) does not encourage a bond between falconer and bird and so this must be forged later. The hack falcon may learn to wander and hunt for herself, both attributes which we avoid by not extending the hack too long. When it comes to tackling quarry the young bird is constantly disappointed and without the falconer's assistance may give up on some species such as rooks and crows and be difficult to enter to them again. While in theory hack is supposed to teach the falcon that she cannot catch pigeons, in my experience it is more likely to teach her that she can catch pigeons, especially if she tears around the countryside and develops the right tactics. These self-hunting problems seldom arise in the first two weeks of hack.

Falcons which are intended for game hawking with waiting-on flights, are required to centralize over the falconer in expectation of invisible game being flushed out. Wandering off, chasing non-game species, landing on the ground or undertaking attack styles other than the stoop, are considered faults. Therefore, on balance, the potential game hawk may learn more bad habits during hack than she gains in terms of physical fitness, flying skills and footing ability. Such skills are probably best learned by an initial intensive training to a swung lure, followed by training in waiting on and then by consistently being served with real game to induce an almost mechanical routine of going up, waiting on and stooping. A strong bond of faith in the falconer serving game is the essential key to all this.

Falcons intended for flights out of the hood at rooks, crows, seagulls, houbara, stone curlews, magpies, larks, and so on are required to undertake a much wider variety of techniques ranging from direct flying attacks to ringing flights, shepherding, stooping, and tail-chasing. Their footing too must be considerably better than the average game falcon's. The falcon needs a knowledge of quarry, of the elements and of tactics which can be learned only through experience of real situations. There is no short cut to producing such maturity. A period at hack benefits such a falcon, especially if she is a slow-maturing clumsy bird such as young sakers and gyrs. We have occasionally had to train some unhacked ones due to having too many birds for the hack site capacity. The difference in their pursuit flights are immediately noticeable.

The first two or three months of a falcon's life are critical. If the bird is not entered and thoroughly wedded to her intended quarryshe may orientate herself more toward the falconer as a parent figure, become lure bound, start to scream and in some cases (particularly malimprints) start to attack him out of frustration.

Most of the above is fairly standard teaching and, as far as it goes, is perfectly true. However, since we have been hacking large falcons on a regular basis we have been able to experiment with the finer points of the procedure. This is what we've evolved: first of all one must breed sufficient birds to make up same-age cohorts, this means 2-6 falcons all hatched within about five days of each other. These are placed in the hack box overlooking a broad landscape. The hack box is well ventilated with a food hatch at the back and an open vertically barred door at the front, facing north or east. Branches or perches around the hack box enable the youngsters to clamber out and to return easily. The fledglings are placed in the box one week before they are capable of upward flight and transmitters are placed on their legs. This gives them enough time to imprint on their surroundings. During this week I kill or drive away as many foxes in the hack area as I can. Mammalian predators are very quick to catch young falcons at this stage, even in the middle of the day. The transmitters are painted in individual colours (including the antenna) and have a life of 60 days. GPS transmitters help hugely in keeping tabs on each bird and also in monitoring their development. At dawn the next day, sufficient food is tied to the hack-board in front of the window to last the branching falcons for three days. The window is then opened from below using draw-cords. The behaviour of the young falcons is monitored by radio and field-glasses from 300 metres or so.

We have found that the best hack period is about 14 or so days. This is long enough for them to become hard-penned and to build muscle and generally develop the sturdy frame which is the hallmark of a hacked bird. Because there are several of them they get flying experience by playing and chasing each other. Also they pull leaves from the tops of trees. But they don't wander far, seldom more than 1-2 kilometres and they don't do much chasing of prey. Occasionally a bird will kill early, in a week or ten days, but normally there is no serious chasing in the first three weeks so the hawks don't get disappointed or too orientated toward self-hunting. But after three weeks the situation can change almost overnight and it is not safe to leave them out. I cannot understand how people reckon to manage 4-5 week hacks. By this time my birds are away almost all the time and although they may come back every day or two they seldom come in at all to the hack-board. I know this because sometimes we put wild peregrine chicks out with ours to be returned to the wild following some incident with their own nests. We always pinpoint all hack birds at dawn, and again once or twice during the day, and again at dusk, making sure that they are roosting up off the ground away from foxes and cats. After two weeks of good flying at hack they have gained most of the flying skills they need, but still need 2-3 weeks to mature mentally. They do not need to be loose at hack during this period, because of the risk that they will disperse; they just need time to grow up mentally.

When the birds come in from hack their transmitters and hack bells are removed. These hack bells are just light bells to help us locate the birds and to identify the falcons as domestic ones in case somebody should be tempted to shoot them. We then turn them away into a big flight pen for a further 2-3 weeks in groups of up to 20 falcons. This completes the mental maturation process; when they come out they are mentally ready for hunting for their own food and no longer have the juvenile urge of looking for a parent figure. They never scream at humans.

On catching up from the big pen, a tail-mount and anklets are fitted and each bird is then made to the hood, fist, and lure, being back on the wing after ten to twelve days, depending on the weather. We prefer to fly them free first at the hack ground. Then, if any of them have a 'technical hitch' they are on their home ground and won't go far. As soon as another bird is put up, the wanderer returns. After about ten days on the wing generally getting fitter and more reliable in the routine of trained birds, they are ready for entering at crows or rooks.

Hacking can be used to teach a falcon to home and some of them pick this up very quickly. All my New Zealanders have been homers but now that we have telemetry to aid us homing tends to be more of a nuisance than a blessing. A falcon on the loose at home can cause a lot of damage before you can get there.

Accipiters are best not hacked except in extreme cases of aggression. Hacking reduces the bond with the falconer which is the last thing one wants to happen with accipiters. Buzzards do not benefit very much from hack; some become confirmed worm eaters and

On our farm in New Zealand in 1976 our nearest neighbours were a mile away in one direction and 60 miles in the other. I put a 40 gallon drum up in a Macrocarpa tree as a nest site and put out a pair of trained falcons. Unfortunately the male was killed at night while incubating the eggs by a polecat that managed to climb up. I found the polecat, terminated his employment and recovered the headless body. Polecats, domestic cats, stoats and weasels were released in New Zealand in the nineteenth century and cause havoc with native wildlife.

Falcons incubating in a nest barrel on the farm 1977.

camp scroungers. They will hang around for years.

I used to enjoy hacking time. I like to be up at dawn watching the youngsters waking up and tipping over in the wind trying to scratch the last fluff off. Then one by one they launch uncertainly into the wind, mounting higher and higher on different levels before turning downwind with panic written all over their faces. A few short days later they are full of confidence, scything around the sky, sneaking between the air molecules as they pierce the wind which before had battered them back down. Even a biologist is allowed a romantic streak. But I don't hack any longer. There are easier and more consistent ways to get better results.

When we are doing conservation projects, often we 'hack back' young or rehabilitated raptors. These can be conditioned to come for food and, having been reared in and imprinted on a particular hack box, may later nest in it once sexually mature.

Later in Marlborough in 2006 we ran a project called 'Falcons for Grapes' in which we established pairs of falcons at permanent hack, breeding in the vineyards, to reduce damage by pest birds. The falcons nested in our nest barrels which we raised up off the ground to deter mammalian predators. One female, Maggie, nested for ten years, watched over by our Project Manager, Colin Wynn.

Maggie (right) feeding a youngster on the food tray.
Photo Colin Wynn

Minister for Conservation, Chris Carter (left) and our project manager, Colin Wynn, watch a falcon drop in for food in our Falcons for Grapes project in the Marlborough vineyards, New Zealand.

We placed food trays immediately in front of the hack box or nest barrel. The tray is white with a bright red rim, about a foot square, of plywood, raised up on a fence post. There are draw strigs on it to attach the food. The youngsters quickly learn to recognise the food tray, and we can then move it. For example, when we wanted the falcons to deter pest birds in the vineyards, we could move the food trays to where grapes were ripening. The falcons quickly got clued up. A male at one site would follow Colin, as he drove to the next site, 15km away, to visit the next food tray.

The main problem was that because the vineyard owners cut down as many trees as they could, to reduce pest birds, the only high perches left were power poles and many of our falcons were electrocuted.

Similarly, if the plan is diversionary feeding, for example of harriers on grouse moors, the food trays can gradually be moved away from vulnerable areas.

5.7 Kitting out and taking in hand

When taken in hand, the bird is fitted out with simple anklets and a tail bell or transmitter mount. Body harnesses should be fitted with enough slack to allow for filling out as she grows. New Zealanders and accipiters are quite hard on their gear and therefore we usually attach the tail mount to the two feathers next to the two deck feathers, one on each side, taking care that the tail can fan fully. This means the weight is spread over four feathers. We may put a sleeve or tape on the tail for the first two weeks until she has settled down. Tape should not be left on for more than a few days because damp feathers may start to rot.

When the hawk is cast we also measure and photograph her properly, and check her identification ring. She is also checked over to be sure she is in full health. Training and handling will inevitably put her under some physical and mental pressure which could fan the flames of a disease. During measuring, the head width can be used as a basis for selecting a hood, either by making one up from Slijper's Canon or by using a blocked hood of that size.

Provided that she can cast through the hood, and is fat and in good health, if she is wild, she is best left hooded quietly for 48 hours. The least traumatic way to man her is to tie her to a perch in a quiet part of the garden where nothing can go behind her and where she can see what is going on from a distance. Most of the day will be spent on the ground, picking at jesses and generally getting used to everything. Go past her (not toward her) every hour or so, putting a tidbit on the block as quietly as you can, and then retire. If she sprawls, or ignores the tidbit, or flicks it away, she is probably still too high to do anything with. If she hops up onto the block and eats it, you are on your way. In the evening, preferably once it is dark, make in as gently as you can, hood her up and put her in her night quarters. Repeat this the next day, and the next, until she anticipates your approach with a positive attitude and starts to take food from your hand. At the same time, stay nearer as she feeds. You can start her on a few tirings, such as a juicy pigeon wing, which she can pull at by her block with you lying near her giving her extra tidbits from the glove. She will then come naturally to feeding on the fist and stepping onto the fist. You have sneaked into her life without confrontation. If, on the other hand, you attempt the old-fashioned approach to manning by putting her on your fist and carrying her about you will give her a lot of negative conditioning at the start which will take further work to cancel out. You also risk injuring her leg and hip joints, her leg scales, and her feathers. Remember, the ends of her long bones are not yet fully fused; she is still very delicate at this stage.

If she had been on her own in her pen you could have reduced her weight gently for a few days before catching her up. But then you would not have known her fat weight or been able to check her health first.

5.8 Manning and early training

If you reared your hawk as an imprint or dual imprint, you can skip this section! She will already be tame, made to the hood and pick up and put down nicely. But if she is a wild one, you will have to go the whole nine yards.

When the hawk was young, she gradually developed a fear response to all unfamiliar objects which were not part of her family routine. She grew up to fear Man and all his works. Manning is the process whereby the handler sneaks himself inside to join the hawk in her world. He may bring some aspects of his own world along with him. The best of course, is if the hawk was raised in such a way that she has no fear of man so that little manning is needed. Accipiters are among the most fearful of hawks and stress for them in captivity can lead to diabetes, aspergillosis and other ailments. Dual imprinting is often the best solution for this group. Nowadays many falconers hard imprint their accipiters rather than treat them like wild birds, and these same imprints are often excellent for breeding through artificial insemination.

Until this barrier of fear between hawk and man is removed, she will not be receptive to further training. To achieve this, as far as possible, anything the hawk has to do with you should be positive experiences with minimal negatives or stress.

The way you handle your hawk on the fist is critical and can easily lead to unintended consequences. Obviously you need to be able to carry the hawk smoothly and give her a comfortable ride. Our falconers learn by riding down to a stream, filling a mug with water, getting back on the horse and galloping home. The mug should be at least two thirds full! On foot, try the same exercise jumping over a wire fence. Consider every bate a failure on your part; try to anticipate and forestall bates and analyse when you fail. If you cannot float your fist for her she will come to hate you, and I wouldn't blame her. If you approach an open doorway, many hawks will bate ahead. You may need to go through backwards until she has steadied up. Can you achieve less than one bate a day?

Always keep the hawk facing the apparent wind: it should be blowing into her face. So, if you are walking along and you can feel the wind on your back, turn your whole hand so that she is facing backwards. If you don't, the wind will blow up her tail and tip her over, which she hates.

To compensate she may turn around on the fist herself and stand sideways or back to front on the fist. From there, if it gets any worse, she will bate. Which is her way of saying your perch is horrible. If you or the wind are travelling fast, she will learn to crouch down on the fist, keeping her head low so that the wind pushes her down rather than blows her off. Keep your arm quite close to your body, not half extended. That way she is sheltered from gusts and your arm will be steadier. Learn to do everything using your spare hand and remember the old car sticker: '*Falconers do it one handed'*.

Especially when your hawk is in training, or with accipiters prone to sudden bates, manage picking her up from a perch and putting her down without at any time having her untethered. Always tie her to your glove before untying her from the perch, and vice versa. Some people find this awkward, or simply want to take short cuts. You might get away with it as an amateur but if you are handling a lot of hawks on a daily basis it is these small details that find you out in the end. We do our best to minimise risks, and there are plenty of them. It's best if you can learn good techniques from the start. Once you have got into bad habits it becomes much easier to justify them than to change them.

With the hawk on the fist, the next step is to feed her. Done well, this looks simple. The decision tree looks like this:

Where is the food? Is it a kill or is it some meat in your bag? If she is on a kill you may wish to reward her right there with the brain or liver. But then you have to pick her up. Never lift the carcass off the ground, always encourage the hawk to step off it onto your hand. So you have to make your hand more attractive (with a pick-up piece of food) and the prey less attractive (by covering it with your bag, with your gloved hand, or with sand or grass.) When crow hawking we pull the wing off an earlier crow, or the one she is on. This comes away with all the pectoral muscle attached. This is then held in the clenched gloved fist with the meat showing between finger and thumb and the wing pointing down below the fist. Similarly a rabbit leg is held in the same way. However you manage it, the hawk must not be able to grab the meat in her beak and try to fly off with it. So the pick-up piece will probably be bigger than the amount you intend to feed. Place your garnished fist over the kill to conceal it and she will transfer her attention to the pick-up piece. Don't move. If she pulls at the meat, rather than let her pull pieces off, just let your fist lift up gently with each jerk. After a couple of attempts, she will realise that she needs to put her foot on the meat to hold it down in order to pull at it. Now you have her with one foot on the fist and one on the kill. Take your time. She may then step up easily or you may need to quietly unpick her claws from the kill and free her second foot. Once she has both feet on the fist don't be in a hurry to lift your hand. Just let her feed contentedly while you thread her jesses back on and secure her with the lines on your glove. Once she is secure, cover the kill with your spare hand and lift her up, not letting her see the kill be removed. So she has a positive experience all the way through and will be happy next time when she sees you approaching her on the kill.

If you are simply feeding her her daily ration, the first step, while she is still hooded, is to take the food from your bag. This is where things go wrong. Do <u>not</u> let her see the meat. Conceal it in your hand and bring it under your glove, feeding it up through the bottom of your clenched fist. The first time she sees the meat should be when it appears like toothpaste between your index finger and thumb. You are holding it for her. It is positioned naturally between her front toes and you have a good firm grip on it. In this way you control the amount she is eating and the speed she is eating it. She cannot bolt whole chunks by pulling it away from you. Nor can she grab the whole lot in her beak and then bate with it. Nor will she develop the habit of trying to foot your hand as it approaches. Your spare hand has always been below your glove where she cannot reach you or even see the food. As she feeds, gently push more meat up through your fist. If she is feeding items like day old chicks, always have the next chick ready in your right hand and push it up through your glove before she has taken the last mouthful of the first chick. If you let her finish the first chick before presenting the second there will be an awkward moment while she is looking around impatiently and seeing your right hand moving. She may bate. Instead, keep the supply of food coming steadily up through your fist and keep a good grip on it by clenching your hand. Watch her

A quick way to make a hawk become bad-mannered.

crop fill until she has eaten the amount you intend her to have. Our management films show how to do this and assess her crop size. If there is food left over inside the palm of your hand, just increase your grip between finger and thumb until no more is visible. So to her, it appears she has finished it, rather than you stealing food away from her. If you are going to put her back onto a perch, do not wait until she has finished eating. Tie her to the perch while she is still feeding so that, as she eats the last mouthful and feaks her beak, she steps gently onto the perch and will see you back away with regret. She will look forward to you coming again next time. Frederick II was a stickler for detail and gave explicit instructions (XLII and XLIX) on how to carry and feed a hawk on the fist, so did Ferreira (p57), and examinations of old portraits show the same method.

There is a school of thought doing the rounds at the moment in which the falconer holds the food between his fingers and the ball of his thumb, with his hand palm upwards. This is the way it is done using an Arab mangala, a cuff that leaves the hand bare. When I ask people why they do this, some say that it makes the hawk more comfortable by having a wider perch. I don't buy that. For one thing, my hand isn't appreciably wider when tilted 30 degrees, and for another, why would it need to be any wider? It is big enough as it is. Another falconer said she did it because her hands are too small to hold the meat. Some falconers have stiff, clumsy gloves with the fingers all gummed up with old food. The big problem with this method is that the falconer has no control of the meat. He cannot place it into his hand without the hawk seeing him do it. He cannot open his fingers to accept the meat because the hawk is standing on them. Often the meat protrudes both above his fingers and below them. He cannot control the meat or the feed rate, so the hawk snatches up a chunk of the food out of the falconer's grasp because he cannot grip it well enough. If he is lucky she will then just shuffle her feet until she has a claw or two on it and then pull it for herself. But what usually happens is that she will pick it up in her beak and try to bate off with it. Or the falconer grabs the meat whilst it is still in the hawk's beak and pulls it back down to the glove, then tries to get it under control, which the hawk interprets as an attempt to rob her. And when the hawk has had her ration, the falconer has no means of removing

Evening training with HRH Prince Muqrin and HRH Prince Bandar Al Saud, while the camels come in.

any remaining meat without openly stealing it. I see this happening time after time, year after year. Bad handling leads to mistrust, bad manners, and on to worse vices. If you don't control the feeding properly, you cannot control the weight properly, and with poor weight control comes poor flight performance or worse still, a wandering hawk…

If you plan to hood her, take your time. Many hawks deeply resent being hooded until they have finished their meal, cleaned their beak, checked that there is no more to be had, and maybe had a nice rouse. Then they are ready to be hooded, not before. But others may be more tricky and it may be that you need to hold the hood over the last bit of food so that she hoods herself when bending down to bite the meat. If you have to do this, let her have a few more mouthfuls once you have closed the braces so that she does not associate it with the food being withdrawn.

Traditional manning was an exercise in overcoming the fear response by using habituation. The hawk was carried around day after day on somebody's fist until it ignored stimuli which at first had scared it. Many trainers didn't even link the habituation with positive reinforcement or appetite motivation. The result was a system which worked but was extremely inefficient both in effort put in and in the quality of the trained bird. It was also needlessly stressful for the bird. Such birds often needed an hour of manning before being flown at quarry and, if lost in the field, were often unapproachable within 24 hours. Modern manning relies less on habituation and more on the creation and reinforcement of pleasant associations of the falconer with food. This system is faster and more positive than habituation which simply teaches the hawk to tolerate situations.

Around a hawk, or indeed any individual, there is an invisible personal territory or no-go area. You can see it operating on the London Underground in the rush hour; despite the crowding, everybody tries to keep slightly away from each other and, to avoid giving threatening signals, they avert their gaze. Staring is hostile. Appeasement displays are frequent: 'Whoops! So sorry!' 'Not at all. Clumsy of me'.

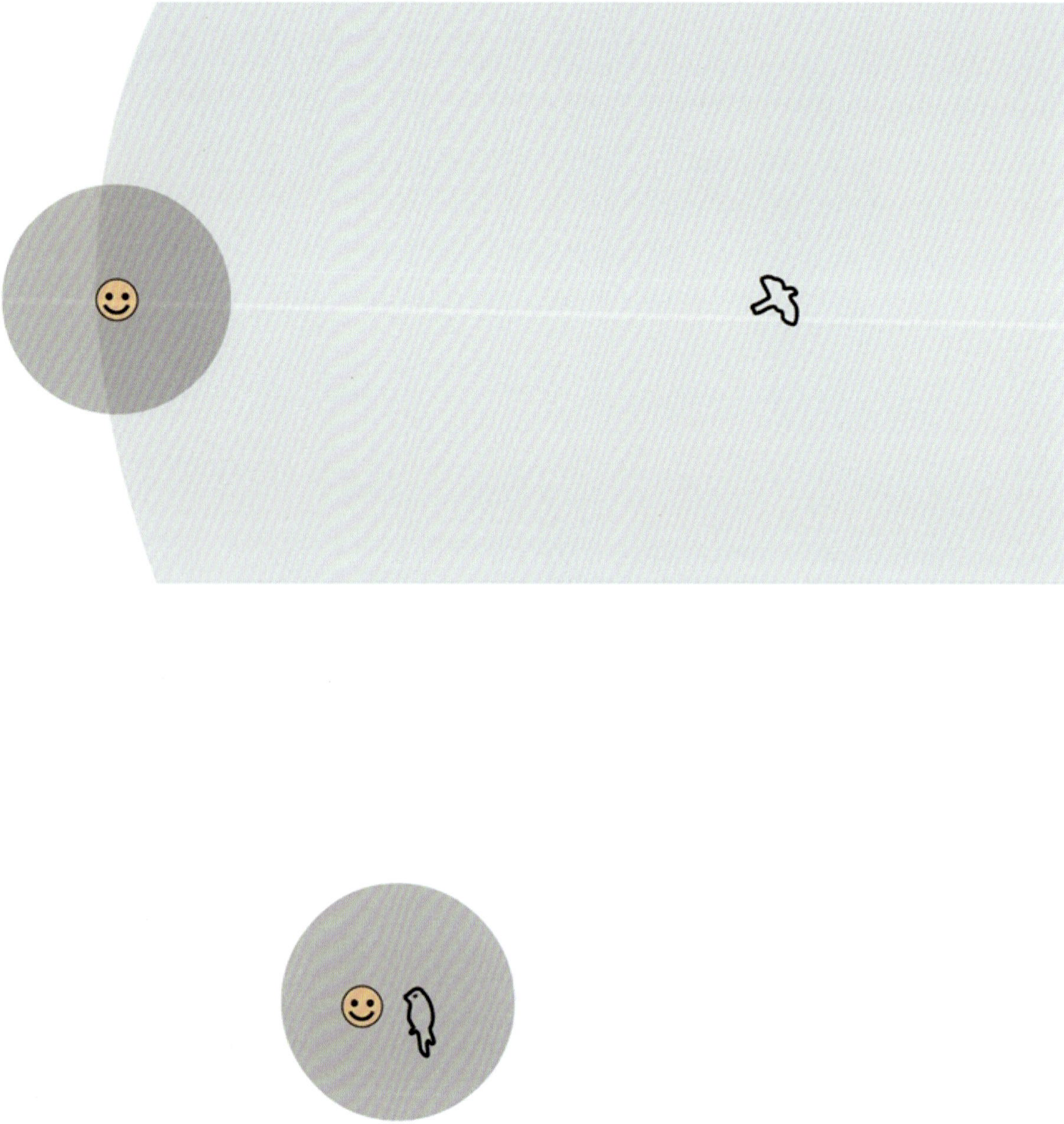

Figure 5.8.1 Top: A shy, wild hawk has a big personal space and avoids man.
Below: A tame hawk is happy to share her space with her trainer.

So it is with manning. If you force yourself into a hawk's personal territory by tying her to your fist she will feel uncomfortable, if not downright afraid. In the wild such intimacy between top predatory hawks (as distinct from kites, vultures and so on) is only tolerated between bonded birds - either a mated pair or parents and offspring. The Harris hawk, which is sociable within her own hunting group, is one of the few exceptions to this; but she will still avoid strange Harris hawks.

So the falconer's first task is to get himself accepted and established in some sort of bond, preferably as a hunting partner and not as a parent which can lead to all sorts of problems. He must give no threat displays, no staring (including mirror glasses or camera lenses), no towering over the hawk, no sudden attack-like movements. He must give appeasement signals: offering food. He must take care not to let himself be associated with any nasty experience for the hawk. Often rather than go into immediate close confrontation he starts off by keeping mainly at a short distance or by coming up to the hawk momentarily to offer her a tidbit and then withdrawing. In this way he works himself naturally and gradually into the hawk's life and trust, forming a bond much stronger than mere habituated familiarity.

Such a hawk will recognize her trainer. She may tolerate any familiarity with him but bate at the approach of strangers. You can tell a person who doesn't understand hawks by the way he walks around the weathering lawn. Insensitive to the hawk's signals of unease at his approach, he comes too close and makes the hawk bate. He is like the dreadful person at a party who insists on pushing up to you and mouthing bad breath all over you. You react in the same way as the hawk does: by trying to escape!

The main use of the hood is to prevent the hawk from getting an unwanted experience or lesson. If you can cut out the bad, negative lessons and give the hawk only good ones, she will train in a very short time. There is no point in having a marvellous training session if the hawk bates herself to bits on the way home afterwards. If you just pop the hood on at the end of the lesson you can carry her home without fuss and leave her in her familiar quarters in tranquillity until she is ready for her next lesson.

At this stage of her training she must be kept tethered in order to manage her. If the hawk is only recently hard-penned, her legs will be in danger from tethering because the ends of the limb bones, the epiphyses, have not yet fused to the shank of the bone. Tethering can exert a side force on the limb, especially at the upper, proximal, end of the tibia, resulting in a broken leg. This is especially true for long-legged species such as Harris hawks, and accipiters which often bate suddenly and awkwardly. They need wide anklets and sympathetic handling, preferably with shock absorbing leashes. Screen perches are a disaster waiting to happen because of the forces put onto the legs. Tethering a hawk before she is hard-penned is inviting trouble. So when tethering, be very gentle, and try to avoid a bate, for physical reasons as well as psychological ones. If the hawk does bate clumsily and cannot immediately get back onto the fist, support her with your spare hand and lift her back on. If you are nervous about this, perhaps because she immediately tries to bite your bare hand, lower your gloved hand near to the ground, so that the ground supports her while she recovers herself. Then consider what went wrong and rectify it. NEVER jerk a hanging hawk back up onto the fist by her jesses. This can not only injure her legs, it can damage her back which, although largely fused, has a weak vertebra in the middle. Nor, as I have seen some vets do, let her hang upside down flapping her wings until exhausted, in a so-called 'stress test'. Some people have even suggested doing this deliberately to teach a bating hawk a lesson. I hope you will be a better falconer than this.

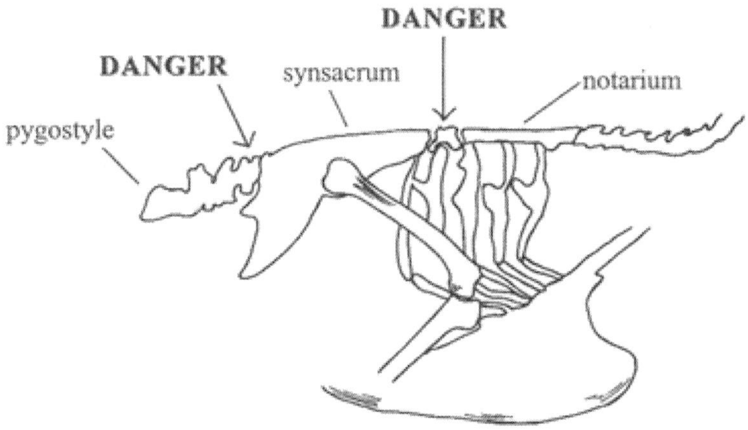

Figure 5.8.2. There are two weak points in the back if the bird is mishandled or flies into something.

Tethering is a necessary evil. Ensure your equipment and systems are safe and in good order. You will need to tether her during her training and while she is flying. But once the need to tether is finished, free loft her. Don't leave her tethered day after day, stagnating physically and mentally. On the other hand, training without tethering is not practical either, whether for hawks, dogs or horses. You need to have some level of physical control and then take pride in not using it. You cannot complete a hawk's training inside a free-lofting room; there quickly comes a time when she needs to come out, travel to a flying area, work on a creance and so on. There have been people suggesting that hawks should not be tethered; this may avoid some physical risk but meanwhile takes no consideration of the hawk's psychological development and other safety elements in managing her. A hawk's place is in the sky, but to reach that happy state she has first got a lot to learn.

Normally I start a hawk's training by hooding her, unless she is imprinted on humans. Hand-raised birds need to be made to the hood while they are still growing; once they are hard-penned they can be very awkward to start to the hood. As soon as the hood is on, I assess it very carefully for fit and if it doesn't fit exactly I make another that does, particularly with respect for comfort for the hawk and for not coming off. The hood must NOT be able to come off. The falconer should of course learn to hood on a made bird, preferably a falcon, not on an untrained bird. If a bird is suspicious of the hood then, whenever you walk up to her with a tidbit, serve it in an old hood. Every little bit of positive reinforcement helps.

Many falconers deem it unnecessary to hood hawks and to some extent they are right. Some hawks travel happily unhooded. But there is quite a lot of difference between having a single Harris hawk or display bird and managing an accipiter or a team of pursuit falcons. These latter birds are hot stuff and cannot be tied near one another or flown in the sight of one another without causing them to bate. A little effort early on to make the hawk to the hood well pays back later in speed of training, ease of transporting and general reduction in stress. I have driven for thousands of miles with falcons fast asleep in their hoods, even when untrained. If they had been unhooded they would have been physical and mental wrecks. Of course if the falconer is bad at hooding, he can cause a lot of damage, but we are talking here about the hawk's needs, not the falconer's inadequacies.

After about a week she should be feeding moderately well on the fist. Her hooding will probably be not very good. Many hawks only accept the hood when a juicy tiring is poking up through the beak hole. Some are awkward to hood when being taken up but accept it more graciously after hawking at the end of the day. If this is your first hawk and you are not a good hooder you may prefer not to hood at all but use a traveling box. At about this stage we usually take the bird for a long walk, hooding or using tirings when necessary, taking the dogs along and getting out into fresh ground. Prior to this her lessons had only lasted a few minutes or as long as one could spin out her meal. This afternoon or two of traditional manning is only done when she is ready for it. After this, it should be downhill all the way. In the space of two or three days she should 'click' and be transformed from accipiter-type behaviour, all nerves and bates, to cuddly merlin-type behaviour, running round you and landing on your shoulder.

Falconers forget that the hawk is getting nasty experiences as well as 'official' lessons. Most falconers cannot carry a hawk in the field without giving it a rough ride on the fist and the bird thus associates the fist with an unpleasant experience and does not perform so well with fist work. Often the green beginner is better than the 'experienced' falconer in this respect. Unless you can really float your fist so that the hawk never stirs a wing, not even when you jump over a fence or ditch, you will have to fight that much harder to overcome the perpetual negative experience the hawk is getting. A comfortable hawk will preen on the fist or put a foot up, sometimes while you're still moving along, and a hooded hawk may go to sleep with her head in her scapulars after a couple of hours walking.

She is ready now for the lure. I usually start with a dead quail on a line. Thrown out on the ground in front of her she should plonk straight on it from a few metres, called off an assistant's fist. I lie down beside her and let her break in, offering tidbits. When she has eaten about half her ration on the quail, I use a small separate pick-up piece to lift her off the quail (covering the remainder with hand, grass, or bag). As I get back to the mews door she should be finishing the last mouthful of the pick-up piece, which comprises the other half of her ration. I have taken nothing away from her. She has only to clean her beak as I put her down, then she is left in peace.

The next day I call her a few times from fence posts in different places (if an accipiter) or from

Figure 5.8.2 The hawk overflies the lure because she is not keen enough.

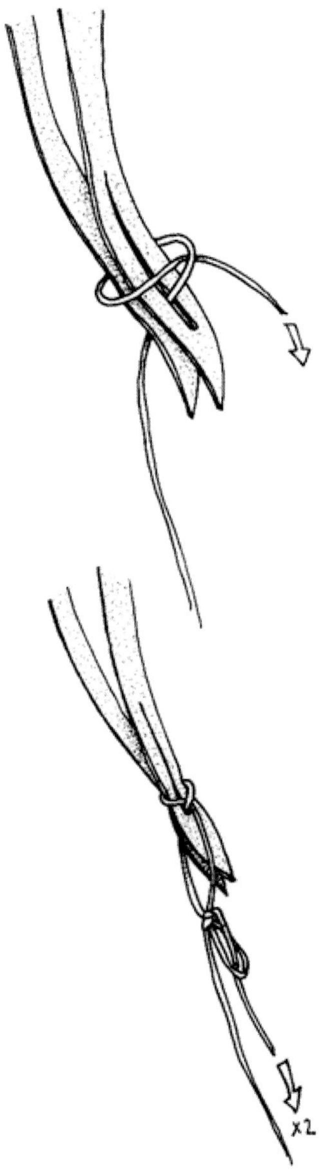

Figure 5.8.3 Attaching the creance to the jesses using a sheetbend and a double falconer's knot.

an assistant's fist (if a falcon) to the fist for small bits of meat, about ten to twenty metres. This is a proper flight, not a glorified hop, but not far enough away for her to start thinking about sheering off. I finish off with a call to the lure which consists of the artificial lure garnished with a good-size chunk of feathery quail, a whole leg for example.

Normally I do most of the creance work to the lure. It is easier to diagnose the hawk's condition by calling to the lure than to the fist. If the hawk starts toward a lure on the ground 5 metres out from you at 90 degrees and then overflies, she is almost certainly not keen enough. But if she overflies the fist you can't be certain that the fear response isn't keeping her from you. The other reason for introducing the lure early is for safety. Once she 'knows' what the lure is, you have some prospect of getting her back if you have a disaster and lose her half-trained.

The creance is attached using a sheet-bend and two falconer's knots direct to the jesses. If you don't use the sheet-bend the line can cut the jesses like cheese-wire, and if you leave the swivel on it will constantly catch and tangle on the smallest thing. The other end of the creance can be attached to a drag weight such as a hawking bag, or even threaded through the sleeves of a coat; anything that will slow the hawk down and stop her, but without a jerk. When we are training a lot of falcons we stretch a 100m line about 30cm above the ground tightly between two posts, described in 3.12. This means that we can call the hawk 100 metres or more without the risk of her getting tangled in any nearby trees. Often she will never realise that she was tied at all.

I try to do as little creance work as possible. The emphasis in training should always be speed of response, not distance. Repeated high-jumping to the fist indoors will quickly hammer home the lesson of a fast confident return to the fist and it is better to get this well learned and then only need three or four days with the creance. I usually fly the falcons loose once they immediately come a full 50 metre creance length and a few years ago I started to wonder why I was training my hawks so quickly. Then I discovered that people had been cutting bits off the creances for other jobs and the creances were getting shorter and shorter! So much for progress.

A typical lesson at this stage would be 2 metres to the fist, 10, 20, 30 metres to the lure, getting a small amount each time, 1/10 crop on last lure call, 2/10 picked up on the fist, hood up, take home, 1/10 crop as she is put away on her perch. She can now stay unhooded when not being trained if she hoods well and if she stands quietly when unhooded.

Once an accipiter is coming promptly 50 m to the lure, swing the lure onto the fist and she will come to that too. I make all species of accipiter, buteo, and falcon to both fist and lure. The falcons are taught to take the lure in the air, using a pole to hold the lure up if necessary for the initial lesson. Later they are exercised to the lure to improve their footing, manoeuvrability, and fitness. Still later, the game falcons can be taught to wait on using the lure. Accipiters also get plenty of lure-work, giving them valuable experience of being picked up off the 'kill'. It is, of course, possible to fly an accipiter without using a lure, just as it is to ride a horse without a saddle, if you are a masochist. To reach the full potential of an accipiter's attack strategies entails starting some attacks from perches other than the fist and you need a lure then to keep control. To fly an accipiter solely from the fist is to miss most of her potential as a hunting bird.

The next day I put the transmitters on. If I can't get 50 metres away from the hawk before she comes, I fly her free there and then. The best place to fly an accipiter free is on open ground with no trees for at least 500 metres in any direction. Teach her to take off and land on the ground and she will do that instead of flying a long distance to reach a perch every time she misses. Put her on the ground or on a stone and call her. Give her 15-20 short (10 metre) calls until you have to run to get away from her.

Call her at oblique angles along walls and fences. Teach her to fly from the fist to a perch you show her a few metres away. Vary the terrain but only introduce high perches that you can't reach when the hawk has proved her reliability. A couple of evenings on fence work should teach her to keep up with you and move on to order. Later on, this lesson is used when hunting through woodland, but first teach it in a situation where you have good control.

In one respect hawks and horses are very similar. There are some people who, when they ride a horse or work a hawk, somehow put her on her toes; they 'energize' her. That animal will become like a coiled spring, poised to do her master's bidding. She is happy, alert, attentive. When the person asks her to do something, she goes into action with vigour. Ten to one that rider doesn't carry a stick and that falconer doesn't need a bag of tidbits. Other people have the opposite effect. The horse becomes a slug, dragging its feet, it gets nappy and needs a stick or spurs to provide any temporary semblance of impulsion. The hawk becomes 'moody' and can only be tempted back with an assortment of bits and pieces accompanied by much whistling, hollering, and general pandering. I'm sure that a

Figure 5.8.4 Calling the hawk obliquely along a fence line.

lot of this is due to differences in actual training and handling methods. But also, behind that, I think the animals can detect the person's natural energy and respond to it.

Most of the hawk's obedience training has been achieved by fast-response high jumps rather than long calls. The outdoor work has been done in a variety of places so that she is accustomed to novelty. There should not be a lot of tedious line work in a 'training field'. Using one place for training can induce superstitious behaviour (see 4.17) and lead to a confused response in unfamiliar surroundings later on.

The moment you are confident that you can get her back it is time to prepare for entering. By entering, I mean catching prey. This may mean wild prey, but equally it can mean robotic prey; in the hawk's mind they are the same. Many falcons are ruined at this stage because people over-train them. Some raptors are very intelligent and it only takes a couple of lessons for them to learn that you are the Great Provider, the source of all free handouts. They will quickly train you to provide these free meals and will soon train you to come to them. Do not fall into this trap! It happens most when hawks are trained while still in their juvenile dependency stage.

Once entered, instead of just tolerating the dog by habituation, the hawk now gets positive conditioning as she links the dog with the point and the flush of game. The falconer must retain the positive link between himself and food and with 'hot' attacking species, such as goshawks and merlins, he may have to stop flying quarry periodically to reinforce this link. Species which are less keen on their quarry must not get too linked with the trainer and the maximum effort must be made to condition the hawk toward quarry. All raptors notice very clearly if the falconer and dog are making a big effort to find and flush game, and the falconer who wanders round the fields nattering to a friend and expecting the quarry to come out with its hands up will soon have a hawk with no interest in killing for herself.

But we are rushing ahead. There are a lot of aspects to consider before the hawk is ready for serious hawking.

5.9 Motivation

Now we're getting into training, you are probably discovering that all this discussion is fine but the reality is that the trainee is not responding as well as she should. Maybe the hawk has gone on strike completely.

It is time to look at motivation.

You can't train an animal to do something until it is motivated to do it. You can't reward the hawk for flying to the fist unless she wants to do it in the first place. Calling to the fist or lure is actually just a simplified version of the Direct Flying Attack (see 8.10) as far as the hawk is concerned. The reasons why a hawk flies to your fist are basically the same reasons why she flies at a rabbit.

Motivation comes initially from inside the animal, whereas training is the outside world working on the animal. The hawk's brain receives messages from her body through nerve signals, chemicals, or hormones. These trigger her into making decisions. Receptors in the crop, stomach, intestine, and liver, monitor what quantity and quality of food is in the digestive tract and the levels of substances in the blood stream. As these levels fluctuate, so the brain is motivated toward the desire to eat and to hunt for food, or, later, to relax after a heavy meal. These pathways are complex and it is not a simple matter of the hungrier she is, the more she wants to hunt. Far from it, a starving bird is likely to be apathetic. Also linked to the hawk's internal drives are external ones, such as the sight of others feeding, or the presence of tempting easy prey, or having a hungry family to feed. Some male Mauritius kestrels we radio-tracked spent up to 95 percent of daylight hours actively hunting when they had young to feed. They themselves weren't necessarily hungry, they were driven by their reproductive instincts to forage for their young.

The neural pathways for hunting for food and for eating are separate. Just because an animal wants to eat, does not mean it wants to hunt. And just because it doesn't want to eat, doesn't mean it doesn't want to hunt. There are many examples of predators, including raptors, which kill more than they want to eat, or which kill something and then hide it without eating it. Although in most everyday examples hunting and eating are linked, this is achieved by separate neural pathways. Therefore, if you want her to do something, you have to try to maximize her motivation and this may require some careful thought and analysis.

When you introduce external training methods, motivation becomes a balance between pleasure and pain. Pain includes punishment and also negative reinforcement, the two are not the same. Punishment is given after the unwanted behaviour and tends to decrease bad behaviour. Negative reinforcement tends to increase good behaviour or decrease bad behaviour, thus the horse increases its effort to avoid the whip. But it only does enough to avoid it, it doesn't give more. An animal which is trained through positive reinforcement on the other hand, will strive to give even more because of the possibility of getting even more reward. This is essentially how we train a hawk; she does something because she wants to, not because she is forced to.

The hawk gets first turn at the dialogue. She must tell you how she feels, what her internal motivation is. So you must be receptive and ask her. Only then can you tell her anything because your message must be tailored to her needs. Be clear in your own mind what your message is. If you are not clear, how can you expect her to be? Each day have specific goals laid on the firm foundation of previous progress. Reach each goal in small steps and get each step right before moving on. Ask yourself: How can I make it simpler for her? The better you get at this question and answer game, the better communication you will have and the hawk will seem to get more intelligent.

Try not to ask her questions that will result in incorrect answers. If you ask and the hawk doesn't do it, either she hasn't understood, or she is saying 'No, I don't want to'. This is establishing a pattern of refusal, which is in fact a form of training, but not for what you want. You must make it easy for her to say 'Yes' and only work from the last point where you got a consistent 'Yes'.The hawk will know when you have lost control, just as a child or dog does. The skilled trainer works right up at the cliff-edge of losing control, but stays in command. The unskilled trainer cannot easily see where the edge of control is; he constantly goes over it, resulting in behavioural problems to be overcome or, worse still, the loss of the hawk.

5.10 The diagnosis and treatment of vices

Vice is unwanted behaviour and as such it has the same roots as other behaviour, ie inherited, imprinted, or learned. First identify the unwanted behaviour, then work out its root. If it is inherited or imprinted, it may be a lot harder to eradicate than if it has been learned. For example your goshawk may take crows and gulls well but may always tail off if they go up more than 30-40 metres (100 ft.). This is a common inherited limitation of goshawks, although with roprey training we can raise this ceiling. Or the bird may scream because it is imprinted on you as a parent figure. You can't change its imprinting but you can change the parent figure; give the bird to another falconer (you may have to pay him to take it!). You can also negatively condition the hawk on you as the food source by flying the hawk at hack or at quarry and minimizing calls to the fist, but this makes falconry difficult. Finally, if the hawk doesn't learn not to scream you will at least have taught yourself not to create a screaming hawk in the first place. Maybe put it into a breeding project using artificial insemination.

We'll look at some common vices, carrying, hood-shyness, and aggression and screaming, in the next three sections and analyze them in terms of behavioural origin and in ways of training to reduce them.

5.11 Carrying quarry

Fifty years ago a friend of mine decided to train a haggard sparrowhawk. She was called *Buttons* because of her bug-eyed look. He waked her and manned her for days on end. After sessions on the creance and flying loose, finally the great day came when he could hunt with her. We beat the hedges and *Buttons* shot off after a sparrow, taking it neatly in the air. She carried it, and that was the last he saw of *Buttons*, a month's work for a three second flight….

Carrying quarry is a vice which can have a variety of roots and is therefore hard to diagnose and cure. Some of the smaller attacking hawks, such a merlins and sparrowhawks, have a strong inherited tendency to carry as a protection against piracy by other predators. You can't blame them. Some species, such as New Zealand falcons, instinctively want to cache food and may carry it away and hide it before coming back to the falconer and carrying on hunting. Raptors that have been reared with jealous siblings soon learn to carry food.

Figure 5.11.1 When she carries she is very expressive. And so are you.

Small, well-advanced, males carry to escape from less-agile, bullying sisters. 'Learned' carriers are easier to cure then 'ethotypic' carriers and it is important to remember that carrying prey to a safe place has survival value both for wild hawks and for trained ones. Many a young raptor has been ambushed on its prey by an eagle, goshawk or fox and there may be times when you wish your hawk had carried and saved herself.

The first steps in preventing carrying are made early on in training. When she is at home on the perch cut yourself a little bowl of bechins: small pieces of meat as tidbits. Every half an hour or so come round the corner into the weathering area and quietly approach her and give her a bechin. Always whistle or make a noise before you come round a corner onto a hawk so that she is forewarned. (I do this with my staff as well.) Soon you should be able to appear and walk openly up to the hawk and give her a bechin in her beak quite boldly, with her showing every sign of eagerness and anticipation. When out lure training, you walk around her and approach from all angles with a bechin. If you attach only a small amount of food on the lure, she will quickly consume everything

Going out to do the horses, I find a scattering of fresh woodcock feathers. Later in the day, working in the woods 600 metres away, I find where the goshawk has settled to pluck it on a fallen larch in a secluded spot, safe from prying eyes.

on the lure and be left standing on a food-less lure with you wandering around with tidbits. You now become the focus of attention. Soon she will bate from the lure back onto your fist as soon as you approach. This should be the pattern for the rest of her days. I do not expect to have to bend right down to a trained hawk to take her up off the lure; she should jump up off it at my approach and I just pull up the lure by its cord and put it away. High jumping too, strongly reinforces this behaviour and quickly gets a wary passage hawk changing into friendly behaviour.

As well as preventive training, it is important to prevent carrying happening and thus reinforcing itself. Don't start getting up to all sorts of strange and rather alarming antics like crawling up to the hawk. Just wander up in a series of oblique diagonals, talking to yourself and not staring or standing high above the hawk.

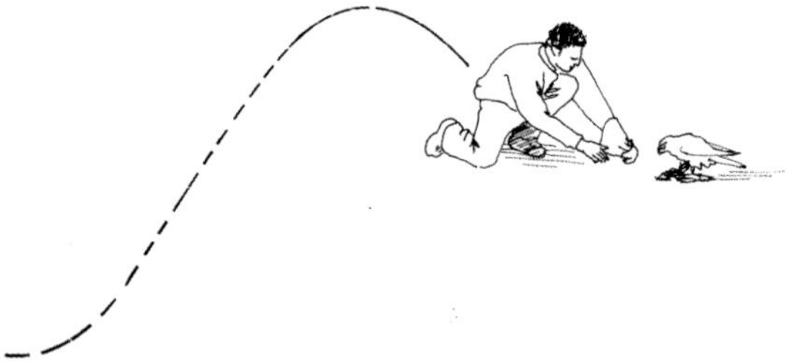

Figure 5.11.2 Do not approach a nervous hawk directly. Come in at an angle with eyes cast down. Try to look bored!

Give her the right signals. She knows you're there. If you crawl up to her, you are like a predator creeping up on its prey. Be open about it and show that you are a non-threatening companion. Beat her at her own game. The hawk is on her hard-earned prey. She sees you as a sister or another predator coming to rob her of prey. First show her that there is no way you want her prey because you've already caught your own: a nice juicy lure or pick-up piece. What's more, your kill is more attractive because you've already plucked yours and there is lots of red meat showing, just ready for eating. Pull a few bits off. Instead of jealousy at your approach, envy takes over. Let the hawk have some of the tidbits and finally let her rob you of your 'kill' before she has broken into her own. Using the links prey/ pick-up piece, or if necessary prey/lure/pick-up piece, let the hawk step onto the fist and quietly remove the kill and lure out of sight into the hawking bag, or cover it with grass or sand. Keep the fist on the ground at the kill spot and let the hawk feed there awhile. Once she is thoroughly settled, walk well away from the kill spot before finishing off. Some hawks, especially Harris hawks, will bate back for the kill spot otherwise, to check. NEVER lift the hawk with her kill off the ground. This sows the seeds of carrying and of attacking the trainer. It is about the shortest cut to disaster.

Figure 5.11.3 Sowing the seeds of vice.

If she is newly entered, thread one glove cord through one anklet and tie the other cord to something heavy, such as your hawking bag, so that she is secured. Settle with her and let her eat her kill, helping her with it and giving her tidbits. I try to spin this out for at least half an hour with a young hawk. Every minute you spend doing this is like putting money into the bank. She is associating you with her success.

In some countries, particularly the Middle East, it is not customary to use a pick-up piece, a piece of meat large enough to be held in the glove to tempt the hawk off her prey or lure onto the glove. It is good psychology to use one, especially with new passage hawks, and helps reinforce the bond and prevent carrying. For some hawks, such as Harris hawks, and especially eagles, removing them from the kill can be tricky. The best way is to throw one or two titbits of food onto the ground nearby and hide the prey while she is going for the food.

If the hawk carries a lot, don't let her break into any of the kills without you being there. You wish to reduce the link kill/food and reinforce the link falconer/food. If the fear of robbery is so great that you can't get near the hawk, then throw the hawk the lure by tying the lure on a 50 metre line and throwing it close to the hawk. Then walk round in a half-circle and tow the lure back until it touches the hawk. If you still can't do it you must have a hard case because I've caught even wild hawks that way, just winding the hawk up with the line before making in. Obviously if you know you've got a difficult one, don't fly her at quarry until you've remedied it. Give yourself a chance to cure the habit on chosen terrain. Otherwise the hawk may come down with the quarry in an awkward place, such as a barbed-wire fence or a corn field, where it is hard to give the lesson properly. This trick is also useful for any hawk which comes down with prey onto thin ice or quaking bog where it is unsafe for the falconer to approach. She can be tempted to grab the lure and then towed gently in to safety. I have had falcons catch crows on cliffs and barn roofs and tempted them to grab the lure on a line, then lowered them down.

Anti-carrying treatment can be very effective and it is possible to teach a hawk to 'retrieve' her kills this way, or to abandon them to come to you, so that the quarry flies away unharmed. This is handy if you are hawking a river which is deep and cold, with the hawk down on quarry on the other side. It is also useful to get the hawk to show herself if on quarry in thick cover. Instead of lying low, waiting for you to pass, the accipiter can be called out or at least made to ring a bell and reveal her position. Some of the small falcons will often cache their prey and then return to the falconer.

5.12 Hood-shyness

Another common vice is hood-shyness. Putting the hood on is a neutral-to-unpleasant experience for the hawk and, unless done adeptly from the start, can give negative conditioning. It is hard too, to link it with positive reinforcement and, if persevered with, the negative conditioning links with all aspects of being handled and the whole relationship is destroyed. Most hawks, a few days after starting, get awkward to hood but settle again after a week or two more. This awkward period is due to the hawk becoming less afraid of the falconer than of the hood, and so she starts to resist.

During early training the hawk is hooded most of the time and gets very comfortable, sleeping quietly and having a peaceful time. The training periods are positive but still stressful and so the hood is a welcome relief. But the hood must be comfortable and it must be put off and on gently and easily, without a lot of menacing waving about. Just roll the hood over its chinstrap onto the hawk's head, using the rotation of your wrist. Your wrists and arms should be held as if you were driving a screw into the hawk's left wing, not in the frontal, sit-up-and-beg position, as if you were placing a pot on a high shelf. This is jerky. I normally close the hood with my right hand as soon as it is on. Then I use my teeth to close the second brace a few moments later.

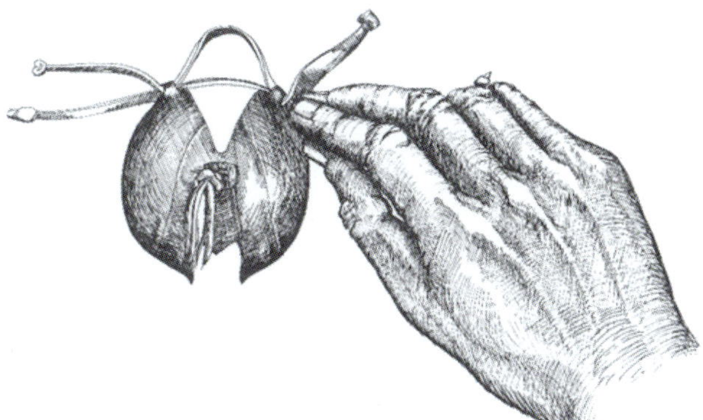

Figure 5.12.1 Closing the braces with one hand.

Usually a hawk which is bad to hood is due to bad technique on the part of the falconer. So blame yourself. Then look at your technique and improve it. Behind every skilled act is a lot of experience and practice and too many falconers think they can run before they can walk. Be prepared to look critically at yourself and improve.

My approach is to put the hood on, come what may. If the hawk shows signs of standing for it, then I do it as calmly as possible. On the other hand, if the hawk doesn't show much sign of staying upright, after two or three bates, I slip the hood on while the hawk is upside down, rather than persist in attempts to get it on the conventional way and reinforce the negative conditioning. This is easily and quickly done: the hawk is already held by fairly short jesses so that her feet are secured. I then raise my left knee so she is draped over it safely. Never let a hawk hang upside down for a moment; it is bad for her back, her leg joints, and her mind. Laid on my knee she may try to bite but it is easy to hook the chinstrap under her beak and flip the hood over her head. The procedure takes no more than a second or two and is much better than entering a confrontation. Should the falconer finally quit without getting the hood on the bird, she learns that if she resists enough she can evade the hood and then you've really made yourself a problem.

If the hawk is chronically, traumatically hood-shy, there are two weapons: habituation and positive reinforcement. Suspend hoods near the hawk on her perch. Soon she stops reacting to the sight of them without associating the unpleasant hood with you. Use your cupped hand, or a huge hood that she can see out of. Let her feed from food inside the hood and stroke her back with your hand. Gradually stroke the back of her head as she feeds, then cup her head with your hand, then come over the front of her head with your cupped hand, then have a big hood in your cupped hand, the plume between your thumb and base of index finger so that it nestles in your palm. These lessons are repeated and become associated with food. When you stop to think about it, there are usually graduations to an activity so that you don't have to do it in one desperate hop. A cutaway mouth opening allows the hawk to see out and is easier to put on. As her tolerance improves, a tighter hood can be used. Positive reinforcement can be given by feeding the hawk tidbits inside the hood and by feeding the hawk while the big, open hood is on, so that she links hooding with food. The hawk can be left hooded at night so that she doesn't link wearing the hood with food, and hence constantly pull at the glove while hooded. You want to link putting on the hood with food.

With really bad hooders, the hawk bates and works herself into a frenzy so that she gives herself no chance to realize that hooding is not too bad after all. The hood becomes a bogey man. The simplest remedy is to put the hawk in a sleeve with her head showing, or sometimes I walk around just carrying the hawk tucked under my arm. Then put the hood on and off many times until you see the nictitating membrane cross the eye as the hood approaches. While she is still sleeved or brailed you can feed her tidbits from the hood, then gradually revert to normal procedures. I find this method more effective and preferable to the alternative of soaking the hawk with water. Spraying with water quietens hawks down and makes them reluctant to bate. It is a useful method especially in a hot climate as it also cools the hawk down.

Some hawks, such as gyrs and New Zealand falcons, are intelligent and hate being hooded at certain times. They often resent it when first being taken up to be weighed and loaded for hawking. After hawking, with a good crop, they take the hood as sweet as pie. They look at you and say 'Okay, you can put it on now.' These hawks are not hood-shy, they just have times when they want to stay unhooded. I usually give them a tiring on the fist, then fit the hood beak-opening over the tiring, then they put their head inside and hood themselves. We all know it is an old trick, but dignity has been maintained.

For all things to do with hooding, read Jim Nelson's book. He has pulled together useful tips from all sorts of very experienced (and patient) people.

5.13 Aggression and screaming

Aggression toward people and screaming both arise from the same root cause and are commonest in raptors imprinted on humans, which have no fear response toward the trainer. In a situation of jealousy, such as on the kill, there is nothing holding back such a hawk from flying at the falconer to drive him away. Another situation is when an entered hawk becomes frustrated from not being served with game and flies at the falconer or his dog. Aggression in the hawk is usually a sign that the falconer has mishandled the relationship somewhere along the line.

The commonest species to become a problem through aggression are red-tails, goshawks and Cooper's hawks. They are large enough to cause real damage and if one attacks a dog or person the victim may be put off hawks for life.

> We once had an old female common buzzard who was brought to us by the police for attacking little old ladies in woods. Probably taken illegally as a nestling and malimprinted, she attacked selected people and ignored others. Out on loan, she chased someone up the garden path. He ran indoors and she flew straight through the window pane of the kitchen door to get him! This same bird was once cast for coping by a young friend called Tom, years ago. In a different place, two years later, out hawking, she saw Tom again and flew through a group of five people to nail him. Tom now has a complex. For some reason, she liked me, and for many years reared baby Red Kites for our Kite restoration project.

As shown in section 4.21, young hawks go through a phase in their development when they become aggressive toward their parents, 'bullying' them to provide food. Normally this wears off as the bird starts to kill for herself and becomes independent. Young falconry birds, particularly imprints, may turn this onto their handler who acts as a parent figure. Anything which produces food, including the hawking dog, may trigger it. As the red-tail is often in inexperienced hands, the young falconer spends longer than normal handling and training the bird (reinforcing the parental bond) and is less successful at producing quarry (diminishing the hawk's independence). As autumn slips by, the hawk begins to orientate inward toward the handler rather than outward toward killing for itself. Aggressive instincts to foot and grab are diverted from the prey to the falconer.

In large hawks, such as the big eagles, this semi-dependent stage is necessarily a long one and it seems that young eagles respond to a certain amount of discipline. Japanese and Asian methods of training eagles usually include some discipline, either physical or social. For example, the falconer holding the eagle is treated with subservience by other people present to demonstrate to the eagle that the falconer is a dominant figure and thus to be respected. Physical methods used with eagles to maintain respect and hence one's own personal safety are based on dominance behaviour in the wild. I will not detail them here because of the risk of misunderstandings and abuse. Handling eagles is an art in itself; talk to experts, such as the Hiebelers.

So what can be done about an aggressive or screaming hawk? Basically she must be encouraged to grow up, to progress through the next stages of development to 'independent' adulthood. First the falconer can reduce the parental bond; he must aim to be a hunting partner, not a parent. Stop flying the hawk from and to the fist. Fly her at quarry and find her plenty of opportunities. Encourage self-hunting. Flying at roprey is an excellent substitute for live quarry and mentally she will react in the same way to it. Call her in to the lure, not to the fist, and don't pick her up until she has had almost a full ration. Do not arrange matters so that the hawk is kept tethered but not flown at quarry.

Handling her by putting her out to weather each day means that you are returning her to dependence. Once the hawking season is over put the hawk in a good-sized moulting pen with a member of the opposite sex (only if both birds are naturally imprinted of course). Let them see people, but not as food providers; let food reach them through a chute or hatch. Provide a nest. If, being young, they don't breed, it doesn't matter. The object of the exercise is to get them to interact as mature adults, possibly even forming a pair bond. More ferocious individuals, such as goshawks and Cooper's hawks, may need to be kept physically separated for their own safety, as do raptors imprinted on humans. But they can still be provided with a nest and allowed to go at least partway through the breeding cycle.

Quite often a young falcon, put out on the weathering lawn for the first time with other falcons, starts to scream a bit. She may have been silent in the hack pen with other youngsters, so what's going on? If one of the other falcons on the lawn is in adult plumage, the youngster is begging at it. It's nothing to worry about. The youngster will quieten down in a few days as she settles into flying.

Occasionally, naturally imprinted birds, domestic-raised by their parents, fall into the hands of inexperienced people and start to scream. These birds are not malimprints. They have had their development arrested at puberty by handlers who have failed to develop their hunting skills and independence. The hawks have simply transferred their allegiance from the natural parents to the handlers. If possible, it is better to have left such birds with their parents for an extra month after being hard-penned and then to get them going at quarry reasonably quickly. It helps too to keep them in daily contact with another hawk, failing that even a mirror may work, to help the hawk retain its identity as a hawk. On the other hand, if the hawk is left in the pen for too long in the autumn she will be past the best period for learning and quarry, and conditions generally, will be more difficult. Although hawking books tell you not to hurry training, in cases such as this there is a lot to be said for pushing through it fairly rapidly with more emphasis on lure-work than on the fist. Dependence should not be confused with obedience. The worst thing that can happen to a young hawk is to be stuck with a falconer who messes about with her all the time, keeps calling her unnecessarily to the fist and fails to expose her frequently and regularly to catchable wild game. The hawk becomes like a shooter who, unable to find real game, becomes frustrated and shoots up road signs instead.

Mantling over food can be extreme in imprints, triggered by somebody watching them. If your hawk is on the lure or a kill she may try to drag it away and mantle over it, but not actually eat anything. She may almost go into a trance. The best option then is to go away. Leave her in peace and seclusion. Maybe just monitor her unobtrusively from 50 or 100 metres away with field glasses. After a while you may see her start to relax and feed. If she is on your fist and won't eat, obviously you cannot move away. The best antidote then is not to feed her a chunk of food on the fist. Instead have it all ready, cut into small pieces in a bowl in your bag. Offer her the bowl and with any luck she will bite a small piece and swallow it, then settle down and eat the rest. Continue in this manner for some days until, bit by bit, she starts to feed normally.

Harris hawks start to imprint at around day 5 and if they are hand-fed for any longer than this, they are liable to scream. Even totally parent-reared Harris hawks often scream during their first autumn, because of their naturally long dependence period. As they become mature hunters, they quieten down, but, being group hunters, they are naturally communicative. Accipiters, on the other hand, become independent very suddenly, about three weeks or so after fledging, and thereafter go about their own business through the woods, silently.

5.14 Further Reading

Bolles, R. C. 1975. *Theory of Motivation*. Harper and Row, New York.

BÖ, O. 1962. *Falcon catching in Norway*. Scandinavian University Books.

Bustamante J. 1993. Post-fledging dependence period and development of flight and hunting behaviour in the red kite *Milvus milvus*. Bird Study 40, 181–188.

Devine, A. 2016. *Peregrine Falcon Dispersal and habitat imprinting (1994-2013)*. MSc thesis. University of Illinois.

Harlow, H. F. 1974. *Learning to Love*. J. Aronson. ISBN 0876681593.

Hiebeler, J. 2006. *Der Steinadler in der Falkenerei*. Panhofer Verlag. ISBN 3-9501123-9-1.

Heers, A.M. 2016. New perspectives on the ontogeny and evolution of avian locomotion. Integr. Comp. Biol. 56, 428–441.

Immelmann, K. 1972. Sexual and other long-term aspects of imprinting in birds and other species. In *Advances in the Study of Behavior. 4*. New York: Academic Press. pp. 147–174. ISBN 978-0-12-004504-4.

Kitowski I. 2005. Play behaviour and active training of Montagu's harrier (*Circus pygargus*) offspring in the post-fledging period. J. Ethol. 23: 3–8.

Macdonald, H. 2014. *'H' is for Hawk*. Jonathan Cape.

Martin, S. 1994. The Positive Approach to Parrots as Pets: Videotape 1. Understanding Bird Behaviour. 2. Training through Positive Reinforcement. Natural Encounters Inc. P. O. Box 68666, Indianapolis, IN 46268.

Martin, S. New Training Thoughts for Falconry Birds. Hawk Chalk XXXIII, no. 2: 44 54.

McDermott, M. 2009. *The Imprint Accipiter II.* Western Sporting. ISBN 978-1-888357-02-08

Norberg, U.M. 1990. *Vertebrate flight: mechanics, physiology, morphology, ecology and evolution.* New York, NY: Springer.

Obrecht, H.H., C.J. Pennycuick and M.R. Fuller. 1988. Wind tunnel experiments to assess the effect of back mounted radio transmitters on bird body drag. Journal of Experimental Biology 135:265-273.

Oliphant, L. 2020. *Developing the Modern Gamehawk.* Prairie Sky Falcons, Canada.

Pearce, J.M. 1997. *Animal Learning and Cognition.* Psychology Press. ISBN 0-86377-433-4.

Pennycuick, C.J., M.R. Fuller and L. McAllister. 1989. Climbing performance of Harris Hawks (*Parabuteo unicinctus*) with added load: implications for muscle mechanics and for radiotracking. Journal of Experimental Biology 142:17-29.

Pryor, K. 1985. *Don't Shoot the Dog!* Bantam Books. New York. ISBN 0 553 25388 3.

Ruaux, G., S. Lumineau, and E. de Margerie. 2020. The development of flight behaviours in birds. Proceedings. Biological sciences, *287*(1929), 20200668.

Sherrod, S.K., W.R. Heinrich, W.A. Burnham, J.H. Barclay and T.J. Cade. 1981. *Hacking: A method for releasing peregrine falcons and other birds of prey.* Peregrine Fund, Cornell University.

Sherrod, S.K. 1983. *Behavior of Fledgling Peregrines*. Fort Collins, Peregrine Fund.

Stevens, R. 1955. *Observations on Modern Falconry*. Privately printed.

Tsukada, Y. and B. W. Agranoff (eds.). 1980. *Neurobiological Basis for Learning and Memory.* John Wiley, New York.

Varland, D.E., E.E. Klaas and T.M.Loughin. 1991. Development of foraging behavior in the American kestrel. *J. Raptor Res.* 25: 9–17.

C6 Physical Condition

This chapter is supported by two films in the Bird of Prey Management Series: *Basic Training* and *Nutrition*.

Young goshawk.
Photo Steve Magennis

6.1 Assessing condition

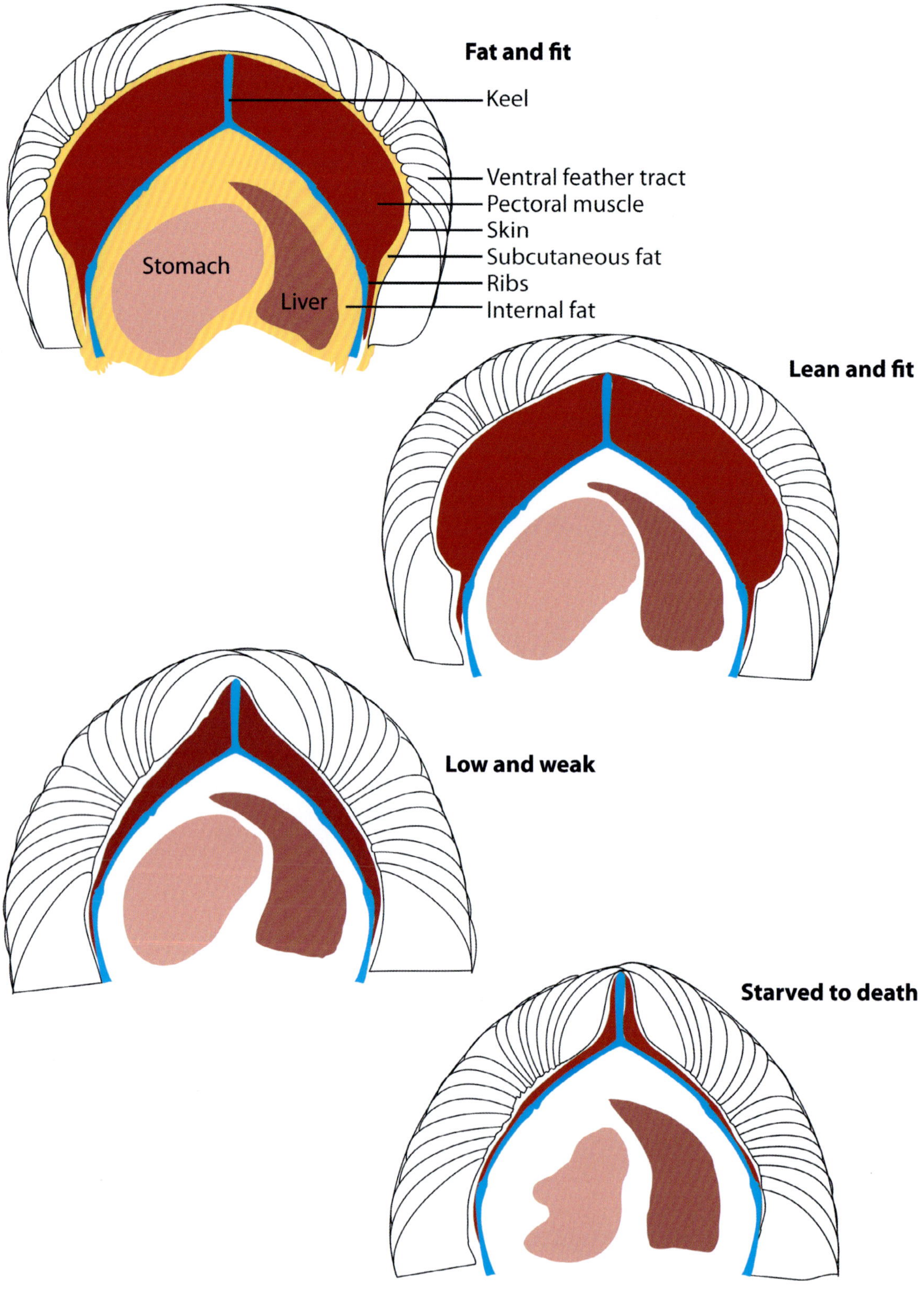

Figure 6.1.1 A section of the chest of a falcon to show the pectoral muscles and fat stores.

So far in training we have assumed the hawk is responsive and in the right physical condition for training. To bring a raptor into condition is the key to successful falconry and requires a good understanding of how the digestive system works in practice and how the hawk metabolizes energy. Starting perhaps with a fat youngster, or a newly moulted hawk, the object is to achieve a bird which has plenty of muscle, little or no fat, a good appetite and which is free of disease. Such a hawk will fly keenly, strongly, and predictably.

There are three ways of judging condition: by feeling the hawk's muscles, by weighing her and by watching her behaviour. The traditional way of feeling the hawk's pectoral muscles is unreliable in that most of a hawk's fat is stored inside the body cavity around the internal organs, over the scapulars and around the sacrum, lower abdomen, and tail. Only the weighing machine can detect it. Although some fat is stored in the pectoral muscles and in a thin subcutaneous layer over them it can happen that an unfit hawk will have wasted breast muscles and yet be very fat. Another hawk may be fit and well-muscled and have no fat at all. Thus you can feel the muscle mass but not the fat. If you can actually feel that your hawk is thin then you have reduced her too much. How can she give of her best if her muscles are wasted? Dr Ken Riddle in *The Art of Falconry from Arabia Westward* gives details of how to assess a falcon's condition by feeling her, but in practice this is very subjective.

Weighing the hawk and keeping a progress chart is a much more reliable method provided that weights are properly understood, recorded, and checked with behaviour on a progress chart. A good counterbalance scales or electronic scales should be used; spring balances are untrustworthy. Keep a 500 gram weight to check the scales each day. Working in grams, once you are used to it, is infinitely easier than in ounces and one tends to work more accurately too. For a hawk of about 900 grams (2 lbs.) I weigh to an accuracy of one gram and pay attention to differences of 5 grams, which is 5/28 or just under 0.18 ounces. This is about the weight of one mute. A difference of 5 grams will normally show a detectable difference in behaviour in a bird of this size, depending on factors we'll examine in the next section.

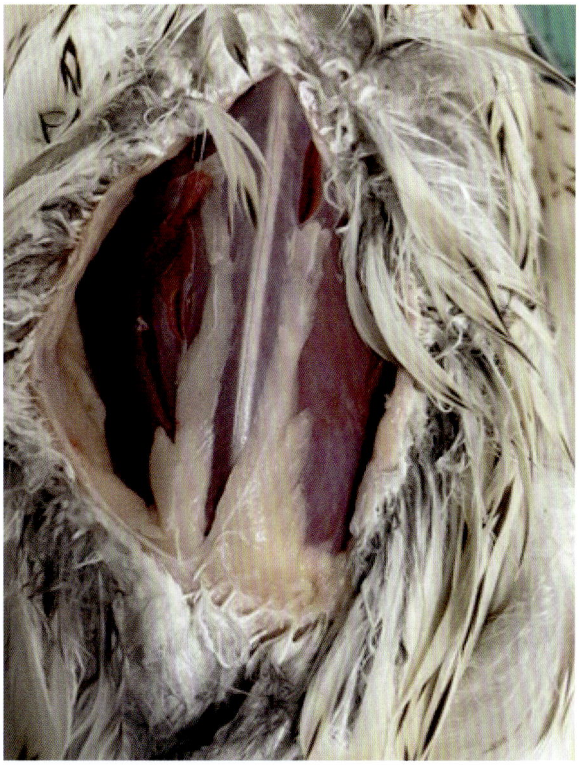

This falcon was very fat with some fat attached under the skin and some overlying the pectoral muscles. But even so it is quite hard to feel this fat in a live bird.

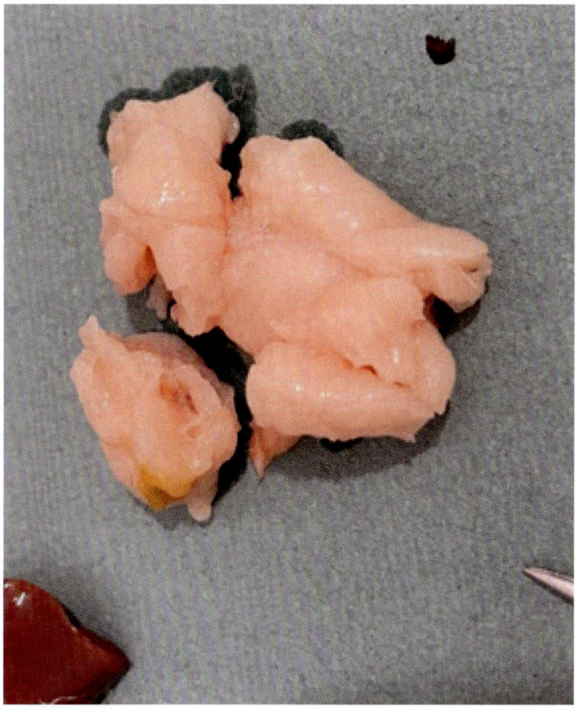

Inside the body cavity, undetectable from the outside, the internal organs may be covered in huge gobbets of fat.

Species	Sex	Ring number	Name
SAKER	♀	10320W	JEDRAN

Date	Time	Weight	Crop	
22/7/95	2300	1210	0	Taken out of pen, measured, tail belled, anklets hood etc fitted after dark.
23 Sun	20 00	1190	0	Hooded all day.
24	20 00	1177	0	Hooded all day.
25	0900	1155	1/10 Rabbit	Unhooded on quiet corner of lawn. Took a few titbits.
26 Weds	0900	1140	1/10 Rabbit.	Sitting up on block well and not bating off when approached with titbit.
27	0930	1130	2/10 Rabbit	Fed on rabbit leg on block with me sitting next to her. Left her to tire on the foot.
28	0900	1125	3/10 Rabbit.	Ate some rabbit on fist indoors. Bit bouncy but not too bad. Good with titbits on lawn.
29	0830	1130	3/10 Rabbit	Busy all day. Fed on fist around campfire in the evening. Hooding well.
30 Sun	1100	1125	3/10 Rab	Stepping up on to fist well + feeding freely. Took her out riding (hooded) for 2 hrs in pm.
31	1030	1120	3/10 Rab	18 fast jumps to fist. Introduced lure. Doesn't want to bath!
1 Aug. Tues.	0930	1118	4/10 Rab	27 fast jumps to fist. 10 metres to lure.
2	0900	1130	3/10 Rab	30 jumps but patchy. 10 metres to lure. Needs to be down a bit.
3	0830	1120	3/10 Rab	45 good jumps. 25 metres to lure. Travelled overnight to Northumberland.
4 Fri	1430	1115	3/10 Rab.	Bit strange for her up here! 30 jumps + 20 metres to lure. Weathering in cold wind.
5	0930	1105	4/10 Rab	Flew loose. One straight 20 metres to lure, 1 call + turn. 40 jumps in mews.
6	0900	1115	4/10 Rab	7 stoops to lure in 3 bouts. Took lure in air on last attempt. 40 jumps.
7 Mon	0900	1120	4/10 Rab	22 stoops in 2 bouts. Got a bit wet so gave 40 jumps after dark.
8	0930	1125	4/10 Rab	35 stoops. 45 jumps.
9	1430	1125	3/10 Rab	Rained all day. 65 jumps.
10	1400	1120	3/10 Rab	Long luring – 500 metres. 30 stoops. Finished with 100 m call to dragged dead crow.
11 Fri	1430	1117	3/10 Rab	About 35 stoops to lures in a cast with Hagar. No problems. Thrown a dead crow.
12 Sat	1500	1115	6/10 Crow	Watch Hill. Hagar took crow in rush patch. Jedran joined flight. No crabbing on kill. Split crow for them

Figure 6.1.2 A training progress chart.

6.2 The components of weight

Let's look at the hawk's weight in more detail. It consists of five components:

a. Base weight. This is the base weight below which the hawk will die. This starvation weight depends on species, sex, age, air temperature, humidity, and on the speed of starvation. A starving Red-shouldered hawk was found by Kendeigh in 1945 to have a base weight of 60 percent of its fat weight. See also figure 1.1.1.

b. Stored fat. This can contribute up to 30 percent of the fat bodyweight, depending on species.

c. Muscle. The muscles throughout the body can vary by up to 20 percent of the fat bodyweight depending on whether the hawk is fit or starving.

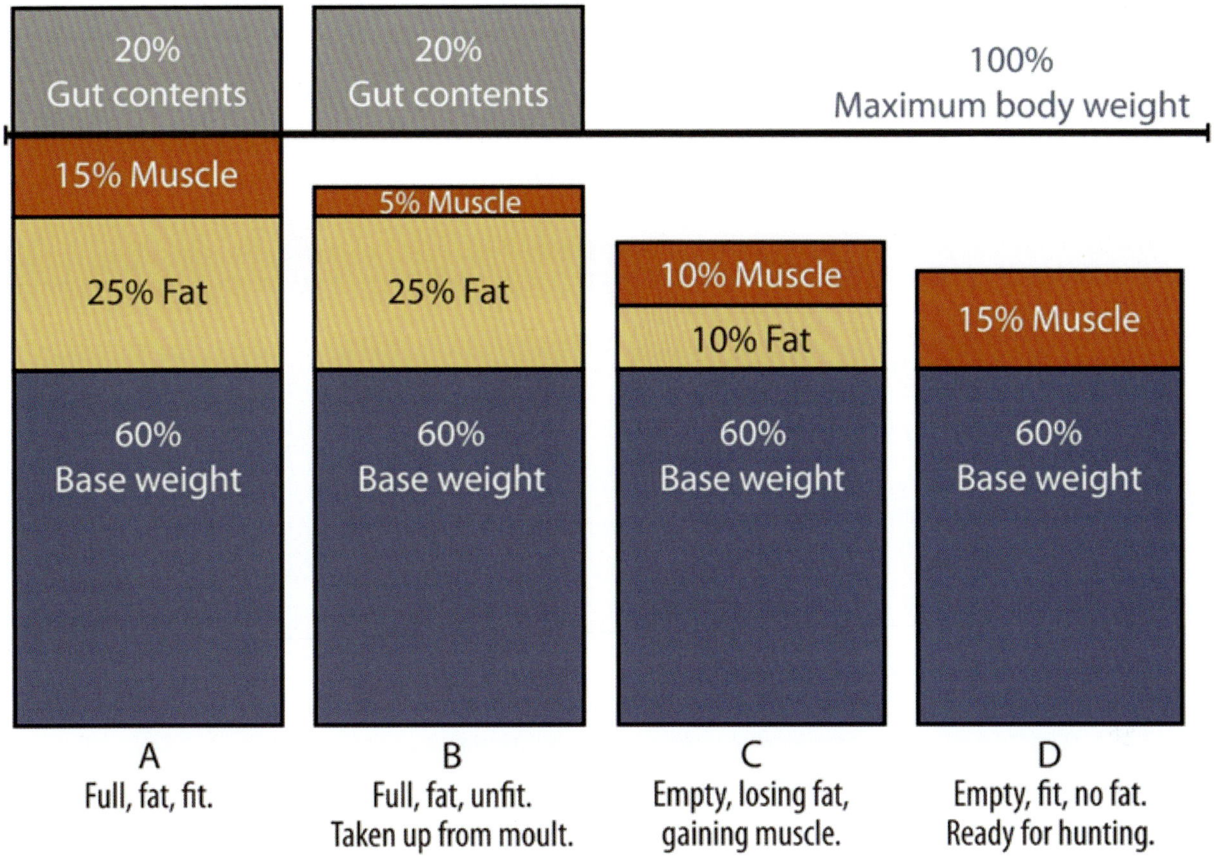

Figure 6.2.1 The components of weight.

d. Gut contents. The maximum weight of the gut contents can add a further 20 percent to a hawk's fat weight and this is why, ideally, a hawk should be empty to obtain a proper weight record. A fresh goshawk pellet weighs 10-12 grams.

e. Equipment. The hood, anklets, bells, and so on should be taken into account and subtracted before calculating percentages but can be left in the total on a daily basis.

These weight components all vary independently from one another and therefore just because a hawk is at a certain weight does not necessarily mean she is ready to fly. The components also vary considerably between individual birds and between species, particularly between large hawks and small ones. No generalized figures can be given but Figure 6.2.1 gives some idea of the variation.

During reclaiming after the moult, the falconer aims to remove all stored fat from his bird and build up the muscles to their full capacity. If fat is left on the bird during the hawking season it will buffer the effect of dieting and appetite controls, resulting in unpredictable behaviour, rather like loose steering on a car. There has been a tendency in recent years to try to fly falcons in as high a condition as possible, in other words, fat. The problem is that fat does not contribute to performance; you would not try to run a dog or horse fat, or an athlete (unless you are a Sumo wrestler). Symon Latham wrote in 1615: '*Remember that a fat Hawke makes a leane horse, a weary Faulconer, and an empty purse: three discommodities befitting every man to beware of.*' Hawks haven't changed since then. If your hawk starts to pull her punches, loses focus or wanders wide, she is probably too fat. She may kill at a long distance and be herself killed by an eagle or cattle, all because of a few grams of fat on board. If you live in the middle of a big empty desert and are happy to follow your hawk by GPS rather than eyesight then by all means, let her wander around doing her own thing, or better still just watch wild raptors. But if you are flying a falcon at taxing

A well-muscled falcon looks solid.
Photo Verity Johnson.

quarry such as houbara or crows, or if you are flying an accipiter, you will very quickly realise the critical importance of hunting condition.

As the fat is lost, there is a reduction in weight and, as muscle is formed, there is a gain in weight. The two processes should be simultaneous and usually result in a net loss in weight because more fat is lost than is replaced by muscle, depending on what type of hawking will be done. Jack Mavrogordato in *A Hawk for the Bush* gives 10 percent as a rough guide and this has become known as the 10 percent rule. It's not a hard and fast rule by any means but it's a start.

The exchange of fat for muscle is done by reducing the intake of energy to below the amount of energy used up. So it is wise first of all to check for internal parasites and treat these before cutting the weight down. The plan is not to eliminate fat (by food reduction) and then build up muscle, but rather to burn off the fat using increasing exercise which at the same time builds up the muscles. Therefore, we need some idea of what energy is being lost and what is being gained.

6.3 Energy losses

The raptor uses her energy in three ways: for her basal metabolism, her existence metabolism, and her productive metabolism. The basal metabolism is the energy needed simply to keep the body functioning and the heart beating. It is proportional to body size; small hawks, such as sparrowhawks and merlins, have a much higher basal metabolism than do large buzzards and eagles. Some individuals and some species, such as the desert falcons, need far less to eat than others of equivalent size, such as peregrines. In survival terms, the energy demands of the central nervous system, the brain and brain stem, take priority. It can only use energy in the form of glucose. The other systems can use energy in other forms.

The existence metabolism is the extra energy needed for the raptor to cope with the physiological effects of temperature changes, wind, rain and so on. Depending on climate, some species need about 15 percent more food in winter than in summer, or if moved to a cold district from a warm one. Every autumn when the frosts come, one hears of trained sparrowhawks dying from low blood sugar fits because their trainers have not made allowance for the effects of the bad weather. As the season progresses the falconer should aim to increase the hawk's weight by at least 5 percent of the early season flying weight. This goes into filling out with extra muscle. Also, by the winter, the hawk should know her business and a little extra weight, especially for the delicate species, is a wise insurance policy. A friend who had cut his sparrowhawk rather finely and then, neglected to give her her second feed on Christmas day because he was celebrating, was horrified to find the next morning that she had lost thirty grams and was in a state of collapse. Only by tube feeding and assiduous nursing was the hawk saved and two weeks were lost before the hawk regained her strength. In such a situation the hawk should not be fed solid food because she won't have the strength to put it over into the stomach; the food will remain undigested in the crop. If this happens the soured crop contents must be flushed out and the crop irrigated with warm water, or, better still 'Reglin' which helps promote motility in the gut. Then she should be tube fed warm electrolyte fluids at the rate of about 20 ml/kg or Emeraid Intensive Care Carnivore. If available she can also be given steroids, B vitamins, iron, and if relevant, antibiotics. She should be kept slightly warm and undisturbed. Once she is perking up, the first meals should be very fine and nutritious, such as bird liver with a little fat, put through a blender.

Conversely, sakers flown in the desert almost live on air. Just the brain, tongue, and liver from a hare is enough to keep them steady; any more and they pile on weight and become unmanageable.

The productive metabolism is the energy required for activities such as hunting, courtship, nesting, and so on. In racing pigeons the energy used during flight is about eight times that for basal metabolism. This is why, if a sparrowhawk is allowed to get low in fat and muscle, an afternoon's flying in cold weather can reduce her blood sugars to a fatally low level. When the body needs energy, the liver converts stored glycogen from a recent meal into glucose in the bloodstream. When the level of glycogen in the liver gets low then stored fat is converted into energy. Unfortunately, in small species with a high basal metabolism, the body cannot convert fat into energy fast enough for the metabolic demand. Thus it is possible for a small hawk to show symptoms of temporary lowness even though she is not near the bottom of her weight range. This lowness can occur in half an hour if the metabolic demands are high and the remedy is to minimize the existence and productive metabolic demands and to supply an easily assimilated form of energy such as glucose or lean meat. As long ago as the sixteenth century the anonymous author of *A Perfect Booke for Kepinge of Sparhawkes or Goshawkes* advised that leaving a sparrowhawk empty in the morning after she has cast makes her poor; a small feed then is necessary to keep her 'lusty'.

Just keeping a hawk awake speeds up her digestive processes and increases her existence metabolism. The Persian author of the *Baz-nama-yi Nasiri* in 1868 noted that a hawk when carried would mute three times for every two mutes when left unattended. Hooding in a quiet place has a similar effect to natural darkness and if a hawk is left hooded on a full crop it sometimes happens that the next morning she still has half a crop left and is not fit to fly.

If we weigh some hawks periodically throughout the 24 hours, and average out their weight loss per day, we can see differences in the rate of weight loss according to what is going on. Figure 6.3.1. shows the weight fluctuations of six raptors over eight days and nights. The shaded areas indicate the hours of darkness. We recorded the ambient temperature which tended to rise in the daytime and fall at night. The hawks were not exercised so they had no major productive metabolic demand and were only concerned with their basal metabolism and their existence metabolism.They were not heavily fat. They were fed at 10 A.M. and their weights recorded before and after the meal and then every three hours.

First we see a spike as each hawk eats its meal, then a gradual loss of weight as the food is digested and excreted. In the first run we fed about enough to keep each bird at an overall steady weight or slightly increasing weight over the eight day

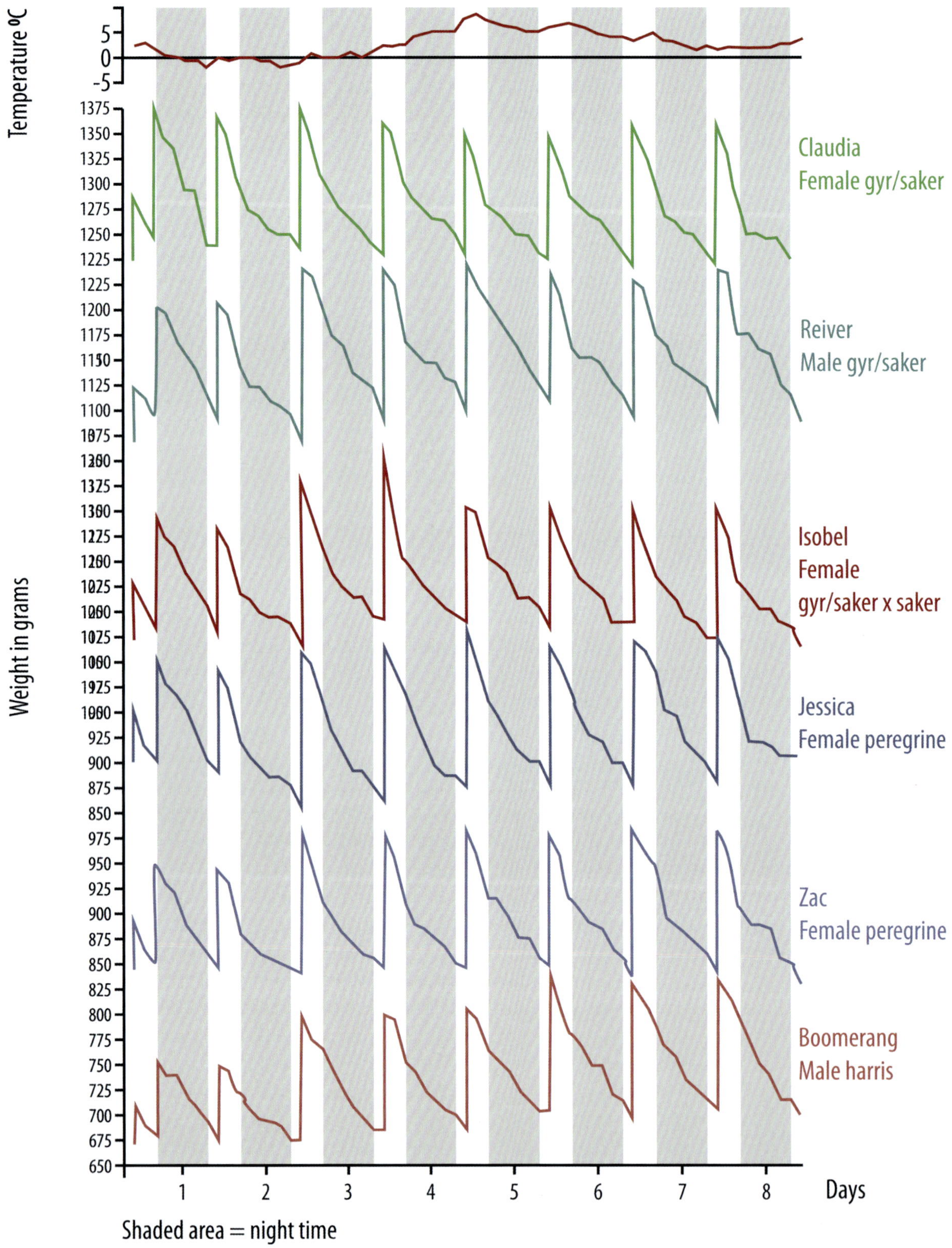

Figure 6.3.1 Changes in weight for six raptors being raised in weight over eight days.

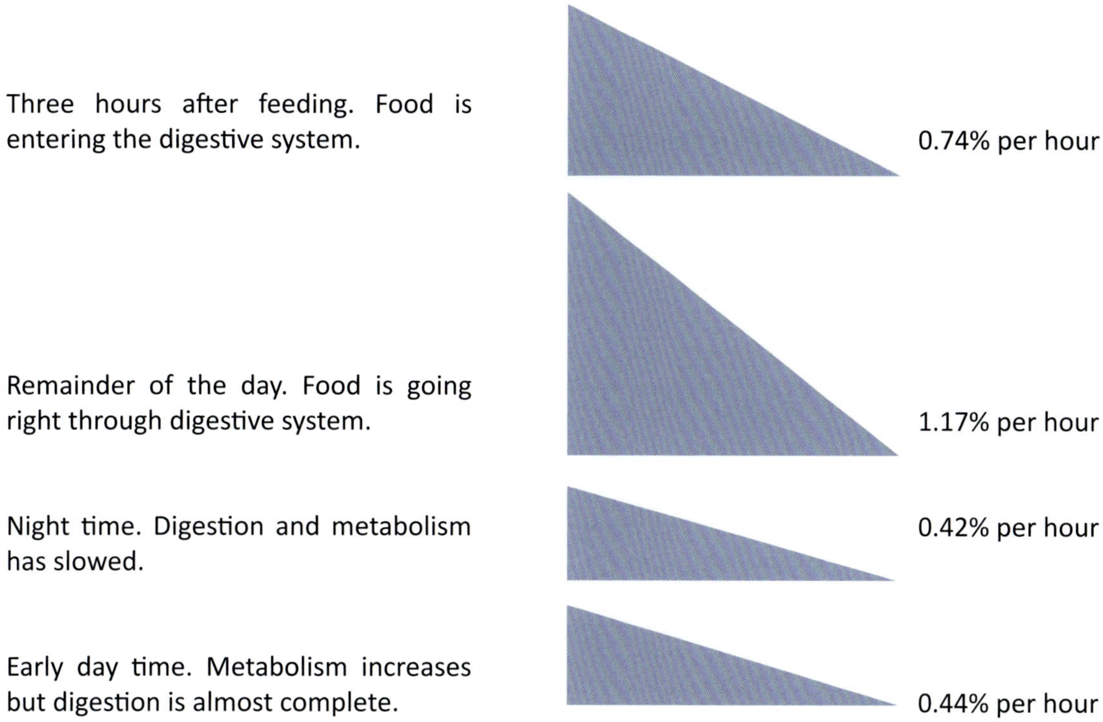

Figure 6.3.2 Mean weight changes for the six raptors in 6.3.1 during the 24 hours expressed as a percentage of their body weights. They are slightly increasing their overall weight. The steeper daily weight losses are simply because they ate more food.

For the first two hours after feeding the food is passing along the digestive system and weight is not lost very quickly, but by the third hour it has travelled the whole length of the gut. Once the first food is fully digested, the hawk is muting frequently, and during daylight she has a rapid metabolism. During this period she loses weight more quickly. On average, weight is lost at 0.74% of total body weight every hour. Once darkness comes, the hawk goes to sleep and digestion continues at a slower rate, usually with fewer, but larger, mutes. There is not much food left in the gut and weight is lost at half the speed as before: only 0.42 percent of total body weight per hour. By morning the hawk is almost empty and does not lose much weight from muting but she is awake and her gut is more active again.

Obviously it is not practical to weigh your hawk every hour. The two important weights to consider are the empty weight just before feeding, and then the weight once fed, in other words, how much you have fed her. This works quite well for a medium-sized raptor such as a peregrine or goshawk whose metabolism requires them to eat about once a day. Small hawks such as sparrowhawks and merlins need to eat more than once a day, and large ones, such as eagles, maybe once every couple of days. In the graphs, I have shown weights as absolutes ie mass. If I had shown each bird as percent of body weight, and if I had included smaller and larger raptors, you could have seen clearly that the smaller ones metabolise much faster than the large ones.

When we are reducing a hawk in weight, we are trying to slowly reduce body fat while retaining or increasing muscle mass. Figure 6.3.3 shows the six birds gradually coming down in weight over six days. The food given is not quite enough for her basal and existence metabolism so for the tail end of the 24 hour cycle she is obtaining energy by metabolizing the last of her fat stores or by metabolizing protein from muscles. Thus the weight loss throughout the 24 hours is by no means a straight line loss, but rather it falls at different rates according to what energy stores are being used at the time and the speed of metabolism.

After eating a meal, the digestion takes about six hours and so for this period energy is being converted directly from the food. This is called dietary energy. Further energy is stored as glycogen in the liver. This supply can be reconverted to energy and will keep the hawk going a further six hours or so. The remaining food is converted into fat and muscle, and some of this fat will promptly be needed to convert back into energy to last the bird for the remaining 24 less 6 (from food), less 6 (from the liver) = 12 hours. So there is a constant ebb and flow of nutrients and energy in the bird's body, a bit like a constant charging and discharging of a battery in an electric car.

In the absence of fat, hawks can convert their own muscle, which is protein, into energy. Proteins are the building blocks of the body and their amino acids are broken down by gluconeogenesis to glucose. Protein yields only about one quarter as much energy as fat and so if

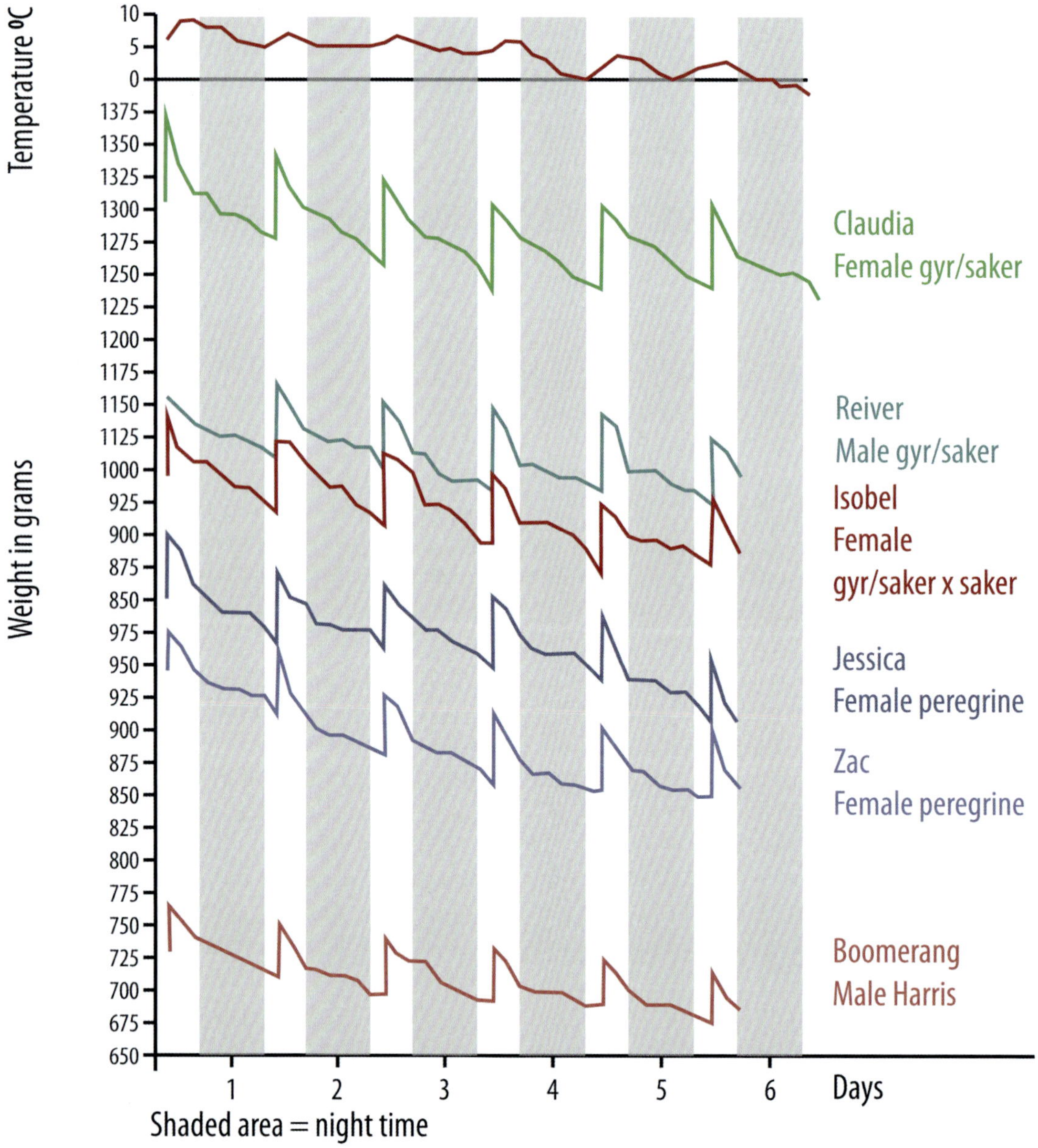

Figure 6.3.3 Changes in weight for six raptors being reduced in weight over six days.

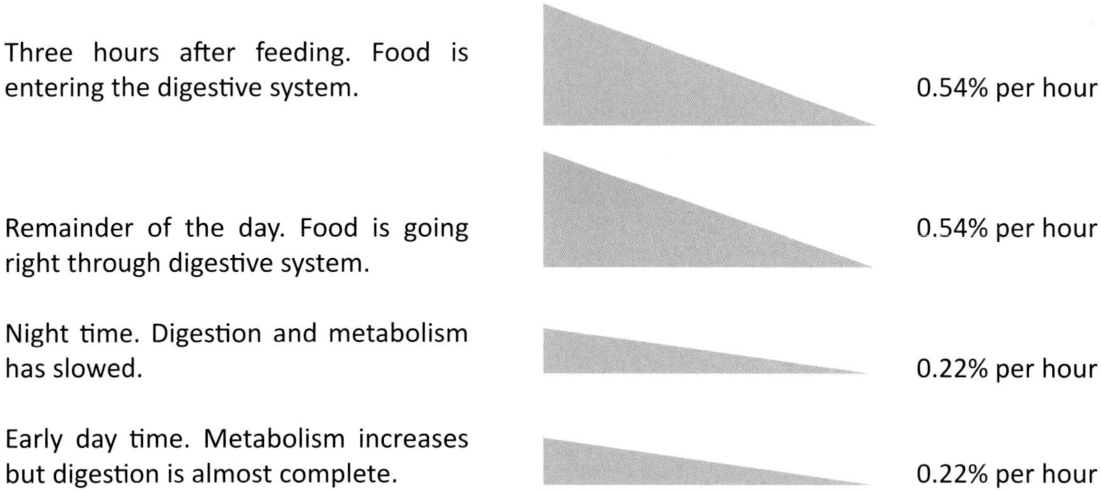

Figure 6.3.4 Mean weight changes as a percent of mean body weights for the six raptors in 6.3.3. The overall trend is a gradual loss of body weight.

a bird depends on burning up her own muscle she loses weight and condition rapidly. This is why, if a raptor is gradually cut down in weight over a two or three week period after the moult, there is a steady weight loss followed by a sudden drop. During the early stages the bird is metabolizing fat into energy and water, which presumably is why the mutes are so watery at this stage, but once the fat is gone, protein is metabolized and four times as much is needed for the same amount of energy. This means you have gone too far and you must boost the meals so that the weight returns to slightly above that at which the sudden drop occurred. The aim is not to remove all the fat but to use it to build up muscle, and the balance, especially in small species, is a knife-edge. Attempts to control the behaviour of an unresponsive intermewed bird, which may have forgotten her conditioned responses, by lowering her weight further, should be resisted. Such a bird does not need starving, she needs training.

It takes time and energy to break down protein into energy, so much so that a small accipiter, in low condition, may not have sufficient blood sugar levels to convert her muscles into useable energy and will become a victim of temporary lowness or hypoglycemia. This is even more extreme than if she had had some fat reserves to call on. Also of course, the last thing you want a hawk to do is burn off muscle tissue. She needs it for flying about with.

Muscles also store large quantities of glycogen which is intended for anaerobic muscle activity. Once this is used up during a sprint, then the muscles need glycogen brought to them rapidly by the blood supply from the liver, as well as having the breakdown products, carbon dioxide, and lactic acid removed.

Another point to bear in mind, especially as it is often the hottest time of the year when hawks are taken up from the moult, is dehydration. Birds adapted to a temperate climate need, very roughly, sixteen times as much water for evaporation to keep cool at 44° centigrade than at 1° centigrade. When a hawk is burning energy in flying or bating, she is heating herself up and must pant to cool down and this loses water through the surfaces of the airsacs. Some lose a lot of water during car journeys or other possible stress situations and even if water is available some may be too wild or stressed to drink. Deep freezing tends to dry out the food and therefore warm tap-water is often better for thawing food than a slow, drying heat. In hot weather, a bath should always be available and the hawk should be kept out of direct sunlight. In cold weather, she should be fed a little extra fat, especially after a lot of flying. Not too much - just a mouthful.

This constant interchange of nutrients for energy is a perpetual process and should be stimulated. It acts at three levels, physiological, physical and mental. Whenever you get the chance, oscillate the energy reserves of your hawk. Don't maintain her at flying weight constantly for two or three weeks at a time. Whenever a day appears when, due to bad weather or other commitments, she will not be flown, tank her right up. Then ease her down just before her next hunting day. This encourages her physiological systems to function rapidly in their task of energy conversion. A good bulging crop will keep her gut in good physical shape and pull at those stretch receptors. And at a mental level she will have more sparkle too. Even with the breeding birds we follow the same principle; raising and lowering their weight stops them getting stale and lethargic and makes them much more active and interactive.

In the short term, within the 24 hour cycle, hawks, humans and horses quickly get into a routine of eating at set times. Horses are very good at training humans to feed them at regular intervals. In their natural life, their grazing patterns tend to reflect weather patterns and are nor tightly clock work, but in a stabling routine horses get very agitated if a feed is not delivered on time. Similarly with hawks, if you delay their feed by a couple of hours, they get keener. Delay it further and this keenness drops off. This is the 24 hour model for motivation, used more than weight control for small hawks. Use it to your advantage but don't let it rule you. One gets people at field meets who refuse to fly their bird until a set time because they have been trained by their bird to follow a routine. A bird the size of a peregrine should be capable of being flown certainly from mid-morning onwards, throughout the day.

The way in which a hawk is reduced after the moult needs to be done carefully. The old writers recommended keeping a fat peregrine or goshawk hooded for the first six or seven days lest in her breathless bating she be struck down by an apoplectic fit. If the hawk is on her own in a flight pen it is a simple matter to lower her by giving her a ration reduced in quantity or quality for a week or ten days before taking her out. As soon as she is at all civilized it is best then to push ahead with exercise. Not only will the work burn off the fat more rapidly, it will also convert it into muscle. It is much better and safer to build up muscle this way than to strip her down almost to base weight and then to try to build up muscle through feeding. The best way to do this with an accipiter is to give her high jumps which not only gives her measurable controlled hard exercise but also sharpens her up mentally and produces a snappy response.

Once the hawk has used up all her fat and is enseamed, the good falconer feeds her up well and flies her hard. The more work she does the more food she burns up, the bigger her appetite, the more toned up her system and the stronger she gets. Being strong and keen she will be obedient and eager to fly. Inexperienced falconers are often astonished at the crops a fit falcon can take. 'Aren't you going to fly her tomorrow?' is the query on seeing the bulging crop. But the next morning there she is, bright-eyed, winnowing her wings, bursting to carve the sky to pieces over your head.

6.4 Energy gains

It is much easier to strip weight off a hawk than to put it on again. The quantities and qualities of food are so hard to judge that inexperienced falconers suffer constantly from their own mistakes.

Different foods have different energy values which can be measured either directly by calorimetry or indirectly by measuring the effects on the hawk's weight. Some values for common foods are listed opposite:

Thus a hawk which required 100 grams of mouse to keep a steady weight would need about 145 grams of day-old chick. Warm, fresh food has a higher energy content than cold or thawed food. Fat in any food puts up its energy value tremendously. If while feeding up your hawk, you stand proudly discussing her performance with a friend and fail to notice that gobbet of fat she bolts down, you will be in for a shock the next morning. We've all done it. Bird fat from

	Kcal/g	Grams equivalent to 100g mouse
Mouse	1.57	100
Small birds	1.31	120
Hare meat	1.19	132
Day-old chicks	1.08	145
Whole rabbit	1.01	155
Beef heart	1.01	155

ducks, moorhens, winter migrants, poultry heads and so on may put your hawk out of condition for several days - a good supplement perhaps during the moult but not during the hawking season.

Different parts of an animal have different food values. The front part of a rabbit is red meat, almost as good as hare, whereas the white meat from the back and hind legs is very poor, almost as bad as washed beef. Small hawks, such as sparrowhawks, need high energy food and if fed on rabbit they cannot digest the food fast enough to maintain weight.

The common diets fed to hawks are dead day-old cockerels, rats, mice, coturnix quail, rabbits, hares, butcher's meat, and assorted small birds and game birds caught by the hawk. Pigeon is popular in the Middle East. Of these only rats, mice, quail, and adult birds provide a complete, balanced diet. Some, such as butcher's meat, are extremely unbalanced and should be fed for only very short periods and never to young hawks. Day-old chicks, rabbits, and hares are subsistence diets and the latter two are not adequate to keep small hawks alive. Subsistence diets require supplementary vitamins and minerals to balance them.

Adult large falconry species, if they regularly receive good feeds of pheasant, ducks, or other fresh kills out hawking, will survive on one of the subsistence diets. But breeders and growing youngsters need a full balanced diet if they are to do well. Many of the hatching problems which have dogged merlin breeders have been traced to diet rather than to technical procedures.

We do not know exactly what the nutrient needs of each species of bird of prey are, nor can one design a perfect supplement because this also depends on the composition of the basic diet. But one should be aware of the different roles of vitamins and minerals in the diet and the problems which can arise if they are over or under supplied (see 2.10).

The quantity of food to be fed depends on its energy content. Also smaller hawks with their faster metabolism need proportionately more food than do larger ones. Sharp-shinned hawks need to consume up to about 25 percent of their own bodyweight in food daily. Medium-sized raptors need about 8-15 percent and golden eagles only 5-6 percent. But of course these figures vary widely depending on the type of food and on the metabolic demands the hawk may have to meet.

A golden eagle can eat about 1500 grams of food in one meal, about six times its daily requirement. Smaller hawks, needing relatively more food, have to eat more often. This means that small hawks, such as sparrowhawks and merlins, can be flown on 'appetite' rather than on weight, if flown daily.

How does one measure the quantity of food a hawk eats? It is all very well to weigh the food or to weigh the hawk before and after feeding but this is hardly practical out hawking. The solution is to measure the crop size. This is the system I use:

- Empty crop: nothing can be felt in the crop and the mutes show no pressure or are green with bile.
- 1/10 crop: nothing can be felt in the crop but you have fed her a certain amount which has gone straight down into the stomach for digestion.
- 2/10 crop: food can be felt in the crop but nothing is visible.
- 3/10 crop: a slight displacement of the feathers makes the food visible.
- 4/10 crop: there is a definite small rounded crop visible.
- 5/10 crop: the crop is larger and more pointed at the base.
- 6/10 crop: the crop is wider but is still pointed at the base.
- 7/10 crop: the crop is now flattened at the base.
- 8/10 crop: the crop is now becoming asymmetrical, building up more on the bird's right side.
- 9/10 crop: the crop is even larger but is not quite tight and the feathers are able to cover it.
- Full crop: the hawk refuses more and seems uncomfortable. Bare skin may show on the front and either side of the crop. The tip of her beak may touch the crop when at rest. She may not tuck her head into her scapulars until some of the crop has been put over.

It is fruitless trying to photograph these crop sizes; there is no substitute for looking carefully and feeling gently with finger and thumb. It is very helpful now and then to cut up the hawk's food into 10 gram rations and feed it bit by bit, noting what the crop looks like with each known weight consumed. We show this in our Raptor Management film on Nutrition. It may seem difficult at first but a skilled falconer can keep his hawk within 1-2 grams of a predetermined weight day after day. Until he is able to feed reliably he will never get on top of the problems which arise from poor conditioning. Also, by trying to train or fly a hawk when she is not fully responsive, disobedience or poor attitudes to quarry can arise and can be hard to eradicate later. Thus the problems pile up.

6.5 Enseaming or preparing the stomach

To get a hawk into flying condition is not just a matter of getting the fat off and the muscle on, although many modern falconers seem to think so. It is also a question of toning up the digestive system.

It is easy to fly a falcon to the lure or an accipiter to the fist. It is not hard to catch things in a mediocre sort of way. But to make a hawk really strong and bold, to take her into the field at the height of her courage, that is the mark of the skilled falconer. Unfortunately it is not so easily achieved or even explained. Inexperienced falconers cannot even recognize it. Certainly the ancients understood it much better than we do.

First, attention must always be paid to the tone of the whole digestive system. Good tone is best achieved by plenty of bulk going through the system. Hunger or appetite seems to be controlled by the hypothalamus via stretch receptors in the crop and stomach and by pH sensors in the stomach. Feeding small crops for long periods seems to allow the crop to shrink, reducing the appetite. On the other hand, a hawk that is frequently fed large crops, even if the food quality is low, will be hungry after only a small crop.

In terms of bulk, casting material such as feathers or fur is of little use because it is returned from the stomach and does not enter the intestine at all. Best then at least once a week to give a good 'blow-out' of rabbit or even, if necessary, of washed meat; anything that will fill her right up without putting her out of yarak.

It is noticeable that goshawks tend to maintain their drive to hunt possibly for an hour after they have taken a full crop. This may be because they are more impulsive attackers than falcons or it may be because it is not until food reaches the goshawk's intestine that hunting is inhibited. Peregrines on the other hand tend to be keen to hunt almost in proportion to the size of their crops. If one's peregrine is over-keen her condition can be adjusted quite well by giving an appropriate small feed.

Figure 6.5.1 Merlin rangle, peregrine rangle and gyrfalcon rangle.

Second, one must attend to the lining of the gut. If the gut is lined with mucus or grease it cannot function properly and the appetite will be poor. Additionally, the cuticle or koilin lining of the gizzard (see section 1.4) is shed periodically, either as a whole or in pieces. Its function is to protect the stomach from digesting itself with its own acids and it is continuously sloughing off and being replaced by new material. If too thick a layer accumulates it can be cleaned off either chemically with 'scourings' or mechanically with rangle and castings.

Typical medieval scourings consisted of pellets made from clarified butter and sugar as a carrier with a variety of additives such as powdered cloves, saffron, rue, rosemary, aloe, nutmeg, and celandine root to make the effect milder or more severe according to need. These recipes used properly had a pronounced effect on the appetite and were not placebos by any means. Saffron for example contains colchicine which is used in the treatment of gout. Eastern falconers still use ammonium chloride or sal ammoniac to loosen and strip off the excess cuticle in the gizzard. This drastic procedure is now fortunately less common; it has killed many hawks.

The more natural way of cleaning the stomach is by using rangle stones and castings. Wild hawks use rangle and I have found it used by Common, Mauritius and Nankeen kestrels, by merlins, New Zealand falcons, at least three subspecies of peregrine, sakers, and gyrfalcons (see figure 5.11.1). I have also found one sample from a wild common buzzard in Scotland and from a wild golden eagle in Utah. There are no records that I know of from accipiters although it would be difficult to find with these woodland species. The author of *The Skoole for a Young Ostringer or Faulkener* considered that rangle is harmful to accipiters and could kill them, but other authors do not agree with him, indeed Bert (1619:78) recommended giving a goshawk stones together with knots of thread from the weaver. It would be surprising if accipiters do not use rangle when buzzards and eagles do. Many other species such as crocodiles, seals, sea-lions, cormorants, divers, fulmars, and petrels also use rangle, some even more extensively than raptors do. The only common denominator appears to be a diet of meat, often with a high fat content.

A wild kestrel, roosting on my bedroom windowsill above the mews, has left a sample of rangle and pellets.

Wild falcons normally take five to twelve stones, preferably rounded but with a slightly textured surface. They fly down to a stream or gravel area where the stones are water-worn and carefully select the size required. This is about 12-16 mm maximum diameter for a 1 kg (36 oz.) falcon, 10-14 mm for a 500 g (18 oz.) falcon and 4-8 mm for a 250 g (9 oz.) falcon. This was determined from several hundred stones. The eight buzzard stones were relatively small at 6-15 mm (mean 8.2 mm) diameter. The eagle stone was 31 X 12 mm. One female peregrine here, deprived of rangle for a month, swallowed 35 stones as soon as she was offered them.

We do not know exactly how often wild falcons use rangle, but field studies and experiments with aviary birds show that some falcons take rangle approximately once per week, others once per month. Some individuals may use it more often, others seldom, if at all. They pick it up at any time during the day and usually retain it overnight. It is taken separately from casting material and the stones are cast up the following day in a group, before eating a meal. Sometimes the stones are covered with a thin film of mucus but usually they are fairly clean. I have seldom found rangle in a pellet. When stones appear in the pellets they are usually the gizzard stones of a prey animal such as a quail or pigeon and are noticeably smaller and sharper than rangle.

The best way to give rangle is last thing at night once the afternoon's casting-free crop has been put over. The stones should all be thrown up the next morning. If the hawk will not take the stones voluntarily pop them into her throat one by one while she is hooded. In practice, the aviary falcons take all theirs voluntarily but the trained birds are given rangle by hand so that I have control over what is going on. It does no harm to leave a dozen stones on the shelf perch.

The best time to give rangle is when the hawk has lost a lot of body fat and is perhaps ready for the creance. Stones can be given every night then for a week or ten days, after which give her good castings. The stones tend to stir up the mucus in the stomach and abrade off any excess cuticle and all this passes on down through the intestine. The castings too will tend to soak up the mucus, but only once the stones have loosened it. Later in the season if the hawk has had a lot of rich feeding and has a sluggish digestion a further course of stones will not come amiss.

The preparation of the stomach is neatly summarized in the following rhyme:

Washed meat and stones will cause a hawk to fly;

Long fasting and great casting will cause a hawk to die.

This whole subject of the stomach is poorly understood by many modern falconers who often blame a poor performance on their birds rather than on their own management. The way in which a hawk flies, assuming she is fit, depends on her inner motivation which depends on the management of her stomach and appetite. The prairie falcon, for example, has a tough constitution and ability to survive when necessary on short and poor quality rations. This makes the management of her appetite that much more difficult and has earned them a reputation among those accustomed to the easier peregrine as being erratic performers. The prairie is a fairly quick-tempered bird which will not tolerate sloppy handling in the same way that a peregrine will and her stomach needs careful attention. In these two attributes she demands more skill from the falconer than a peregrine.

Whereas the goshawk will perform adequately despite quite varied treatment, the less rapacious hawks such as the buzzard and the kestrel need very careful management if they are to chase with any enthusiasm. Unfortunately most of the beginners who handle these species do not understand this aspect well enough to realize their hawk's full potential and often find the experience rather frustrating and disappointing.

Wrong feeding has caused many deaths. Examples include feeding food which has 'gone off,' food which contains lead shot, and food which is lacking in essential nutrients with a result that the young chicks get rickets. If you are in China and feed your goshawk only on yak meat, don't be surprised when she loses the use of her legs. Food high in fat leads to high blood cholesterol causing arteriosclerosis. This is common in breeding females and is preventable by using a low-fat diet and giving some exercise. Even the way the food is fed causes accidents: food tossed to a tethered hawk on a high perch, the hawk drops it and bates herself to death trying to get at it. Or food thrown to a hawk in view of another bird, causing mass bating and feather damage. Or food pulled away from a hawk so that it punctures the sole of its foot with a hind claw. Or food fed to a chick which has not yet put over its previous meal, thus leading to sour crop. Or food with long or sharp bones in it, causing a perforated crop or a blocked gizzard. Every small detail in falconry is a trap waiting for you.

6.6 Appetite and weight control

As we have already seen, small hawks need a reasonable crop of food every day. They cannot go without food for several days as eagles can. This means that, even if they are quite fat, they become hungry at some time each day for their meal and this daily hunger is used in the 'appetite' system as motivation for the bird to fly in falconry. It is not a question of starving the hawk, but simply of ensuring that the time of day when she is hungry and ready for her meal coincides with the time at which you intend to fly her.

Under a given set of conditions, a merlin may need 3/10 crop free of casting material in the morning and 9/10 in the afternoon to maintain a steady weight. Because productive metabolism is greatest during the day, most of the energy drain occurs then. On a time basis one can calculate that 1/10 crop provides for only just over one hour. Thus if given 4/10 crop instead of 3/10 crop the merlin will be in optimal hunting order about one hour later than she was on the previous day, or, if fed 2/10 crop one hour earlier, all other variables staying the same. This is the 'appetite' system of conditioning. It works well, but only within strict limits. The bird must be near her 'fit not fat' weight, and if the meal is delayed for more than two hours or so the bird's hunger and responsiveness may actually decrease.

To be successful with the appetite system one must be clear about the time at which the hawk is fed up. When food is fed to a hawk it is stored first in the crop from whence it is transferred or 'put over' bit by bit into the stomach for digestion and from there to the intestine for absorption. A goshawk or peregrine with an empty gut, with no pressure in the mutes and often with green bile present, will put the first food right through in 2-3 hours and produce a good high pressure mute. Now if you hawk all afternoon and start giving odd bits of food at 2 P.M., finally feeding up at 6 P.M., the effective time of feeding was 2 P.M. If altogether you gave 5/10 crop it doesn't matter much whether you give 1/10 at 2 P.M. and 4/10 at 6 P.M., or the whole 5/10 at 2 P.M. The extra 4/10 will only be stored in the crop while the bird is digesting the first portions. If, on the other hand, you start flying at 2 P.M. but the hawk gets no real feed until 6 P.M. then obviously feeding-up time was 6 P.M. The late feed-up time means that the hawk will have lost more or gained less than it would have done and yet will have less keen an appetite at 2 P.M. the next day. This is the falconer's eternal dilemma. On afternoons when it is pouring with rain one hangs around wondering whether to feed up now, so that the hawk will be ready the following day or to wait, hoping the rain will clear up in time for a brief excursion before dark, even at the risk of a poorer performance the following day.

Larger hawks are usually flown on a combined appetite-weight system. More attention is paid to the weight of the bird than to her appetite. Thus a goshawk may take hares at a certain weight but refuse them if she weighs 50 grams more. A similar situation applies to peregrines and rooks.

Figure 6.7.1 A keen, strong European sparrowhawk, in yarak.

Watching the falcons weathering can pick up on fitness.
Photo Verity Johnson.

6.7 Symptoms of high and low condition

Watching falconers with their birds, it is obvious that many are unable to tell with any certainty what condition their birds are in. They are insensitive to or unaware of the signals which the hawk is giving them. More than one falconer has starved a hawk to death because he has believed a faulty weighing machine rather than the symptoms

A fit falcon will bathe vigorously.

before his very eyes. Also it often happens that by the time a hawk is first weighed after being taken up for training she has already lost quite a bit of weight and so the falconer has no fat weight on which to base calculations for an approximate flying weight. It is crucial therefore to be able to see from a bird's behaviour what her condition is despite wildness or imprinted tameness which may mislead you.

When a hawk is fat and fit she is very strong, and may well fight hard when cast in the hand. She easily gets breathless and in the field may be lethargic and unresponsive, or she may fly off a long distance without stopping. She may ignore quarry and, if the lure is thrown out to her, she may overfly it. She may refuse to feed on the fist or take a bite or two and flick the meat away contemptuously. She will bate strongly, bouncing back up afterward and glaring at the falconer with hot breath, or she may even refuse to stand on the fist at all.

A fit hawk is strong to fly but does not easily

Mummio is a black resin, a mixture of mineral and organic substances from the mountains of Central Asia. The Indians use it in Ayurvedic medicine, it being mentioned in Charaka in the first century AD and was used in European human medicine in the 16th century, possibly brought back in the Crusades. It was mentioned by Latham and Bert for use in falconry, then seemingly forgotten until British falconers, such as Stephen Biddulph and Delmé Radcliffe, were taught about it again by local falconers in pre-partition India. It is known as *Mummio, Mumiya, Shilajit* (Urdu: 'Sweat of the Mountain'), *Baragshun* (Mongolian: 'Mountain Juice'), and *Kaotun* (Myanmar: 'Blood of the Mountain'). I found it in the Karakorums in the Hunza, where it is called *Nerasto*. It seeps out of the rocks like tar in the summer and can be purified by slowly melting it in the hot sun. In Kirghistan I met a man who had accidently chopped off the top joint of his finger, leaving it hanging by a flap of skin. He dipped it every day in mummio and it healed back together perfectly.

Mummio in western Mongolia.

Mummio is sold in apothecaries throughout Asia as capsules or pills and is used as a fortifier, for broken bones and bruising, as an antiseptic, diuretic and emollient. Russian research demonstrated its properties as an antiseptic and as an anticoagulant, and for burns. The Russians used it for body building. It is also used as a sex potion, for exhaustion and for aching joints. The Russians use mummio from Tajikistan but I also found it throughout Mongolia and the Altai.

get breathless. A really fit goshawk should be able to take pheasants in the air in their first flight and should seldom give up the pursuit of a rabbit or pheasant while she can still see it. If grounded she will get up and have another go. Unless poorly manned, the fit accipiter will come to the fist and bind but not overfly it. Similarly a fit falcon may, through ebullience, overfly the lure once or twice but still take it seriously and soon hit it hard. The fit falcon in still weather should manage 60-70 stoops to the lure in two bouts of 30-35 without resting on her wings, looking away, perching, or panting. At home she should perch with one foot up and feathers slightly fluffed, both eyes open, and watchful. She may winnow her wings for exercise and, when resting,herwing-tips will be over, not under, her tail. When feeding she will tear up her food strongly and impatiently.

A low falcon will not fly so strongly and may give up when flying quarry after one or two hundred metres. The pectoral muscles will be noticeably wasted. She may not be particularly responsive in the field but when she does comeshe will not overfly the fist or lure. Some birds may be maniacally keen and aggressive. Young birds may start to scream, mantle, attack the handler or dogs, and drag food away in a mantling shuffle. However, after one or two greedy bites, the low bird may be slow or reluctant to feed, standing jealously over the food but not actually pulling at it. On the perch the feathers are more fluffed out, the eyes are oval and although one foot may be lifted it may still rest on a knuckle. The wings may droop below the tail. Bating is weak and moth-like.

A very low hawk, just before she dies, will be fluffed out, standing on both feet, wings and under-tail coverts drooping. Eyes are semi-closed and sunken, and one of them may close completely occasionally. The bird appears unusually tame and dozy and when offered food may refuse to eat or may pick at it a little without swallowing. She cannot be relied on to feed herself and will need tube-feeding. She has difficulties recovering from a bate and is reluctant or unable to fly. As her blood sugar levels drop furthershe may lack co-ordination, have fits and finally die.

Of course such a hawk may well be suffering from some disease and so, together with these general signs of lowness, there may be symptoms of the disease, such as digestive, respiratory or parasitic disorders. If there is the slightest question mark over the bird's health the very first rule of good management is to *get the bird's condition up* and stop flying her. A strong hawk can often fight her way through disease even if untreated, but a low hawk, even with the latest medicines, may quickly slip beyond hope of recovery.

6.8 Factors which can mask condition

Although a hawk may have been carefully brought to the optimum physical condition, there are various factors which can offset these efforts. Obviously if there is high level of an undiagnosed parasite or a disease, such as aspergillosis, the hawk will be unable to do her best. So if for some reason the hawk seems below par, her health should be checked. A bird well-gone with aspergillosis for example may show no symptoms at all except a lack of stamina; while she is functioning anaerobically she appears fine, but after an initial effort she is unable to continue aerobically.

Physical or mental stress can mask conditions. A newly imported hawk which is not acclimatized may require extra rations and careful treatment. Wild-caught hawks which are insufficiently manned may be too stressed to perform properly although in the right physical condition. Even tame hawks get spooked in a strange place or with unfamiliar sights.

Biological urges such as migration, territorial disputes or homing may put a raptor right off hunting. If things aren't going too well my New Zealand falcons fly off home in disgust!

Malimprinting can cause a hawk to show deceptive behaviour. A mantling, screaming imprint can put on such display that one is convinced that she is at death's door. It can be very difficult to judge the condition of such hawks accurately; they can quickly go from one extreme to the other.

One afternoon I took five sparrows with a slightly malimprinted New Zealand falcon who broke the neck of each one of them then flew off and cached its corpse, returning after a few minutes to continue hunting. Similarly with goshawks; often when a river has separated me from my hawk on her quarry I have called the hawk off the prey back to the fist or lure. On one occasion the quarry, a moorhen, flew off unharmed. On two other occasions my goshawks have taken one prey in each foot. These experiences occur because the behaviour centres for hunting and for eating are separate. We don't know yet what part of the brain triggers searching and attacking behaviour, but the appetite is controlled through the hypothalamus. Thus it is possible for a hawk to be keen to hunt but not to eat, as for example when a parent bird has young to feed. On several occasions I have found kills of wild falcons and harriers with a casting beside them, indicating that the kill was made before the hawk had cast. This dissociation of hunting and feeding is relevant to multiple kills. Once a hawk is thoroughly made, after fifty kills or so, I often do not feed any of the kill at all if the flight was an easy one.

Wild hawks have daily cycles for hunting and trained hawks behave similarly. Most perform well in the early morning, less well during the middle of the day and best at the last half hour before dark, especially in winter. Sometimes the change of behaviour can be quite dramatic. On the other hand, if you feed your hawk, or horse, by the clock, you can quickly find you have created a rod for your own back.

Hawks, and dogs for that matter, seem to work better after being kept home for one or two days. Bert (1619:41) recognized this and Harry McElroy calls it 'frustrational conditioning'. Thus if one is planning to fly at a field meet, rather than flying the hawk every spare minute beforehand, it is usually best to rest her on the day before the meet starts and feed up early in the day. Also if one is wanting to introduce the hawk to an unfamiliar species of quarry the best time to try it is during the first week of the hawking season after the moult. That is when she is at her most open-minded.

Some old intermewed game hawks, who have flown with mechanical precision for several seasons, will go well almost regardless of what condition they are in. Then you can relax and enjoy yourself!

6.9 Further Reading

Acharya, S.B. 1988. Pharmacological actions of Shilajit. Indian Journal of Experimental Biology 26:775-777.

Barton, N.W.H. and D.C.Houston. 1993. A comparison of digestive efficiency in birds of prey. Ibis 135:363-371.

Baz-nama-yi-Nasiri: *A Persian Treatise on Falconry*. 1908. Translated by D.C. Phillott. Bernard Quaritch. Facsimile Eryr Press.

Berners, J. 1496. *The Book of Saint Albans*. Wynkin the Worde. Facsimile 1966 Abercrombie and Fitch, New York.

Bird, D.M. and S.K. Ho. 1976. Nutritive values of whole-animal diets for captive birds of prey. J. Raptor Res. 10(2): 45 – 49.

Clum, N.J., M.P. Fitzpatrick and E.S. Dierenfeld. 1997. Nutrient content of five species of domestic animals commonly fed to captive raptors. Raptor Research 31(3):267-272.

Dierenfeld, E.S., C.E. Sandfort and W.C. Satterfield. 1989. Influence of diet on plasma vitamin E in captive peregrine falcons. Journal of Wildlife Management 53(1):160-164.

Dierenfeld, E.S., N.J. Clum, E.V. Valdes and S.S. Oyaruz. 1994. Nutrient composition of whole vertebrate prey: a research update. Proc. Assoc. Zoo Aquaria Conf., Atlanta, GA USA. 20.

Duke, G.E., 1987. Gastrointestinal physiology and nutrition. In Giron Pendleton B.A., Millsap K.W., Cline K.W. & Bird D.M. [Eds.] *Raptor management techniques manual*. Nat. Widl. Fed., Washington, DC USA. 262 –269.

Forbes, N.A. Raptor Nutrition. Honeybrook Foods.

Fox, N.C. 1976. Rangle. Journal of Raptor Research 10:61-64.)

Gaunt, A.S., R.S. Hikida and J.R. Jeh. 1990. Rapid atrophy and hypertrophy of an avian flight muscle. Auk 107, 649–659.

Gessaman, J.A. 1987. Energetics. In Giron Pendleton B.A., Millsap K.W., Cline K.W. & Bird D.M. [Eds.] Raptor Management Techniques Manual. Nat. Wild. Fed., Washington, DC USA. 262 –269.

Kirkwood, J.K. 1980. Maintenance energy requirements and rate of weight loss during starvation in birds of prey. In Cooper & Greenwood [Eds.] Recent Advances in the Study of Raptor Diseases. Chiron Publications, Yorks., UK. 153 - 157. 37.

Kirkwood, J.K., 1985. Food requirements for deposition of energy reserves in raptors. In : Newton I. & Chancellor R.D. [Eds] Conservation studies on raptors: Proceedings of the ICBP World Conference on Birds of Prey, 1982: 295-298. Cambridge: International Council for Bird Preservation.

Lavigne, A.J., D.M.Bird and J.J. Negro. 1994a. Growth of hand-reared American kestrels I. The effect of two different diets and feeding frequency. Growth, Development & Aging 4: 191-201 40.

Lavigne, A.J., D.M. Bird and J.J. Negro. 1994b. Growth of hand-reared American kestrels II. Body composition and wingloading of fledglings fed two different diets. Growth, Development & Aging 4: 203 – 209.

Marsh, R.L. and R.W. Storer. 1981. Correlation of flight muscle size and body mass in Cooper's Hawks: A natural analogue of power training. Journal of Experimental Biology 91:363-368.

Rosser, B. W. C. and George, J. C. 1986. The avian pectoralis: histochemical characterization and distribution of muscle fibre types. Can. J. Zool. 64: 1174 1185.

Taylor, R.L., S.A. Temple and D.M. Bird. 1991. Nutritional and energetic implications for raptors consuming starving prey. Auk 108: 716-719.

UFAW.1989. *Animal Training: A review and commentary on current practice*. Universities Federation for Animal Welfare. ISBN 0-9007-67-66-9.

Van Zyl, A.J. 1999. A comparison of basal metabolic rates and thermal conductance in the Common Kestrel at different latitudes. Unpublished Ph.D. Thesis, University of Cape Town.

C7 Flight Training

This chapter is supported by a film in the Bird of Prey Management Series: *Fitness Training*.

A young gyrfalcon getting fit.
Photo Mark Williams.

7.1 Fitness training

What do we mean by fit? Fitness implies suitability for a job. Thus a well-muscled body builder might be fit for weight-lifting but be hopeless at the marathon, and the super-fit marathon runner would be pathetic on the weights. The top endurance horses are gaunt, lean-looking beasts, far removed from the sleek well-muscled show-jumpers. The best pursuit falcons are the endurance birds, while the game hawks are the rounded, heavily muscled ones. Feathers may hide her body but you have to learn to undress her with your eyes.

Muscle needs glycogen (for energy) and oxygen brought to it by the blood, and lactic acid, carbon dioxide, and heat removed from it. During exercise, the blood can increase its transporting capacity by up to sixty times. There are two main types of skeletal muscle fibre: red muscle which is good for endurance work and has a high capacity for using oxygen, and white muscle which is associated with power and sprinting and has a low capacity for using oxygen. Within these broad groups are further types of muscle fibre, not all of which come into use every time the whole muscle contracts. Of these there are two main types: the slow-twitch muscle fibres which have a slow contraction time and a good capacity for using oxygen; they can therefore work continuously without fatigue. The second is the fast-twitch muscle fibres which have a fast contraction time. These are the ones most involved in performance work. Some of them have a high capacity for using oxygen and can manage sustained work over long periods, the others have a low oxidative capacity and fatigue rapidly but are capable of the most strenuous fast work.

Thus for most of the daily activities, the hawk uses mainly her slow-twitch fibres. When she sets off on a long flight, perhaps a hunting patrol, she uses mainly the high oxidative fast-twitch fibres, and when she bursts into an attack on prey she uses her low oxidative fast-twitch fibres.

All these fibres are scattered throughout the muscles in varying proportions. Harriers and short-eared owls for example, are likely to have a high proportion of high oxidative fast-twitch muscles to sustain them while they quarter up and down, assiduously searching for prey. Accipiters, which depend on a powerful sprint, are likely to have a high ratio of low oxidative fast-twitch fibres.

Although the ratio of fibre types depends on the species and the ways they hunt, fitness training can increase the oxygen-using capacity of the fibres, enabling them to avoid fatigue longer. Fatigue is when the muscle can no longer continue at peak performance due to one of the factors becoming limiting. It may not have enough glycogen, or enough oxygen, or it may not be able to get rid of surplus lactic acid or surplus heat. It may also get dehydrated.

When the hawk is not flying hard, the blood can bring sufficient oxygen for the muscles to work aerobically, using the glycogen stored in the muscles themselves as an energy source. If this stored glycogen is exhausted, it will take her 1-3 days to replace it. Thus, if she has done a hard day's fast flying, she must be given one or two very quiet days before being again asked for peak performance. If you try to get her to fly to the point of exhaustion two days in a row, she will not manage much at all on the second day.

When the hawk has to fly really hard, she will be using her fast-twitch fibres and soon the blood cannot bring enough oxygen for the muscles to work aerobically. They go into oxygen-debt and function anaerobically, resulting in a build-up of lactic acid. The acid eventually will cause the muscles to seize up. If there is not too much acid, a little rest or gentle exercise is sufficient to clear it and the hawk can continue again until its glycogen runs out. Anaerobic respiration is only about 1/12 as efficient an energy-user as aerobic respiration. It is crisis management. Thus, lactic acid build-up limits the fast performance of the hawk's muscles in the short term, causing fatigue, whereas glycogen depletion causes longer-term exhaustion.

Fitness training aims at encouraging more mitochondria to improve glycogen utilization and at increasing the blood supply to speed up the transport of materials and delay the onset of anaerobic respiration and the resulting build-up of lactic acid. It also improves the capacity and blood supply of the lungs, giving better oxygen uptake. The improved blood circulation also helps dissipate excess heat, which is a limiting factor for gyrfalcons flying in warm climates.

To do this requires exercise which must be of the same type as the performance ultimately required of the hawk in hunting. Thus the accipiter needs fast, hard exercise which will increase the power of her muscles, resulting in more thrust and more speed, and which will increase the distance she can sprint before fatigue sets in. The crow-falcon, on the other hand requires hard, sustained work so that she can keep going at peak performance for ten or more minutes. Fitness is not so critical for waiting-on flights, but exercise is needed to improve internal motivation which will maintain discipline in responses. Without it, the game falcon becomes a fat, lazy prima donna. Harris hawks need to improve both their general fitness and their sprint; obviously a gentle country walk does nothing to improve sprinting.

You need to be able to measure how much exercise the hawk does, and what effects it is having on the bird. This is not easy. On the one hand, the hawk is motivated to fly by hunger, or the desire to chase and so on, on the other hand she is limited by the pain of fatigue. Hawks tend to govern their own intensity of training, but the best ones will overwork themselves. Although it is possible to monitor heart rate, respiration rate, body temperature, and blood lactic acid levels under research conditions, this is not feasible for the average person with a free-flying raptor. Instead you can only 'read' her behaviour, watch for her beak starting to open, and see how well she performs in a given amount of exercise.

As in any animal, youngsters become fit more quickly than old ones, especially old ones which have never been properly fit before. Interval training gets hawks fit more quickly without overworking them. In this way you can exercise two or three falcons at a time like council workers (one working, two resting). You can go round and round the cadge, exercising each bird in turn, then resting her while you do the others, then another bout, and so on. This gives her time to clear any lactic acid, reducing fatigue but increasing the amount of exercise. She should be only partially recovered before the next exercise period. As soon as her breathing returns to normal, she is ready again.

7.2 Exercise

There are several ways to exercise raptors: calling off and long-luring which are good for the slow-twitch muscles, chasing a towed lure or climbing to a lofted lure, which is good for both types of muscle fibres, high-jumping and exercise luring which are good for the fast-twitch muscles, and flying at quarry or roprey which uses the fibres which the hawk actually needs for hunting.

In the early stages, or when daylight time is short, the hawk can be given high jumps indoors. The hawk is placed on a low padded perch (not its normal block) and called steeply upward to the fist for a tidbit. Seventy to a hundred of these jumps can do a peregrine a lot of good because she finds it quite hard, but an accipiter can do it almost without using her wings at all. They can manage 200 jumps or more. Buteos are easily disgusted with it and may refuse to cooperate unless the trainer uses special techniques. The problem is that if you give the hawk a tidbit every time she jumps, she gets more and more tired and less and less hungry. As she gets fitter and has to do a lot of jumps, then all those tidbits soon amount to a full crop.

Steve Layman has developed a good system for the high-jumping technique. First you need a safe room, free of obstructions and perches, with the window covered over. On the floor is a deflated basketball stuck to a metre square of carpet. The hawk can land on this safely without damaging feet or feathers and learns to centre on it as part of the routine. The hawk is taken up as for hawking, weighed, brought into the jumping room and allowed to settle down until she is relaxed. The initial plan is to shape the behaviour we want. Once a firm foundation of routine has been established then high numbers of jumps will become possible later on.

So to start with, the hawk is placed on the ball and then asked simply to jump up onto the fist for a tidbit which is hidden in the glove. She cannot see it from the ball. She must jump up, as an act of faith, and then see and eat the tidbit. Then she can be placed back on the ball. Get a rapid repetitive up and down behaviour established. Sometimes, as she is eating the tidbit on the glove, place another tidbit on the ball. She will quickly turn around and fly back down onto the ball. Soon she is flying fast up to the fist and down to the ball. If she misses the ball and lands on the floor, wait a moment until she regains the ball before calling her to the fist. As she comes to the fist, raise it high above your head.

It may take several days to establish this behavioural routine. Psychologically this is an attack sequence and she must be encouraged to use energy and commitment. Just use sequences of 5-6 calls to start with. Do not attempt prolonged sequences and risk refusals. Quit before she starts to lose interest. By the time she is doing about 75 good fast jumps on a 1:1 ratio, i.e. being rewarded every jump, start to skip rewarding her now and then. As you have not been showing her the tidbit, she never knows until she has arrived, whether there is one there or not. You are now on a variable ratio of reward. Gradually reduce the numbers of rewards until you reach the point where she is getting about three rewards for 10 jumps, but these are on a random basis. Sometimes she gets two in a row. Sometimes she has to do six jumps for nothing. It is the same for her in the wild and she is mentally prepared for disappointment. The hunting success of young wild raptors is usually below 10% and in adults seldom much above 30%. If you have managed to shape her behaviour well, instead of becoming disappointed, she will be inspired into maximum effort. She will start to get really fast and aggressive in her approach and, as you get into high numbers, you will need to raise her flying weight sometimes by 10 to 20 percent. You can also relate food quality to performance. Have two tubs of food, one of washed rabbit, one of quail. When she is slow, give her rabbit. When she is fast, give her quail. Don't try for maximum numbers every session; sometimes finish early and feed all the remaining tidbits at once. Thus she is encouraged to make maximum effort immediately and cannot outguess what will happen next.

High jumps have long been used by eastern falconers and are increasingly being used in the west. The system has many advantages. It does not lower a game-falcon's pitch. It can be used indoors, in the dark, and in the rain. A few high jumps are a good way to warm a hawk up before proper flying. Jumps can easily be quantified and used as part of a controlled fitness program. Goshawks, for example, are not fit to hunt until they can manage 80-90 jumps. I used to high jump my birds over twenty years ago, but gave up because I could never get much beyond 40-50 jumps. But when Steve showed me the technique of rewarding on a variable schedule then 200-400 jumps became possible. This has a tremendous psychological impact on the bird and helps to build up stamina and endurance. The only problem is that I get bored.

Calling off from posts or trees is a common way of getting an accipiter or buteo fit; unfortunately this can be very time consuming. It is certainly a good way of getting the falconer fit! I think the best time for this is for the first two or three days after flying an accipiter loose for the first time. It helps build confidence in both hawk and falconer. Much more than this risks orientating the hawk too much toward the falconer and not enough toward quarry. It does nothing for the fast-twitch fibres which the hawk will need for its first attempts at quarry. To show much effect, the hawk needs to go a good distance, for example, if you take a Harris out on a horse for a five-mile ride, throwing her off and calling her in all the way, she will start to improve in general fitness. But a slow potter from tree to tree for a mile or two is really neither here nor there.

Pursuit falcons are often trained by long-luring, that is, called to a lure from about 500-1000 metres in a straight line. The hawk flies at her own speed, which is not top speed. Although this does help get them fit to some extent, the main reason for doing this is a mental one. It is teaching the falcon that when the hood is removed, she should look out and be ready to start at something a long way off. Secondly, it teaches her to come in to the lure from a distance. Long-luring does not improve flying and footing skills in the same way that the exercise lure does. It is the precursor of the Telwah competition, which is flown on a 400 metres level track.

Trailing is a progression from calling-off and long-luring. Basically the falcon is called to a lure held by a falconer sitting in the back of a vehicle.

The lure is trailed on a long line from a strong fishing pole. As the hawk comes in, the vehicle starts to move, the lure speeds up and becomes airborne, held out to the side of the vehicle away from the slipstream and dust. The line is wound in until the lure is only a few feet from the rod with the hawk close to catching it. The vehicle can then proceed at a steady, fast speed so that the falcon is compelled to use her fast-twitch muscles in order to keep up. The trail can go for a measured distance at a measured speed and thus the falcon's performance can be monitored until it reaches a peak. It is better to use a sandflat for this in the desert, rather than an English country lane! Trailing does not improve flying skills much but is good for experienced birds. Written records should be scrupulously maintained so that none of the falcons are worked over their performance limit, otherwise they will be set back several days.

An alternative to this is towing the lure behind a radio-controlled tow-plane. Delta-winged EDF planes can be purchased relatively cheaply and a servo installed to create a release mechanism. The lure is attached by a light line to a parachute which trails behind the plane. When the falcon takes the lure, the parachute is released, and the hawk comes down holding the lure. The pilot then has to land the plane while watching where the falcon is landing. This is an excellent way of getting the falcon fit and emulates the stage when the parent carries food in front of the young falcons, then drops it for them to chase and catch. It doesn't teach the young falcon what the prey looks like, or how to catch by the head, or good footing skills, but it certainly gets the falcon fit. Do not be tempted to over-do it. Do not believe people who tell you 20 minutes flat out is OK; it isn't. Falcons, like thoroughbred horses, will race beyond their physical endurance and kill themselves or destroy their own muscles through lactic acidosis.

Releasing lures, parachutes etc from a UAV is illegal in most countries, which is a reason why we use roprey. Beware also of any plane or quadcopter with unguarded propeller blades. Many falcons have had severe or fatal injuries from these. Although your intention, in theory, is that the falcon doesn't go near the blades, in reality - 'stuff happens'.

For accipiters, buzzards and eagles, a lure can be towed on a line on a straight or zigzag course, with the line pulled by a motor. The problem with these ground lures is that they soon get very bedraggled in wet grass, and it can be tedious rigging up everything. I'd rather tow the lure on a 50 metre line behind a horse or quad bike. Someone energetic on a mountain bike would do nicely. You can shout encouraging words to make him go faster…

The exercise lure for a falcon is an excellent way to give her a good pipe-opener and can be used on all non-hawking days except when it may cause her to lower her pitch or to become lure-bound. However, an unfit falcon will not reach or hold a good pitch for long, so one is on the horns of a dilemma here. Personally, I find that a good workout to the lure (about 50 stoops in two lots of 25) in her home exercise field, does not lower a game-hawk's pitch out in the hawking area. Also, it does make her more responsive to the lure. I have seen game falcons in some countries which would ignore the lure and only come down to live game. Such a bird is half trained and half lost. The crow falcons are given a lot of exercise to the lure and also we try to get them fit in a different way. They must be capable of very enduring flights, they need plenty of red muscle but nothing surplus. They need to be able to go on and on and on. And then when you finally think they will have to give up, they've got to find that little bit more and finish the job. Game falcons, on the other hand are more like body builders, they need plenty of weight on board to strike a devastating blow. Fitness is important but not so critical. So as far as possible they are flown at maximum weight, but not to the extent that they become disobedient.

Hard flying at quarry every day will eventually bring the accipiter to the peak of fitness. But to attempt this before the bird is at least half fit will over-face her and lead to other problems, such as lack of confidence. In Britain maybe only about one in fifty goshawks are flown fully fit. If a hawk is flown less than three days a week of good, hard flights at quarry, she will not be fully fit and supplementary fitness training will be needed. For a goshawk, a hard flight is a sprint of at least 100 metres. Pheasant flights of 200-300 metres will force the gos into lactic acid production and will gradually improve her fitness, whereas a 20 metre dash at a rabbit doesn't tax her system to its limit at all. Accipiters flown only at weekends do best with about a 100 high jumps on non-flying days. Ropheasants are great for exercising goshawks when time is short.

Buteos and Harris hawks do not even have the explosive anaerobic sprint of the accipiter and for them fitness training is essential. Because of the mental attitude of the Harris, the falconer flying her tends to have the same, laid back, casual approach. This is a pity because it is possible to get a much better performance with the Harris than most people seem to be content with.

For waiting-on flight training, the lure can be suspended below a kite, balloon or a quadcopter. Quadcopters have more or less superseded kites and balloons nowadays because they are less weather dependent and less hassle to rig up. Basically the lure is suspended and the falcon unhooded and allowed to fly up to it. Either during the climb, or on successive days, the lure is raised higher until the falcon is fit enough and used to climbing to 300 metres (1000 ft) or so.

7.3 Assessing soundness

Before a hawk is flown loose, whether for falconry or for rehabilitation, she must be got fit. It is not a clever position to be in, when a hawk is:

a. on her first flights loose

b. at her lowest weight

c. unfit

d. unentered.

How long will she survive if you lose her? Careful consideration should always be given to hawks which have not learned to fend for themselves.

She may be sound in wind, but is she sound in limb? A rehab bird especially may have received an incapacitating injury and it is vital to find out if she is fully fit again before hacking her out. There are two easy ways to assess her soundness. First she must be exercised until she is quite tired; only then will a weakness become apparent. Then, just like trotting up a lame horse, she should be flown in a straight, level flight directly toward and away from an observer filming with a video camera.

The film is then played back on slow speed or freeze frame. Look closely at the wing beats. Are both wings exactly in unison? Do they beat to the same depth? Are they level? If one wing is working slightly harder than the other it will skew the mid-line of the body, causing her to yaw. The hawk will try to compensate for this by tilting her tail slightly. Can you see the tail out of line?

The second way is to return the tired hawk to a quiet pen and watch her without her being aware of your presence. Preferably use closed circuit television. Does she droop her wings, and does she do so evenly? While it may be acceptable to release to the wild an unsound buteo or other 'searching' hawk (see chapter 8), any species, such as accipiters and falcons, which rely on their flying ability to attack agile prey need to be 100 percent fit. No compromises.

Owing to hostility towards falconry in Australia, rehabilitators have largely been prevented by State regulations from using falconry free-flying techniques to rehabilitate raptors. Instead the regulations require flight pens to be used, ranging from 4 metres long for small raptors to 6+ metres long for eagles. Chasing a raptor up and down, or round a doughnut-shaped flight pen, is by no means adequate to get it fit. Nor will it teach an inexperienced young bird to hunt. It is a shame that in the 21st century there is still such a basic misunderstanding of raptor physical and mental fitness, even in a western country. Fortunately more and more re-habbers in Australia are getting the skills needed for free-flight exercising. When we take out our young falcons from our 30x30 metre hack pens we estimate they are no more than 25% fit and this is very quickly revealed when we go on to advanced flight training in chapter 9.

7.4 Lures and lure theory

In the late fourteenth century Chaucer wrote '*With empty hand men may none hawks lure.*' But

Figure 7.4.1 Dragging a rabbit lure for a young Harris hawk.

the lure has more uses than simply for calling the hawk back. It follows therefore that the lure takes a variety of forms and is used in different ways and circumstances.

The main uses for the lure are:

a. to call the hawk back

b. to exercise her

c. to improve her footing ability

d. to teach a falcon to wait on

e. as a dress rehearsal for quarry

f. to call her back when she is afraid of you

g. as a more visible way of calling her back from a distance

h. some falcons prefer to hit a lure rather than to land on the fist

i. to make an aggressive hawk less orientated toward the falconer

j. to give flying displays for the public.

a) To call the hawk back is the commonest use for the lure. I normally introduce the lure as soon as the hawk is coming a metre or so to the fist. Some hawks are reluctant to come any distance to the fist and they can often be induced to overcome this by calling to the lure. Soon the lure can be raised up to the glove and the hawk begins to come to the fist with no difficulty. The lure should be garnished with food to start with and this should be firmly attached. By the time the hawk has been flying free for a week or two I no longer garnish the lure except in particular circumstances. Garnishing it is a tedious chore and can lead to problems in the field.

Some people do not use the lure for accipiters or buteos, regarding them as birds of the fist. I do not follow this line of thinking; the lure is far too useful to leave at home. Newly-trained goshawks are usually good to the fist and even return to it after a miss. But it is not long before such perfection fades, and very often during the season there will be times when the lure gets you out of a scrape. My goshawks have not always been models of obedience to the fist, especially when baulked of a kill, but with the lure to help me I have never had to wait for more than a few minutes at most and I expect them to come in from a good distance. An accipiter is best given some reward whenever she comes a distance whereas the Harris, at the other end of the spectrum, can usually be flown without lure or reward. Merlins, Aplomados and New Zealand falcons are happy to use people's heads as perches and will fly from one to the other when beaters are lined out looking for quarry.

Sometimes, in an emergency, it is impossible to find a clear area of ground where the lure would be visible and approachable. Therefore falcons should be taught to take the lure in the air before they are trusted loose. We use a pole lure, but if you use a hand lure you can wind the lure line spirally along a stick so the lure is either at the end, like a feather duster, or on a short length of line, as on a fishing rod. In this way the lure can either be held static in the air, just off the ground, or it can be trailed about a bit. The traditional Japanese lure is of paper, at the end of a stick, with no cord.

Falcons normally get the idea after one or two lessons. When calling a hawk back to the lure from a distance, never twitch it away at the last minute. She may fly straight on or next time she might refuse it. Lose her trust and you lose her. Remember Shakespeare: *Dare a falcon when she's in her flight and ever after she'll be haggard-like.*

A traditional Japanese paper lure.

When calling in a hawk the lure should be presented so that it is easy for the hawk to take, the glide path should not be too steep and the ground should be clear. If the hawk does not come promptly do not keep calling her, you are simply teaching her not to come. Nor go toward her otherwise she is teaching you to come. Failing to come is almost never a physical problem; she should easily be capable of flying the distance. Rather, the problem is that she lacks motivation. She doesn't want to come sufficiently to fly the distance. On the other hand, as with a shy dog, if you start to walk away and leave her behind, she may come twice the distance without being called at all. The more you pander to her whims the worse she will get.

Never forget the use of time. Often if the hawk is disobedient, it pays to use the politician's approach and do nothing at all. Rather than lay out your wares of assorted lures like a street vendor, just ignore her. Just let time do its work. After a while she will get restless and you will have the upper hand again. Young hawks, for all their waywardness, are extremely dependent on the falconer. Almost every incident of disobedience or 'moodiness' in a hawk is traceable to poor technique on the part of the falconer. Although it takes some patience and forbearance there are times in the training of hawks, and of horses too, when the implacable disregard of time can have a profound effect. The difference is that while the minutes or even hours tick away, by not calling the hawk at all she is made to wait for you. You call the tune. She is the one who is waiting, not you. She begins to realize that if she does not come promptly she may have a considerable wait before she has another chance, just as she would with real quarry. Thus the falconer 'tames the shrew'.

Many problems stem from poor techniques in picking up from the lure. This can lead on to difficulties at the kill too. When she has landed on the lure, give her a moment, then, if she looks up at you, just flash your pick-up piece in your gloved hand. This signals that you are an opportunity rather than a threat. Maybe tweak the lure line to keep her interested in the ungarnished lure while you approach. Step on the line as you move in, holding out your garnished fist. This prevents her from lifting off with the lure, especially if it is windy. Hold your fist at beak height and at the same time hold the lure down on the ground with your right hand. She should bite at the food. A settled bird will simply hop onto the fist and leave the lure. But she might want both! Just take your time. Keep the lure held down on the ground. Every time she bites the meat, 'give' a little with your left hand so that she cannot pull food off. Soon, in frustration, she will put one foot onto the glove so that she can pull the meat. Be patient and keep the lure held firmly down on the ground. You may have to tactfully unpick a claw embedded in the lure. She cannot see the lure because it is directly underneath your glove. Whatever you do, do not let her lift the lure off the ground. Eventually she will put the other foot up and scramble upright on the glove. Still take your time. Keep the lure properly concealed with your right hand and then lift her up, keeping her from looking back and seeing the lure. Once you are upright, put the lure away in your hawking bag which is on the opposite side of your body from the hawk and out of her sight. Then secure her with jesses or tie cords and let her have her ration. She will quickly learn to do all this very smoothly and you may not even need to bend down to the lure; she will simply leave it to come up to your fist. Out hawking she should also pick up from the kill without fuss. Of all the basic skills in falconry, picking up is probably the most often done poorly.

Once the hawk is fully trained, the recall or retrieve lure can be used. It is a simple lure with about four feet (1.2 m) of line and no stick (figure 3.10.1). In olden days it was carried looped over the shoulder like a climbing rope. The problem is that the lure is visible, rather than hidden away in a bag, so a hawk can see it and come in and snatch it.

Pippa Scott-Harden picks up one of our crow falcons at the kill.

b) Falcons can be exercised to the lure by a variety of methods. The technique commonly used in Britain works well and has been passed down from generation to generation. It is described in detail in the next section. The exercise lure should be capable of being struck hard without injury to the hawk. There should be no metal on or near it and it should weigh about one third of the weight of the hawk. For crow falcons we often use two crow wings, both right wings or both left wings. They are dried and the bony parts are dipped in black latex to seal them against the rain. Placed together, they spin like a propeller. The falcon makes a stoop and pass attack at it. For accipiters, buteos, and eagles a stuffed synthetic rabbit or fox skin dragged along serves well but in a wet climate needs frequent changing. They tail-chase and make a direct flying attack, rather than a stoop and pass. For exercise purposes they can easily be trained to use a rubber foam lure.

c) The falcon's flying and footing ability can be improved by luring especially with the pole lure. An unhacked young falcon, newly hard-penned, on first being flown, will be incapable of landing properly let alone hitting any prey accurately. A week or ten days lure flying will improve fitness and teach the falcon the use of her wings and feet; a prerequisite of serious hawking. Without it a falcon used for waiting-on flights at game may take a long time to develop good footing ability and will seldom fly like a wild bird.

Like so many things, more of it is not necessarily better. A poor hawk will soon become lure bound if over-lured. After ten days or so, the hawk should be entered at quarry without further delay. The rook or crow hawk can continue to be exercised to the lure on days in between hawking quarry, or better still, flown to roprey. Care is needed to maintain the balance of attractiveness between quarry and the lure. The best falcons get too quarry-orientated and the poor ones too lure-bound; both faults must be remedied as soon as noticed. If a falcon shows signs of becoming lure-bound, then fitness can be maintained by high-jumping until she is catching quarry better.

Accipiters do not improve their footing much with lure work and they certainly do not appreciate being made to miss the lure. It is best to get straight into real hawking with them.

d) Careful and imaginative use of the thrown lure can teach a falcon a great deal toward waiting on. Poor use of the lure, when the falconer does not understand clearly what he is trying to achieve, will soon ruin the falcon. With the thrown lure you can teach a falcon to stoop on command, only throwing the lure at the last moment at the bottom of the stoop. It is soft enough so that she can hit it hard without hurting herself.

e) Often the young hawk needs to know what the quarry looks like. She needs to develop a 'search image.' By placing a dead rabbit in some rushes and then twitching it out on a 70 metre creance the young buzzard or goshawk can be taught to start at rabbits. After a brief chase she can be allowed to 'kill' it and eat some of it. Following a couple of dress rehearsals, as realistic as possible, the hawk is ready for the real thing.

Falcons intended for game seldom need a dress rehearsal; most game species are virtually irresistible to them. Pursuit falcons may need more careful entering and we will look at this in the next three sections.

f) Sometimes hawks, especially accipiters, get jumpy and are reluctant to return to the fist. Throwing out the lure may bring her down; the falconer may even have to move away. If the situation has deteriorated to this extent the falconer must analyse the problem and remedy it because such a hawk is half lost.

g) All raptors usually come much further to the lure than to the fist. Also a swung lure is much more visible than a raised fist. A trained falcon can see and come in to a pole lure from 3-4 kilometres. Remember that a lure swung slowly and in a twitching fashion is much more enticing than one swirled at high speed, however anxious one may be. In some countries, people still use live lures, but a flapping artificial lure is just as effective in attracting a lost hawk, with none of the ethical problems associated with live lures. Remember, if your hawk has chased prey out of sight and you set off after her, get out your lure and swing it. This is easier done with a pole lure. She may not have killed and will come back in to the lure, saving you a lot of walking. The GPS signal may have shown her as static, but that doesn't mean she has definitely killed.

h) The fast-flying aerial falcons, particularly peregrines, are reluctant to come in to the fist once they have got speed up. They prefer to truss the lure in the air and flutter down with it further on. Failing a lure, a cloth cap on a string, or anything similar, will often bring her down. This trait of falcons is probably the reason why the lure was originally invented.

i) The way in which the lure can be used to reduce aggression by making the hawk less falconer orientated is described in section 5.13.

j) The exercise lure is commonly used to give flying displays for the public at centres or at regional summer shows. There are also mechanical lures which tow a dummy along a zigzag route, like a jinking rabbit.

The actual construction of the lure depends on its use. The best lure for attracting a hawk or for a dress rehearsal is a dead quarry on a string or a flapping dummy. Such lures are not suitable for every day use and indeed are not always desirable. If the hawk refuses it then you are in the proverbial cactus. You have nothing left to offer. For everyday use, I prefer a simple lightweight foam or feather lure which, although similar in many respects to the quarry, is still quite obviously a lure. The hawk is familiar with it and responds to it. If, for some reason, she is a bit uppity or jumpy and refuses the lure, I can still use dead prey or a flapping dummy as a last resort.

Lures made of dried wings do not do well in our British climate. Left in the hawking bag by neglectful falconers, they soon smell and become prime sources of bacterial infection. Bumblefoot and enteritis are just around the corner. For big lures I use moulted goose feathers.

Some falconers advocate the use of a heavy lure to prevent carrying. I tried this quite thoroughly when I was young with two sparrowhawks and a merlin and it made no perceptible difference. To prevent carrying requires a more fundamental approach to the problem than this (see section

7.7). Hawks easily distinguish between light and heavy quarry and act accordingly.

There is one further kind of luring which is used in the waiting-on flight. This is when the falconer waves his glove or hand to a falcon raking away from the point. Both he and the falcon need to be quite clear that the meaning of this signal is for the falcon to centre over the falconer, not to come down to the falconer. Try to keep this signal unambiguous, for example, whistle or call differently and perhaps wave or swing your glove in a different movement to luring. Some people use a hat with a fluorescent lining, and then turn it over as a signal to the falcon.

Whatever you do, don't lure her or you will lower her pitch, or in the old words, she will 'abate her gate.'

7.5 Lure training for pursuit flights

The initial training of a pursuit falcon is the same as for any other type, but once she is flying free she must be brought to a peak of fitness and appetite. She can be exercised by a mixture of high jumping, lure flying to the exercise lure, long luring, trailing or hunting roprey.

Lure flying to the hand lure takes some skill. The falconer must be able to place the lure in the air in the right position at the right moment, and this takes coordination. He must also be able to 'read' the falcon as she flies, to judge how motivated she is and how tired she is, and he must be clear in his mind what he is trying to teach or to achieve.

The exercise lure is described in 3.11. If you garnish it, use a good joint of rabbit or similar, strong so that it will not come off, and not too much, so that if it did come off, she would not get a full crop from it. Never use day-old chicks on the lure; either they come off, or they spray yolk, leaving you looking like a politician after a bad campaign.

There are two basic moves in exercise luring. Other moves are just variations on these. They are the left turn and the right turn, the left being faster and easier.

First wind up the lure line until only a short length is left. Always hold the stick in your gloved left hand (assuming you are right-handed). Hold out the lure in your right hand as in figure 7.5.1. This is how much line you need. Now hold it as in 7.5.2. Hold out your hands as though you were holding a tray of food, elbows loosely in at your sides, but without being stiff. The line should slide over your right hand like a pulley when you move your hands apart. When your hands are together, the lure should just clear the ground. When they are apart, it should come up near to your right hand. When you need to, you can use your thumb on the line as a brake.

Now start to swing the lure so that it comes forward, underarm past your ankles and then backwards over the top past your head, in the vertical plane. Swing it smoothly and as slowly as you can, but keeping the line straight at the top of the circle. Practice this for a while and walk around swinging it until you are thoroughly at home with the movement.

Next put your right foot forward as in figure 7.5.3, and slide the lure out in front of you as far as you can, letting your hands come together as you do so, to give maximum reach. Do this many times until you can shoot the lure out and land

Figure 7.5.1 Measuring out the right length of line for the exercise lure.

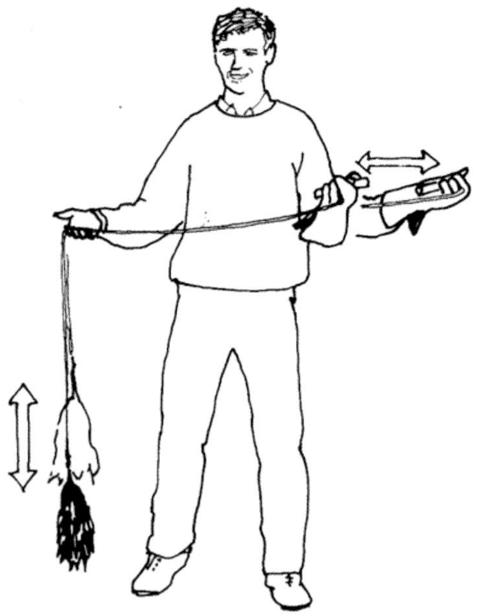

Figure 7.5.2 As your left hand pulls away, the line should run through your right hand as if on a pulley.

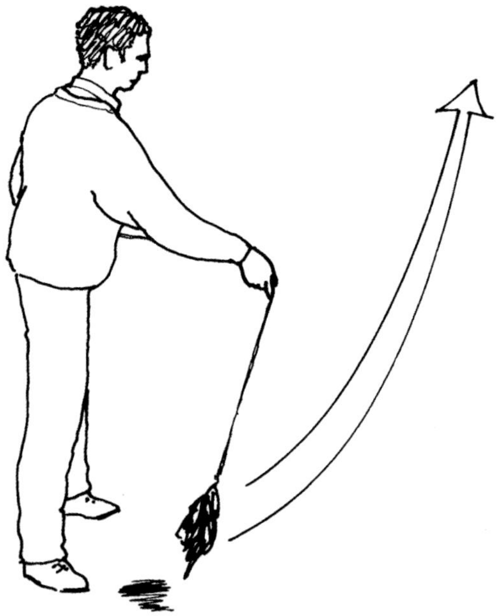

Figure 7.5.4 As you pull it back, pull your hands apart, letting the line pulley through your right hand.

Figure 7.5.3 Put your right foot forward and shoot the lure out onto the ground with your hands coming together. Then pull it back.

it exactly on a target spot, like a fly fisherman casting for a rise. The line should stay straight at all times. There should be no slack and no bounce back or jerk.

When you're getting slick at this, it is time to practice the left turn itself. Shoot the lure out onto the ground and then pull it back, low, just above the ground. Turn left as you pull, and bring your hands apart, sliding the line through your right hand. This gives the lure extra speed as in figure 7.5.4.

This movement is shown again from the side in 7.5.5 and 7.5.6. Keep turning and bring your right foot round until you face the opposite direction (figure 7.5.7 and 7.5.8). Keep the lure swinging and you are now facing the other way with the lure rotating at your right side, as in your original exercise. The falcon of course would have gone past and you are looking up watching her as she climbs away from you. Pretend you are watching her go up; she turns on a wingtip, and starts to come back in toward you. You have to prepare to turn left again. So put your right foot forward as before, and shoot out the lure. This time be confident, don't hesitate, before the lure has had a chance to hit the ground (figure 7.5.9 and 7.5.10), pull it back and turn left as before. Keep

Figure 7.5.5 The same movement as 7.5.3 seen from the side.

Figure 7.5.6 The same movement as 7.5.4 seen from the side.

Figure 7.5.7 As you turn left, bring your right foot forward and follow through with the lure.

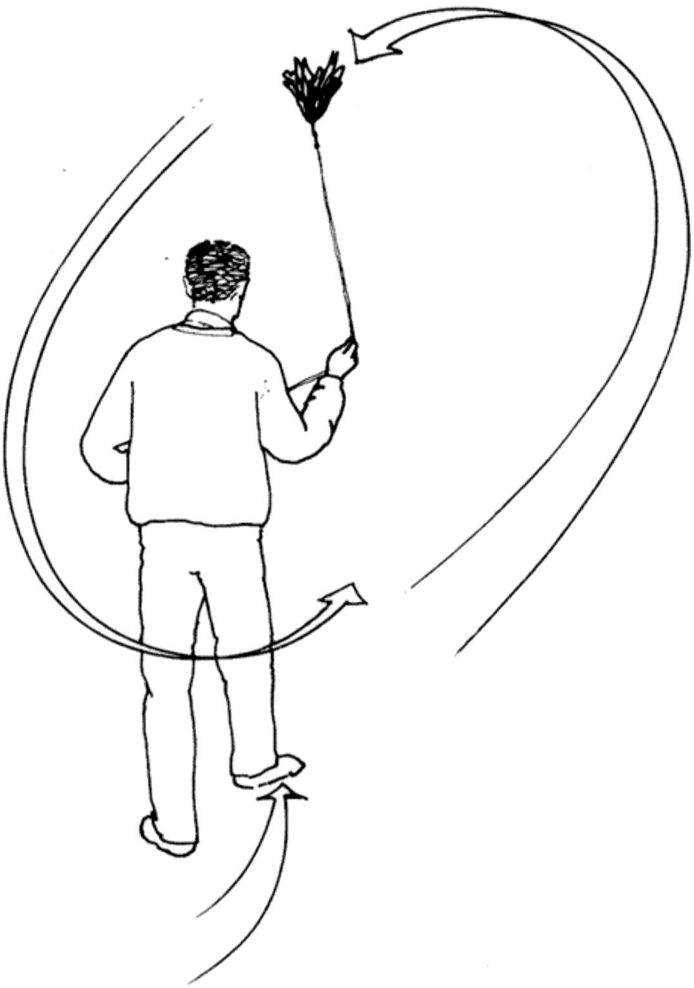

Figure 7.5.8 The lure comes backward over your head, circling on your right side. You are watching the falcon on her outrun.

practising. Swing the lure a few times, then right foot forward, shoot out the lure, then turn left. Keep going until it becomes smooth, controlled, and second nature. When you do it with a falcon you will not have time to be thinking about arms and legs. You must constantly watch the falcon. You have to time it so that at the exact moment she comes in, you shoot out the lure in front of her beak, then pull it back, keeping it no more than four finger widths in front of her as she goes past. If you do it well, her legs will come forward trying to catch the lure. This means she believes she almost has it. If you misjudge it, either the lure will be too far ahead of her, in which case she will go through the motions but will start to slacken. Or she will catch it and you will have to let her down with it.

Figure 7.5.9 The same movement at 7.5.3, but faster so that the lure does not touch the ground. This is what the approaching falcon sees.

Figure 7.5.10 The same as 7.5.9 seen from the side. This is a fast version of 7.5.5.

When you are happy with your left turn, try the right turn. This time have your left foot forward as in figure 7.5.11. Shoot out the lure, then pull it back, sweeping it along low down, turn right and separate your hands all at the same time (figure 7.5.12). Again, you should end up facing the other way with the lure swinging at your right side as before. You may have a spot of bother with this. The trick is to keep the lure below the level of your right hand as you turn, then, as you start to swing it, it will naturally circle in the right direction. When you've got the hang of it, speed up until you can present it to an imaginary falcon coming toward you at Mach 3.

Now get a friend to stand in front of you. His hand is the approaching falcon. Get him to hold it high, low, left, or right. When you shoot out your lure, you should be spot on target. The friend now calls out 'left' or 'right' and you should be able to turn either way without strangling yourself or getting a black eye.

You will notice that when you turn left as in figure 7.5.6, the line, as it pulleys through your right hand, does not do so at a sharp angle, so your right index finger does not get so sore. But when you turn right, as in figure 7.5.12, it almost

Figure 7.5.11 The right turn. The left foot starts off forward and swings round as you turn.

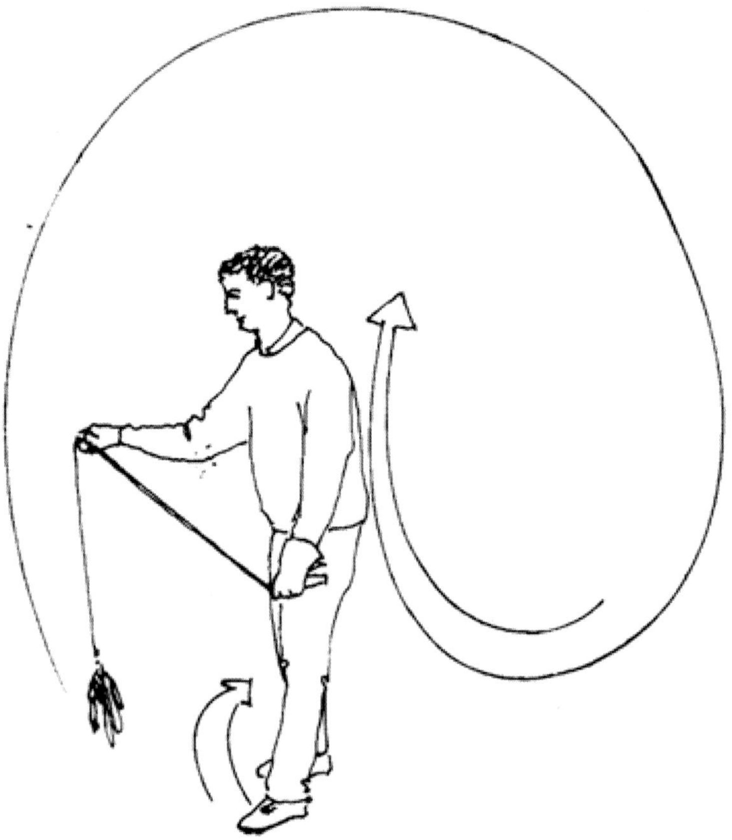

Figure 7.5.12 Turn right round with the falcon and you will end up back in position 7.5.8.

cuts your finger in half. This limits the speed with which you can turn right.

When you exercise a slow falcon, such as a lanner, saker, or kestrel, you have plenty of time and can easily turn right or left. But when you fly a big fast falcon such as a gyr or gyr hybrid, when she is hunting fit, she is traveling too fast for a right turn. Also, you must indicate to her in advance, which way you plan to turn, otherwise, if you dither, she may hit you and knock one or both of you out. It is the difference between a racing car and a family runabout.

Also with slow falcons you can start doing fancy twiddles and high passes. This is not wise with a fast bird; she will carve you up. Keep the tricks for displays.

To make a falcon exercise hard she must be keen to catch the lure and must believe in her own abilities. If the lure is twitched away no matter how hard or well the falcon tries, she will tend to be half-hearted, waiting for the falconer to throw up the lure signalling the end of the exercise. The falconer must make her believe that if she tries harder she can catch it. If the falconer lets the bird 'best' him now and then in a good effort, and finishes in such a way, the falcon will work hard and improve. It is the same variable schedule of rewards as that used for high-jumping or for calling off. Without this incentive, many falcons make large shallow circuits and rest in between stoops. This gives them time to clear lactic acid. The falcon should be encouraged to work well in the vertical plane, stoop steeply and downwind and to circle tightly keeping up the pressure. Failure to achieve this will show in its poor performance in the field, especially in pursuit flights. The best ones stoop like yo-yos. Most of the gyrs come over your head and then stoop vertically on you, a move which is hard to evade and liable to leave you sitting in a cow-pat.

I've described hand-luring in some detail because it is basic training for a falconer. In reality, for most of our luring, we use pole lures which are far more effective, versatile and easy to use. The movements are very similar and again, assuming that you are right-handed, the left turn is easier than the right.

The young falcon will not fly at top speed because she is frightened of crashing. At the bottom of the attack she has to pull out and this puts a tremendous strain on her pectoral muscles, and it is hard for her to judge her windspeed, groundspeed and margin for error. We have had gyrs hit the deck and somersault five metres (15 ft.). Soft ground and thick grass is handy! She must gradually be given more and more confidence and higher levels of difficulty and she will start to fly faster and faster until, as the Americans say, she's really cruising. Gradually you can introduce cross-wind stoops and down-wind stoops. Always let her beat the lure on a good effort. You can increase her accuracy and steepness by flying her in clearings in trees or amongst trees, but again, make sure the trees have soft branches. Never fly her with your brain in neutral and always concentrate on what she needs to achieve next. Low wing-loaded falcons cannot stoop hard down-wind because the strain on them is too great at the pull-out. These are the sort which have to circle to climb. The powerful falcons can naturally stoop hard down wind and always stand on their tails to climb. This is in their breeding. As the falcon gets more confident you will see her flaring her tail open less and less often. Young ones do this to balance out little mistakes, but older falcons only use their tail when it is really needed. A fanned tail creates drag and reduces speed.

A common problem with young falcons is that, after the first approach, she circles and either hangs over your head, or lands nearby. If she does this, immediately hide the lure (properly) and walk away, into the wind. If she is on the ground, try to walk away at least 100 metres or so. Then, when you show the lure, she should come in with sufficient speed to throw up high and stay airborne. As a beginner, you will be concentrating one her approach and pass. That's fine, but you also have to plan what happens after she has passed you. You need her to be travelling at speed and up wind, or at least cross-wind. That will enable her to climb out and gain height ready for another approach. If you leave her travelling low and downwind, she will be in trouble and may land. You will find out why when you try piloting roprey and do these manoeuvres yourself. Later, when you both have more confidence, you can try some down wind work.

Thus the exercise luring has elements of teaching flying skills, developing obedience and reliability, and of physical fitness training. Always keep up the discipline of using the lure, maybe once a week or so. In tandem with the lure training, we are using the roprey and this is so much more versatile that it is tempting to neglect lure work. But there will be times out hawking when the falcon has lost her prey and you need to take her down promptly to the lure before she checks off at something.

So far she has looked inward to you for dead food. For pursuit flights, you must teach her to look outward for live prey. Most of this is done with roprey, but when she is ready for entering, obtain some dead prey, such as a crow which you've been keeping in the freezer for just such an occasion. Attach a 30 centimetres cord to its legs, with a little toggle like a dog retrieving dummy. Break both wings so they flap freely and attach some good food securely to the neck. Go out to a strange place and give her a quick call to the ungarnished fist so that she realizes she is free. As you approach a blind hedge or a bit of cover, the crow will flush out from the other side and fall on short grass. She should flip over the top and nail it. All this happened because you had a hidden friend following your instructions who threw out the crow when he heard your signal whistle. He stays hidden until you reach the falcon on the crow, where you make a fuss of her, preferably for about half an hour. It does no harm to tweak the cord a bit when you first arrive, to make the crow appear alive. You can repeat this a couple of times in completely different places.

Once she has had a dead crow in her foot and is confident enough to start at it without hesitation, she is ready for a dragged crow. We attach the dead crow to a 100 metre creance and hide it in a tussock. The lure man is on a horse, in a vehicle, or even just running. The falconer is 100-200 metres away from the lure at right angles to the direction of drag. When all is ready there is an agreed signal, the falconer unhoods the falcon and holds her up. The 'crow' then flushes from its tussock. The falcon starts after the crow, cutting across the corner so that it always sees a slightly side view of the crow, not just a straight-away view. The horse or vehicle is now moving fast so that the lure bounces 2-3 metres into the air as it goes along. The falcon comes up into a tail-chase and binds to the crow, and the lure man drops the other end of the lure line to prevent the falcon being dragged. He slows down and comes back, picking up the end of the lure line for a while, jerking it as if the crow was still alive. This makes the falcon hang on tight and teaches her to go in strongly, not timidly. If she is flying confidently at roprey and done her basic manoeuvres, she is now ready for entering at wild prey.

Similar methods can be used for young goshawks and Harris hawks. They will not stoop repeatedly at a swung lure, but they can get good exercise on a ropheasant, and for ground game they can chase a dummy bunny lure, either towed or on a motorised line.

7.6 Teaching a falcon to wait on

There are many excellent books available now on waiting-on flights. These authors are dedicated, consistent and pay attention to detail. They also have good access to prime land and plenty of game. Without these things, forget waiting-on flights. Take up dominoes instead. Most of their books describe methods and recipes that work for them and have stood the test of time, but not all of them examine the underlying principles of how to do it.

Those species which search for prey naturally from waiting-on are much easier to teach than those which do so less frequently. Of the large falcons the peregrine, lanner and the Black falcon are the most aerial. The gyr, saker, prairie, and lugger are not usually quite so aerial and may be more difficult, but many take to it very well and are easily equal to a good peregrine. The New Zealand falcon and Aplomados will wait on but tend to be like an indignant wrestler who can't wait until the bell goes before running over and tackling his opponent. Any prey at all from a frog to a hare will cause them to dive. The common and American kestrels hover well and are probably the easiest of all to teach to wait on.

The Red-headed falcon in Pakistan.
Photo Hyder Jaffer.

The Turumti or Red-headed falcon *Falco chicquera* and the merlin prefer to zoom around, power flying. Although they will come overhead for brief periods, they are not very keen on waiting overhead and need to be served promptly.

Individual birds vary greatly in their tendencies to go up. Some go up immediately and can hardly be tempted down. Others can hardly be induced to go above head height. The easy ones are no problem but the difficult ones require careful treatment to get them steady and to keep them steady. What follows is intended for those with a difficult bird.

The first problem is to teach the bird to go up and the second is to teach it to wait there, in position. The third problem is to teach it what to do when it comes down. We teach the answers in reverse order. Compared to some of the other species, such as the prairie falcon and the New Zealander, and especially to the accipiters, the peregrine tends to be a poor footer. It is exasperating to see a peregrine wait on and stoop beautifully only to lose the quarry through poor footing.

Therefore the first thing to attend to is the falcon's fitness, flying ability, and footing ability. Hacking or taking a passage bird largely solves these problems, but let's assume we are dealing with a newly fledged unhacked captive-bred bird. Once the falcon is flying loose, I normally give

Pallas's Sand Grouse *Syrrhaptes paradoxus*. Although wild sakers take them, they are not an easy prey for trained ones.
Photo Andrew Dixon

her about a week of stooping to the exercise lure. Soon she will be making 6 or 7 passes at the lure. If she starts to pant or to set her wings or looks as if she might land, I let her catch the lure. This is done as far as possible by me being a little too slow so that the falcon believes she caught it through her own efforts. After she has had a few bites on the lure she is coaxed on to the fist with a pick-up piece and the lure very carefully and discreetly removed and stowed in the hawking bag. The pick-up piece is then removed into the palm of the gloved hand until only a small bit is left showing.

This makes the falcon believe that she herself has consumed the meal, not that it is being taken from her. When this final bite has gone, the remainder is removed carefully without her seeing it and she is put on the post or cast off, ready for another session. Some falconers do not like to use posts because the falcon may develop a habit of landing on them, in which case an assistant is needed.

Soon she will manage say 60 stoops in two sessions of 30 each at the ungarnished lure. Apart from teaching her to fly and getting her fit, this improves her footing and her picking up from the kill. It also improves her discipline and, importantly, teaches her to look inward to the falconer for quarry rather than outward. It also gives you an idea of what kind of hawk she will be. The best go up and down like yo-yos with a good hard downwind stoop and a minimal interval between stoops. Low flying, up-wind passes and big circuits are faults. I do not regard cutting the lure on the ground without binding to it as a sign of style. Rather it is a fault, particularly of unhacked peregrines, which will result in the loss of downed quarry unless it is cured. Flying the falcon in a small field with trees around it makes her stoop more steeply and forces her into tighter, more controlled, circuits. By letting her catch the lure only on good driving, down-wind stoops a lot can be done to improve her style. Once she knows she can do it she will repeat it, but to start with a lot of youngsters are a little afraid of hurling themselves about too much.

Many falconers believe stooping to the lure will lower her pitch and if you continue to do so once hawking has started this may occur. But a week or two during her early development is a good investment for the future. In many ways it equates with the days when the young falcons chased their parents until food is dropped for them to catch.

A waiting-on falcon will normally be trained for a game species such grouse, pheasants, partridges, sand grouse or various duck species. We use roprey designed to mimic these chosen quarry because you need to develop her search image or fixation on the chosen prey, and perhaps more importantly, not let her hunt other species. Otherwise she will fly at check and constantly disappear all over the district. Some people swear by using live pigeons, which is legal in some countries. But pigeons will lead your falcon away and I have spent many hours in different countries waiting for falconer friends to come back from searching for their falcons. Apart from the risks to your falcon, it is a time-waster and poor falconry. On the other hand, some falconers are terrified that their falcon will put down on a fence post or other low perch. You see their falcon try to land on a branch of a tree and she is so inept that she almost falls off. Don't worry that landing might ruin her pitch. Obviously wild falcons are landing all the time, yet they can fly high!

Concentrate on getting her flying and footing skills as good as you possibly can, and for that we use chases at roprey. I cover this in 9.15. The aerial chases can go up to 300-400 metres (1000-1400 feet) altitude. This will also teach her about using the upper air and the importance of height dominance. She will realise that during these aerial 'dogfights', whenever she is below the prey, she cannot attack, that first she must climb above the prey, and the more dominance she has, the longer, steeper and more effective her stoop. It also breaks through any 'glass ceiling' she may have in her mind, so that she is confident working high in the sky. Game hawkers are obsessed with their falcon's pitch and have used lures towed by kites, balloons and multicopter drones to make the falcon climb. This does indeed teach her to climb and get her fit, but psychologically it teaches her nothing about aerial dominance, nor does it teach her footing skills. On the contrary, it teaches her

Learning aerial dominance.
Photo Stuart Grimwood.

to look upwards for her prey and that all she has to do is climb steadily up and it will sit there while she snatches it off the line, like picking fruit. With the roprey, if she is slow climbing, it will climb higher, so she learns to climb quickly, and when she has reached the same level, the roprey doesn't sit there waiting to be caught; it will take evasive action, which means diving and twisting, and the falcon learns the advantages of height and of controlled, accurate stooping and footing.

Once she has reached this stage you are almost there. Essentially you have turned her into a passage falcon, but one that should not check at other quarry and is focussing inwards towards you to produce prey, rather than scanning distant horizons. If you cast her off without familiar prey visible, she will circle and gain height because she knows that being low down puts her at a disadvantage whatever might happen. Take her to a good place with lift and she will go up. You can flush the roprey automatically from the ground when she has reached some height. It will be near you, but not from you. She will start waiting over you, but will be ready for game to flush nearby. If she stoops well then let her 'kill' first stoop. But if she is poor or out of position, then avoid her attack and start to climb high. Now her only choice is to work hard climbing until you allow her up to your height and then stoop. Work her hard in the vertical dimension. Once she is climbing and stooping well, extend the time before you flush the roprey. Maybe walk a little way. If you have companions, teach her to mark you and not them, by using distinctive clothing or a coloured hat that she knows.

This is teaching her to wait in expectation of her quarry. Don't overdo it. This teaches her to look in, to hold position, and to wait. This is just as important as height. Shout 'Ho' when the prey flushes. Soon she will start her stoop on your shout, before she has seen the game.

There is no point, indeed it is dangerous, to get the falcon up high without her having a clear idea of why she is there and what she has to do next. Waiting-on needs operant conditioning. First we have to get her up in the air before we can reward her to encourage her. Although the best young falcons will go up in a flat calm, it is much easier to get the young falcon up in rising air. Where can we find such air?

In the warmer parts of the world, the morning sun heats the bare ground so that mushroom-shaped bubbles of warm air (thermals) start to rise. These are exploited by the soaring raptors such as vultures and eagles. Thermals occur here in Britain on good warming days but they are seldom frequent enough or strong enough to be relied on. The more usual source of rising air is to look for a place where horizontal wind is diverted upwards. Wherever there are hills or obstacles the windward side will have rising air. I prefer to use hill lift rather than thermals for two reasons: with hill lift you can predict exactly where it will be even before you leave home and you can clearly define the shape of the lifting air and its ceiling. By looking at your falcon's position you know what lift she is experiencing and, if she goes too high, she will go beyond the lift and come down. She will thus automatically tend to take the easy option and hold position. Thermals, on the other hand are not easy to predict before leaving home, or to locate, and often their ceiling is very high indeed. There is a real danger that the young falcon will be sucked up out of sight and control, coming down in a hazardous place some distance away, without learning the specific lesson that you were trying to reinforce.

A little care in selecting the right place makes a world of difference. For young grouse hawks in Britain, often it is late July and the falconer is having to train in the lowlands before moving up to the open moors in August. So things may not be ideal, but it is important that you train in a place with clean lift and not a lot of prey around for her to check off at.

If the falcon is being used for partridges it is a wise investment to breed and put out partridges in fields with suitable holding cover and a good prevailing wind direction. Leave a pen of call birds to stop the others straying. Using a pointer, it is then possible to get reasonably reliable set-ups for your falcon in the early part of the season and during this tricky stage in her development. Simon Tyers describes this in detail.

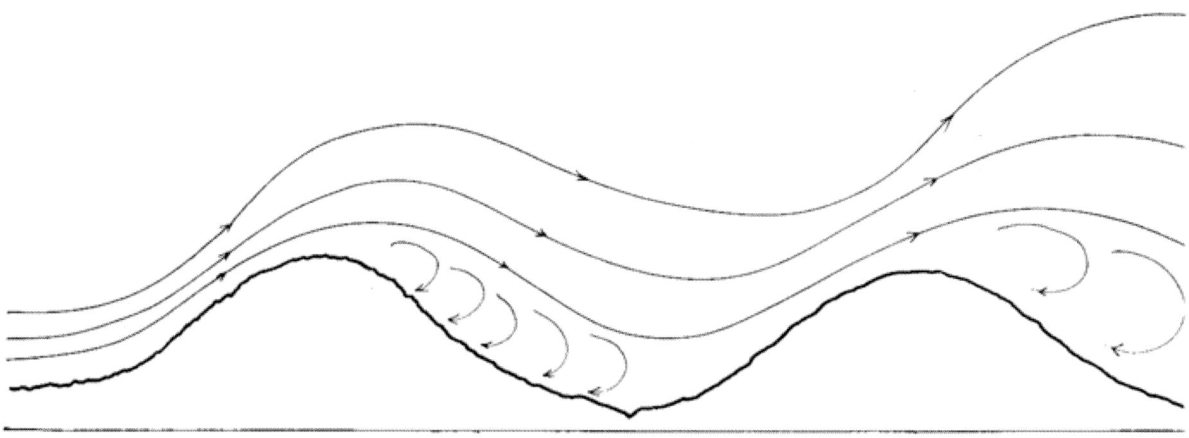

Figure 7.6.1 Lift and sink areas in the hills.

If the transfer is being made direct to grouse then you must find grouse on a slope with some lift (figure 7.5.1). Do not even bother to run the dog in areas of no lift or in a lee; if you find a grouse there you can't fly your young hawk so leave those areas for another day when the wind is in a different quarter, or use a more experienced falcon. To fly a young falcon on a lee slope at this stage is inviting trouble. I checked out the air on my usual training slope in an ultralight. On a normal day the lift was 600-800 feet per minute (180-240 m/min.). Just by opening her wings for one minute, the falcon would reach 500 feet. On the lee slope behind, it was dangerous. The air was descending at the same rate, which is faster than the ultralight could climb. If we flew into that area we would inevitably be driven into the ground, and so would a young falcon.

Where wind hits the hill, it will tend to flow around and channel up the side valleys. But if the hill presents a broad front, or is at the head of a basin, the wind will be forced over the top, giving excellent orographic lift. A hawk is almost bound to go up here, indeed there may well be several wild hawks already up there. It's practically an avian flying club! On the lee slope the air is moving downward and may be turbulent, the hawk is forced to the ground. If quarry is found in this place (as it often is, sheltering from the wind) the hawk must either struggle through the lowering air or reach a windward slope and come back over. With an inexperienced hawk it is better to carry her to a lifting place before unhooding then, once she is well up, run back to the point keeping the hawk higher than the lowering air.

Falcons are as lazy as humans. They do not use the stairs if the elevator is working. Anybody who has studied wild falcons knows that they have their lift spots, usually windy gullies or faces. They simply glide across to the lift and go straight up, either like an air bubble in water or in tight lazy spirals. Then off they go hunting.

If you fly your falcon often in the same place, she will soon learn the lift spots and go and use them even if this means going some distance away. Let her go. When she has got enough height she will come back over. Don't try to impose your own earthbound mentality onto her concept of aerodynamic space. Only when there is no lift will the falcon slog upwards in powered flight. It is tiring and cannot be maintained for long. You are asking a lot of a young bird in those conditions.

Having got the falcon in position over a point on a windward slope the grouse must be flushed uphill and downwind, straight into the dog's face. There is some advantage here if the dog will lie or 'set' right down. Come around the point making a big circuit until you are facing the dog. Early in the season provided the falcon has over-flown the grouse and is approximately in position, the grouse should hold unless cover is sparse. You can make some small noise to let the grouse know your position. As the falcon comes overhead, preferably slightly up wind, rush into the grouse. I use a light stick to help beat the heather patches. As the grouse rise, or just before, if you are confident of an immediate flush, shout 'Ho!' The falcon will start to turn and stoop, lining up on the grouse. If you have served her well, the grouse will be flying uphill, downwind and the falcon will have a downwind stoop. Most probably the grouse, if not hit cleanly, will hit the ground to evade the falcon. A good falcon will turn and take it on impact, or will go back up waiting to be re-served.

What you don't want happening is for the grouse to put in and for the falcon to land on the ground nearby. Otherwise the grouse will spring off again leaving the falcon (and you) with a slogging tail-chase.

If the grouse and falcon are both down on the ground separately, do not approach because you may spring the grouse. Just wait for a while. Then lure the falcon out and once she is in the air hide the lure so that she goes up. If she is tired and won't go up and you think the grouse is an old one and likely to flush at any moment, then it is best to abandon it rather than risk a tail-chase. If you decide to for whatever reason to keep going, then you must keep the pressure on and not waste time. If the falcon is near the grouse, run in to about 50 metres from the grouse. Then lure the falcon but make her miss the lure so that she turns to come in again. As she throws up and begins to come back in, run on with dog at heel and flush the grouse.This is called flying 'at the toll'. Either the falcon will hit it, or she will tail-chase it or she will put it in again. It depends on the position and height of the falcon, the timing and direction of the re-flush and the heart of the grouse. Alternatively, the grouse may not flush at all in which case you know it is thoroughly intimidated. If you try to flush it the dog will catch it. If you pick it up in your hand and throw it, it will tumble to the ground and hide rather than fly.

In such a situation the best solution really is to leave the grouse. Failing that, one can hood up the falcon and wait for 10 or 15 minutes after which time the falcon may go up well and the grouse flush cleanly. With a young bird, without bagged game, you must decide whether to stick to a very strict discipline of only going for a kill from a perfect set-up, or of getting her a kill so that she understands the reason for the whole exercise and appreciates the limits of her killing cone (see 10.9). She can learn from her failures, as well as from her successes, but the one thing you don't want to do is dent her confidence. Probably the best compromise is to achieve a few kills to start with, in the best style you can, and then gradually tighten up the discipline until she is only allowed one stoop at each grouse and if she fails, is taken down with no reward. The ultimate aim is to establish a pattern of stereotyped behaviour. You may be on a moor with plenty of young grouse. On the other hand you may have just worked for two hours just to get a point on a barren pair of old birds. Decisions…!

Often the young falcon will miss the first grouse, tail-chase for a while, then return at a good height. Now is the time to flush a second grouse from the covey under her, if you have a good dog. If all goes disastrously wrong and you are left with the falcon faithfully waiting on but no game to serve her, either run the dog on looking for fresh grouse or, especially if the hawk is young, shout 'Ho' and throw out the thrown lure. At this critical stage in her development, success is important, it is the foundation for confidence, and then style.

Purists will say 'This hill lift is all very well, but I want my falcon to learn to mount in dead air over flat ground.' In operant conditioning you must get the bird to do something first, then reward her. In other words you train her, link by link, from the end result, forward. First you get her flying, footing, centring, and stooping to the thrown lure or roprey organized. Then you build forward from this. You get her to go up and appreciate the advantage of this position. Then she will want to go up. Then you take her to a less lifting place where she must work harder to get up. Until in the end she will mount in a powered climb, knowing it will bring its reward. What you mustn't do is take her to some dead air and expect her to climb for no reason. She may do so, in which case you are lucky. But if she skims around low down, you are establishing a pattern of refusal. You are actually training her not to mount. The skilled falconer is measured not by his best bird, but by what he can achieve with these worst ones. Obviously don't pick a bird with Telwah racing bloodlines, bred to fly a foot above the ground.

Later in the season it may well be hard to find the time, the good weather and the quarry to continue with waiting-on flights at game and the falcon gets unfit and goes to pieces. Some falconers stoop their falcons to the lure for exercise at home in between hawking days. This works for some falcons but lowers the pitch of others and the latter will have to be grounded until next season. Only a few falconers have the time and quarry to continue all season with waiting-on flights. For most, it is an addictive holiday.

Buteos, eagles, and even accipiters can be trained to wait on. Buteos, Harris hawks and most eagles learn it easily because it is part of their natural search methods (see 8.4). The training of these species is not complete until they hunt well from waiting on. Most buteos and eagles are poor in the tail-chase, whereas in the dive from high up they are reallyspectacular. The principles of teaching them to go up are the same as described for falcons, although of course they will not stoop to a swung lure. If you have a Harris hawk, teach her to wait on. It will open a new world of possibilities for you.

7.7 Entering

Game falcons and accipiters do not normally need any entering. But some of the less rapacious buzzard species and pursuit falcons which must chase quite challenging prey such as crows, gulls or houbara may need a gradual introduction to their quarry.

When entering, you must first get the hawk properly fit, with a good digestion and appetite. She will have been introduced to dragged or thrown dead quarry in a dress rehearsal so already has the makings of a 'search image.' Maybe she has been flown at roprey and built up confidence in flying and footing. She knows what to look for. The moment that you judge she is ready for entering, feed her a small ration early in the day so that the next day she has a 28 hour hunger and has lost a little weight (perhaps 10 grams on a 900 gram bird). She should not be low, but this is the lowest weight that you will ever fly her at. You must achieve maximum motivation without losing strength. Take her out and try to give her some good easy slips or flushes. If she has already been flying hard at roprey, and if you have piloted the roprey to match the style of flying of the real live prey, she should have no difficulty with some easy slips. Your job as falconer is to make sure that she is successful, and then build on the success. I always say, when we are taking out a young falcon for her first day's hunting "OK, we're looking for a crow that's blind in one eye with a wooden leg". One year, first day out, we did actually kill a crow with a stump for a leg!

If, despite all your best efforts, you have a problem with this, take her to an area where her prey is common, such as a rookery or a rabbit warren (depending on the hawk) and put her in a tree or on a vantage point. Sit down in a quiet place one or two hundred metres away and let her start self-hunting. Some hawks just need a little time to get used to everything and see their quarry. If she flies away, don't call her back. Follow her, if necessary using the radio. Keep going until dark. The moment she kills, run in and help her. Make sure the quarry is dead, secure the hawk with a cord through one of her anklets tied to your hawking bag. Open up the brain and neck and then sit down about twenty metres away for about an hour, she may want more, possibly the rest of the afternoon. Don't rush her. Every moment that she sits with her own kill a little voice is saying to itself in her brain 'Brilliant! So this is what life is all about! I must do this again!' If she is getting stuck up with feathers or fur, open up the carcass. Don't let her pig-out, let her have just enough so that she is flyable again the next day. Don't use a pick-up piece; cut off a wing or leg and walk her home with that.

If it is getting dark and she still hasn't killed, throw out your dead prey from behind a screen and try to make the best of a bad job. But don't screw her weight down lower and lower.

Once she is entered, do the same the next day, but it should all come quicker and easier. Soon you will be able to start slipping her properly at things, with reasonable confidence that she will go for them.

It is important that you think positive at this time and make sure that one way or another your hawk is successful. We try for hours to find a crow which is just young, or moulting, in an easy place. You must find any off-peak quarry which will just give your hawk an extra chance. The hawk too will sense if you are really trying. Most hawks are willing to make several attempts to catch prey, but if they are unsuccessful, then they lose heart and refuse. They must kill long before this, otherwise they will be half-hearted for a long time before becoming properly wedded. We try to enter the bird either on the first day, or at latest, the second. She will get special treatment for at least the first five kills and won't be pushed to make doubles until she has killed 10-30 singles, depending on her commitment. If possible the first five kills should be made without a miss. After that, the level of difficulty can gradually be increased.

In some countries, bagged quarry is used or misused. The ethics of this are discussed in 11.10. Bagged quarry is a short cut which often makes falconers lazy. When rehabilitating a bird which has no self-confidence, it may be necessary to arrange easy prey for it to get started. Without it, the bird could never manage to become independent in the wild. But the skilled falconer makes every effort to get his bird hunting genuine quarry without the need for substituting bagged game.

Some falconers are very vocal if their prized falcon so much as creases a feather, and yet will cram a pigeon or bagged game into a bag, or worse still deliberately maim it in some way. Surely we have a responsibility to respect all life, and while predation itself is a natural process, there is no need to cause unnecessary suffering or to engage in unnatural practices. Any animal in captivity should be looked after properly and responsibly, and hunted prey should either be killed quickly and cleanly, or escape unharmed.

7.8 Further Reading

Abs, M. (ed) 1983. *Physiology and Behaviour of the Pigeon*. Academic Press, New York.

Anon. 2011. *Guidelines for the Rehabilitation of Birds of Prey*. National Parks and Wildlife Service, New South Wales. ISBN 978 1 74293 165 4.

Aschoff, J. (ed.) 1981. *Handbook of Neurobiology*. Plenum, London.

Beebe, F. L. 1992. *The Compleat Falconer*. Hancock House, Surrey, BC. ISBN 0 88839 253 2.

Beebe, F. L. and H. M. Webster. 1994. *North American Falconry and Hunting Hawks* 7th edition. Denver, CO.

Bert, E. 1619. *An Approved Treatise of Hawkes and Hawking*. Richard Moore, London.

Bolles, R. C. 1975. *Theory of motivation*. Harper and Row, New York.

Boni, B. 1994. *A Tale of Two Passagers: A Treatise on the Passage Red tailed Hawk and Passage Female Copper's Hawk*. Privately printed. Tucson, Arizona.

Bower, G. H. (ed.) 1980. *The Psychology of Learning and Motivation*. Academic Press, New York.

Dewsbury, D. A. 1978. *Comparative Animal Behaviour* . McGraw Hill, New York.

Edmunds, M. 1974. *Defence in Animals*. Longmans.

Ewart, J. P. and D. J. Ingle. (eds.) 1983. *Advances in Vertebrate Neuroethology*. Plenum, New York.

Farner, D. S. and J. R. King. (eds.) 1974. *Avian Biology*. Academic Press, New York.

Ferster, C. B. and B. F. Skinner. 1957. *Schedules of Reinforcement*. Appleton Century Crofts, New York.

Fitzsimons, J. T. 1979. *The Physiology of Thirst and Sodium Appetite*. Monographs of the Physiological Society, no. 35. Cambridge University Press.

Garatinni, S. and E. B. Sigg. (eds.) 1969. *Aggressive Behaviour*. Excerpta Medica, Amsterdam.

Halliday, T. R. and P, J. B. Slater. (eds.) 1983. *Animal Behaviour*. Blackwell Scientific Publications, Oxford.

Hinde, R. A. 1970. *Animal Behaviour*. McGraw Hill, New York.

Hinde, R. A. (ed). 1972. *Non verbal communication*. Cambridge University Press.

Holtz, P and R. Naisbitt. 2000. Fitness levels as a determining factor in the survival of rehabilitated raptors released back into the wild – preliminary results. In Lumeij, J.T. *et al.* (Eds) Raptor Biomedicine III. Zoological Education Network. O-9636996-1-X.

Huber, F. and H. Markl. 1983. *Neuroethology and Animal Physiology*. Springer Verlag, Berlin.

King, A. S. and J. McLelland. (eds.) 1981. *Form and Function in Birds*. Academic Press, London.

Krebs, J. R. and N. B. Davies (eds.). 1984. *Animal Ecology* (2nd Edition). Blackwell Scientific Publications.

Lorenz, K. 1965. *Evolution and modification of behavior*. University of Chicago Press.

Marler, P. R. and W. J. Hamilton. 1966. *Mechanisms of Animal Behaviour*. John Wiley and Sons.

Marler, P. R. and J. G. Vandenburgh. (eds.). 1979. *Handbook of Animal Neurobiology*. Plenum, New York.

Marshall, A. J. (ed.) 1960. *Biology and Comparative Physiology of Birds*. Academic Press, New York.

Martin, S. 1994. The Positive Approach to Parrots as Pets: Videotape 1. Understanding Bird Behaviour. 2. Training through Positive Reinforcement. Natural Encounters Inc. PO Box 68666, Indianapolis, IN 46268.

Martin, S. 1994. 'New Training Thoughts for Falconry Birds' Hawk Chalk XXXIII, no. 2: 44 54

Mavrogordato, J. G. 1960. *A Hawk for the Bush*. H. F. and G. Witherby, London.

McElroy, H. 1977. *Desert Hawking II*. Privately printed.

Oakes, W. C. 1993. *The Falconer's Apprentice: A Guide to Training the Passage Red tailed Hawk*. Eaglewing Publishing, Roy, UT. ISBN 1 885054 01 7.

Parry Jones, J. 1994. *Training Birds of Prey*. David and Charles, Devon. ISBN 07153 0142 X.

Pedley, T. J. (ed.) 1977. *Scale Effects of Animal Locomotion*. Academic Press, London.

Phillips, J. G., P. J. Butler, and P. J. Sharp. 1985. *Physiological Strategies in Avian Biology*. Blackie: Glasgow and London. ISBN 0 216 91780 8.

Pryor, K. 1985. *Don't Shoot the Dog!* Bantam Books. New York. ISBN 0 553 25388 3.

Stein, B. E. and M. A. Meredith. 1993. *The Merging of the Senses*. MIT Press, Cambridge, MA. ISBN 0 262 24036 X.

Stevens, R. 1955. *Observations on Modern Falconry*. Privately printed.

Tyers, S. 2021. *The Specialist Falcon*. TJ Books Ltd. ISBN 978-1-8383843-0-2.

Upton, R. 1990. *Falconry: Principles and Practice*. A. & C. Black, London. ISBN 0 7136 3262 3.

Weiskrantz, L. (ed.) 1985. *Animal Intelligence*. Clarendon, Oxford.

C8 Hunting Strategies of Wild Raptors

A New Zealand falcon makes a direct flying attack on a bellbird (*Anthornis melanura*).

A hunting goshawk.
Photo Steve Magennis

8.1 The components of the hunt

We have now reached the stage at which the hawk is ready to be taken into the field to hunt wild quarry. She is healthy and mentally and physically ready for action. But what can we expect from the bird and will our standard falconry techniques let her realize her full potential? To find this out we need to look in detail at the ways in which raptors go about taking prey in the wild, unaided and unhindered by the falconer.

In any hunt for prey there are two basic parts: the search and the attack. In the attack itself there are often two parts: the first part in which the prey is still unaware that it is being attacked and therefore makes no effort to escape and the second part when the element of surprise has been lost and the predator and prey must contest for supremacy.

The information in this section is based on existing scientific literature and on my own observations of approximately 3,100 hunts by 53 raptor species. Some of these hunts were by wild birds, either casual observations or radiotracking studies, some were by birds at hack, a large number were by trained birds being radio-tracked and allowed to do whatever they wished, and some were by trained falconry birds under more controlled conditions. In the last thirty years most of my hunting has been with trained falcons at Carrion Crows (*Corvus corone*) covering about 14,000 hunts. Of the five methods of study, radio-tracking tame birds gives the most complete understanding of each situation, while other methods tend to be more fragmentary or artificial. Studying untagged raptors hunting in the wild is fraught with difficulties and one is forced to focus on places where hunts can easily be seen. Probably Dick Dekker's studies of peregrines hunting over mudflats or at urban rock doves provide the largest dataset, but even these are skewed towards open terrain. Trying to see complete hunting sequences of untagged wild solitary forest raptors is simply not feasible. If nothing else, the presence of an observer disrupts the element of surprise in ambush attacks. The nearest I have come to it was following tagged wild Mauritius kestrels hunting *Phelsuma* geckos.

Hunting occurs whenever a predator searches for, or attacks, prey. During the searching phase the raptor is attempting either to locate prey which already is, or could become, vulnerable, or it is attempting to place itself in an attack situation in anticipation of prey becoming vulnerable. An attack may be opportunistic and not preceded by any searching. Both search and attack may use more than one strategy.

The main hunting methods used by birds of prey include:

Search strategies

1. Still-hunting.
2. Fast contour -hugging flight.
3. High searching hovering, poising, prospecting, and soaring.
4. Slow quartering.
5. Stalking.
6. Listening.
7. Flushing from cover.
8. False attack.

Attack strategies

1. Direct flying attack.
2. Indirect flying attack.
3. Tail-chase.
4. Glide attack.
5. The drop, the dive, and the stoop.
6. Deception flights.

There are more, such as hunting in pairs or groups, or semi-co-operatively such as when Eleonora's falcons *Falco eleonorae* hunt incoming migrant passerines while still out at sea. But these are the main ones.

The many species of raptor which catch prey show a broad spectrum of hunting methods. At one end of the spectrum are the 'searching' hawks, such as the vultures, kites, harriers, many buzzards and owls, and kestrels. These tend to concentrate on prey species such as small rodents, reptiles, amphibians, and invertebrates which are relatively helpless, widespread, easily captured, and which make little or no attempt to escape when attacked. Most of the prey are found singly but if in flocks, such as of termites, they still make little or no attempt to escape. This means that the searching hawks themselves tend to need smaller hunting ranges relative to their size than 'attacking' hawks and can exist at greater densities. Some are heavily dependent on one or two staple prey species, such as voles, and their own breeding performance follows the numbers of the prey. The extreme searchers, such as the carrion feeders, do not need to make an attack at all, because their prey is already dead.

The attacks of the searchers tend to be simple and short - a mean distance of 15 metres for small hawks and about 25 metres for larger ones. Their most frequent forms of attack are the direct flying attack, the drop, and the dive. If they tail-chase at all it is only for short distances. Because prey are small, attacks tend to be more frequent and have a higher success rate than those of attacking hawks although the biomass captured per unit time is proportionately similar. The wing-loading of searchers is relatively low, enabling a slow, careful flight, emphasizing the importance of the searching phase of the hunt rather than the attacking phase. The size difference between the sexes is small. In general male are about 75-105 percent of the weight of the females.

Between the searchers and the attackers are a variety of intermediate forms, such as the red-tailed hawk. Some of these intermediate species tend to behave more like searching hawks for the first year or so of life but gradually become more active in their attack strategies, particularly during the breeding season.

At the other end of the spectrum are the 'attacking' hawks. These include most of the accipiters, most of the falcons except the kestrels, the Micrastur forest falcons, the hawk-eagles, and some of the eagles. A high proportion of the attacks by juveniles include tail-chases but adults more often use less energy-expensive surprise attacks, such as the indirect flying attack and the glide attack. Attacking species make more complex attacks and the mean attack distance, although proportional to the size of the raptor, tends to be longer than those of searchers. Sometimes in falcons and eagles it exceeds 4 kilometres. The quarry are usually alert, agile, and relatively large vertebrates, sometimes larger than the raptor itself. Prey normally shows active escape behaviour and may live in flocks which aid the detection of predators (see 7.4). The wing-loading of the attackers is relatively high and the sexual dimorphism is pronounced. Males are only 50-70 percent of the weight of the females.

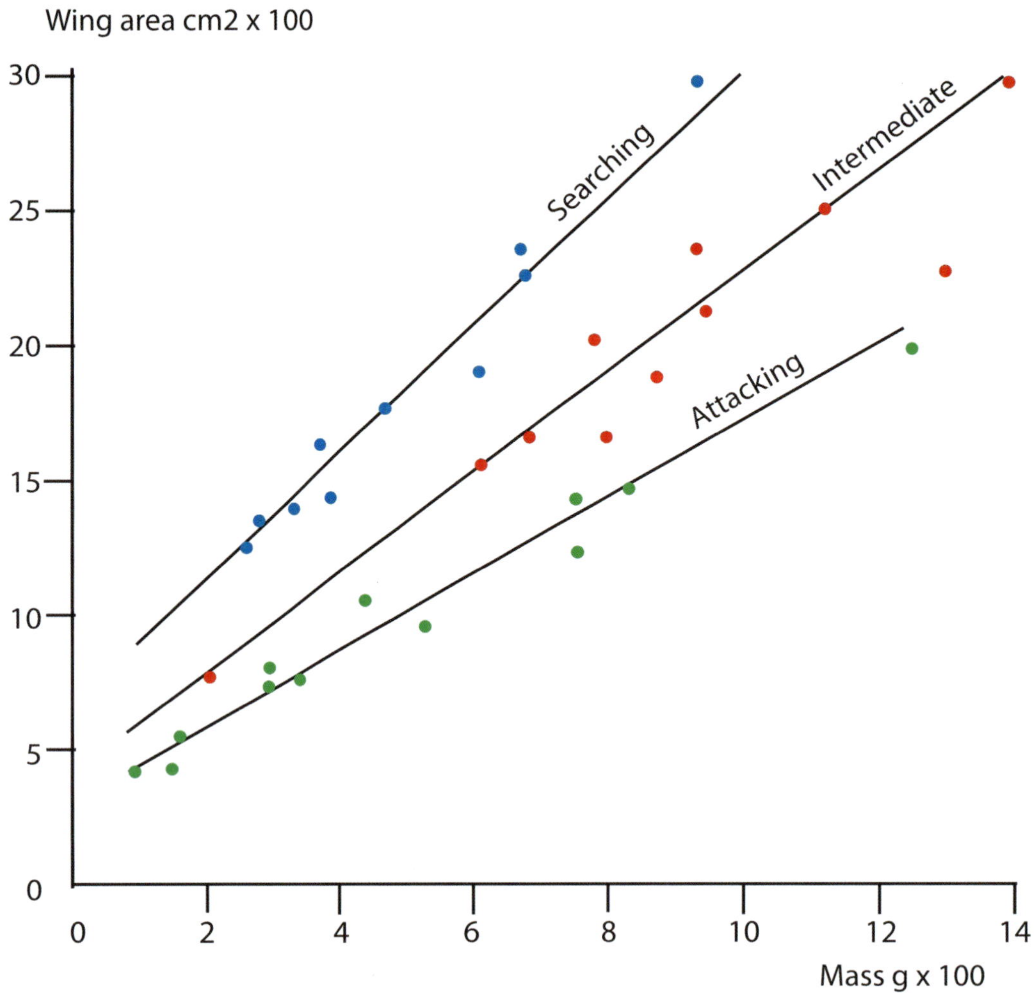

Figure 8.1.1 Wing loading reflects hunting habits.

Searching raptors (blue)

Montagu's Harrier *Circus pygargus*
Pied Harrier *Circus melanoleucus*
Hen Harrier *Circus cyaneus*
African Marsh Harrier *Circus ranivorus* (male)
Pallid Harrier *Circus macrourus*
Hen Harrier *Circus cyaneus* (female)
Honey Buzzard *Pernis apivorus*
Black Kite *Milvus migrans* (male)
Western Marsh Harrier *Circus aeruginosus*
Red Kite *Milvus milvus*

Intermediate raptors (red)

Black-winged Kite *Elanus caerulus*
Australasian Harrier *Circus approximans* (male)
Mountain Buzzard *Buteo oreophilus*
Australasian Harrier *Circus approximans* (female)
Red-shouldered Hawk *Buteo lineatus*
Red-tailed Hawk *Buteo jamaicensis* (male)
Rough-legged Buzzard *Buteo lagopus*
Common Buzzard *Buteo buteo*
Long-legged Buzzard *Buteo rufinus*
Red-tailed Hawk *Buteo jamaicensis* (female)
Jackal Buzzard *Buteo rufofuscus*

Attacking raptors (green)

Sharp-shinned Hawk *Accipiter striatus* (male)
Merlin *Falco columbarius*
Sharp-shinned Hawk *Accipiter striatus* (female)
European Sparrowhawk *Accipiter nisus* (male)
European Sparrowhawk *Accipiter nisus* (female)
Cooper's Hawk *Accipiter cooperii* (male)
New Zealand Falcon *Falcon novaeseelandiae* (male)

Cooper's Hawk *Accipiter cooperii* (female)
New Zealand Falcon *Falco novaeseelandiae* (female)
Northern Goshawk *Accipiter gentilis gentilis* (male)
Peregrine *Falco peregrinus*
American Northern Goshawk *Accipiter g. atricapillus* (male)
American Northern Goshawk *Accipiter g. atricapillus* (female)

The main physical advantage of the accipiters, hawk-eagles and micrasturs, all of which hunt in or near forests, is the sprint. Small accipiters can sprint only 50-100 metres but are very agile and dexterous. Large accipiters can sprint for about three times this distance if they have to. Their attack strategies are designed to place the hawk within sprinting distance before the prey starts to escape.

The main attack strategies of the typical long-winged falcon are the stoop and the tail-chase. The attacks are designed to place the victim inside the killing cone (see section 8.14) for a dive or stoop. Failing this, many species such as the merlin, saker, and the gyrfalcon, can manage an arduous tail-chase.

Buzzards and harriers are not very effective at sprinting, stooping or tail-chasing and therefore, as searching hawks, concentrate on less agile prey which are more easily caught. The emphasis which different raptor species place on each hunting strategy is examined more closely in section 8.17.

Different studies use slightly different terminologies, but all the permutations tend to involve the core strategies listed above, sometimes in more refined or derived forms. Raptors will also steal prey from other raptors (kleptoparasitism) and this can be a major factor for their survival. Young female sakers in their first autumn rely heavily on robbing male sakers for food and this habit exposes them to the baraks, or noosed dummy prey, used by falcon-trappers. Smaller raptors which are very vulnerable to being robbed on their prey, or even being killed, are understandably reluctant to stay long in the open with their prey. Instead they will quickly lift off and find a more secure spot to eat, maybe under a bush.

8.2 Still-hunting

Still-hunting is probably the most universal type of searching used by birds of prey because it requires the least effort. There are two sub-categories: 'perch and search' and 'waiting in ambush.' In the former the hawk perches for a time at a suitable vantage point and looks and listens for prey. Small accipiters and owls usually search a radius of only 10-100 metres but large falcons and eagles can detect and attack prey over 4 kilometres away. Should the hawk fail to spot any vulnerable prey it may fly to another perch to examine a fresh area. Ideal perches for still-hunting have a good view of an area holding vulnerable prey and some concealment for the hawk either in the form of foliage or a disruptive background or simply sufficient height or distance from the prey for the hawk to remain unnoticed.

As the hawk gets hungrier it becomes more active and tends to change perches more frequently-every two minutes or so. It may also launch attacks at poorer opportunities. More aerial species, such as falcons and harriers, usually abandon still-hunting as they get hungrier and use more energetic search methods, still-hunting intermittently while resting. On windless days, kestrels and buzzards which would otherwise be hovering or high searching (see 8.4), often find it easier, in energy terms, to hunt from a perch, even though this may restrict the area they can cover.

Forest species, such as accipiters, hawk-eagles and probably the Micrastur forest falcons, sometimes deliberately wait in ambush in a concealed position near places such as waterholes and food sources where prey congregate and become exposed. When the prey moves into a vulnerable position the hawk bursts from concealment, taking the prey by surprise. Such hawks seem fully aware of the importance of concealment and I have at times watched a New Zealand falcon deliberately keeping behind a tree limb to remain hidden until the best moment, watching intently with just half its head showing round the side of the trunk.

Peregrines too will wait persistently overlooking gorges where racing pigeons and migratory birds habitually pass. Some individuals have gained a reputation for seeming to know the pigeon racing schedules! Because the attacks launched by the large falcons and eagles are often very long - over one kilometre - the field observer tends to underestimate the importance of still-hunting in these species. The falcon that stoops from 200 metres like a bolt from the blue may have actually started the attack from a perch 2 kilometres away. It may also have deliberately waited for the quarry to reach a certain vulnerable point before intercepting.

8.3 The fast contour-hugging flight

In this kind of searching, the hawk flies rapidly 1-2 metres above the ground, forest canopy or water, or along hedgerows and wood edges, to take prey by surprise. Speeds of over 55 kilometres per hour are achieved, usually by flapping flight, although eagles in hilly country can glide round the contours on fixed wings. It is expensive in energy demands but the gain in surprise offsets this.

The substrates hunted over vary considerably, from moorland and tundra hunted by merlins and gyrfalcons, or to hedgerows hunted by sparrowhawks. Several genera hunt in this opportunistic way for birds or insects just above the forest canopy and accipiters hunt in a similar fashion below the canopy in open forest.

Birds are the commonest prey taken in the fast contour-hugging flight; flushing too late, they are snatched in the air. Small mammals are usually overlooked and in any case the hawk is traveling too fast to stop safely. This type of search is used primarily by diurnal raptors with high wing-loadings; it is presumably too hazardous for nocturnal species. I have twice seen young New Zealand falcons whose total flying experience to date I have known, break into a complete

Figure 8.3.1 A European sparrowhawk hunts along a hedgerow.

200 metre contour-hugging sequence without any apparent initial learning attempts. They did it quite spontaneously, so presumably it is instinctive.

It is difficult to interpret attacks launched from this kind of search because often a failed attack consists of only a slight deviation from the raptor's already erratic path. Many attacks are aborted. Also, if the hawk already knew the prey was there, then the flight was not a contour -hugging search but an indirect flying attack (see 8.11). Both can look the same unless you have been tracking the hawk for some time beforehand and have seen the full sequence.

8.4 High searching

When high searching, raptors use a variety of flight styles to maintain height in order to search the air and ground below. These styles include hovering, poising, prospecting and soaring.

Hovering is most commonly used by kestrels, some kites, and occasionally, some buteos; these species usually have long or forked tails to give added lift and control. Many species which have difficulty in hovering, such as large falcons and some buteos, sometimes poise, momentarily hanging in an upcurrent to pinpoint prey.

When prospecting, the raptor flies steadily at a height of 10-100 metres on a basically direct course. Although it has not actually located prey at this stage it may be aware of the presence or approximate location, or probable presence of prey on its flightline, perhaps because it has found prey there before, or in similar areas. If the prey is detected, the patrolling hawk may attack immediately or it may overfly and then return with a low surprise attack from cover. Accipiters usually make short 100-1000 metre prospecting flights whereas large falcons often make long patrols covering a lot of country, sometimes using cloud as cover.

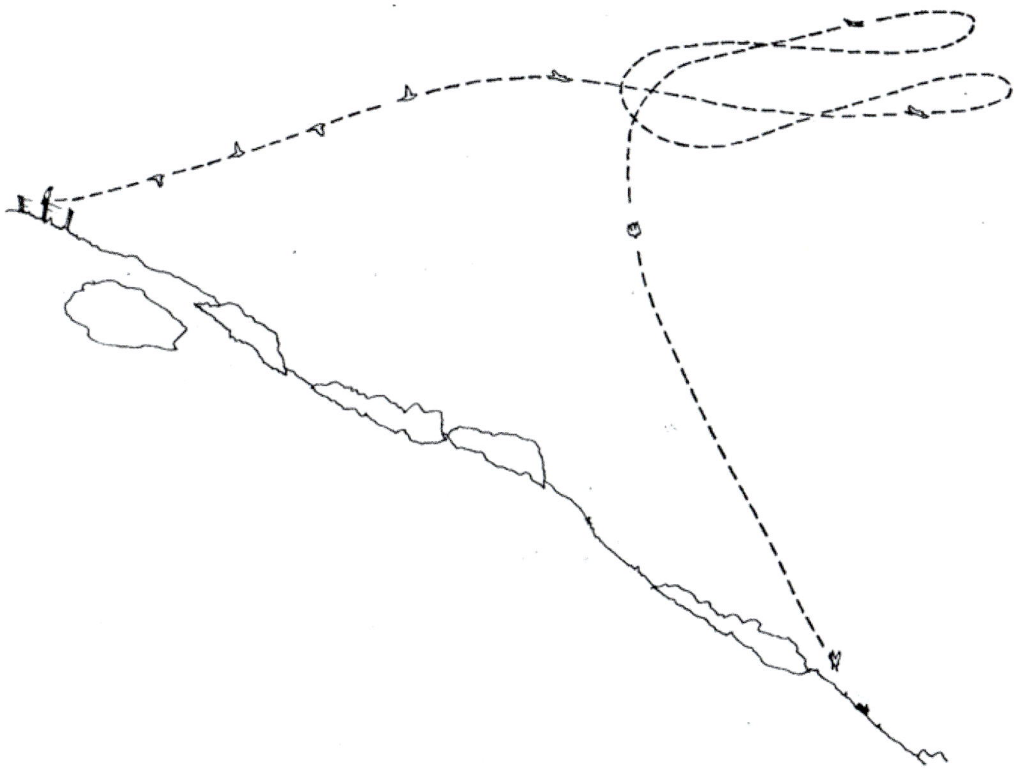

Figure 8.4.1 shows a hunt by a goshawk in a rough hilly area which she has hunted many times before. She knows that there are usually rabbits out feeding among the scattered bushes and bracken but she can't see them from her low perch on a fence post. There is a little breeze coming up a slope so she flies up until she has caught the wind, then slope soars effortlessly for a few turns, while scanning the clearings below. She sees a rabbit which hesitates a moment too long and in a steep fast dive she takes it, just outside its hole.

Soaring searching is commonly used by vultures, eagles, buteos, kites, and some falcons, particularly in habitats where the air is warm and lifting. Indeed, the geographical distribution of most of the vultures is dictated by the presence of lifting air for a major part of the day throughout the year. The soaring bird covers a roughly circular path on open wings but may use flapping flight for short periods when moving from one thermal to another. Falcons and eagles sometimes hunt co-operatively, one bird flying low shepherding up prey which succumbs to an attack from the partner soaring above. Groups of soaring Eleonora's falcons form a curtain ambush above migrating passerines running the gauntlet on sea crossings between the Mediterranean islands. Raptors will also soar above flushing agents such as herds of game or stock, ground predators, other raptors, fires, shooting parties, trains, cars, agricultural machinery, and motor boats. It is this instinct that falconers harness for waiting-on flights.

High searching allows a raptor to examine thoroughly an area which, due to an absence of perches, could not be searched by still-hunting. In the wild it is often less successful than still-hunting because there is less element of surprise. Also the more strenuous forms are very energy expensive.

8.5 Slow quartering

This is most often used by raptors with low wing-loadings such as harriers and short-eared owls *Asio flammeus*. The bird beats along slowly at ground speeds usually below 35 kph (22 mph), 2-10 metres high, with prolonged steady flapping

Figure 8.5.1 A harrier hunts reed beds by slow quartering.

flight, looking and listening for prey. The ground is searched carefully for small mammals, invertebrates, carrion, nests, and the young or incapacitated of more active species. Most prey are caught on or near the ground by a simple direct attack.

Hawks with high wing-loadings are incapable of slow quartering. They cannot fly slowly enough without stalling except in hilly terrain where up-currents make slow, low flights feasible. As shown in 1.14, juveniles of some species have lower wing-loadings than adults, allowing them to quarter to some extent.

8.6 Stalking

Strictly defined, stalking is a search strategy only if the raptor is unaware of the location of the prey. In practice the hawk usually seems to know or suspect the prey is present but seems unsure of its exact location. For example, a kestrel may notice a movement in a tussock and fly down to make a more detailed search on foot. Or the hawk may be seeking out prey which it has just chased into cover and lost.

Many species of accipiter, buteo, and other genera stalk prey on foot, particularly when invertebrate prey is abundant. Wakeley found that a male ferruginous hawk made 18 percent of its hunts entirely on the ground and some individual common buzzards in Britain become earthworm specialists. Juveniles often eke out a subsistence living this way until they learn to catch more active quarry. David Hancock reports nesting bald eagles walking the exposed mudflats and eel grass, poking with bill or feet to attack invertebrates and small stranded fish.

When the prey is small and relatively helpless, this is perhaps better called 'gleaning,' whereas true stalking is a way to approach alert, elusive prey.

Forest-living accipiters and falcons make aerial stalks, flitting quietly from tree to tree preparing to attack a crowing pheasant or a quacking duck. Male gamebirds displaying in the forest often lure in more than they bargain for! In contrast to the indirect flying attack, the aerial stalk consists of slow moves calculated to pinpoint the prey's position without the hawk itself being detected; the hawk can then unobtrusively move into a position ready to launch an attack. The stalk is therefore a search strategy.

Harry McElroy's passage Cooper's Hawk on a Scaled Quail *Callipepla squamata*. *Photo Harry McElroy.*

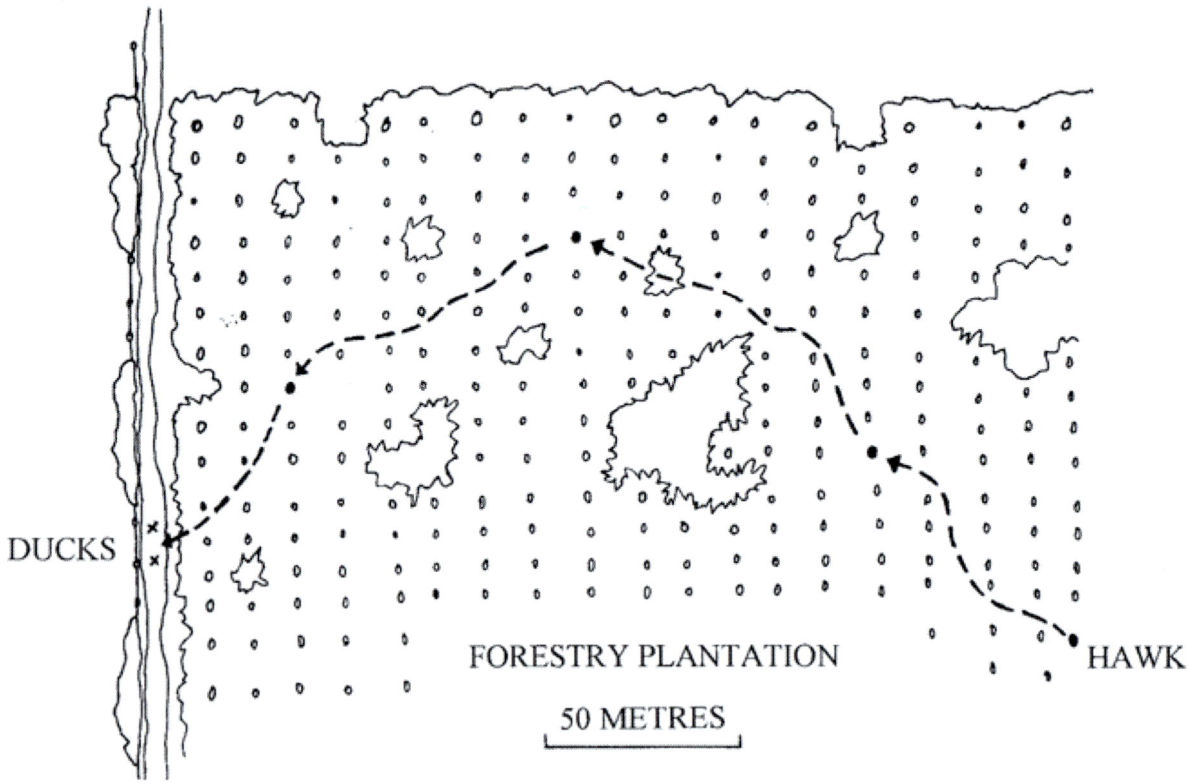

Figure 8.6.1 A goshawk hears some ducks in a distant stream. He flies quietly from tree to tree. When he has pin-pointed the ducks, he attacks.

Figure 8.6.1. shows a stalk made by a young male goshawk called 'Tom'. It was in pre-telemetry days and he wore three bells. We were in a forestry plantation when in the distance we heard some mallard quacking. Tom moved softly and inconspicuously through the denser parts of the wood, avoiding the clearings. I ran quietly below him on the thick pine needles. Somehow he managed to fly and land without making more than a few chinks from his bells. He stopped three times to get his bearings, waiting for the quacking to resume. From his last perch, 50 metres from the ducks, he could tell exactly where they were. He burst out from the trees in a direct flying attack (see 8.10) and bound to a duck before it had got a metre into the air.

I have seen the same pattern of behaviour when hunting with changeable hawk-eagles *(Spizaetus cirrhatus)* and Humayun Taher reports almost identical behaviour in India of changeable hawk-eagles successfully stalking junglefowl.

8.7 Listening

Listening is usually used in conjunction with the more leisurely search methods such as still-hunting, slow quartering, and stalking. Owls of course are specialists at this, but diurnal raptors have excellent hearing too and will locate prey by sound alone. While studying spruce grouse and blue grouse, Hancock has several times had goshawks appear overhead when he was holding young grouse chicks giving distress calls.

One of our New Zealand falcons at hack discovered a sparrow's nest by the cheeping of the chicks inside: it tore the nest to pieces and ate the contents. Several times I have seen New Zealand falcons attempt to attack twigs or wire fences squeaking in the wind. Result: one very puzzled falcon!

8.8 Flushing from cover

This is a search strategy because the aim is to move the prey into a vulnerable situation rather than to attack immediately. Often a hawk will flit

Figure 8.9.1 A peregrine stoops straight through some rooks in a false attack to search out weaklings.

around in the branches of a tree attempting to push quarry out into the open where it can be attacked. Accipiters chasing woodland birds often 'bounce' bushes trying to flush the quarry. They slam into the side of the bush and then immediately fly up, ready to pounce if the prey flushes from the other side. Sometimes they get quite frustrated.

Large falcons, having seen or driven prey into cover, also often swoop violently over the spot attempting to flush the quarry out in panic. Sometimes they will work in pairs; Humayan Taher reports a male red-headed merlin in India flying low and flushing small birds and doves which were then actually attacked by the female.

8.9 False attack

The false attack is not easy to identify in the field. Usually it takes the form of one or more rushes which are not pressed home but which search out the more vulnerable individuals in a flock of birds such as pigeons, gulls, ducks or crows. Once such an individual has been singled out the attack begins in earnest.

False attacks are also used to scatter groups or to shepherd prey away from cover. Once driven into a vulnerable position the real strike is launched.

I saw a flight by a New Zealand falcon in Britain at a scattered group of crows about 600 metres below us. The falcon was in very strong condition and slightly overweight for crows. After watching them in full view from a fence post for several minutes, she launched into a glide attack (see 8.13) off the side of a hill slope. She came in on them at such a terrific speed that she could hardly have stopped if she had tried. She shot over their heads and 'bounced' them all up off the open pasture into the air, thoroughly alarmed and intimidated. Immediately she threw right up and winged over straight down onto the nearest crow which was then in a totally vulnerable position, a metre or two off the ground and struggling to get air-borne. The first pass was a false attack, a shadow punch to be followed by the real body blow.

Similarly, the experienced falcons flown at crows already in the air, often put in a devastating first stoop before throwing up high above the prey. Such a show of power takes a lot of the fight out of the prey, making it easier for the falcon. In this sense one could say that the prey has been mentally 'wounded' even though not physically touched.

The difference between flushing from cover and a false attack is that in the former, the hawk is physically unable to launch an attack until the prey is exposed, whereas in the false attack the prey is exposed but the hawk wants to do some more work on it before actually attacking. It is difficult to distinguish sometimes between false attacks and play. Often a falcon will stoop at prey but not actually hit it, although it obviously could have done so. The reasons for holding back may be because the falcon is young and playful, or not very hungry, or because the prey seemed a little too strong for it, or because the prey was with protective companions. The prey too will sometimes seek to play with the falcons. We used to have a flock of racing pigeons at home which every year, when the hack falcons across the valley had been flying for about a week, flew over

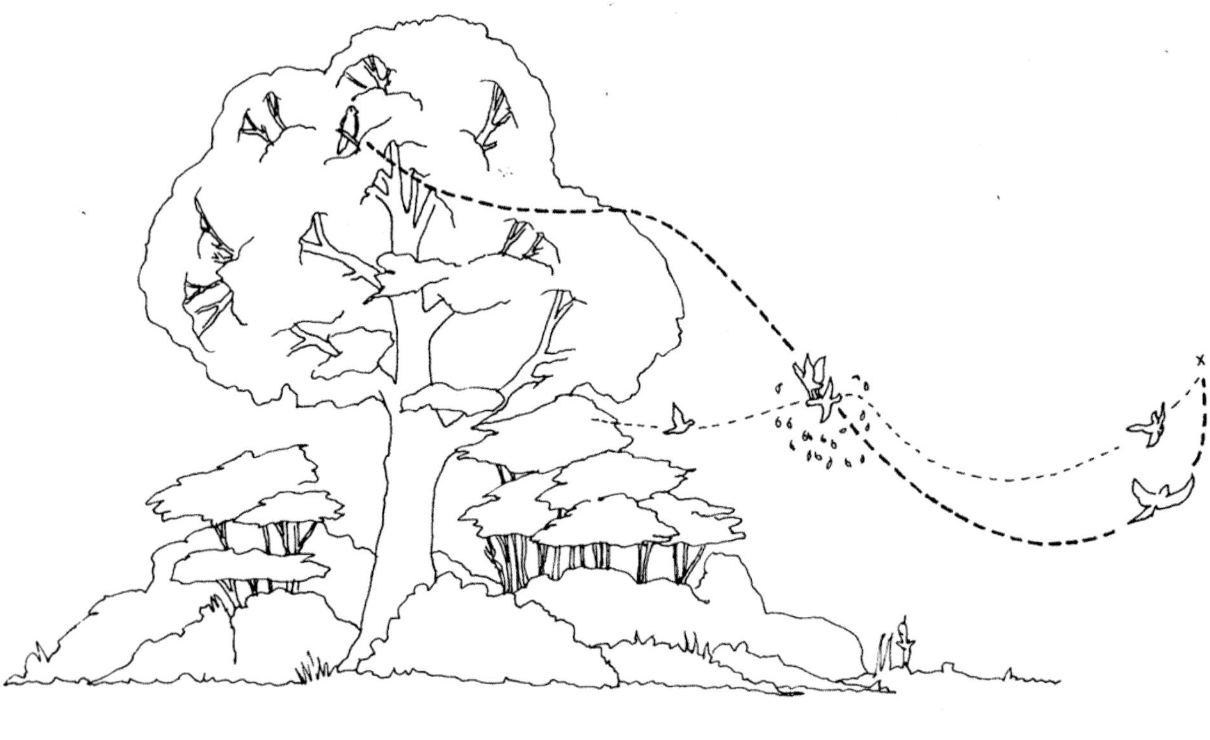

Figure 8.10.1 shows a wood pigeon which has the misfortune to fly past a tree containing a goshawk. The hawk, without hesitating, launches a direct flying attack, hits the pigeon but fails to bind because so many feathers come out. The pigeon struggles to regain momentum but the goshawk piles back into the attack and, after a tail-chase of just a few wingbeats, successfully catches the pigeon.

In figure 8.10.2 A northern goshawk makes a direct flying attacked on a pheasant.

within about 50 metres of the soaring falcons, as if tempting them to chase. Perhaps they knew that the falcons are still young and that they would be safe for a week or two yet.

8.10 Attack strategies: the direct flying attack

This is the simplest form of attack and the first to be attempted by the young hawk. It flies directly toward the prey in an attempt to catch it. Alert prey may promptly start to flee so that the attack then merges into a tail-chase.

Invertebrates and less agile prey are normally taken in a simple direct flying attack and this is commonly used by owls, harriers, buteos, and kites. It is also used when contour-hugging. Of necessity it is a short attack when launched against agile prey, a quick burst of sprinting. It is the commonest attack to be seen by people flying trained accipiters from the fist at flushed game.

The direct flying attack has four usual outcomes: either the prey is caught, or it flees and the hawk tail-chases, or it escapes to cover, or the hawk abandons the attack. It takes many attacks for the young hawk to learn the maximum distance at which it is feasible to launch a direct flying attack at each particular species of prey. While a moorhen out in the open is extremely vulnerable to a direct flying attack by a goshawk, wood pigeons can seldom be taken on such attacks longer than 50 metres. Because there is no surprise element, the prey normally starts to flee immediately, and then it is a question of whether or not the hawk can fly it down. The longest direct flying attack I have seen by a goshawk was by a female at an escaping lesser black-backed gull 400 metres away: she outflew it after a further 200 metres. Falcons of course attack at much longer distances, although the attack generally becomes more complicated rather than a direct flying attack.

Only when the young hawk learns the abilities of the various prey species, the significance of cover and of concealment, and its own limitations in the direct flying attack, does it start to avoid

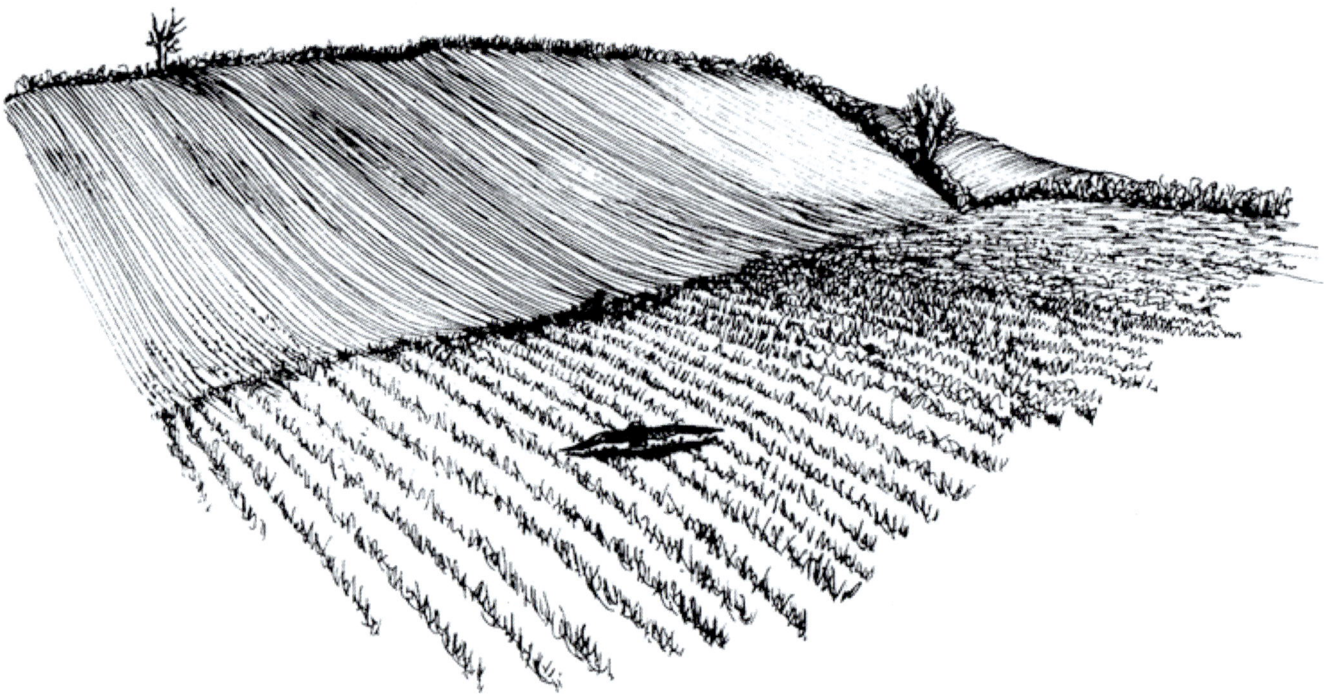

Figure 8.11.1 A goshawk making a ground-skimming attack on a covey of partridges out on some winter wheat. The hawk manages to get within about 10 metres before the partridges detect her and explode into action.

chases where the chances of success are small and attempt more advanced strategies.

8.11 The indirect flying attack

This is a surprise attack and can be very effective. Before starting, the raptor knows exactly where the prey is, not just its approximate location as in aerial stalking. The hawk makes use of cover or irregularities in the ground to conceal the initial part of the attack and thus the route is indirect. The whole flight is made at speed, with vigour.

In the open the hawk may skim the ground, wingtips brushing the surface. To the uninitiated this may look like a direct flying attack but this is not so; using the small undulations of the ground, the hawk often succeeds in getting right into an unwary flock and this makes all the difference. A direct flying attack would have flushed the quarry off right from the start and would have faced the hawk with a hopeless tail-chase. Watching such an attack out in a clean open field one can hardly believe that the prey cannot see the approaching hawk but if you put your head down flat against the ground at pigeon's eye level, differences of elevation of just of few centimetres suddenly become significant. Dr Steve Sherrod has even seen peregrines and bald eagles using the ocean swell to conceal these surprise attacks.

This ground-skimming attack was recognized in old falconry when it was called flying to the 'querre,' probably from the old french 'curee,' a flat skin or hide. The original 'quarry,' from the same derivation, was parts of a deer's offal laid out on the flat skin for the hounds.

The longest ground-skimming attack I have seen by a goshawk was 1100 metres at herring gulls on Dornoch Links in Scotland. She reached within 50 metres before they flushed and she spurted straight up into them. The longest I have seen by a falcon was about 300 metres by a New Zealand falcon at a magpie. Indirect attacks using obstacles as cover are often much longer; 600-700 metre attacks are quite common for New Zealand falcons.

Figure 8.11.2 A goshawk makes an indirect flying attack by using the river for concealment to attack moorhens feeding out in the open.

Maybe the hawk will use a bush to hide its approach, or a sunken lane or a riverbed. This low flight along a river was known as flying 'at the raundon or randan,' possibly after the light boat of that name which used to skim along rapidly. Figure 8.11.2 shows a male goshawk which has spotted some moorhens feeding out on some grass on the opposite bank of the River Eden in Scotland. Any obvious move would have sent the moorhens scuttling for the riverside cover, so instead, the hawk drops quietly from the fist and skims along the surface of the river. When he reaches the point where the moorhens are, he twists quickly up and over the top of the bank, giving them a nasty surprise.

Indirect attacks seldom diverge more than about 45 degrees from the direct route and use a fast flapping flight from the start. Aerial stalking, on the other hand, may diverge 180 degrees and is an unhurried, stealthy approach, seeking to pinpoint the prey first. In the indirect attack the raptor relies on its original estimation of the prey's position and if the prey has moved a few metres, or if the hawk has misjudged, then the attack may be too far off target for the hawk to make good the error in the final sprint in the open. Hawks use the indirect flying attack once they have learned to recognize when certain prey are too far away or too near cover for a direct flying attack to succeed.

Fran, Noel Hyde's haggard 11 year old female Australasian harrier takes a mallard in the water. *Photo Noel Hyde.*
And drags it ashore.

Page and Whitacre (1975) found that 81 percent of 343 attacks by merlins hunting shorebirds were either ground-skimming or glide attacks which gave the advantage of surprise. Only 14 percent were at birds already on the wing and none of the aerial stooping tail-chases they watched were successful. They noticed that even kestrels were successful in catching shorebirds by ground-skimming.

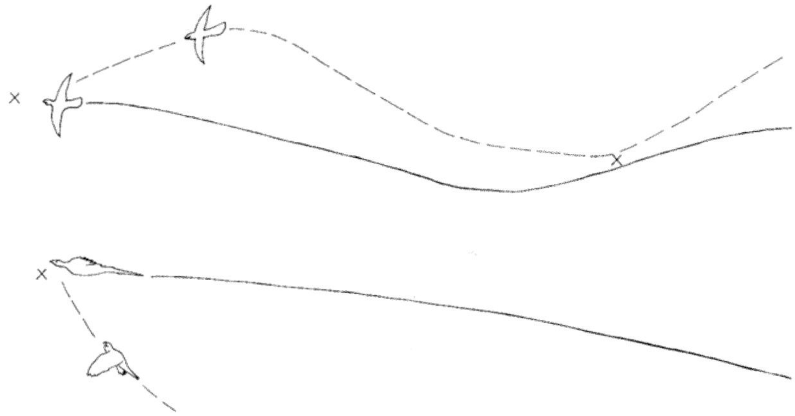

Figure 8.12.1 Top: The falcon tail-chases in a series of shallow attacks. Below: The accipiter follows below and approaches from the blind spot.

Figure 8.12.2 A Cooper's hawk tail-chases a woodcock (*Scolopax rusticola*).

8.12 The tail-chase

Tail chasing is when a raptor pursues a fleeing prey and attempts to catch up with it. Accipiters usually follow on a bird's blind spot, below and behind, then swing upwards in a short burst of acceleration to truss the quarry (figure 8.12.1). But styles of tail-chasing vary tremendously and the most that can be said is that the more experienced hawks tend to throw up high if their attempt to grab the prey fails whereas young hawks usually make a total commitment and if this fails, give up.

If the quarry flies up high into the air goshawks usually abandon the chase: young hawks at about 10 metres and experienced ones at about 30 metres up. However, fit, determined hawks occasionally far exceed this, especially in pursuit of favourite quarry. A male goshawk of mine once made a glide attack from about 300 metres at a flock of 40 mallard on a barley stubble. They flushed as he closed and he went up chasing them over a river valley to a height of about 100 metres and about 800 metres from me. At this point I was unable to distinguish which bird was the hawk until a duck dropped out and dived for the river with the hawk close behind it. The duck found sanctuary in the river and the hawk had caught a moorhen by the time I arrived. Such flights, unfortunately, are the exception.

In the tail-chase, the prey has some control and may attempt to ring up, that is to stay above the raptor and outfly it. Low wing-loaded birds which can climb easily use this strategy: skylarks, rooks, gulls, herons, short-eared owls, even hoopoes, and there are undoubtedly many others. In the struggle for height-supremacy strict tail-chasing may be abandoned and the two contestants may circle up in opposite directions, seemingly unrelated to each other. If the falcon succeeds in getting above the prey this must then dive for cover as quickly as possible and risks being taken either high in the air or on the descent. Such ringing flights are very energy-expensive and have a low success rate; falcons often seem to start them just to try the quarry out. After about 100 metres any weakness in the quarry will show but if it should be strong and buoyant most falcons, sensibly, abandon it.

Old falconers used to say that merlins fly 'cunning' in their second season. All this means is that the merlin has had sufficient experience of gruelling tail-chases to learn that there are easier ways to catch prey, such as the glide attack and the indirect flying attack.

8.13 Glide attack

The raptor usually launches into a glide attack from a perch and attempts to reach the prey unforewarned. Although a few initial flaps may be made to clear foliage, the raptor more often slips off the perch without a single wingbeat and then glides in a fairly flat, less than 30 degree angle, toward the pre-located prey. As airspeed increases, the hawk slowly closes its wings until the last few metres, then suddenly opens its wings and tail to reduce airspeed. It may also rock from side to side either to baffle the prey or perhaps due to the difficulties of rapid braking.

The raptor then slams into the prey on the ground or after a brief tail-chase. Using slow-motion film of goshawks, Goslow showed that, on striking, the feet were thrown forward so that they struck at a speed 15 percent faster than the pelvis.

The glide attack is fundamentally different from stooping or diving and even when the entire attack is not observed it is usually possible to identify it. The glide attack is launched at initially non-fleeing, unaware, agile prey. Whereas a high overhead direct flying attack might suffice for a hare sitting in its form, a rabbit out feeding near a hedge is more likely to bolt for cover on detecting the hawk, and this is where the glide attack meets with success. It is a surprise attack, relying on the inconspicuousness of the head-on profile to prevent detection, rather than using any natural cover for concealment. Some speed is sacrificed in order to maintain this small profile and most of the distance is covered at speeds less than the hawk's flapping flight speed at that flight angle. If the hawk is detected in the later stages of the glide, it may accelerate using a few pumping wingbeats into a direct flying attack. Usually the hawk is not noticed until it is a few metres away and making braking movements. At this speed of approach the hawk is committed to attacking a small area and the prey may react quickly enough to dodge out of the danger zone. One rabbit so attacked

Figure 8.13.1 A common buzzard makes a glide attack on an unwary rabbit.

by a female goshawk of mine did a backward somersault one metre in the air, the hawk passing harmlessly underneath. One of us was surprised, one was frustrated, and one smiled. It's a common trick for rabbits and hares.

Un-belled hawks have a relatively high success rate against agile prey by using the glide attack. It is seldom used by young hawks, possibly for three reasons: they tend not to recognize immobile prey as living but require a trigger, such as a sound or movement, to stimulate an attack. Second, young birds seldom have the self-confidence to attempt attacks at relatively long distances. Thirdly, young hawks have yet to realize the significance of self-concealment.

Glide attacks by falcons can be very long indeed. A peregrine that scores across the sky from one side of your field of vision to the other may be making a glide attack. Over distances of up to 2-4 kilometres, even at a shallow angle of only 5-10 degrees, the falcon can build up tremendous speeds and even though the quarry, such as pigeons or waders, takes flight well ahead, the falcon is hard to evade.

Some experienced falcons learn to combine this technique with other methods. The hunt may be started from a crag or high point at quarry on the ground 2-4 kilometres away. The falcon pumps out high for the first 500 metres or so, climbing hard, then sets her wings into a glide attack. Within 500 metres of the quarry, having approached thus far undetected, the falcon is close to her terminal velocity and, instead of going straight in at the prey and risking it flying out ahead, she may shoot out over the prey so that within seconds the prey is covered by her killing cone and is set up for a stoop.

As well as having a high success rate, the glide attack is very energy-effective. It is used a great deal by experienced birds of prey, more than was realized before the advent of radio-tracking.

Because the glide attack is in the open and yet is a surprise attack, it does not work as well on flocks as on singletons. Outlying members of a flock see more of a side view of the approaching hawk and are more likely to detect it and give the alarm. On flocks, the hawk is better off with a concealed approach such as an indirect flying attack or, in the case of a falcon, by coming overhead and getting the flock within her killing cone. Page and Whitacre (1975) found that when merlins performed glide attacks and ground-skimming indirect flying attacks on shorebirds 25.6 percent of 343 hunts were successful when the prey was alone. But on flocks of 2-10 birds only 6.9 percent were successful. As flock size increased the shorebirds got more confused and on flocks of 11-49 birds, 8.3 percent of attacks succeeded. On flocks of fifty or more, 21.4 percent of attacks succeeded. Overall, a singleton was 3.2 times more likely to be caught than one in a flock.

Large raptors weighing over 300 grams tend to use glide attacks more than small hawks do. This is because they have heavier wing-loadings and launch attacks at longer distances. Smaller hawks do not build up sufficient speed over their short attack distances. During a radio-tracking study of wild Mauritius kestrels, which have an average attack distance of 13 metres, only two out of 63 attacks were glide attacks and their mean distance was 25 metres, easily the longest attacks attempted. In a comparable study of wild New Zealand falcons 15 percent of all attacks were glide attacks. In the three species, Mauritius kestrel, New Zealand falcon and northern goshawk of which I have studied the hunting behaviour quantitatively, the glide attack always had the highest success rate.

Probably the main reason why so few falconers do well with buzzards, including often the red-tail and sometimes even the Harris, is that they misunderstand the nature of their aggression and the ways they prefer to hunt. They treat them as second-class goshawks not realizing that the buzzards are completely different physically and mentally. Whereas goshawks have hair-trigger reactions, an explosive take-off, good acceleration and a rapacious disposition, buzzards are slow, calculating and incapable of rapid, powered flight (see 1.17). Any form of taxing tail-chase is impossible for them. Their hunting is based on searching and their preferred attacks are the dive and the glide attack. Pitted against agile prey, their only hope for success is from a height advantage, preferably with the benefit of surprise. Although experienced common buzzards will take rabbits well, they need to do so on their own terms and will often glide over seemingly vulnerable rabbits to smash into the one selected as the best bet - usually young, sick or unsuspecting.

Once one accepts the buzzard for what it is and helps it to do its own thing, instead of trying to force it into the goshawk mould, then success will come. The successful hawk will be less frustrated and less likely to attack people.

The above applies to a lesser extent to the red-tail, the Australasian harrier *Circus approximans* and to the young Harris. Older and more experienced birds should be capable of more 'classy' flying, but when they get really experienced some start to show their limitations and refuse flights which they would have taken on in their hot-blooded youth. Don't we all?

8.14 The drop, the dive, and the stoop

Figure 8.14.1 A redtail drops straight onto a snake.

These three attacks are all launched from above at an angle steeper than 30 degrees to the horizontal.

The drop, or parachute, is used for short distances; in a big downward hop, the hawk simply opens its wings and drops from a post or low branch onto easy prey. During the whole descent, normally no more than 2-3 metres, the head and wings are held back and the legs are thrust forward ready to catch the prey (figure 8.14.1).

The dive is longer than the drop; usually between five and thirty metres. The dive of the kestrel or of the osprey is typical. During the descent the head leads, as in the stoop, but before impact the head goes back and the feet come forward to grasp the prey. Although the dive looks like a stoop there are three main differences: first, the bird is not traveling as fast as it can. On impact it must stop, or almost stop, and so it descends only at a rate at which it can seize the prey without injuring itself. Often it will hesitate during the dive and realign itself. Secondly, on impact, the diving bird's head is behind its feet, out of the way of damage. Thirdly, on hitting the prey, the hawk attempts to bind to it and to maintain its hold (figure 8.14.2).

The stoop, on the other hand, is usually executed with all the speed that the hawk can muster in the circumstances; nothing is held back. During impact, the falcon's head is leading and the feet are held open on each side of the keel. The strike is made with the claws and the undersides of the open feet, the semi-folded legs cushioning the shock, shoot out at the last moment to extend the falcon's reach if the stoop is slightly wide of the mark. I had this well-drummed into me during fieldwork on wild New Zealand falcons who struck me some hundreds of times on the head and back during nest visits. Analysis of slow motion film by a number of workers has confirmed it and when one considers the falcon's anatomy there is no other part of the body capable of striking such a blow. The upper surface of the toes shows bruising after quite a slight blow and could not be used offensively.

Figure 8.14.2 A common kestrel dives onto a vole.

A male peregrine which I flew one season, a rather reckless hard-flying bird, bruised both his hind toes on the ground in a mis-timed stoop. He then started to 'miss' his grouse, shooting straight through the coveys without hitting a bird. It only slowly dawned on me that he was 'pulling his punches' and after about three weeks, as his feet improved, his stoops started to connect again.

The stooping falcon is not attempting to bind to the prey but to kill or disable it with a glancing blow and then to continue onher flight, turning to recover the prey afterward. Occasionally a large falcon hitting small prey will simply gather it up and keep on going. I have a vivid recollection of a male prairie falcon flown from a pitch at pipits. Having put a pipit back in after a short sharp flight, we re-flushed it for him. He stooped past my shoulder, smacked it hard right in front of me, and just left behind him an empty space. The pipit seemed to have disappeared into a time warp! The falcon landed some distance further on to consume his meal.

When the quarry is nearly as large as the falcon, the stooping bird cannot bind to the quarry; she could not survive the impact. However, if the stoop is a shallow one and the prey is flying away at a good pace, and particularly when the falcon has to avoid dropping the prey into water or forest below, she will often do her best to bind.

Figure 8.14.3 A peregrine stoops onto a red grouse.

Each falcon has a slightly different style of stoop, some have shallow scything stoops, some have steep yo-yo stoops; the most dazzling perhaps is the flickering corkscrew stoop or canceleer. One cannot hope to cover all the permutations, instead I will discuss what happens in simple, straight-forward stoops.

Of course speeds vary between individuals, the larger sexes and species traveling faster. With some stooping falcons one can almost hold a conversation as the bird is falling, other falcons seem to slip down through the air in the blink of an eye, as if all air resistance had momentarily disappeared. Ken Franklin's peregrine reached 389 kph (242 mph) in a stoop, starting from 5,100 metres (17,000 ft). But from the 120 metres (400 ft) in the Hunt Race, an average stoop is about 200 kph (124 mph).

8.15 Deception flights

Some people believe that one can tell when a raptor is making a serious attack. This is not always true. Sometimes an experienced hawk, seeing prey out in the open with no cover around and knowing that a direct or indirect flying attack, or even a glide attack, is not feasible, may make a deception flight in full view of the prey. Usually the prey is already aware of the hawk's presence so there is no point in the hawk concealing herself. All she can conceal is her intentions. Even to the experienced observer, unless he knows the hawk, this is undetectable until the end. I have had two hawks which made several of these attacks. Essentially it is a question of either getting the victim inside the killing cone or within sprinting range.

The first hawk was a wild-caught adult female changeable hawk-eagle which I flew at brown hares in the big open fields of Fife in Scotland in 1970. A typical example: I carried *'Oddjob'* to the edge of a field and let her scan it for squatting hares, too well concealed for me to see. After a few moments she took off and started to climb in a rather stately way until she had reached about 70 metres (200 ft). I thought she was heading back to India. Suddenly she flicked over and dived hard. As she did so, a hare bolted from its form. Before it had gone five metres, the hawk-eagle hit it in a vertical dive which left the hare unconscious and the bird shaken on the ground nearby. During the season she repeated this performance several times and it was quite clear that both she and the hares had been aware of each other throughout. Although she would sometimes catch hares by outflying them she preferred to use this technique and make life easier for herself.

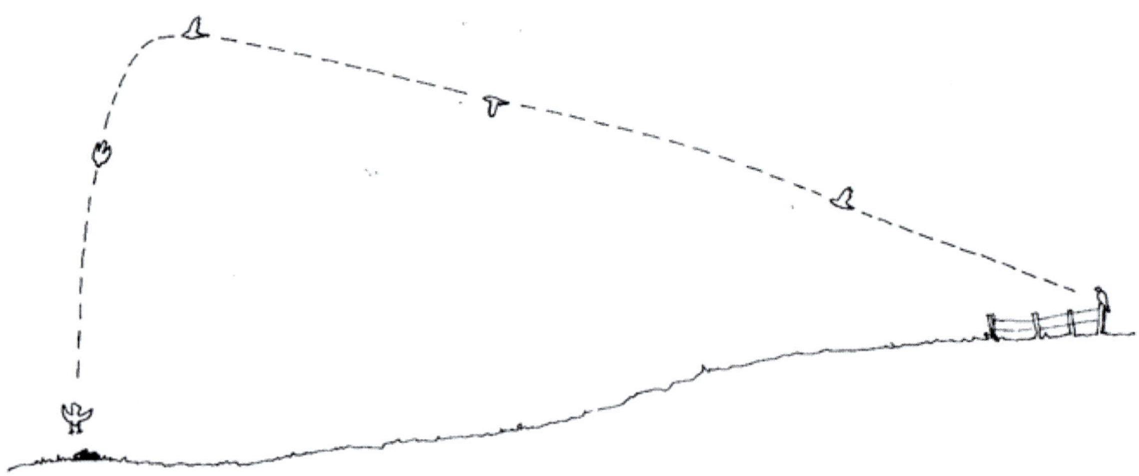

Figure 8.15.1 A changeable hawk-eagle spots a squatting hare, launches out high and lazily, then dives into an attack.

Another hawk was a New Zealand falcon called *'Rebel'*. In her first season, after about thirty kills, she became adept at killing crows. On several occasions I showed her crows out in the open in full view with no height disadvantage 500-800 metres away. There was no way she could get near them using a 'conventional' attack, they would have made off ahead of her and reached cover or at least given her a long and strenuous flight. So instead she would launch out 45-90 degrees from the crows steadily gaining height until she was 70-120 metres up. Then, almost casually, she would drift over the crows which always stayed on the ground unless she had tried it on them before. Once she had them in her killing cone it was a simple matter for her to make a hooked or corkscrew dive, binding to one on or near the ground. As an example or energy-effective predation in the face of apparently uncatchable quarry, it was hard to beat.

Such flights remind me of how, as boys, we used to catch hares and rabbits by walking past them in their seats and casually spiral in on them until within striking distance. The victim is never approached head-on and never has a break-point where it makes a decision to bolt. It seems to think you are not attacking - until it is too late.

8.16 Hunting in groups

Some raptors join forces to hunt. This may just be a loose aggregation of mixed raptor species cashing in on a temporary abundance of prey, such as near a fire. This 'group foraging' occurs among many of the 'searchers.' Some, such as the zone-tailed hawk *(Buteo albonotatus)* are wolves in sheep's clothing and mimic turkey vultures *(Cathartes aura)* to disguise their approach to prey. Or it may involve several individuals competing for the same prey, such as when Eleonora's falcons all go after the same passing passerine over the sea. Or it may involve actual cooperation between individuals. In this case they are usually related in some way-they may be a mated pair or family group and in Harrises they may also be cousins or 'helpers.' In this case the prey is normally shared between participants.

Cooperative hunting helps the participants to catch prey which they could not so easily catch on their own. Dean Hector founds that Aplomado falcons killed in 45 percent of their joint hunts but only 21 percent when solo. When it is a family group hunting together, the gain may be not so much in food, but in knowledge. It is part of the training process for the young birds.

8.17 The ways in which experienced raptors hunt

Young birds of prey have many or all of their hunting patterns genetically implanted in them. They can instinctively undertake a complete hunting sequence although they have never been in such a situation before. But this is only a part of the story. Just as the young bird can fly instinctively and yet needs experience and practice before she becomes a skilled flyer, so also does she need experience of actual hunting for her abilities to develop fully. Thus, in the first year of life, 60 percent or more of the young birds may die, largely through factors related to poor hunting success. Yet experienced adults in the same period may suffer a mortality of only 5-20 percent.

Looking at these figures, some falconers have suggested keeping wild-taken raptors in captivity for their first year and then hacking them out to the wild, thus circumventing the high juvenile mortality. This of course fails to take into account the underlying reason for the mortality which is not lack of age, but lack of experience.

An experienced adult can survive in an area where prey availability is perhaps only 25 percent of that needed to support a young bird. We know this from flying both experienced and inexperienced birds in the same area. The inexperienced bird tends to search a smaller area, be less able to detect and assess attack opportunities, and has poorer flying skills to be successful in the attack. When he has a mate and chicks to feed, an adult male may have to provide for three or four times his own needs and may spend 95 percent of his daylight hours during the breeding season actively hunting. The juvenile raptor faces a hazardous apprenticeship before it can reach this level of achievement. A wild peregrine, by twelve months old, may have experienced 1000-2000 hunts; a falconry bird would be doing well to achieve this in five years.

As hawks get older they diversify their strategies and often switch from one method to another. They learn to judge accurately the reactions and escape distances of their prey and learn to recognize quickly when a quarry is vulnerable to an attack. Different species emphasize some hunting patterns more than others. It will be a long time before we can quantify these strategies for all species but figure 8.17.1 is an approximate outline for a few selected, mainly falconry species. The morphological adaptations for different flying abilities are discussed in section 1.18.

	Still-hunting	Contour-hugging	Hover/poise	Prospecting	Soaring	Slow quartering	Stalking	Listening	Flushing from cover	Direct flying attack	Indirect flying attack	Tail-chase	Glide attack	Stoop and pass	Dive and bind	Attack static ground quarry	Pursue into cover	Continue pursuit inside cover	High aerial chase	Hunt cooperatively
Peregrine	C	R	N	O	C	N	N	N	R	C	O	C	R	C	C	R	O	N	C	C
Gyrfalcon	C	C	O	C	C	R	O	?	O	C	C	C	O	C	C	C	C	R	C	O
Saker	C	C	O	C	C	R	R	?	O	C	C	C	C	C	C	C	C	N	C	O
Prairie falcon	C	O	O	C	C	R	R	?	O	C	C	C	C	C	C	C	C	N	C	O
Lanner	C	C	O	C	C	R	R	?	O	C	C	C	C	C	C	C	C	N	C	C
Lugger	C	O	O	C	C	R	R	?	O	C	C	C	C	C	C	C	C	N	C	C
New Zealand falcon	C	C	O	C	C	N	O	C	C	C	C	C	C	C	C	C	C	C	C	O
Red-headed merlin	C	C	O	O	O	N	R	?	C	C	C	C	R	C	C	C	C	R	C	C
Merlin	C	C	N	R	R	N	R	?	C	C	C	C	R	O	C	C	C	R	C	O
European hobby	C	R	R	R	C	N	N	N	N	C	R	C	R	O	C	R	N	N	C	O
Common kestrel	C	R	C	R	C	N	O	O	R	C	R	O	C	R	C	C	C	N	R	R
Sharpshin	C	C	R	C	O	N	C	C	C	C	C	C	O	R	C	C	C	C	N	N
European sparrowhawk	C	C	R	C	O	R	C	C	C	C	C	C	O	R	C	C	C	C	N	N
Cooper’s hawk	C	C	R	C	O	R	C	C	C	C	C	C	C	R	C	C	C	C	N	N
Black sparrowhawk	C	C	R	C	O	R	C	C	C	C	C	O	C	R	C	C	C	C	N	N
Northern goshawk	C	O	R	C	O	R	C	C	C	C	C	C	C	R	C	C	C	C	N	N
Changeable hawk-eagle	C	O	N	C	O	R	C	C	C	C	C	C	C	R	C	C	C	C	N	N
Common buzzard	C	R	C	O	C	R	C	R	R	C	R	O	C	R	C	C	C	O	N	R
Red-tailed hawk	C	R	R	O	C	R	O	R	R	C	R	O	C	R	C	C	C	C	N	R
Ferruginous hawk	C	R	R	C	C	O	O	R	R	C	R	O	C	R	C	C	C	O	N	R
Harris hawk	C	R	R	O	C	R	O	R	C	C	R	C	C	R	C	C	C	C	N	C
Golden eagle	C	C	R	O	C	R	R	R	R	C	O	C	C	R	C	C	C	N	N	R

8.17.1 The hunting strategies of various raptors. C=common, O=occasional, R=rare, N=never.

A young saker checks out the town pigeons in Ulan Baatar.
Photo Stig Olsen

8.18 Further Reading

Anderson, M. and R.A. Norberg. 1981. Evolution of reversed sexual dimorphism and role partitioning among predatory birds, with a size scaling of flight performance. Biological Journal of the Linnaean Society 15: 105-130.

Anderson, M. and J.R. Squires. 1997. *The Prairie Falcon*. University of Texas Press.

Abu Dhabi Sports Council 2011. *Falcon Eye Abu Dhabi 2011*, www.youtube.com.

Alerstam, T. 1987. Radar observations of the stoop of the peregrine falcon *Falco peregrinus* and the goshawk *Accipiter gentilis*. Ibis 129, 267-273.

Bentson, S.A. 1971. Hunting methods and choice of prey of Gyrfalcons (*Falco rusticolus*) at Myvatn in Iceland. Ibis 113:468-476.

Bird, D.M. and K.L. Bildstein. 2007. *Raptor Research and Management Techniques*. Hancock House. ISBN 978 0 88839 639 6.

Buchanan, J.B. 1996. A comparison of behaviour and success rates of Merlins and Peregrine Falcons when hunting Dunlins during winter. Raptor Research 20:130-131.

Cade, T. J. 1982. *Falcons of the World*. Cornell University Press. ISBN 0 8014 1454 7.

Cramp, S. and K. E. L. Simmons 1980. *The Birds of the Western Palearctic*. vol. II. Oxford University Press. ISBN 0 19 857505 X.

Cliff, D. and Miller, G. F. 1996. Co-evolution of pursuit and evasion II: simulation methods and results. In From Animals to Animats 4: Proceedings of the Fourth International Conference on Simulation of Adaptive Behaviour (ed. P. Maes, M. Mataric, J.-A. Meyer and J. Pollack). Cambridge, MA: MIT Press.

Cochran, W. W. and Applegate, R. D. 1986. Speed of flapping flight of merlins and peregrine falcons. Condor 88, 397-398.

Craighead, J. 1940. *Hawks, Owls and Wildlife*. Peter Smith. 084460562X.

Cureo, E. 1976. *The Ethology of Predation*. Springer Verlag, New York.

Dekker, D. 1980. Hunting success rates, foraging habits, and prey selection of peregrine falcons migrating through central Alberta. Can. Field Nat. 94, 371-382.

Dekker, D. and Lange, J. 2001. Hunting methods and success rates of gyrfalcons, *Falco rusticolus*, and prairie falcons, *Falco mexicanus*, preying on feral pigeons (rock doves), *Columba livia*, in Edmonton, Alberta. Can. Field Nat. 115, 395-401.

Dekker, D. 2009. *Hunting tactics of Peregrines and other falcons*. Hancock House. ISBN 9780-0-88839-683-9.

Destin 2012. Slow motion raptor strikes. In Smarter Every Day 38, www.youtube.com.

Domenici, P., Blagburn, J. M. and Bacon, J. P. 2011. Animal escapology II: escape trajectory case studies. J. Exp. Biol. 214, 2474-2494.

Fox, N. C. 1981. Hunting behaviour of trained Northern Goshawks (*Accipiter gentilis*). P121-133 In R. E. Kenward and I. M. Lindsay (Eds) Understanding the Goshawk. Proceedings of the IAF 29 September-1 October 1981, Oxford, UK

Fox, N.C., B.I.M. Fox and T. Bailey. 1993. The hunting behaviour and predatory efficiency of the Mauritius kestrel (*Falco punctatus*). pp. 136-143 *In* M.K. Nicholls and R. Clarke [eds.]. Biology and conservation of small falcons: Proceedings of the 1991 Hawk and Owl Trust Conference. The Hawk and Owl Trust, London, England.

Gillies, J. A., Thomas, A. L. R. and Taylor, G. K. 2011. Soaring and manoeuvring flight of a steppe eagle *Aquila nipalensis*. J. Avian Biol. 42, 377-386.

Gonzales, G. 2010. Falcon catches dove on board flight video. www.youtube.com.

Goslow, G. E. 1971. The attack and strike of some North American raptors. Auk 88, 815-827.

Haak, B.A. 1995. *Pirate of the Plains*. Hancock House. ISBN0-88839-320-2.

Hamerstrom, F. 1986. *Harrier: Hawk of the Marshes. The hawk that is ruled by a mouse*. Smithsonian Institution Press. ISBN 0 87474 538 1.

Hedenstrom, A. and Rosen, M. 2001. Predator versus prey: on aerial hunting and escape strategies in birds. Behav. Ecol. 12, 150-156.

Howland, H. C. 1974. Optimal strategies for predator avoidance: the relative importance of speed and manoeuvrability. J. Theor. Biol. 47, 333-350.

Jenkins, A.R. 2000. Hunting mode and success of African Peregrine Falcons: Does nesting habitat quality affect foraging efficiency? Ibis142:225-246.

Kenward, R. E. 2006. *The Goshawk*. T & A.D. Poyser. ISBN 978-0-7136-6565-9.

Lee, S. H., Pak, H. K. and Chon, T. S. 2006. Dynamics of prey-flock escaping behaviour in response to predator's attack. J. Theor. Biol. 240, 250-259.

Leonardi, G., A. Longo and G. Corpina. 1992. *Ecology and Behaviour of the Lanner Falcon*. Privately printed.

Marcstrom, V., R.E.Kenward and M. Karlbom. 1990. *Duvhoken och dess plats in nature*. AB Trycksaker, Norrkoping. ISBN 91-630-0008-3.

Mecholsky, N. A., Ott, E. and Antonsen, T. M., Jr 2010. Obstacle and predator avoidance in a model for flocking. Physica D 239, 988-996.

Monneret, R.J. 1973. Techniques du chasse du Faucon Pelerin (*Falco peregrinus*) dans une region de Moyenne montagne. Alauda 41:403-412.

Moran, D.T. 1996. *Life on the Wing: Adventures with Birds of Prey*. Round Table Press. ISBN 0-9655519-0-3.

Nahin, P. J. 2012. *Chases and Escapes*. Princeton, NJ: Princeton University Press.

Newton, I. 1986. *The Sparrowhawk*. T. & A. D. Poyser Ltd. ISBN 0 85661 041 0.

Orton, D. A. 1989. *The Hawkwatcher: Adventures among Birds of Prey in the Wild*. Unwin Hyman, London. ISBN 0 04 440140 X.

Page, G. and D. F. Whitacre. 1975. Raptor predation on wintering shorebirds. Condor 77:73 83.

Palleroni, A., C.T. Miller, M. Hauser, and P. Marler. 2005. Prey plumage adaptation against falcon attack. Nature 434:973-4.

Potapov, E and R. Sale. 2005. *The Gyrfalcon*. T & A.D. Poyser. ISBN 0-300-10778-1.

Quinn, J.L. and W. Cresswell. 2004. Predator hunting and prey vulnerability. Journal of Animal Ecology.73:143-154.

Ratcliffe, D. 1980. *The Peregrine Falcon*. T. & A. D. Poyser Ltd. ISBN 0 85661 026 7.

Schipper, W. J. A. 1973. A comparison of prey selection in sympatric Harriers (*Circus*) in Western Europe. Le Gerfaut 63: 17 120.

Village, A. 1990. *The Kestrel*. T. & A. D. Poyser Ltd. ISBN 0 85661 054 2.

Treleaven, R. B. 1998. *In Pursuit of the Peregrine*. Tiercel Publishing, Wheathampstead.

Wakeley, J. S. 1978. Activity budgets, energy expenditures, and energy intakes of nesting Ferruginous Hawks. Auk 95:667-676.

Walter, H. 1979. *Eleonora's Falcon: Adaptations to Prey and Habitat in a Social Raptor*. University of Chicago Press, Chicago.

C9 Rofalconry and Racing

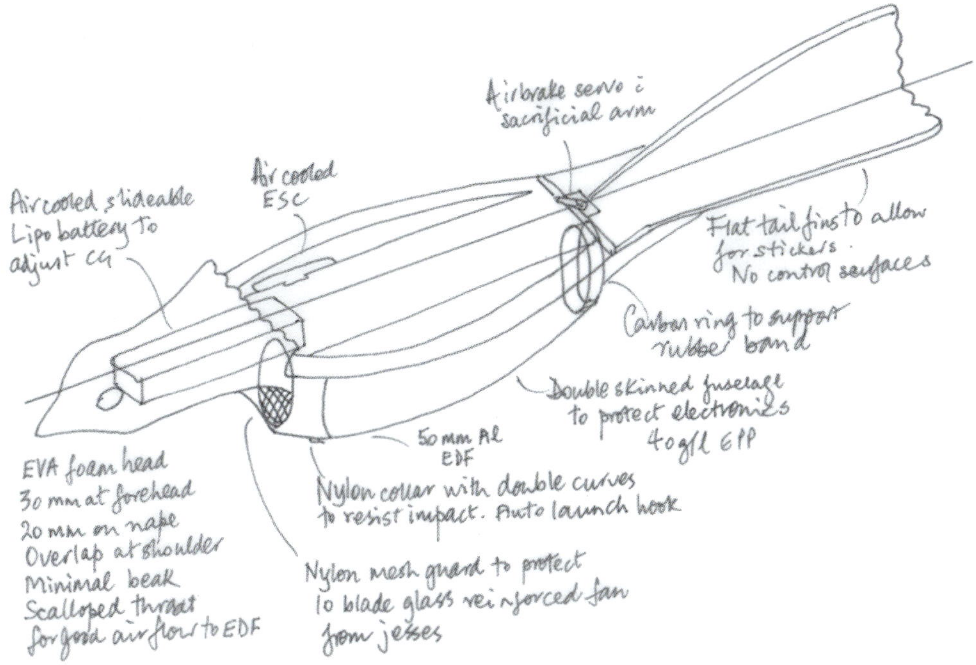

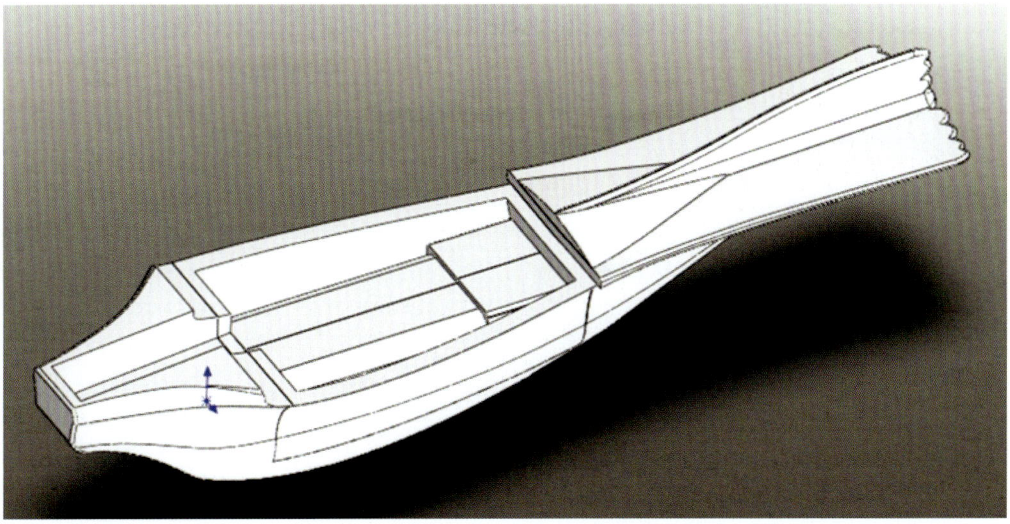

This chapter is supported by several short films and video manuals based on www.rofalconry.com for training with roprey, and www.vowley.com for racing. As the sport is fast developing, expect to see more films being posted and also watch some of the social media.

Becca, a young female gyr/New Zealand falcon, defies all rules of physics and anatomy.
Photo Stuart Grimwood.

9.1 Introduction

While Falconry is the art of hunting wild prey in its natural state using a trained raptor, in our pressurised modern world, natural habitats and wild prey are not always available. So the choice is either to give up and go and play computer games, or to seek substitute prey. Robotic prey fill this need, and open the door to further scope, such as racing. Although rofalconry could be likened to hunting a drag with hounds, or shooting clays as a substitute for the real thing, in reality it is a lot more than that. The roprey is not simply dragged along the ground, or catapulted up into the air. It is flown by a pilot, and that is a whole skill and sport in itself. The RC flying community is actually a lot bigger than the falconry community, with many clubs and events of their own. When you have tried piloting a roprey and experienced a falcon coming in to kill you, you will realise that this can be just as much of a thrill for the pilot as it is for the falconer. On top of that, the two sports are interactive; the falcon 'thinks' she is chasing real prey, and any spectators soon get captivated with the aerial chase as if it was real. And of course rofalconry supports falconry by improving the performance of the hunting birds, and by enabling them to hunt a substitute on days when time or quarry are in short supply.

9.2 History of roprey

Making dummy prey to use in training is an old idea. Sir John Chardin in his Travels in Persia in 1686 noted: '…and the Way to teach them, is, by tying a bit of Flesh to the Head of one of those flea'd Beasts and stuff'd with Straw, which they lay on four Wheels, and keep moving as the Bird is eating, to use him to it.' Others before me have dreamed of a radio-controlled prey bird. In 1979, Ronald Stevens saw a lad flying a radio-controlled plane and conjectured "Would it not be worth someone's while to design a model aircraft with feathers on it to simulate a bird? Flying to the lure would have a vastly different meaning then, and what spectacular high flying it would encourage. Would it not be a short cut to persuading a hawk to wait on? But of course there would be a risk of hawks becoming wedded to this quarry." Later he went on to try a flapping bird model sent to him by Henry Tyler Swain, flying his kestrels at it.

Later, in 1964, Frank Beebe commented on the advantages of a controllable flying lure, but he had no way to construct one.

I started on this in 1994, consulting people like Colin Pennycuick and Mark Fuller, but the technology simply wasn't available then. My spec was:

Wing span: 94cm

Wing area: 1500 cm^2

Weight: 550g

Level flight speed: 15 m/s

Endurance: 5 minutes.

Propulsion: Ducted fan (not exposed propeller).

Construction: Able to survive being repeatedly caught and 'killed' by falcons. Surface covered with material for talons to grip. Colour black. Holder for food reward on 'head'.

Radio control: Manoeuvrable enough to take evasive action. Motor stoppable.

I later somehow got hooked on the idea of flapping wings. Certainly they make the plane very attractive to hawks. But the penalty is that they are very fragile, hard to control and incapable of the aerial speeds and manoeuvres needed. We could only really test this out by flying the models with falcons and seeing how they reacted. It turned out that while flapping wings are very attractive for getting a young raptor started, after an initial catch or two, flapping wings make no difference. Instead they become a liability.

There are three main flapping wing designs:

- Wings that power flap and make the model fly.
- Wings that flap using servos but do not contribute to flight power.
- Wings that flap passively once the model has airspeed.

The first method is great for studying the theory of bird flight but the mechanisms are fragile and often have exposed hard parts that could injure the falcon. The flight performance of ornithopters cannot provide what is needed in falcon training and, because the fuselage moves up and down, it does not provide a stable platform for a FPV camera. So after initial trials we abandoned that option.

We tried the second option on our first model, the V1 Robara. But again the mechanism was prone to failure and the wings were liable to break during extreme flight manoeuvres. So we use it only in early training of Houbara falcons and then switched to the Sports wing which is fixed and therefore stronger.

Our second model, the V2 roprey, has a tough one-piece fixed wing with an embedded w-shape carbon spar. In early training we found you don't need a flapping wing at all. To make the model more attractive, any fluttering will do, and the simplest way to achieve this is to glue some shredded black polythene bin liner to the tail or trailing edge of the wings. This is virtually weightless and does not interfere with flight performance, but is very attractive to young raptors. After a few catches this is no longer necessary and we remove it. Most falcons never need it.

So our second model, the V2, designed by Remy Van Wijk and me, is a much simpler device. It has been specifically designed as a prey for hunting falcons. It's the result of many years of research and development carried out by a team of experienced falconers and technicians, as well as many thousands of flights by several different species of falcons in different terrains and climates. We have made and flown many prototypes, continually changing and improving the model. It is still evolving and in use around the world.

The advantages of using roprey:

- Unlike live prey, the behaviour of roprey can be carefully controlled to allow the falcon to learn and succeed at the desired pace.
- Roprey can be tailored to match a huge number

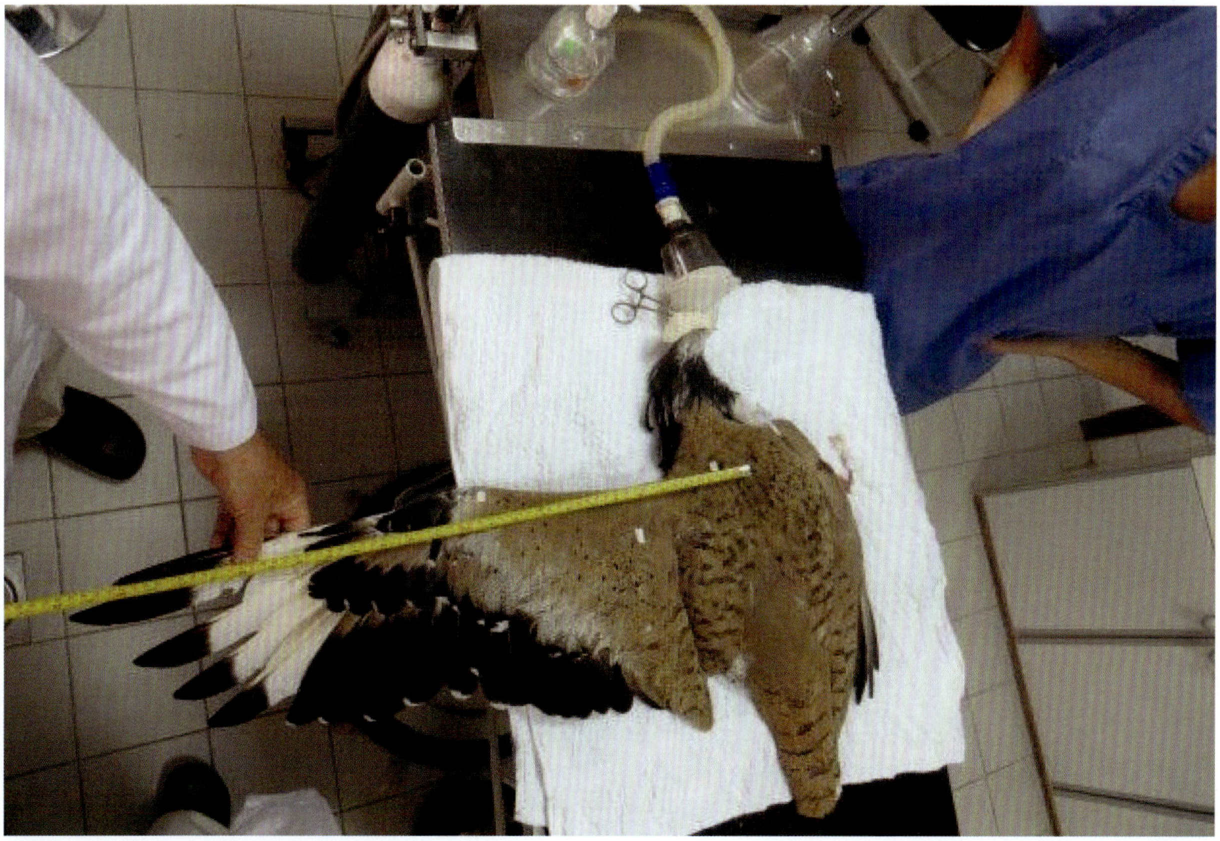

Measuring a Houbara wing.

Prototype V1Robara, 2014.
Photo Linda Wright.

of real prey species and so are excellent for developing a falcon's search image, a great preparation for hawking live prey.

- Falcons become hugely motivated to chase roprey, and those falcons with a tendency to abandon the lure in favour of something more exciting to chase rarely pull off from a roprey slip.

- With instruments such as altimeters, airspeed meters and timers, it is possible to put a falcon through a carefully prepared programme of physical and mental development. Such a falcon can gain a year's hunting experience in a month.

- For falconers who do not have access to good hunting grounds, rofalconry allows flights to be undertaken in one chosen spot without the need for many square kilometres of hunting area. Also friends can bring their falcons along too and everyone can enjoy flying their falcons together at the same venue. In countries such as Japan, and in western Europe, where the land is heavily developed, this is a huge advantage.

- Not only can falconers enjoy rofalconry individually, it is possible to hold competitions together, creating a whole new sport in its own right.

- For flying displays, rofalconry is the best way to show what falcons love to do best – chasing prey. Instead of tame displays calling falcons to the lure or fist, it is possible to watch them hunting high in the sky as they do in nature. One quickly forgets that the prey is not real.

- Rofalconry allows falcons to hunt prey every day without any danger of reducing the numbers of wild prey. It is a totally sustainable sport and has had great conservation benefits for species such as the Houbara bustard in the Middle East.

- Rofalconry does not use wild prey, so there are no welfare issues and roprey can be hunted at any time of year. Even where it is illegal to hunt wild prey, the falconer can still fly falcons in rofalconry.

- Roprey are now being used in rehabilitation of raptors. Both orphaned raptors and injured raptors need proper fitness training and practice hunting skills before they can be returned to the wild.
- Rofalcon predator models can be used in pest control at airfields, landfill sites, vineyards and areas of valuable crops, to chase away pest birds.
- Roprey can be flown from moving cars out in the desert in high-speed hunts across the dunes. Total adrenaline rush!

9.3 The construction of the roprey

After years of experiments with lures, kites, drones, prototypes and so on, it was clear that I needed to design a prey that:

- Looks the same as the real prey.
- Behaves the same.
- Has similar flight performance characteristics.
- Is controllable at all times by average people.
- Can fly high and freely in the sky.
- Can handle strong winds, rain, heat and fine sand.
- Is safe to catch.
- Does not require a release and parachute.
- Can withstand repeated attacks.
- Is modular so that parts can be easily replaced.
- Is easily repairable.
- Does not require a special runway but can be launched from the hand.
- Can be auto-launched from the ground away from the falconer.
- Has a 'come home' facility.
- Is safe to use near people.
- Is legal to use.
- Can be mass produced at an economical price for the tiny falconry market.
- With standardised models that would allow for competitive events.

This was a tall order and we have had many frustrating dead ends and expensive mistakes. Only in the last few years have advances in materials technology and electronics become sophisticated enough for us to achieve what we wanted. Now we have tough and light EPP (Expanded Polypropylene) foam which is rigid enough to fly with, tough enough to withstand impacts and light enough for the falcon to hit. And we have miniaturised electronic speed controllers, receivers and stabilisers, electric ducted fan motors and LiPo batteries suitable for our flight specifications. We can even use FPV (First Person View) if we wish. But bitter experience has shown that the more technical and complicated you get, the more hitches you will have, so yes, while technically we can provide various fancy add-ons, usually the good old KISS method is best. (Keep It Simple, Stupid!). Being hit hard by a raptor and then crashed to the ground is not the most friendly treatment for a sophisticated technical gadget. Very few items can withstand this treatment. A female gyr can deliver a devastating blow and things must bend and absorb shock to dissipate the impact. The safety of the raptor is paramount for us, so if anything breaks, we want it to be the model, not the bird. But never forget that this is an extreme sport. If our falcons were humans, we would be wrapping them up in all sorts of safety gear. So the model has to be able to take a beating, to be hit from above, below, on the extremities, on the head, to be clawed, bitten and smashed head first into the ground. Today, tomorrow, and all season. The only way to do this is to have mass-produced standardised parts that can be easily replaced on a modular system. Some parts, such as the head, are deliberately sacrificial because this is where the falcon catches and bites. So the EVA head foam is easily and cheaply replaced as needed.

Our Wingbeat factory in Wales, with the warehouse behind.

Other parts, such as the EPP body and wings, will take collision or claw damage. These can all be glued back together usually in a few moments in the field and, once damaged beyond repair, easily replaced.

The first model we made, the V1 Robara, designed by Robert Musters, was based on an adult male Houbara and although it worked, and we produced it commercially, it had several design faults which we could only find out by trying it in the hands of different pilots and falconers. After two years of flying it, we had learned a lot and we decided to invest in a V2 model based on a whole new design configuration. For this I recruited Remy Van Wijk to do all the technical aspects and the CAD drawings, while I contributed the conceptual elements and the falconry inputs. The V2 airframe is now the standard design on the market and is adaptable to many different uses.

Most of the component parts are moulded in China and we designed and built a factory and warehouse on the farm in Wales to do all the construction work. Going from one-off hand-made

The warehouse.

Production lines.

The shipping room.

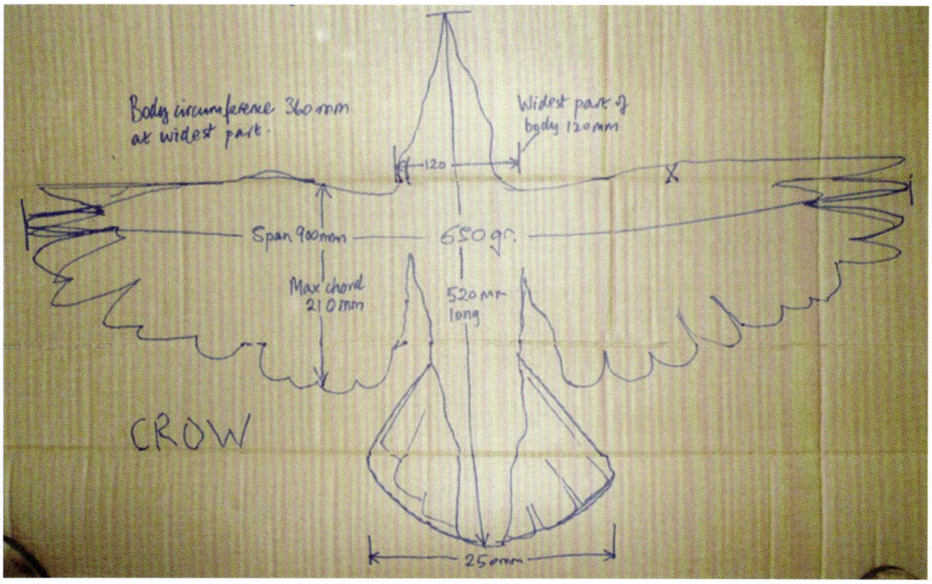

models to standardised commercial production is a major step financially and logistically and the international falconry market is tiny.

I wanted a model that was medium-sized, a 'general purpose' bird that could imitate a wide range of prey species or even raptors. We traced dimensions from an adult crow caught by one of our falcons in Northumberland, as a starting point. It could become anything from a small houbara to a pheasant, seagull or duck.

The roprey is not just an adapted model plane. It is specifically designed to be caught by falcons. So its propeller is hidden away inside the body as an EDF (Electric Ducted Fan). The EDF is 50mm with 10-12 glass-fibre reinforced fan blades to make it as strong and as silent as possible. The fan can be changed if broken or the whole EDF is removable. It is shielded by a fine nylon mesh which is also replaceable, to prevent any trailing jesses getting into the fan.

The front end has a nylon collar extending as a cup all around the body. This spreads the forces of a frontal impact and protects the battery inside. On the outside is a moulded head made of EVA, a tough light foam (used in making croc shoes). The whole head end is thus well-padded, reducing the risk of injury to the falcon or to humans and it is mandatory that the head is properly padded for any public event. The beak is of a basic crow shape but can also be bent to form a raptor beak, flattened to form a duck beak, or shortened to make a grouse beak. The upper surface of the neck is thick foam that extends back over the shoulders to absorb the impact of a hard-stooping falcon. Around the neck is a red elastic loop with a knot. You pull the knot to slip a piece of food under the band. If the falcon finishes the food before you arrive, she will often pull at the red knot rather than the EVA.

The head must be of a certain robustness to pad the battery and to survive multiple strikes. On the chest of the bird the collar has a little streamlined hook. This is used for auto-launching from the ground. From my physical model prototype, Remy then makes the CAD drawings and we ensure that the object is capable of being released from a mould, with sufficient release angles and no overhangs, and with sufficient material thickness. The density and thickness of the foam and the Shor number must be just right so that the head is not too flimsy but is still within our weight limit.

The fuselage of the bird contains all the gizmos. It is double shelled, so that the outer shell absorbs the impact, rather than the forces being directly transferred to the sensitive electronics. The receiver is tuned to the pilot's transmitter by pressing a tiny button and receives the signal via a short aerial. It controls the motor speed via the ESC or Electronic Speed Controller, which is air-cooled from the engine. The autopilot is mounted centrally and links to a GPS which enables the model to either loiter over your head, or to come home if drifting away. But the main function of the autopilot is the stabiliser which can be set to

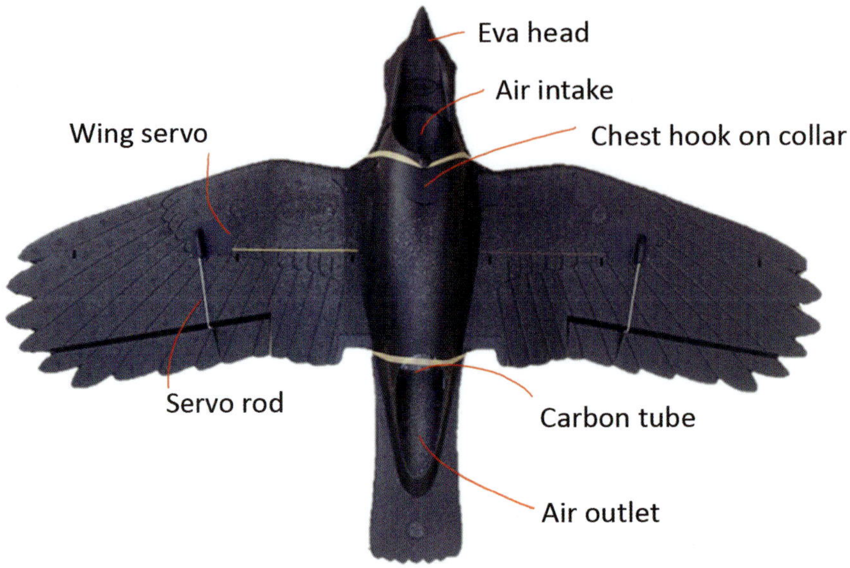

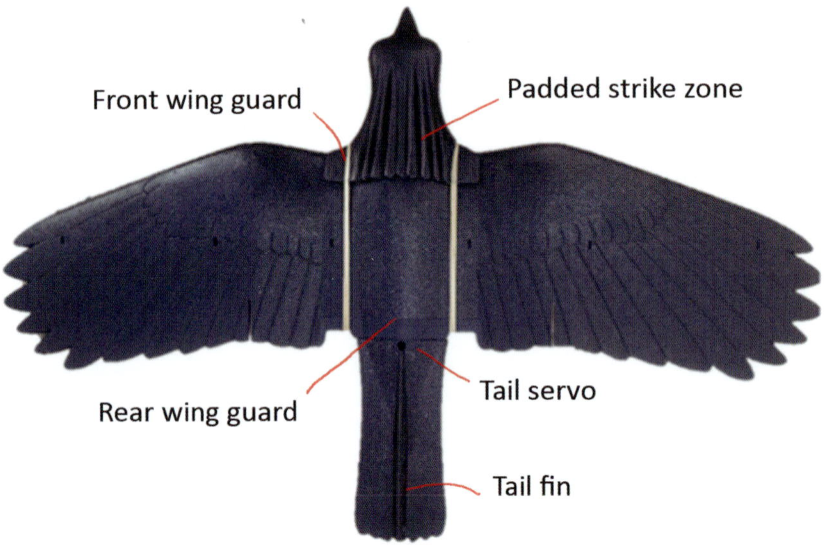

Taking a houbara head as a starting point.

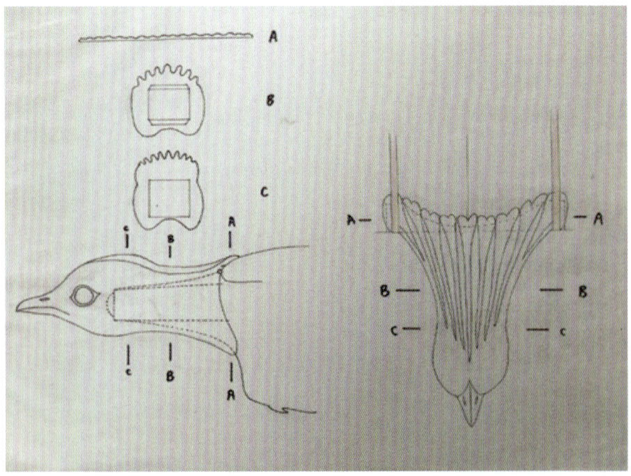

I create drawings of the outside and inside of the head.

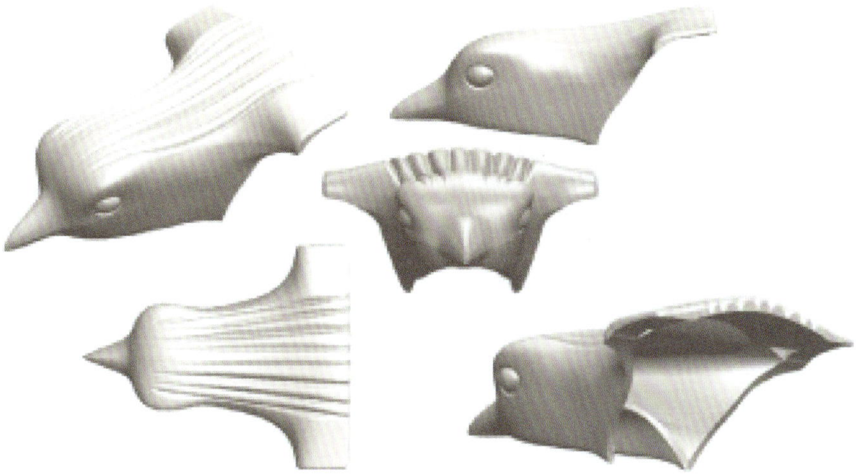

Remy converts these into CAD drawings.

We 3D print some prototypes to see what they will look like and whether they fit properly.

We have the heads moulded in EVA foam in China and paint them.

Lots of heads are painted, and some are made as different species.

different levels of control. Usually we configure it so that if the plane rolls more than 15° it will take over the controls to bring it back towards level. Similarly if the nose pitches up or down, it will level it out. A switch on the transmitter enables you to engage the stabiliser. For aerobatic flying, you need full control, but when things get extreme, for example in pulling out of a stoop, you may need to flick on the stabiliser to get you out of trouble. The stabiliser is easily adjusted to make it less stable if needed. The model is designed so that as new gizmos become available, you can upgrade.

At the root of the tail is an aperture for a servo to release the wing. The tail itself has a vertical and two horizontal fins but no control surfaces. Young falcons often grab the tail so we have removed all flaps. If the tail is badly broken, there is a fine vertical line marked on the fuselage, where it can be cut off and replaced. Below the tail is a carbon ring to take the pressure of an elastic band if fitted.

The wing is in one piece with a 'w' shaped carbon spar running the whole width. This enables it to have some dihedral on the inner wing, conferring some stability, but not on the outer wing. A degree of instability is needed for aerobatics. The inner wing is quite deep and cambered to give lift at low speeds, when there is less drag. The outer wing has the elevons which have a simple pinched hinge which is easily repaired if damaged. The elevon is controlled by a wing-mounted servo protected by a fairing, and a springy push rod to absorb shock and avoid injury to the falcon. The wing tip is emarginated and the tips can get broken. This does not affect flight performance but we can easily fit EVA foam rubber wing tips. This means we can make the wingtip in different shapes, for example, more pointed like a peregrine.

The leading and trailing edges of the centre of the wing are protected by nylon guards. These enable strong rubber bands to be used to hold the wings on. The centre of the wing is thickened to absorb impact and a feather pattern is imprinted into the EPP foam for more realism. The wings and body can have printed stickers on them for houbara, karrowan or cock pheasants. All other models are hand painted using alcoholic inks. We dissolve the ink powders into 95% isopropyl alcohol and then add 25% artists varnish to improve adhesion. The 40 g/l EPP foam is moulded in black for crows and in white for all species that need painting.

Although all the parts are modular and replaceable, all the foam parts are repairable. What looks like a total wreck of a crash can be repaired in a few minutes using the superglue and accelerator provided. For rocrows, which we know are going to get a hard life, we usually tape up the wings first, using a lightweight duct (gaffer) tape. Repairs can be done in the field. But when we get a chance, we take the models into the house or factory, clean and dry, then sand off any surface glues or vent pimples. Hot glue makes a better repair than the super glue (which can get brittle). Other glues such as neoprene or rubber glues (Evostik) can also be used, the main thing being to spread it well and thinly and hold the parts firmly together while setting. To make an extra strong job, put a thin trail of hot glue along the smooth surface of the repaired break and immediately cover with duct tape, pressing the glue flat with the blade of a knife to spread it out widely and thinly. Wings go on almost indefinitely with careful repairing. We have some now four years old. Just be careful not to add weight more than absolutely necessary.

The model is designed so that its centre of gravity is just below the wing spar. If you stall it, the nose will drop and enable you to regain airspeed and recover. The CG can be moved by adjusting the position of the battery, by inserting some foam to move it back. Rubber bands can used to hold the wings on, giving a degree of 'give' in a crash. These are not needed for the airbrake. When launching, hold the body at the CG and throw horizontally with ¾ throttle. Do not be tempted to throw upwards! The model cannot gain sufficient airspeed, will pancake in the wind and will stall.

All through the prototyping process we were constantly measuring performance parameters. It has to be forgiving to pilot yet capable of some quite intense manoeuvres. With software, we can assess what is going on.

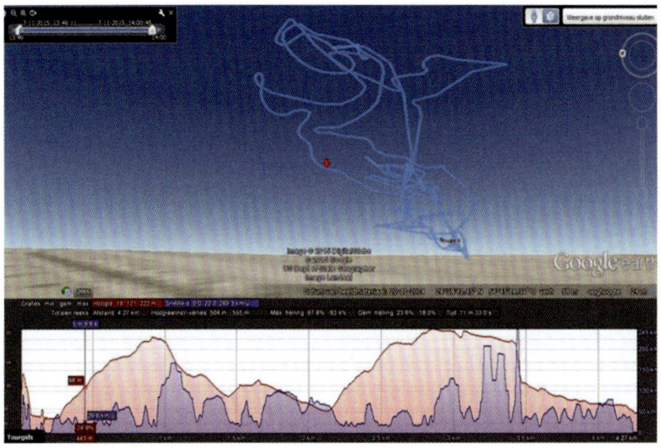

The red arrow can be traversed along any part of the flight path and the graph below shows the altitude in pink, the airspeed in purple, and the track distance travelled on the bottom line.

The transmitter supplied with the model is very basic. As you become a more experienced pilot you will need a programmable transmitter with more adjustments available to you. For aerobatic flying you need exponential controls which are soft at low levels but become more active at the higher end of the throw of the control lever.

Different parts, such as the fan, can get broken and can be replaced, so can the motor itself, without too much difficulty or cost. The internal electronics do not usually take damage because they are protected inside the double-shelled fuselage. There are no wheels; the model is hand launched and no special airstrip or landing surface is needed. The result is a model that, despite taking punishment, can be repaired and have parts replaced indefinitely, a bit like my old hawking knife that has had three blades and two handles – philosophical question: is it still the same knife?

Although the V2 model is based on a crow in shape and size, it can easily be transformed into another species. Species with complex feather patterns such as a Houbara, Karrowan or Pheasants, have photographic stickers. Other species, such as Herring Gulls, Mallard, Red or Grey Partridges, grouse, Magpies, Herons etc are hand painted using very lightweight dyes. Sometimes different head shapes are used, or feathery plumes added. Extending on from our various prey species, we can also make Peregrines and Goshawks and use them as deterrents in pest control.

9.4 Legal considerations

Kites, balloons and Unmanned Aerial Vehicles require various licences to fly. Especially if you plan to fly above 120 metres (400 ft) or in controlled airspace, most countries require you to be registered and to supply a NOTAM. Usually the craft must have your name and registration number on it. Also in most countries, including the UK and UAE, it is illegal to release an object from a UAV, which precludes the use of releasable towed lures. There are also restrictions on how close you can fly to private houses and buildings, and to members of the public or spectators. Check out the regulations in your own country. Usually they are not too onerous but there is always someone ready to do something stupid and that attracts more and more legal controls.

9.5 Flying grounds

Although rofalconry can be carried out on ground that is too enclosed for real hunting, the selection of good quality flying ground will help you achieve the highest quality flying. Essentially, the bigger and more open the ground, the better. In theory, as long as the pilot ensures that the roprey is caught less than tree height, flights can be carried out in the middle of a small clearing. In the real world it's advisable to give as much room horizontally as we do vertically. So, if we are flying to 400m altitude, we want 400m of open space in every direction. This helps to ensure that if the roprey is caught at great height, the falcon still comes down in a safe place. All the usual hazards should be avoided – fences, electricity wires and pylons, bodies of water, etc.

Consider how the wind is behaving at your flying site. If the wind is blowing over a wooded area just before reaching you, the air will be turbulent and very unpleasant to fly in. If you are at the bottom of a sloped field with the wind blowing down it, you will find yourself in a down draft and the wind will fight you. Wind can either be "dirty" or "clean", turbulent or laminar. Laminar wind, regardless of strength, allows for smooth controlled flying for both the pilot and falcon. Beginners of both should be flying in pleasant conditions until both gain experience. It's also important to consider the airspace you are flying in. Airspace use in some areas is restricted, often due to the proximity of airports or other airspace users.

9.6 Basic piloting

The first thing you must realise is that piloting a fixed-wing radio-controlled plane is a whole new discipline for you. Don't just take your model out of the box, chuck it in the air and then wonder why you can't fly it! Just because you have flown a quadcopter doesn't mean you are a pilot. There is a huge amount of basic stuff to learn. As a falconer, you may be up for that and find it complements your understanding of flying falcons. But if you are not up to it, find the nearest ten year old to do it for you. Although I did hang-gliding and micro-lighting in my younger years, I did not properly take up RC flying until I was in my sixties and wanted to develop roprey. Learning any new skill is not as easy as you grow older, especially as you need good eyesight and quick reactions. But it is possible, and I have been able to train some dozens of falcons this way, but I will never be good enough to be a race pilot. Still, I have enjoyed the challenge! Contact people in your local RC flying club. They will help you over the first basic hurdles.

The first thing you must do is turn your mind inside out. You are down here controlling a model that's up there. You have to put your mind into that model and keep it there. When the model is flying away from you it's easy. You steer left and it goes left. But when it is coming towards you, the model while steering left, heads to your right. It's all back to front! So the least painful way to train your mind for basic flying is to get a flight simulator. There are plenty on the market, but get one with a transmitter similar to the one you will later use. Most people fly in Mode 2, that is with the throttle on your left thumb and the steering controls on your right. Select a plane option similar to the roprey. We have made it very simple, you have no rudder or tail controls, only flaps on each wing, called elevons. These combine elevators (which make you go up and down) with ailerons which bank the plane, making you go right and left. With a simulator you can crash as much as you like while you train your brain and hand and eye co-ordination.

Once you have got a bit of a feel for flying on a simulator, choose a still day and go with a pilot friend and let the pilot set up the model and launch it. Once it is about two 'mistakes' high – two or three tree heights or so – you can take over and start basic flying, staying roughly in the upwind quadrant from you and out of the sun. As your confidence improves you can practise launches and landings.

You will soon start to appreciate that the sky is not just an empty space up there, it is an active fluid medium. You cannot easily see the air moving, but you will feel it soon enough once you start piloting. The model is of necessity a compromise in design. It cannot change its wing shape and configuration like a bird can. There are times when you will need to fly really slowly, even stationary in the air. See if you can do it. And other times when you need a good turn of speed – how fast can you go? There will be times when you need a shallow glide angle, and times when you need agility and aerobatics. And there are times when the model must become a pheasant, or a seagull, looking and flying very differently. As for the air, you will discover wind shear, turbulence, pockets of thermals, hill lift and – the bane of everyone – down drafts.

If you go up a little in still air and then cut the engine, you can glide. The rocrow is not very efficient at gliding. With a ratio of about 8:1, it is about the same as a Cessna. So for every metre it descends, it can fly forwards 8 metres. This compares with a paraglider at about 3:1 or a high aspect ratio performance glider at 30:1. Perhaps we should have made the rocrow lighter so that it would glide better? Actually, weight doesn't really change the glide angle, just the speed you go down it. A Boeing 767 has a glide angle approaching 20:1. From 30,000 feet, without engines, it could still make it for about 96 miles. Who would believe a huge chunk of metal weighing 186 metric tons could glide like that? And if we'd made the rocrow with long thin wings like a glider it wouldn't be capable of aerobatics.

It glides because, while gravity is pulling it vertically downwards, the wing shape wedges against the air, pushing it forwards. Once it has airspeed, air flows past the wings, which are cambered so that the top surface is more curved than the underneath. The air flowing over the top of the wing has further to go than the air going below it, which means that the air pressure above the wing is less, and below the wing is more. The result is to lift the wing up.

If you hit some air that is rising faster than you are descending, then you can go upwards, so that instead of gliding down, you start to soar up and gain height.

Alternatively, if you turn on your engine, this will increase your forward airspeed, giving more airflow over the wings, giving more lift and more flight control. So you can play around with gliding and using the engine, and explore the sky a little to find where air is going up or going down. If you bring the nose up too high you will lose airspeed and start to stall. With hardly any airspeed you will find your elevons don't do anything either. It's the same with a boat; the rudder only works when you are under way. As long as you have a little height, the rocrow will look after you. The stalled nose will drop, it will dive a little way, gain air speed, and off you go again. Next time you will anticipate it and start 'dolphining', going up and down, but soon you will get the hang of it and fly level. Flying level in still air will enable you to 'trim' the model so that even with hands off it will continue level flight.

Most of the wind is moving horizontally and you will find yourself dealing with a headwind, a cross wind and a tail wind. When you are flying into a 10 kph wind at 20 kph your groundspeed is only 20-10 = 10 kph. You have plenty of control and you can climb easily. When you turn down wind your ground speed is 20+10 = 30 kph. You feel you are going too fast, so you throttle back. When you bring your airspeed back to 10kph and the wind around you is doing the same, you hardly have any lift or control. So you try and turn (which makes it worse…) then panic, then crash. We've all done it.

Next time you go downwind you decide to stay high. Give yourself plenty of space. That's when you discover wind shear. It's quite a bit stronger up there. Before you know it, you are going out of sight. You can't see your orientation. Are you coming or going? Too late, the rocrow has disappeared over some distant trees. Pity you forgot to put a transmitter onto the tail mount attached to the ESC for you… So you start your 'walk of shame' and trudge off to look for it. Mark the line it took and when you get near the spot, rev the engine. You may find it up a tree or in a cornfield by hearing the engine sound. For those who have lost their rocrow at sea, the good news is: it floats well!

But maybe you had the presence of mind to hit the GPS come home switch. If the wind is not too strong and you keep your height, it will slowly turn the model and bring it back. Provided that you found satellites and set it on your location before you started. If you've left it on the factory setting it will head towards Wales! But maybe things aren't that bad and you can still just see the model. How can you tell which way you are facing at such a distance? Try making a sharp turn right. If it goes right it is still going away. If it goes left, it is facing you. Simple really, if you'd had time to think about it.

You need to practice. Increase the thrust downwind, especially turning. Climb into the wind. Find a bit of slope with some lift to help you along. But then you will find that the lift is very localised. If you move out of the lifting area you may go over the crest of the hill and hit a roller or down draft which will unceremoniously crash you before you've realised what happened. You need to explore your new environment, the element that your falcon is living in. There are plenty of books and films out there to help you. Kermode is worth a look.

If you have a friend with another rocrow, fly together and chase one another. Instead of flying gingerly you will get caught up in the dogfighting and find yourself flying without thinking about it. Try a few stoops at him. Not so easy if he is close to the ground is it? Now you know why grouse stay low. Downwind low stoops are pretty terrifying. Now you know why falcons stooping to the lure pick their shots. Try finding some strong thermals. They are hard to find in Britain, especially in winter. Now you know why we don't have vultures.

9.7 Getting started with roprey

A bad pilot makyth a puzzled falcon,
A weary falconer,
And an empty glue gun.

Old rofalconry saying.

We have looked at all the basic procedures you need to understand to bring your hawk physically and mentally in good condition to the point of flying free and entering at prey. We've also looked at the methods different species use to find and catch prey in the wild. Just as we wouldn't expect a peregrine to fly through brambles chasing a rabbit, we wouldn't expect a goshawk to ring up after a seagull. Even so, the young raptor will probably need more advanced training to help her identify wild prey, know how to attack it, and know what to expect when the prey tries to escape. Some of these flights, such as waiting-on flights at game birds, require a certain amount of choreography, as well as operant conditioning, and this can be tricky.

In recent years people have used kites, balloons, quadcopters and tow planes to encourage their falcons to climb higher or become fitter. These devices are all methods of towing a lure for the falcon to catch. They mimic the stage when the parent released prey for the pursuing young. They do not look or behave like real prey. In some countries where it is still legal, falconers release live prey, such as pigeons or gamebirds, to teach falcons to wait on overhead and gain high pitches. Apart from being very haphazard, this poses an animal welfare problem and, when posted on the internet, a PR problem. Wood pigeons are widespread in western Europe. Any falcon trained using domestic pigeons develops an eye for pigeons and as soon as she gets any height it is inevitable that she will see pigeons and check away to chase them.

When training our young falcons to hunt crows, there is a difficult gap between getting the falcon fit and ready, and actually chasing and catching real crows. Some falcons, either by bad luck in initial slips, or through being heavily mobbed on their first crow kill, give up and refuse to chase crows. Crows can be very intimidating to a young falcon. They can also fly well and know plenty of escape tactics. So what I was looking for was some kind of artificial crow that I could control, to make it as easy or as hard as the young falcon needed.

Although in falconry we tend to recognise two basic styles of attack – the pursuit flight in which the falcon starts below and behind the flushed prey, and the waiting-on flight in which the falcon stoops at the flushed prey from a position of aerial dominance, in nature falcons use both methods. Some falconry flights also employ both methods, for example, in magpie hawking, or in the *anwerten* flight at crows. Accipiters and buteos also use both techniques, but often in confined spaces.

As more and more of us use domestic-bred raptors, which have no hunting experience at all, our challenge is to teach these youngsters how to become as proficient as wild adults. During the past fifty years I have tried out most of these techniques, such as wild hacking, tame hacking, make hawks, thrown lures, dragged lures, hand lures, pole lures, lures pulled from cars, spinning lures, flapping lures, lures lofted on kites, balloons, multicopters, tow planes, bagged prey, and flapping dummies, with varying degrees of success and frustration. All have their limitations.

I was watching a saluki race the other day. Two dogs are run against each other as in coursing, but instead of a hare, they chase a dummy made of polythene bin-liner. It is on a 700 metre line which zigzags around pulley pins along the course and is pulled by a motor. So as the dogs chase it, it keeps zigzagging as it goes around each pulley. The things is, it zigzags regardless of the position of the dogs. Of course they swerve to follow the new track to the next pulley, but the dummy does not actually react to the approach of the dogs. With roprey the dummy moves away from the falcon just like a real prey and the falcon responds accordingly. For spectators it appears just the same as a falcon chasing real prey; the two interact in an aerial ballet. It is surprising how important this difference is. The dummy appears simply as an automated object, whereas the roprey appears imbued with a life of its own.

9.8 Developmental targets for the young raptor

Getting back to first principles, what are we trying to achieve?

First, remember that the transition from dependence to independence is not easy. Two thirds of young raptors don't make it through their first year. Catching their own quarry every day is no simple feat. Our newly-fledged raptor needs to learn and develop both physically and mentally, while at the same time avoiding undesirable traits, such as lack of focus, raking off, checking at other prey and so on. These are the attributes I'm trying to develop in my young falcon:

- Confident flying ability in all gears and in all situations. Pinpoint accuracy in flying, landing and using wind shear, able to stoop hard down a strong wind near to the ground. Able to climb both in a straight climb but also knowing when ringing is more efficient. Able to find lifting air. Able to confidently negotiate obstacles including understanding wires and fences.

- Excellent footing ability. Knowing to strike at the head whenever possible. Knowing which parts of the body, such as tail and wing feathers, are not worth catching. Knowing the likely last-moment moves of the prey to avoid being footed and how to counteract them. When the two dots converge in the sky they shouldn't come apart. Being able to do a devastating knock down with an accurate stoop.

- Physical fitness. In falcons, able to sustain a 500 metre (1650 feet) climb without missing a wing beat. Good aerobic as well as anaerobic respiration. Ability to maintain a sustained attack for up to ten minutes. Carrying no surplus fat. In a goshawk, able to sustain a high-speed chase for 500 metres.

- Launching into immediate flight the instant the hood is removed, without hesitation or delay, climbing and planning her attack as she goes, and evaluating the skill, power and ability of her prey.

- Not breaking off the hunt or skipping a wingbeat.

- Willing to take on prey on passage, even with a 200 metre (660 feet) height advantage.

- Flying only at her chosen prey species, not checking at other species, and not changing from one selected prey individual to another mid-hunt. She must lock on and stay focused on a single target.

- Strategic ability. Able to take advantage of the wind and sun to place herself at an advantage for attacking the prey. Not slavishly tail-chasing. Appreciating the relative attack climb angles of herself and the quarry, both into the wind and down wind. Fooling the prey into making a bad move. Making feinting stoops to drive down the prey or put it in a bad position for a second, killer, stoop. Planning a stoop so that if it misses, she can instantly command good height and position advantage rather than being thrown wide and giving the prey opportunity to escape. Out-performing the prey in a tight turn down wind. In the case of a waiting-on falcon, the ability to attain aerial dominance without being able to see the hidden prey, to position herself well in relation to the falconer and the wind, and to be confident enough to wait until she is served.

These are the main things I am looking for in my hunting falcons. Of course if you are rehabilitating an orphaned falcon to the wild you may want her to hunt a variety of species, not just one. Or if you are doing flying displays for the public, you may just want your falcon to chase around the sky; the public will not know any better. But if you are a serious falconer, and want your domestic-bred falcon to fly like a haggard, then these are your main criteria.

I'm not going into an exhaustive critique of all the different methods, from hacking to kites to drones etc. If you have tried them, ask yourself: which of the six attributes I have listed does each method or training tool develop? Which does it fail on?

9.9 Training with roprey

A young raptor is a clean sheet. What you put on this sheet is critical for her whole life. Choose a roprey species that is the same as that which you intend to hunt. The raptor needs to develop a 'search image' and therefore we provide a choice of roprey species. Common ones are crows, houbara, karrowan, pheasants, gulls, ducks and grouse. If you want to, you can paint your own model. It is important that the model looks and behaves like the real prey and of course you must fly it just as if it was your chosen species. Pheasants do not ring up 500 metres into the sky!

Before you introduce your young raptor to roprey you must complete her basic training: coming to fist or lure and so on, and you must keep this up as refreshers all through her life. In parallel, while she is completing her creance

Australasian Magpies *Gymnorhina hypoleuca*

Roducks in the Wingbeat factory.

work, you can show her a roprey model on the ground with food attached to its head. Because the model is only on the ground and not flying, you can put a reasonable chunk of attractive feathery food on the head. Stand just a few feet away to start with so that she sees the food. You can twitch it or tow it on a line and rev up the engine. This can be the last 'lesson' of the day and you can feed her up on it, just like finishing on a kill out hawking. By the time she is stooping to the lure or in the case of an accipiter, flying free to the fist, she will know the roprey and associate the engine noise with a call, just like a dinner whistle. Always encourage her to stay on the roprey until you arrive and reward her with a pick-up piece of food. If she starts to leave the prey to come and find you, this leads to all sorts of difficulties, not least finding the abandoned prey afterwards.

The next step is to get her to catch it in the air. Attach your roprey to your pole instead of a lure, so that you can hold it just off the ground and run the motor. Even while she is still on a creance she can catch it and feed on the ground with it. You can put plenty of food on at this stage. Remember to go for a clean slip. The first thing your raptor should see when the hood comes off is the roprey, and she should instantly start. Out hawking a one second delay could make a big difference to the outcome.

Around this stage of development, the mother falcon will fly holding prey, tempting her youngsters to chase her, and then she releases the prey. From there they start to chase prey for themselves. We can emulate this by towing a lure behind a tow plane. Any simple EPP delta wing hobby RC plane is easily converted for towing. It needs an EDF of about 65-70 mm and a simple servo release mechanism. After a few flights on this the young falcon quickly gets into aerial chasing and can graduate on to roprey.

A tow plane is not necessary for falcons to learn to chase. Most can go straight onto roprey. The raptor must be flying loose for this stage. Stand the pilot upwind and 3-4 metres in front of the raptor. Hold the roprey so that the head and food is visible. Make the food as feathery and attractive as possible, but not heavy and not so much that if the raptor pulls it off she will be overfed. You can use the roprey as a towplane for the first flight or two by attaching a couple of quail wings on a short line to its tail. Then unhood her and let her see the model. The moment she launches after it, the pilot launches the model into the wind in a gentle flight to enable her to catch it, come to the ground and eat her reward. You can repeat this another time or two on the same day. Usually you can extend the distance quite quickly. If you have a slope with a nice lifting breeze, this makes it easier to fly slowly and entice the raptor.

The 'conker' crow held off the ground.

Flying slowly is a great skill to master. As well as looking nice, it's really helpful for starting young or inexperienced raptors. Flying slowly is made easier by wind, as well as by ensuring the model isn't carrying any extra weight. Wind means that the aircraft can fly slowly, and you can even hover on the spot in a gale. Another way of helping the model to fly slower is by making it as tail heavy as you can get away with. This causes the nose to pitch up and the model can be flown slower, almost at the point of stall. Be careful though, an excessively tail-heavy model is very difficult to keep in the air. The best way to move weight towards the tail is not to slide the battery all the way in. This takes weight away from the nose and shifts it further back in the air frame towards the centre of gravity. You can use a small square of foam pushed in front of the battery slot to stop it sliding forwards. You can also add weight at the base of the tail. The airbrake servo cover is secured with a small amount of hot glue. In this recess, you can add up to 30g of weight. We find that putting in one or two 2 pence coins, each weighing 7 grams, works perfectly, but anything can be used. Securely replace the plastic cover, and go for a test fly. You can add and remove weight to your liking.

The basic model weighs 650g. Although it can take a pay load of up to 50-70g, this must be centred at the Centre of Gravity or you will unbalance it. If you put half a quail on the head it will crash! Just put a small amount of food, such as a day-old chick leg, under the red neckband when you wish to fly. Never put any more weight on than you have to. The instinct to chase is very strong in raptors. So you can add lightweight things to your model to make it look more attractive. For example, you can glue long tail feathers to the cock pheasant for a goshawk or Harris Hawk. The model has no control surfaces on the tail so that this works fine. For crows, we just shred some black bin liner and glue it onto the trailing edge of the tail or even the inner wings. It flutters like a lure without destroying the aerodynamics of the model in flight or affecting its CG. Don't add any bits and pieces on the aerofoil areas such as the wings and back. This disrupts the airflow, destroying lift, and causing a crash. For white models, such as Herring Gulls, use white bin liner polythene. Polythene is better than feathers because it does not absorb water and add weight.

Be careful, due to the nature of rocrow use, each model soon becomes individual, so spend some time tinkering about and getting to know it.

After one or two sessions like this, the raptor is getting keen and you need to fly the model as close in front of her as you can – just a few metres. This will make her lock on and get really focussed. Then with a falcon, you can start to do more extended tail-chases which also start to get her more fit. By this stage she will have graduated beyond the need for any flappy bits and you can remove them. She may be catching by the tail or wings to start with. As she improves, she will learn to go for the head and a skilled pilot will tilt the model up at the last moment so that the head is presented for an easy catch. Also, as far as possible, ensure that the ground below is soft grass for an easy landing. As a pilot, you must always fly like a prey bird. NEVER fly towards the falcon, always turn away. Never behave as if you are attacking the young falcon. Don't try fancy stunts like multiple barrel rolls. Don't fly upside down. Think and behave like a real bird. Through all these stages the pilot can fly with the stabiliser on.

The flight performance of the rocrow is more powerful and more manoeuvrable than that of a falcon. We have moved the balance in favour of the pilot, to make it easier for you. But with increased power comes responsibility. You need to fly in such a way that she thinks that if she just flies a little bit harder, she can catch you. If you just lead her around the sky in a hopeless tail-chase she will not have her heart in it and we have seen a lot of falcons in the Middle East chasing lures behind powerful tow planes that just go through the motions or, worse still, are over-flown and get muscle cramps. The roprey is designed to fly for up to five minutes on full power and up to ten minutes on half power, so that the falcon cannot be over-flown.

As a pilot your job is to build her confidence, strength, endurance and footing skills. As she improves, she will develop strategies to get above you and stoop. The higher the flight goes, the more advantage she has because piloting the rocrow at 200-300 metres altitude gets more difficult whereas the open sky is her domain. Once she is starting to stoop and hit the model hard, use a foam head that covers the impact area over the shoulders to reduce the risk of impact injuries.

And if she is a big falcon catching the roprey high in the sky, then use the airbrake to bring her down gently. The airbrake works by dislocating the wings from the body so that all lift is lost.

Training with a Rokarrowan.

It will bring both falcon and prey down almost vertically. If you are training to fly crows or gulls on passage, do not unhood until the roprey is 100-200 metres high, so that she expects to have to climb immediately before coming to terms. If you disappear in cloud, keep the engine running and the falcon will find you by the sound. Gently drop down in a curving flight. Although you can 'surrender' to the falcon when she is just starting, once she is chasing well arrange things so that sometimes she catches you after a hard flight but other times she catches you almost immediately. Encourage fast climbing by sometimes letting her beat you in the climb at 100-200 metres.

Don't fly your falcon at two roprey at the same time. A good falcon should lock onto her prey and never leave it, no matter what. It is a common trick for a pair of crows to keep distracting a falcon and take a stoop each, until the falcon tires or they have reached cover. The falcon must learn not to change targets or to check at fresh ones. Later, when you are flying a falcon at flocks of crows or gulls, she must be able to single one out and stay on it. Similarly, if you are flying in an area where there are pigeons, crows or other birds flying about, your falcon must stay locked on to the roprey. If you want your raptor only for flying in demonstrations, do not let her enter at live prey. You can even spray your demo roprey fluorescent pink if you want, to keep her from looking at real prey.

If you are flying a wild-caught falcon who knows how to hunt real prey but not the roprey, go through the same step by step introductory process. Many falcons, even wild ones, come in and chase the roprey. If you want them to foot the roprey, tempt them in still with a little height, maybe 30 metres or so. Then cut the engine and glide the roprey, presenting the head and rocking the wings. Many falcons cannot resist the temptation and take you down. The unfamiliar engine noise had been putting them off. You can also tow a feathery barak on a short cord behind the model so that it looks as if it is holding food. Wild raptors find this hard to resist.

The first couple of months, for both the falcon and pilot, can result in a subtle struggle for dominance. Initially, as pilot, you tend to be in control. You can fly around until you decide that the falcon has done enough. But once the falcon gains some confidence, and some skill in the sky, she begins to take control of the situation, catching the pilot in situations where he was convinced he would get away.

The balance goes back and forth repeatedly. For a few consecutive days, the pilot has a good run, twisting and turning, having full-length flights involving a climb, a descent and even a low-level tail-chase resulting in an intentional self-sacrifice. But then the tables turn, the falcon has a great run, making the pilot feel completely at her mercy. The falcon climbs after the roprey as expected, but she then puts in a burst of speed to close the gap, turns upside down and binds to the head of the prey at height! She begins to predict your movements. You turn one way but she's already committed to the stoop in anticipation. You both dive, and as you pull up, she keeps going, only to keep her momentum, throw up even higher than her quarry, to drop on it completely unexpectedly.

The power balance between your model and your falcon is critical. When we made our first model, the robara, we were careful not to over-power it. With the standard motor it is very well balanced to the ability of the falcon. However, pilots in the Middle East complained that there was not enough power to get them out of trouble, and so we provided a more powerful motor option available as an upgrade. When we designed the V2 rocrow we built in more power as standard, which moves the balance in favour of the pilot. If you are not a very good pilot you can use this extra thrust to get you out of trouble, either to avoid the falcon or to avoid a crash. However, if you are a good pilot you have so much power that you can out fly and out climb any falcon, and by a good margin. The problem with this is that raptors very quickly, almost instantaneously, recognise the strength and power of a prey animal. If just for one moment you hit the throttle hard and let the falcon see you have more power, she will treat you differently than if she thinks you are 'weak'. She will probably chase you on the basis that eventually you will give up and in the end allow yourself to be caught. She will tail-chase and follow on and to the uninitiated eye she looks good. But she is not learning to dominate and do slashing stoops like a haggard. You see the same effect with lure flying. Some people stoop a falcon to the lure and then at the end throw the lure in the air and shout 'Ho!'. It all looks very pretty and for sure the falcon has had good exercise. But basically she has just been waiting for you to throw the lure up. It is fine for falcons just doing displays. On the other hand, if you lure her in such a way that she can catch the lure at any pass, even the first one, she will make more effort because she believes that by trying harder she can beat you. This is the kind of falcon we want to develop for hunting.

When you launch the rocrow for a new falcon, just to start with, hold it in front of her, showing the meat on the head and as soon as she bates for it, give her a gentle flight and let her catch it. Once she is doing this, lengthen the distance of the slip until you are 50-100 metres apart at the launch. The reason for this is, first to teach her to take on longer slips (which she needs to know as a hunter), and second so that the pilot does not have to use full power to escape from the close and accelerating falcon. This would betray how much power he has. Once the falcon is taking on 100 metre slips, slip the rocrow and start to climb before unhooding the falcon. This teaches her to take crows on passage. Once she is fit and confident, you can wait until the rocrow is 200 metres high before unhooding the falcon.

With the rocrow, because it has more power, it is easy to make the mistake of letting the falcon realise that you have so much power. Therefore when training young falcons, we can reduce the thrust available to 70% on our programmable transmitters. You cannot do this on the standard transmitter supplied with the rocrow, so you must be disciplined! When it comes to competitions we have to ensure that the balance is right between pilot and raptor otherwise the pilot can just make the falcon look silly. At 70% power, the rocrow's acceleration, top speed and climb rate are reduced and are in balance with most falcons. The falcons recognise this and each day becomes more and more determined. Instead of plodding along behind you, tail chasing, she starts to mount and dominate you, aiming to get a hard shot at your head and take you down. The rocrow still has lots of manoeuvrability and you can jink away from most stoops but in the end she will catch you, on her own terms, not on yours. This means she may catch you high in the sky. On the other hand, if you just want to do demos, don't worry about all this. Just let her chase you and get yourself caught in front of the crowd. By using the extra power available, you can put off being killed until the exact moment you are in a good position.

Once your falcon is entered and catching real prey, you can look at her style and on thc non-hunting days use the rocrow to improve it. For example, some falcons have a 'glass ceiling' and break off from a pursuit when they are only 50 or 100 metres high. With the rocrow or Robara you can improve this day by day until she will go as high as you want. Other falcons may be poor footers. They follow slavishly from behind and try and catch you by the tail. You see this a lot with lures on tow planes. With a lure behind a plane, the falcon has no choice but to take the trailing object because the 'head end' is actually the plane. With the rocrow, keep jinking and using height so that she learns to dominate you from above and stoop only at your head. Crows and gulls, if they don't escape to cover, escape by getting above the falcon and keeping out of trouble. Your falcon needs to appreciate this and when she stoops and misses, always throws up in such a way that she keeps the height advantage and can stoop again.

Wingbeat makes several prey species, from crows to gulls to ducks, and even game birds. They are essentially the same model but painted differently. We do this so that your raptor develops a proper 'search image' for her prey species and becomes wedded to it. But search image is not just about what the prey looks like, it is also about the way it behaves. Always remember to fly your model according to the species it represents. A seagull for example has a low wing loading, is very buoyant, can manoeuvre well but does not have good acceleration or fast level airspeed. So adjust your power and fly like a gull. A duck on the other hand builds up its airspeed and has a fast level flight, and can jink but does not do clever aerial strategies like a crow does. So if you want to be a duck, then fly like a duck. Grouse and pheasants have a good acceleration, don't fly up into the clouds and have a poor aerobic performance. Often the falcon must learn to stoop and bowl the grouse over when it is only a few metres off the ground. Of course, with the game birds, you can lead your raptor in a gently curving flight so that the 'kill' is not too far away from you. Saves you a lot of walking!

These are all elements that you need to study and understand. A skilled pilot can really prepare and improve his falcon until she is hunting like a haggard. And by giving her the kinds of flights she needs he can keep her going and improving throughout her life. If, on the other hand, you just give her tail-chases, you will end up with a falcon that is OK for demos, but totally lacking in style and technique out hunting. The commonest faults we see with pilots are: trailing around with the falcon twenty or thirty metres behind, and not using the vertical dimension enough. You can be using any of the sky up to about 300m (1000 ft) for general exercise and put in stoops and dives, not just horizontal work.

9.10 Climbing

In real falconry, or in the Hunt Race, flights are won or lost in the climb. The falcon needs to be fit and motivated, but she also needs to learn how to gain height rapidly. The quarry species vary in their flight characteristics and the falcon needs to know what she is taking on. Crows for example are all-rounders; they climb well end-on and by ringing, but they are also quite fast in level flight and unless the ground is very open and the air very still, will prefer to race towards cover if they have the chance. Gulls on the other hand have a lower wing-loading and are more inclined to climb, particularly as they will not put in to cover on the ground as crows and rooks do. The smaller gulls are more agile and will often simply keep dodging the falcon until she is exhausted. Houbara are fast end on and climb at a steady angle rather than in rings, with the flight going a long way horizontally. Houbara are not as agile in avoiding the falcon as gulls and crows are. In the old days when herons were hawked in Holland, the best time of day was often in the evening when the wind had dropped and the herons would ring up, creating an aerial contest overhead, rather than a long cross-country downwind chase.

Game birds such as pheasants, partridges, grouse, francolin, tinamous and so on do not climb and are normally flown with the falcon waiting on overhead, or by accipiters in direct pursuit. Ducks are also flown the same way but can climb out high and go a long way rapidly. They can quickly lead the falcon out of sight and into all manner of dangers, such as predation by eagles, or simply being lost a long way off. The climb rate of the game falcon is less critical to success; she can take her time and the prey are unlikely to flush prematurely once they have seen her overhead. The game hawker's difficulty is in persuading his falcon to take a good pitch when there is nothing there to encourage her and it is an act of faith.

Starting with a young falcon which knows the roprey and is capable of about 25 good stoops to the lure, the first step is to give her the simplest slip possible. Hold the roprey 2-3 metres in front of her and launch. Allow her to catch it in a low-level tail-chase for a few days, ensuring her confidence builds well. Once fixated on her quarry, you can lead her at a shallow angle. Her straight-line speed will dramatically increase over the first 2 weeks. Initially she isn't convinced about going at it all guns blazing, but after catching the roprey half a dozen times she will begin to put the burners on. You can extend the slip to 50-100 metres or more if you intend to hunt with her, and you can have the prey in the air already when the hood is removed.

In still air the falcon will climb steeply to start with, at about 4 m/sec, depending on her power and motivation. Birds with shorter and deeper wings have better climb rates and acceleration to start with, but those with longer narrower wings are faster if they have the space. You can encourage this initial burst to be steeper and longer by flying just in front of the falcon, and by 'accidently' getting caught.

After this initial steep climb, the falcon will settle down to her preferred climb rate and this will be affected by the position of the prey. If the prey is right in front of her, maybe 3-5 metres or so ahead, (Figure 9.10.1 position A) she will be fixated on it and do her best to directly catch it. You can tow her around the sky following your track slavishly, but if you climb too steeply she will not be able to manage it and will be forced to break off. Obviously if the prey is slightly below this maximum climb angle (position B) she will trade off climb rate for horizontal speed. If the prey is too far above the maximum climb angle (position C) she may break off and ring or try another tactic. If the prey is too far ahead (position D) she will decrease her effort and may go her own way. At no stage should she miss a wingbeat. In climbs of 100-300 metres altitude, expect average climb rates of 1-2 m/s. The best falcons achieve up to 2.5 m/s but weather heavily influences performance. The reality is that although the roprey and falcon can achieve good climb rates for a few seconds at a time, over the entire climb the average rate is more modest. With a moderate wind the falcon

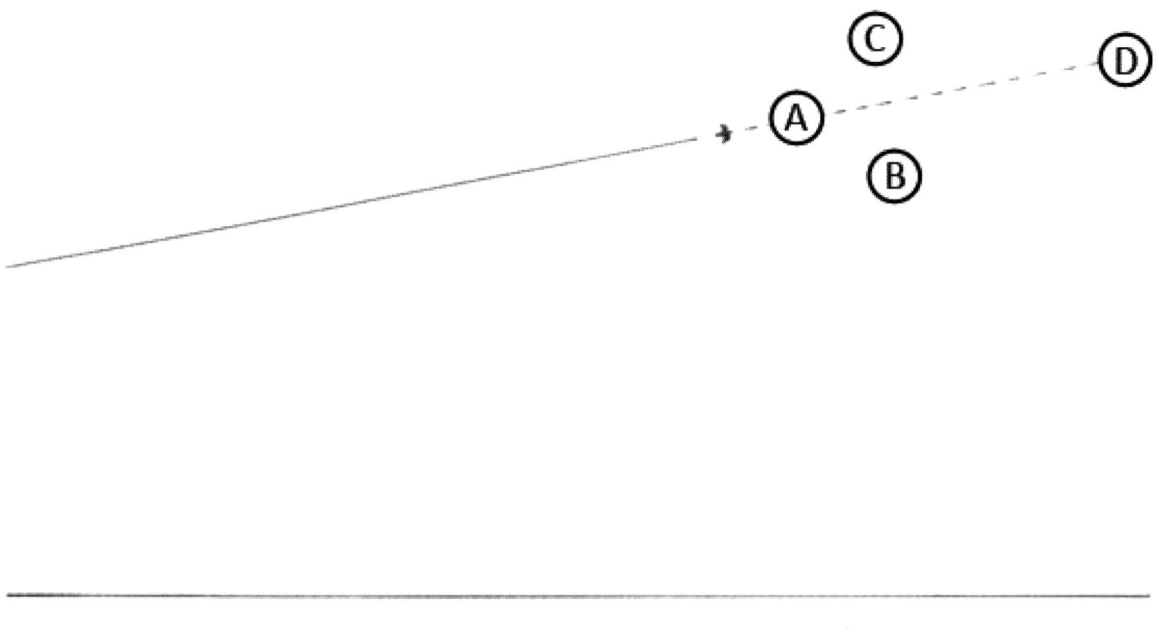

Figure 9.10.1 Roprey positions in the climb.

can climb well, but on a still day times will be much slower, sometimes barely half the rate. The main fault I see with pilots is failing to appreciate the importance of these positions. Often pilots fly 20 or 30 metres ahead, or even more, just zooming around. The falcon is left to helplessly try to follow, constantly having to re-align on this haphazard target. This is especially obvious with tow plane pilots who literally tow or troll the lure around the sky with the falcon often 50 metres away. This maybe fine if all you want to do is exercise the falcon, but it is not improving her in any other way. As well as not appreciating the importance of the position of the roprey relative to the falcon, pilots are often scared of being caught. Spectators will tease them. Or they may have it in their heads that the falcon is going to be given a five minute flight whether she likes it or not, like a dose of medicine. The pilot must improve his flying skills, possibly aided by improved autopiloting systems. He must be able to place the roprey where he wants it at all times. But don't be afraid of being caught; on the contrary, this is a great stimulus for the falcon. A variable ratio of reinforcement is far more encouraging for her than a fixed repertoire.

Although falcons will chase the prey steadily up into the sky and around, she will learn more and work harder if you vary the chase. Rock your wings, twist and turn a little. Never fly towards her and never reveal that you can zoom away from her if you want to. She must think you are a vulnerable quarry struggling for its life, and if she tries just a little bit harder, she can catch you. Also do not fall into a routine of the same kind of flying every day, only allowing her to catch it near the end so that you do not have to run too far to retrieve her. If you make everything predictable, she will do some predicting. She will take it easy. Get yourself caught at any stage of the flight and she will stay eager.

As the climbing portion of the flight is extended, you open your options to carry out more exciting manoeuvres. Some falcons naturally climb along the same route as their quarry, following it in the turns. Others climb wide, looking for advantageous lift in likely parts of the terrain. The climbing style can be influenced by the pilot. Keeping the roprey close but just out of reach tends to draw the falcon along a similar path, while opening up a larger gap can often get the falcon thinking about other methods of gaining height. When you are going for height in a wind, it is tempting to climb straight into the wind so that both falcon and prey go up together a few metres apart. It is a slog. You need to teach her to understand wind shear.

A sailing boat can sail into the wind without power. It cannot fly straight into the wind, but a well-performing boat can tack at 40 degrees off the wind, and by tacking, make headway into the wind. Of course it cannot flap its sails to develop thrust in the same way that a falcon can beat her wings. Yet that boat can go into the wind! So it is with the falcon, with one proviso: whereas the sailing boat can only operate from side to side, the falcon can go up and down into the wind as well. So, when she is facing the wind, she faces a 'no-go' zone, a cone of maybe 40 degrees around the point head wind. Outside this she can gain speed, using the camber and shape of her wings. The falcon works in three dimensions – she can go up and down as well as side to side. But in this dimension she has another force acting on her: gravity. So her vertical cone is distorted downwards by her glide angle.

In the boat sailing into wind, we say the sail is 'close hauled'. That is, the sheets are pulled tight to make the sail as flat as possible. A narrow, flat wing is the tool for this job, and we see it in peregrines and birds such as petrels and shearwaters who use windshear for power. Going across the wind on a beam reach, the sailing boat softens the sail forming a deep camber. Falcons can change their wing shapes rapidly, folding back primaries sometimes and emphasising the secondaries, other times, decreasing the inner wing and extending the outer wing. The patagium stretches tight as the wing extends, flattening the curve, and then the camber increases as the wing draws in but disappears completely as the carpus is pulled close to the body. The young falcon must learn that, instead of churning head into wind using up power and energy like a motorboat, she needs to exploit the wind like a sailing boat.

To teach her this, you must fly the model in a 'sailing' style, never heading directly into wind but always angling into it. Of course you cannot change your own wing shape, so you are less

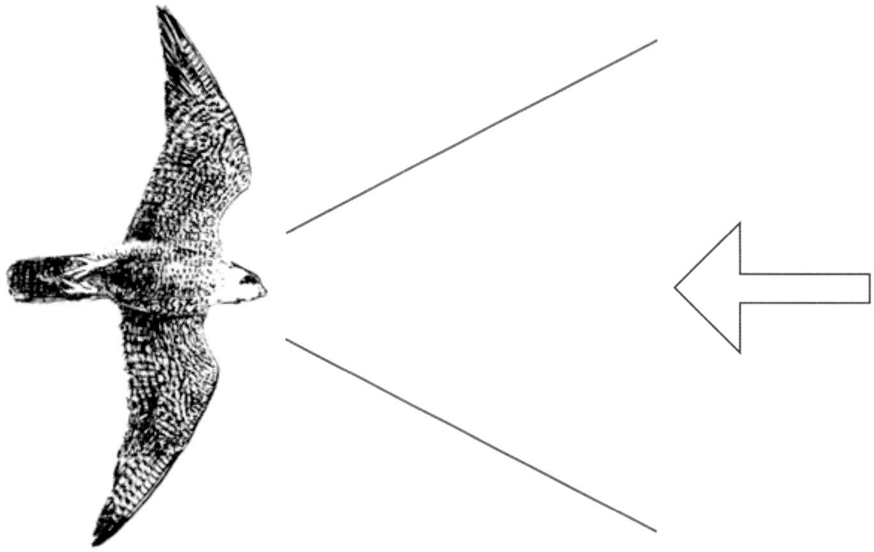

Figure 9.10.2 Heading into wind the falcon (seen from below) cannot fly within this cone without flapping her wings.

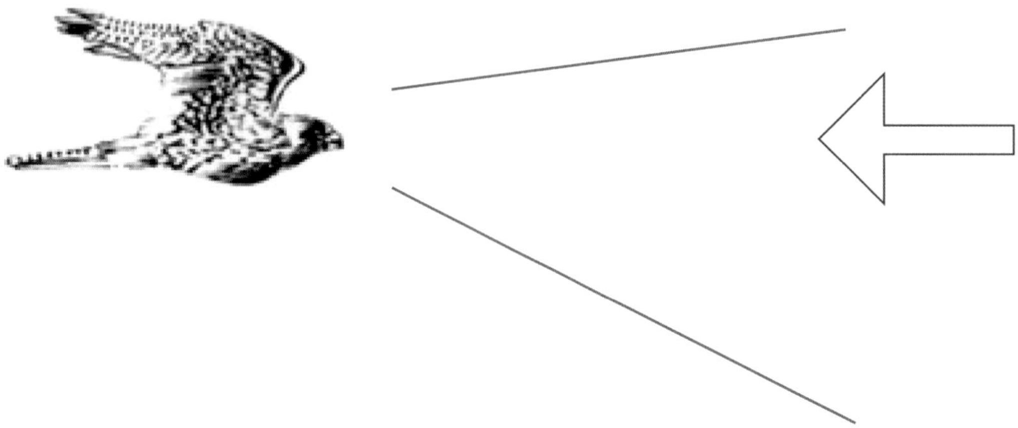

Figure 9.10.3 In the vertical dimension, gravity distorts her cone downwards.

efficient. As she approaches you, you can angle to the side, banking as you go. Or you can drop your nose and the wind, while driving you down, will also propel you forward, into wind. Then you can raise your nose and swoop upwards, gaining height but also some forward distance. She will try to follow you, and in doing so, will learn the advantages. The sailing boat has a keel which 'grips' the water and to some extent counteracts the sideways force on the sail, helping to 'squeeze' the boat forward like an orange pip. The falcon too can twist her tail in relation to her wings. The roprey has fixed vertical and horizontal fins for the same purpose.

Soon she will fly like a wild falcon, playing with the wind dynamically instead of slogging away. She will make it look effortless. As she makes fewer 'mistakes' she will not need to keep flaring her tail to compensate (which creates drag). Later you can find your own separate ways up into the sky, seemingly independent of one another but in reality linked by the invisible thread of a deadly contest. And if the wind is gentle, you can both climb in rings and she will learn this too. It is easy to teach her.

After a month or two, she should be flying like a haggard. The day will come when instead of enticing your falcon around the sky, you suddenly find you are flying for your life. The tables have turned. Instead of planned catches, the day comes when she is catching you out. You make a mistake. You try to turn downwind too close to the ground. Bam! You are dead!

High ringing flights are hugely enjoyable, but being caught all the way up there isn't always that much fun. In suitable terrain, high altitude binds can be great, and can give a falcon a reward for good climbs, but being caught up high in unsuitable terrain can allow the falcon to land somewhere dangerous. She will likely descend with the wind, turning into the wind as she reaches tree height.

To avoid an unintentional catch, don't fly straight above your own head. This makes judging the height difference between the falcon and the roprey very difficult. Instead try to keep the flight a little to the side of you and upwind. This makes it easier to judge the horizontal difference between the falcon and the roprey. Some people allow the flight to drop down wind. This is the road to ruin. A point is reached when you cannot fight your way back home against the wind. The falcon is happy; she is right there with the prey, but you are trying to judge things from an increasing distance away. Maybe you get disorientated and cannot see if you are flying back, or going away. If this happens, do a sharp turn to reveal which way you are headed. You learn to judge, by her wing shape and beat, whether she is still climbing up to you or whether she is level with you. Alternatively you can put altimeters on both falcon and prey, with a voice read out on the two heights. But remember, the more complicated you make things, the more can go wrong…

Although multicopters have been used for several years for climbing flights, the use of roprey for the same purpose is more favourable. The propellers on drones are accessible to the falcon, despite the various guards that can be fitted, and toes have been injured. Drones are also limited in their use. The falcon spends time and energy getting up there, just to hang onto a lure and fall to the ground. Once the falconer starts to enjoy the vertical spiral stoop at the end of a rofalconry flight, everything else seems a bit tame.

9.11 The Airbrake

The airbrake is designed to reduce the ability for falcons to carry the roprey. It is especially useful when you are doing high flights or flying in enclosed areas where you need to land them close by. Once the falcon has caught the prey by the head, flick the air brake switch. This dislocates the wing from the body and brings them down at a steep angle. If the falcon has caught the prey by the tail or wing, you do not need to activate the airbrake. She cannot fly forwards with the prey dangling. Do not deploy the airbrake when the falcon is not holding the roprey – you will have a spectacular crash!

We tried all sorts of airbrakes during prototyping. We avoided parachutes because they are heavy and complicated and because they destroy the illusion that the model is a real prey. The ones we use now are the lightest and most effective so far. The original airbrake springs the wing up at 45° to the fuselage. This brings them down at a smooth angle of about 45°. But some falcons learn to take the roprey by the head and then biplane with it for some distance.

The newer airbrake is designed to overcome this problem and is more catastrophic. It blows the wing completely off the body, holding them together only with a short cord attached to the wingspar. This is long enough to allow the wing to disconnect, but not so long that the wing will rotate and tangle. It brings them down vertically. Once the wing servos have been connected, the front of the wing is fitted under the collar. It is locked from moving from side to side and can only move up or down. On closing the wing, the servo at the base of the tail is engaged to hold down the trailing edge. This airbrake is handy for getting the bird on the ground quickly during a Hunt Race.

We supply some spare servo arms with each airbrake kit. This small plastic arm is designed to break in an impact to reduce strain on the gears inside the servo. The arm is easily replaced by removing the small screw. Metal servo gears last longer than nylon ones.

9.12 Stooping

Young falcons are afraid of stooping. I am too! You need to practise this with them, first in very shallow descents, then steeper ones, then vertical corkscrews or canceleers. Pulling out at the bottom is the tricky bit for both of you, and things can go horribly wrong.

When first stooping to the pole lure, the young falcon will always attack into the wind. That is the safest option. Gradually you can encourage side wind approaches and in gentle winds, a downwind approach. As she gets more confident, she will attempt downwind stoops and to promote this, first ensure that the pull-out from the previous stoop is going to put her in an upwind position. Then use a good pole and present the lure well up in the air so that she can get a good shot at it and has room to pull out afterwards. If she takes it, allow the bungy cord and the pole to lower her to the ground quickly but safely. Do not let go of the pole or it could skewer her. But of course, even with a pole, there is a limit to what you can teach with the lure. With the model, all these exercises can be done well up in the sky without risk of hitting the ground.

The Hierofalcons, gyrs, sakers, lanners and luggers, and also prairies, like to hang over your head and do a vertical stoop on you. This move is difficult to counter and you may end up getting dizzy and falling over backwards, a bit like the guy who was put in a barrel and told to pee in the corner. But with the pole lure it is not hard to sweep the lure wide from you and high, so that she can get a good shot in. Do not discourage her from this; it is a very deadly move when hunting real prey, leaving the quarry very few options and undecided what to do.

Once the falcon has graduated to roprey, she needs to be coaxed into the stoop to start with. First climb to a reasonable height, maybe 100m, and bring her up close behind you at about the same level. Then drop the power to approximately 25% and start diving, not too steeply. Once the nose is pointing downwards, turn the power off completely. This can take some practice, as well as some confidence as you may be worried about being caught high up. Try and tease the falcon into a stoop, attempting to draw her down with you, rather than firing the roprey towards the earth. Letting her catch it in the stoop will build her confidence. Remember that young falcons are sometimes clumsy, and a reluctance to speed into the ground is understandable when you're still not familiar with how the brakes work. Each time increase the descent speed and angle until she can do a true vertical.

If you dive too steeply and too fast for a young falcon, she may not follow you. Then you find yourself in the tricky position of being close to the ground with the falcon in total dominance, like a game hawk. She can splat you any time she likes. It is a strategy she needs to learn, and it is fine for waiting-on flights. But for pursuit flights, if she allows the prey any quarter like this, the chances are it will find its way to cover quickly. The crow hawks know the advantages of height and dominance, that is mainly how a flight is determined, but equally they know that they have to keep the pressure on, or they will lose touch.

In a fast vertical stoop, you need to pull out into the wind. If you try to pull out downwind you need to pull out early otherwise the ground will come up faster than you had anticipated. If your stoop is aligned towards or away from you, you cannot judge if it is vertical or not. So you need to arrange your stoop so that it is side on to you and also head to wind.

When the roprey pulls out from a stoop, with the falcon close behind, there is a point where, if you have got insufficient flight speed in the throw-up, she will simply snatch you out of the air, catching you in the stall. Try to maintain speed and throw up sideways so that the falcon goes the other way. Until you are an experienced pilot, always switch on the stabiliser near the bottom of the stoop to help you pull out cleanly. It is very easy to get disorientated and it all ends in tears. I did a 400m (1300 feet) vertical stoop once with the V1 roprey which had no stabiliser, straight into a sand dune. Our Egyptian pilot, Khaled, who saw the bits fly into the air, still suffers from PTSD as a result.

In little stoops, of 100 m or so, the falcon does not reach terminal velocity. Most stoops will be in the 150-200 kph (93-125 mph) range. To attain faster stoop speeds, you need to start

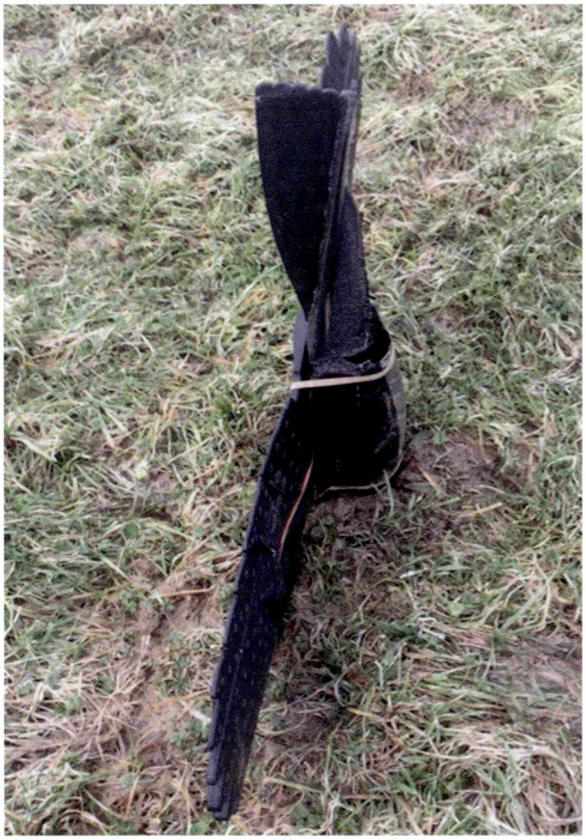

The rocrow withstood this stoop without any damage!

higher, at 300-400 metres, or have a heavier body mass. Although Ken Franklin's female peregrine reached a record stoop speed of 389 kph (242 mph), she started at almost 4,572m (15,000 feet). With roprey the highest you can see to fly is about 500 metres (1650 feet).

Tucker concluded that in high speed stoops the pressure on the eyeballs was so great that falcons, instead of stooping straight at the prey, maintained their target at an angle of about 40 degrees to one side so that they could use the maximum acuity of the deep fovea of one eye. This curving stoop was also examined by Richard Sale. We also

looked at prey tracking with Caroline Brighton and the Oxford Bird Flight Study Group. I have not noticed a one-eyed, curved stoop. In my experience the lateral or deep foveas are used for scanning and detecting things, whereas during the stoop the problem is not in detecting the prey, but in judging relative distances, and this requires stereo vision. To examine this properly we need GPS tags and barometric altimeters on both raptor and prey, sampling at at least 10 Hz. Stereo vision is so important at this stage that I am trying to design for myself a simple device to improve my depth of field perception. The canceleer, or spiral stoop, is most likely used to confuse the prey and to stay on track to hit it. Falcons are perfectly capable of stooping straight and vertical, but as they approach the prey it is inevitable that both trajectories curve.

If the falcon takes the prey high up, she will drift down with it, sometimes very slowly, but sometimes alarmingly fast. If the ground is hard and stony she could easily be injured. We had a young female gyr/peregrine once in Morocco who took on a houbara high on passage and after about 4km got up to him and took him. They came down so hard that both were killed stone dead on impact. The houbara was a banded captive-bred cock bird, released nine years earlier. After one or two hard landings, most falcons learn to let go just before impact, turn and then land on the prey on the ground.

9.13 Footing and tactics

The falcon must learn how to deal with agile prey. For example, a young falcon may make a good effort in the initial rush or stoop, then miss, and throw up widely into a bad position. This gives the prey a vital second or two to improve its position. It may use the time to climb or to race towards cover. The falcon must learn the advantages of height dominance and to throw up vertically after a stoop, ready for the next one. If she finishes the stoop below her prey, she will have a slog to re-engage, and if she goes wide, she may take some distance to get back in touch. A crow will get on the blind spot of a young falcon and it looks as though the crow is chasing her. In reality the crow is hiding behind her and gaining time to get closer to cover. Unfortunately you cannot do this manoeuvre with the roprey because the falcon can hear your engine. Or the crow will get into the falcon's no stoop zone, just below her, preventing her from turning over and lining up a stoop. Some of these tactics are brief and you may not notice what is going on if you do not understand the dynamics. But half a second is vital in this deadly game.

The pilot on the ground, with limited depth perception at a distance, cannot easily see exactly where his model is. You notice this when doing a dogfight between two models. You do a magnificent swoop at your opponent, only to realise that actually his model was ten metres further back than you thought. So, from where you are standing, it looked like a near miss, but for anyone viewing from the side, they can see you were hopelessly out. Fortunately this is not a problem normally, because it is the falcon doing the attacking and you only have to dodge.

During the race for height in the ringing flight, after a few rings with both falcon and prey closely matched, she may break off the pursuit and fly away two or three hundred metres. You and the prey may think she has abandoned the chase. The prey stops climbing so strenuously, waiting to see if the falcon has really gone. But watch her wingbeat. She is still climbing at a good climb angle, but on a straight line. Now she is curving round. She is coming back in and now the crow is below her maximum climb angle. She powers back across the sky, still climbing, and the crow cannot regain the height advantage that it squandered. Its only chance is to stoop for the ground as quickly as it can, with the falcon, a teardrop, on its tail.

Another outcome of the ringing flight is when the crow is well above the falcon and makes a break for it. We had a flight like this recently with a gyr/New Zealand falcon called *Herself.* After two rings the crow had a good start on her and decided to break for trees about 800m away. She followed underneath him, close to his blind spot and about 30 metres behind, both of them racing hard. But she knows this game; instead of striving for aerial dominance she stayed below the crow, preventing him putting into cover. So the flight went over Brieredge Farm trees, over Hesleyside, over the North Tyne River, climbing all the time. Eventually the crow had no choice. He had to come down and she could just foot him as he tried to brake his speed putting in to cover. A 3.2 km (2 miles) straight chase.

A clever falcon, such as a New Zealand falcon, may not do any of this. Maybe the crows are on the ground 300-500 metres away. If she flies towards them, climbing, they will head off immediately and maybe make cover before she can close with them. Instead, she flies away at 90 degrees to them, seemingly uninterested, but gaining height steadily. The crows stop feeding and start to eye her nervously. Then she drifts almost casually over them. Now they are in a bind. They are completely covered by her killing cone and whichever way they flush they are in trouble. So, with their heads drawn into their shoulders, they stay put. She canceleers down, preventing them deciding a direction to flush in, and simply nails one where it stands. Totally unsporting, but energy efficient predation! I did this once with a falcon called *Rebel* on a 600m slip, with Gordon Robinson and Dick Treleaven on Dartmoor and they didn't know what to make of it. On the other

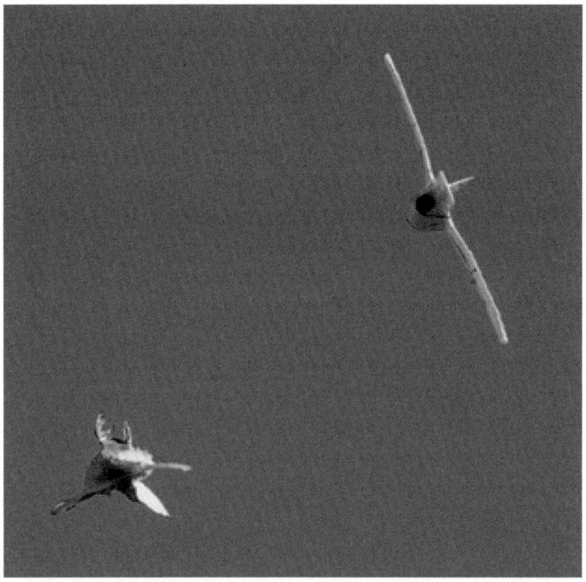

hand I have a ten year old peregrine/saker called *Flamenco* who will break off a tough flight to stoop at and kill an easier crow below her. This is annoying; your falcon should stay locked on through thick and thin. Merlins are notorious for refusing tough ringers as they grow older. You cannot blame them, and it is up to the falconer not to show them easier options, which is easier in theory than in practice.

You can teach her some of these tactics using the roprey. Sometimes the falcon must do the action first, either by instinct or by chance, and then you can reinforce it using operant conditioning. Some of it is a question of battling dogfights day after day in strong winds as she develops her skills in

Photo Stuart Grimwood.

turning adroitly and accurately. As she improves you will see her tail flaring open less and less. She will keep it tight closed because it creates drag, and after her first moult her new tail will be shorter.

9.14 Promoting Head Binds

The falcon should take the roprey by the head rather than by the tail or wing. It's more suitable if she's to be used for real falconry; it's more exciting and dramatic, it's safer as she can control her descent easier, and she's less likely to cause damage to the aircraft. Young falcons often start out as "tail tappers". They catch up to the roprey and lunge out a foot to hit the tail, either holding on or tapping it so that the aircraft falls to the ground. This is not a detrimental thing for the young falcon to learn, and it aids her confidence, but it is only a step on the way.

At low altitudes, approximately tree height, it can be difficult or impossible to recover airspeed for the roprey if the falcon hits the tail and points the model towards the ground. However, when you are taking the falcon higher, once she has tail tapped the roprey, you can recover and pull up without crashing. The model can then be flown underneath her giving the falcon a good view of the head with meat attached.

Another useful technique is to lift the head of the roprey slightly when the falcon is coming in level behind it. By reducing the power and pulling back on the stick the roprey can be put into a stall, where for a second or two, the model is almost stationary with the head higher than the tail. A skilled pilot can time this to perfection and the falcon, with a great view of the head, flies over the body to bind flawlessly.

9.15 Waiting-on Flights

If you want to develop a falcon for waiting-on flights, first train her normally until she has done all the basic lessons of climbing, head binds and so on. She should be happy to climb to 200 metres or so without missing a beat. From there, in the dogfights, you can start to impress on her the benefits of aerial dominance. She will start to appreciate that, when she is below you, she is in for a slogging climb, but once she has got above you, she can stoop; you are at her mercy. Soon she will develop a pattern of stooping, but be careful. If you are still well above the ground – say 50 metres or so – she can stoop vertically

A young gyr at Vowley races goes for the tail.
Photo Stuart Grimwood.

A juvenile female Australasian harrier grabs a pukeko.
Photo Mike Williams.

into you and give you a real hard wallop which could injure either of you. Make sure the head and shoulders of the roprey are well padded. Give her plenty of practice at stooping and always twist and dive at the last moment. This will make her stooping more accurate, and it will take the sting out of the impact.

Himself. Male gyr/NZ falcon.
Photo Stuart Grimwood.

By then she is ready to learn to wait on. Find a good slope with some lift. Have your prey hidden, maybe on a ground launcher. Hood her off. She will fly around wondering what to do. With any luck she will start to climb a bit, maybe just 100 metres or so. Give her a bit of time to sort herself out. It doesn't matter if she strays out a little way, you can always fly under her. Then serve her by flushing the roprey. This will help you set up a pattern, which is the key to making a game hawk. Being able to flush prey every day, when and where you want it, is crucial. Work in a different place each day. If you don't want to use a ground launcher, keep your roprey hidden under a cloak, then throw as needed. You can decide whether to reinforce a good stoop, or to make her work hard if she is out of position and not stooping well. You can also fly low, skimming the ground like a grouse, so she has to work out how to tackle it. Once the game season has started, you will still have odd days when you cannot go hawking. Use the roprey then to keep her up to the mark. Game falcons often don't get properly fit because they climb at their leisure and don't actually have much active air time. Extra weight that is solid muscle is just what she needs.

9.16 Accipiters

For shortwings you can either position the pilot behind some cover, or use the automatic launcher. Keep the prey close in front of the hawk and curl the flight around so that it finishes near you. As your hawk improves, you will find that you can get her to fly much higher and further than you have ever seen her go before and she will be able to tackle strong pheasants and ducks with confidence. Finally you have a way to get an accipiter really hunting fit!

9.17 Hunting Real Prey

Once you have your raptor chasing and catching roprey, the transition to live quarry is often seamless. The raptor is already set into the routine of chasing something flying when the hood comes off. She knows what to expect, what quarry looks like and what she can expect from it. She knows that if she puts the effort in, she can catch it.

Occasionally we drag a dead prey on a creance, or suspend it on a pole lure, and let her catch it, twitching the line so that it appears live and she will try to bite its neck. Make a fuss of her as if she was on a real kill. Now she is ready to go hunting without further training. Find her some good slips and reward her on the kill.

9.18 Rehabilitation

Roprey are now being used by serious rehabilitators in several countries. Active raptors, whether youngsters or survivors of injuries, need to learn to fly, become fit, and demonstrate their ability to survive before being released to the wild. Chasing them up and down aviaries is no preparation for the wild. There are still some rehabilitators, especially in Australia, who are so anti-falconry that they prefer to release unfit birds rather than face flying them properly in a soft release.

On non-hunting days the falcons can stay fit by catching roprey.
Photo Verity Johnson.

The best method we have found is to fly the raptor at roprey as normal, emulating the type of quarry she will take in the wild. When she is flying strongly and skilfully, with no hint of old wounds, fly her in the release area for some days. This will flush out any resident raptors who might resent her presence. Then attach the roprey on top of a post in the open, with food tied to it. Let her feed up on it and replace the food daily until she no longer comes for it. Other predators will be suspicious of the roprey and refuse to go near it, so if the food is disappearing you can be pretty sure it was her.

9.19 Pest Control

As well as producing roprey models for falconry use, we have developed Ropredators for bird deterrence. We are able to produce models that mimic any medium to large sized raptor, with Peregrines and Goshawks proving to be the most popular.

Raptors have been used in bird deterrence since at least 1967, and have proved to be effective in many circumstances. Wingbeat models aim to emulate the natural hunting of raptors, including the high speed stoop of the falcon. However, we also want to improve upon some of the characteristics of the raptor. Ropredators can be focused on one area, can never be fed too much, don't run out of energy, and don't fly off into the distance only to kill a gull and leave you with a long walk.

For pest control we can add flapping wing tips to the Roperegrine. These simple polythene or Xray film extensions give the wing shape a more pointed profile and flap passively once the model has airspeed. During the stoop they make a scary Brrrrr noise. But our tests so far with corvids and gulls tend to show that the flapping wings do not make much difference. What is important is the flight of the predator. If the predator makes moves towards the prey or attacks it, then the prey flees. On the other hand, if the predator just flies around, with or without flapping wings, corvids and gulls may well come in and mob it. This makes the problem worse, not better. Flying LOS (Line of Sight) limits the pilot to a radius of operation of around 300 metres maximum. This may be sufficient for a runway or land-fill site. For longer chases, the pilot needs First Person View (FPV) and the model must be capable of carrying it. We have done this with the Roperegrine but need to do more trials before offering it as a commercial option. We can also do an autopilot feature in which the Roperegrine flies on a predetermined route using waypoints, but although this allows coverage of a larger area, such as a field crop, orchard or vineyard, it does not include attack, so you need to intersperse it with LOS attacking flights, and it is complicated to set up. Patrol flights which do not include attacks aimed at specific prey, with or without flapping winged models, quickly trigger mobbing, or habituation.

A flight time of 10 minutes gives the pilot plenty of time to clear large sites of pest birds. The battery can be changed within 30 seconds and the predator can be airborne again. Small models can be seen at 200m altitude, the large model twice as high, allowing the pilot to take a commanding position in the sky. The smaller model is particularly versatile. It can be flown low and slowly, or can tear up the sky with repeated slashing 200 kph stoops.

Wingbeat Ropredators can be operated by anyone with some practice. It is easy for beginners to fly and does not require specialist staff. We have had ten year-olds flying with confidence in ten minutes. They can be flown in weathers that are unsuitable for live falcons, up to wind force 6 (48 kph/30 mph). They can also be flown from a moving vehicle allowing you to cover a huge amount of ground. Light rain is no problem and they can cope with fine sand.

These models have been tested in many thousands of flights. It is not unusual for us to carry out over 50 flights a day with a single rocrow when in the peak of the falconry season. The model is both the cheapest and safest available. Designed to be hit at high speed by a precious falcon, it takes minimal damage when flown into an obstacle – it can bounce rather than break. Designed to be smooth with no protruding or moving parts, it can safely be caught by hand. The model is also hand launched and so can be deployed immediately anywhere without the need for a runway. In contrast, model falcons that are ornithopters depending on flapping wings are very fragile, very expensive, almost impossible to repair, can only fly in 20 kph winds and require skilled pilots.

The model is designed with replaceable modules. Any parts that are broken can often be fixed on site with minimal down time. Parts can be quickly replaced without having to send the model in for servicing. A flight box is provided ensuring that all your parts, spares and batteries are kept secure. Our models have been flown in a temperature range of -10 to +40 degrees Celsius giving them an operational range much greater than that of a live falcon.

Several extras can be added to Ropredator models to make them better suited to a particular application. For example, a Rofalcon being used at an Air Base can have an autopilot fitted which allows you to pre-programme several routes. The model can be instructed not to fly over a sensitive area, such as a runway. A Rogoshawk flown at a land fill can be fitted with a First Person View (FPV) camera that streams live to the pilot. Pest birds can then be pursued at a distance, and roofs can be checked for nesting birds.

The use of live falcons hunting roprey at a site is also beneficial. Prey can identify whether a predator is actively hunting or just cruising around. A falcon chasing a rocrow is certainly hunting. So, instead of flying a trained falcon at seagulls on a landfill site, which can risk the flight going some distance away, it is possible to fly the falcon at robotic prey right amongst the gulls. This scares the gulls who see it as much as if they watched a real hunt.

9.20 Racing

As prey became scarce in the Middle East, hunting was actually banned in many regions. Yet the countries were committed to protecting their cultural heritage of falconry through UNESCO and the supply of domestic falcons was sustainable. Every evening the falconers would go out into the desert training their falcons by calling them to the lure or *telwah*. They might spend all winter doing this and hardly ever go hunting. So people started timing their birds to see how fast they were, and the sport of Flat Racing was born.

In the Flat Race or Telwah competition, the falcon flies 400 metres to the lure against the clock, breaking a start beam and a finish beam on the way. The falcons fly low, using ground effect to help their speed. This has two problems: one is that the event is not visual enough to be a spectator sport. The other is that falcons are selectively bred to fly low. This is not a good attribute for a hunting falcon.

As Flat Racing evolved, we were also developing roprey for falcon training and it soon became clear that roprey could be used to test the abilities of hunting falcons. It starts with a falcon being unhooded and chasing a robotic prey into the sky, climbing as quickly as it can until it reaches a pre-set height, usually 120 metres (400 ft). Once it has reached the set height, the pilot dives the prey. The falcon must stoop and pursue the prey, catch it in the air and land on the ground with it inside a 100 metre square landing area. The fastest falcon wins.

The falcon races suit all falcon species and can be divided into class categories depending on demand. The races must have clear start and finish end-points that can be accurately timed without dispute. They also allow electronic recording for post-flight analysis in case of any uncertainty. A falcon race shows a progressive top score as the event unfolds, so that competitors can see what time they must beat in the event.

Flat racing at Vowley.
Photo Stuart Grimwood.

Falcons waiting at the President's Cup, Abu Dhabi.

A gyrfalcon in the Flat Race at Vowley.
Photo Stuart Grimwood.

While Flat Racing tests the level speed of the falcons, Hunt Racing offers different elements:

- Hunt Racing tests the strength and fitness of the Falcon in the climb, its speed in the stoop and the accuracy of its footing of the prey. While a skilled pilot can help the Falcon to demonstrate this, ultimately the competition is designed to measure the ability of the Falcon as a potential hunter rather than the skill of the pilot.

- Hunt Racing provides an exciting and varied spectacle for viewers so that it is a fun event to watch and attend, thus maximising day attendees and demand for TV or film recording and hence sponsorship.

- Hunt Racing offers an element of uncertainty. This means that the recorded times of finalists can be several seconds apart, rather than milliseconds, and that an 'outsider' has a sporting chance of winning.

- Hunt Racing allows flights to be run in about six minutes, or ten per hour. This enables a significant number of competitors to participate in one day. Flat Racing can run up to thirty falcons in one hour.

- The Hunt Racing area, being limited to 100 x 100 metres, allows clubs to participate even if they do not have access to big open spaces.

- Hunt Racing provide different levels of difficulty so that, for example, breeders can showcase young novice falcons flying at the beginning of the season for buyers to evaluate them. It also provides tougher levels of competition for more skilled or more powerful falcons.

- Hunt Racing supports the tradition of real Falconry by having some events that are only open to proven hunting falcons, and by testing skills that a Falcon needs to be a real hunter. This is to avoid a division between 'competition' falcons and 'hunting' falcons, especially for breeders developing bloodlines.

- In Hunt Racing the Falcon must unhood cleanly and fly without hesitation the moment that the hood is removed. This is what we expect of a hunting Falcon.

- Hunt Races always take place within 100 metres of spectators and cameramen. This enables film crews to set up cameras in a predictable way for filming action.

- At larger events, the time clock and the GPS track of the falcon, can be shown live on screen direct from the timekeeper's receiver.

- While the body and wings of the prey must all conform to the same factory-made format, the colour or species of the prey are optional to allow individual preferences of each falcon, especially those that hunt. The motor and other parts of the roprey are optional to encourage progress and development in the performance and reliability of the models. Making an engine more powerful does not necessarily confer an advantage, if the falcon itself is not powerful. If the prey gets too far ahead of the falcon, the falcon may be de-motivated and slow down. Also a more powerful motor could drain the battery more quickly, potentially leaving the pilot in difficulties at the end of the flight.

- The model prey provides a platform for team colours and logos, and for logos from sponsors. These logos will be fully visible in photos and films of the event. This provides an incentive for sponsors to support a falcon or team.

The Hunt Race starts with the Judge checking the roprey, that it complies with specification and is safe with the foam head, and that the falcon is fitted with a barometric altimeter tag accurate to 10 cm. Pilot and falcon enter the Hectare, the 100 x 100m landing arena. They choose a spot facing the wind for launching. When ready, they raise their hands, the first horn sounds and the clock

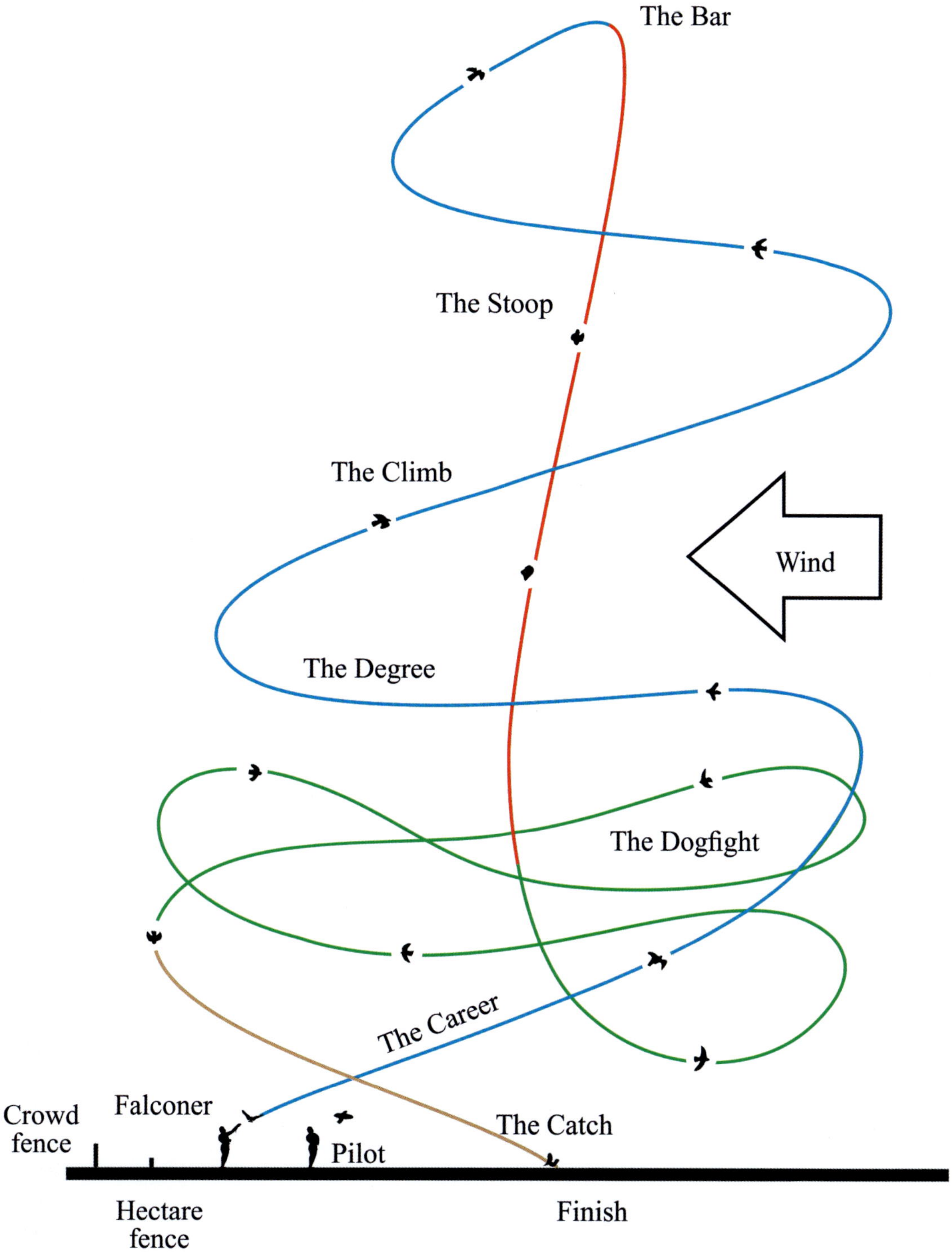

Figure 9.20.1 The Hunt Race.

The Pilot, Dylan Freeman, launches the roprey.
Photo Stuart Grimwood.

starts. The pilot must coax the falcon up to the height bar, the altitude set for the event. This is usually 120 metres (400 ft). The race is usually won or lost in the climb and the falcon must be fit and experienced. Climbing at the optimum angle into the wind the pilot can usually only reach 60 metres or so before being obliged to turn across or downwind.The upwind climb is known as the 'career' and the downwind leg is the 'degree'. This loses precious seconds before the falcon can turn and perhaps reach the bar height. The wind is stronger up there and the last few metres may be painfully slow. Once the falcon reaches the bar the second horn blows and the pilot dives, with the falcon stooping on his tail. A good fast vertical stoop can save some seconds and as the falcon descends past 30 metres altitude the third horn goes. The pilot is now in a difficult position. He must throw up immediately to avoid hitting the ground but he has a determined falcon right behind him who could easily catch him near the stall. So he must maintain airspeed and twist away, trying to win himself some space. He must avoid being caught for thirty seconds during the Dogfight. This takes place near the crowd. The pilot cannot simply fly off, trying to keep away from the falcon, because he must be back near the Hectare after thirty seconds. The clock is being counted down and the Pilot must try to line up low and near the arena ready for a quick catch. The fourth horn goes after thirty seconds of Dogfight and the Pilot must get the roprey caught and both falcon and prey on the ground as soon as possible. If he comes in up-wind there is a danger that the falcon may grab the roprey and carry it out of the Hectare over the white tape. This is 30 cm above the ground and 5 metres beyond it is the crowd barrier. So there is a strong chance she will just tip over the tape and stop at the barrier, incurring penalty points.

The record for the Hunt Race is currently held by *Swallow*, a male gyr/peregrine, bred and trained here, with a time of 90 seconds. Of course a lot depends on the weather. A gentle steady breeze is best. Rain or strong wind is no good and a flat calm is a slog. As each year goes past, people are upgrading the technology and techniques. Measuring wind strength and air density will become part of it, like recording the ground conditions for a horse race.

The Falconer, Mary Craney, unhoods and slips the falcon.
Photo: Stuart Grimwood.

The climb. *Photo Stuart Grimwood.*

The second horn starts the stoop.
Photo Stuart Grimwood.

The third horn starts the Dogfight. Male gyr/peregrine *Swallow* lines up for another attack.
Photo Stuart Grimwood.

Pilot and Falconer have a tense moment at the end of the Dogfight.
Photo Stuart Grimwood.

The fifth horn goes as the falcon touches the ground with the roprey.
Photo Stuart Grimwood.

Contestants 2020.
Photo Adam Barnard.

9.21 General Racing Rules

As racing became more popular the rules became standardised. The General Racing Rules cover both racing disciplines:

1 The Trainer is the person acting individually or leading a team of people who have trained one or more falcons together throughout the training period.

2 The Team is the Trainer and support crew who have trained one or more falcons together for a race on behalf of one or more Owners.

3 A Team cannot include two independent Trainers.

4 In the Flat Race the Falconer is the Team member luring and retrieving the falcon.

5 In the Hunt Race the Falconer is the Team member slipping and retrieving the falcon.

6 A Team can engage one or more Pilots but a Pilot cannot fly for more than one Team in any given race.

7 The Owner is the Owner of the falcon at the start of the race.

8 An Owner can own more than one falcon and a falcon can have multiple owners.

9 The Owner of the falcon is the entity who wins the prizes.

10 The Breeder is the person or stud that hatched the falcon.

11 The falcon must have a means of attaching a GPS tag supplied by the race management.

12 Each entry on the race card must include: the falcon's name, ring or microchip number, species (including hybrids), gender, age, the Falconer's name, the Trainer's name, the Owner's name and the Breeder's name. A wild-caught falcon shall be listed as 'wild-caught' under Breeder's name. For Hunt races, the Pilot's name must also be included.

13 Open Races are open to falcons only, of any species, gender or ages.

14 Novice Races are open to falcons not previously raced competitively.

15 The Contestants must provide their own insurance for any claims whatsoever against them.

16 The Organisers shall not be responsible for any injury or loss to any falcon, howsoever caused.

17 The falcon must carry a non-removable official registered ring or microchip identifying it as the individual described in its legal papers.

18 If offered for sale the falcon must have any legal documents required to make the sale.

19 The Owners' Racing Colours may be displayed on hoods or the uniforms of participants. Pilots and Falconers may wear sponsorship liveries, logos or racing colours.

20 The Order of Flying will be declared one hour before the start of the competition drawn from a hat by the Timekeeper. Wherever possible no two falcons from one Team will be flown consecutively.

21 If the weather changes during the course of the competition, no allowance will be made to any competitor except that the Timekeeper may decide to temporarily halt the competition or cancel it altogether because conditions are unsafe, for example low fog, storm or heavy rain.

22 The wind strength at two metres must not exceed Force 5 i.e. 38km/hr, 24 miles/hour or 10.7 m/s and the temperature must not exceed 28°C.

23 The winner is the falcon with the lowest time score.

24 The Timekeepers must have no declared interest in any of the competitors.

25 If the falcon wanders off, the Timekeeper will decide whether or not to fly the next falcon.

26 If a lost falcon is still not recovered, the Trainer of the next falcon has the right to refuse to fly because of risk to his own falcon. In this case his falcon will be moved to the last position in the Flying Order.

27 All filming and media rights belong to the organisers of each event.

9.22 Flat Racing Rules

These rules are in addition to the General Racing Rules.

1 The Flat Races can include events for different ages or species of falcon and for different classes of competitors.

2 The race distance is normally 400 metres along a smooth, level track.

3 The start and finish lines are infra-red or radar beam gates connected together to record timings to 0.001 seconds accuracy and screened at 0.01 seconds accuracy.

4 The falcon must pass through both start and finish gates.

5 The start gate is 5 metres wide and the Falconer must stand no closer than 5 metres from the beam and no further than 20 metres from the beam.

6 The finish gate is no more than 30 metres wide and 30 metres high at the beam.

7 The Falconer can call the falcon using only the lure (Telwah) and audible calls, and must stand behind the finish gate.

8 The dress of the Falconer can be any type specified by the race Organisers.

9 On unhooding the falcon, the Falconer must keep his hand static. Attempts to launch the falcon from the hand will disqualify the entry.

10 A falcon which fails to start after two minutes from being unhooded will be disqualified.

11 If a falcon fails to complete the course within three minutes it will be disqualified.

12 If the falcon wanders off, the Timekeeper will decide whether or not to fly the next falcon.

Tom Spink slips one of our falcons at the King Abdulaziz Falconry Festival.

9.23 Hunt Racing Rules

These rules are in addition to the General Racing Rules.

1 Any qualifying falcon may be entered into 'Open classes.' 'Restricted classes' limit entries to falcons of a specified class.

2 The feet of the falcon must be in natural hunting condition, that is to say, with sharp claws and fully functional toes. Fitting beads to the claws or any device or surgery that prevents the falcon from having full use of its feet will disqualify the falcon.

3 The Pilot is on the same side as the falcon and can be an in-house Pilot or a freelance Pilot, the same as jockeys in a horse race.

4 A 'Team' at minimum consists of the Falconer who slips the falcon and the Pilot. It may also include the Owner or Owners and the Trainer, if different.

5 An Owner can have falcons in more than one Team.

6 The Falconer, Pilot and Trainer can act only in one Team during a given Race.

7 The number of falcons entered by one Owner in a race may be restricted by the Organisers.

8 The Pilot can fly for more than one falcon in the same Team.

9 The Pilot can fly for more than one Team in Qualifying races, but is restricted to one Team in Finals.

10 The roprey is supplied by the Pilot or the Trainer, not by the competition Organisers. The EPP wings, fuselage and tail of the model must be unmodified factory models supplied by Wingbeat. The motors, electronic controls, batteries, heads and airbrake springs can be modified from factory issue. The transmitter can be any model. Any feathers added to the model must be non-detachable and must not extend more than 5 cm beyond the end of the tail. No modification will be allowed if, in the opinion of the judges, the model risks injury to the falcon or spectators.

11 All models must have properly padded heads to prevent injury to the falcon or spectators.

12 The models may carry any amount of logos.

13 In the event that the model malfunctions for any reason, this is the fault of the competitor, not the Organisers.

14 While a race is in progress no person may distract the falcon in any way, including showing a lure or roprey, or feeding a falcon, within 300 metres of the Landing Square. If a competitor files a formal complaint of distraction, and it is upheld by the Timekeeper, the competitor will be allowed a re-run and the distractor, if a member of a competing team, will have his or her falcon(s) disqualified.

15 If the falcon wanders off and the Pilot runs out of battery and is forced to land, the falcon is declared 'lost' and is disqualified. The Timekeeper will then consult the GPS location of the falcon and decide whether or not to fly the next falcon.

16 In the event of any cheating, irregularity or uncertainty the Timekeeper's decision will be final.

17 The standard Landing Square is 100 x 100 metres marked out by corner posts and white tape 30 cm above ground level so that the falcon cannot go underneath it. Outside the tape is a 5-metre-wide clear zone separating the tape from the crowd barrier. This zone must be kept completely clear at all times.

18 The falcon must be fitted with a standardised GPS tag with a barometric altimeter transmitting the height reading down to the Timekeeper in real time. This tag is supplied by the race Organisers.

19 The classes will be 120, 200, 300, 400 and 500 metres altitude for the top height.

20 The model can be any colour.

21 The model can include an altimeter allowing the Pilot to know the height of the model.

22 Only the Falconer, the Pilot and the Timekeeper are allowed inside the Landing Square.

23 The competition starts with the Falconer and the Pilot standing inside the Landing Square facing the wind. They can be any distance apart.

24 The Pilot can be in any position in relation to the falcon as long as both are within the Landing Square.

25 The braces on the falcon's hood can be open but the hood must be on, with the Falconer's free hand at his side. When the Falconer and the Pilot are ready to slip, they must both raise their hands to signal to the Timekeeper.

26 Once both hands have been raised, the Timekeeper sounds the start horn and starts the clock. The Pilot launches the roprey and the falcon is unhooded.

27 If the Pilot has a mis-launch, one more attempt is permitted provided that the falcon has not left the Falconer's fist. Once the falcon is in the air the flight is underway and cannot be repeated. In the event of a mis-launch, the clock continues running from the initial horn time.

28 The Pilot can do anything he likes to encourage the falcon to reach the top height as quickly as possible and use any part of the sky, including outside the Landing Square.

29 When the falcon reaches the top height, the Timekeeper sounds the second horn.

30 The Pilot must then immediately dive to encourage the falcon back down as fast as possible. When the falcon has descended past 30 metres altitude the Timekeeper sounds the third horn. At this point there is a break of 30 seconds during which time the roprey can do anything it likes but must not get caught. The reason for this is to reduce the risk of the falcon being injured by hitting the roprey in a hard stoop, and to provide spectators with the excitement of an aerial combat at close quarters, demonstrating the flying skills of the falcon.

31 If a falcon touches the roprey at any point after the third horn the pilot may ditch the model any time after the fourth horn.

32 Thirty seconds after the third horn the Timekeeper sounds the fourth horn. The falcon must catch the roprey in the air in the shortest time possible after the fourth horn. The fifth horn sounds at the moment the falcon touches the ground holding the roprey. This is the Finish.

33 The winner is the falcon with the shortest total time between the first (start) horn and the fifth (finish) horn. The logged time includes the mandatory 30 second chase period between the third and fourth horns.

34 If at any time the falcon touches the prey but does not bind to it, the hunt can continue.

35 If the falcon catches the prey and lands on the ground with it before the third horn, the falcon is disqualified.

36 If the falcon catches the prey between the third and fourth horns and lands with it before the fourth horn it incurs time points equivalent to the number of dogfight seconds remaining, times ten.

37 All parts of the model must be clear of the ground at the moment that the falcon catches it.

38 The Pilot may use the airbrake, or reverse thrust, but nothing external to the model, such as a parachute.

39 If the roprey hits the ground without the falcon having touched it, the falcon is disqualified.

40 If the falcon hits the roprey to the ground but does not bind to it, the clock continues until the falcon has landed on the roprey on the ground.

41 The decision as to whether or not the falcon is holding the roprey on landing is made by the Timekeeper who may consult action replay footage.

42 If the falcon carries the roprey and lands outside the Landing Square, the falcon incurs 15 seconds time faults.

43 If the landing of the falcon with the roprey on the ground outside the Landing Square cannot be seen by the Timekeeper the landing time shall be determined from the GPS recording.

9.24 Further Reading

Akos, Z., M. Nagy, and T. Vicsek. 2008. Comparing bird and human soaring strategies. Proc. Natl. Acad. Sci. USA 105: 4139-4143.

Bradbury, T. 1996. *Meteorology and flight: a pilot's guide to weather*. Bloomsbury. ISBN 9780713644463

Brighton, C.H, K.E. Chapman, N.C. Fox, and G.K. Taylor. 2021. Attack in naive gyrfalcons is modelled by the same guidance law as in peregrine falcons, but at a lower guidance gain. Journal of experimental biology 224(5).

Cosgrove, B. 2013. Microlight Pilot's Handbook. (8th edition). Crowood Press. 978-1847975096.

Franklin, K. 1999. Vertical Flight. NAFA Journal 68-72.

Hardaswick, V. and K.L. Christopher. 2011. *High Flying Gyrfalcons*. Western Sporting. ISBN 978-1-888357-13-4.

Hoey, R.G. 2010. Exploring bird aerodynamics using radio-controlled models. Bioinspiration & Biomimetics 5(4):008.

Kane, S.A. and M. Zamani. 2014. Falcons pursue prey using visual motion cues: new perspectives from animal-borne cameras. Journal of Experimental Biology 217: 225-234.

Kermode, A.C. 1989. *Flight without Formulae* (5th edition). Longmans. 978-0582026988.

Pareja-Obregon de los Reyes, M. D. 2013. *Cincuentaanos de cetraria*. Egondi Artes Graficas. ISBN 978-84-616-6930-1.

Pennycuick, C.J. 2004. *Modelling the Flying Bird*, (Elsevier, ed. 1, 2008), Vol. 5.

Ruxton, G.D and D.C Houston. 2004. Obligate vertebrate scavengers must be large soaring fliers. Journal of Theoretical Biology. 228, 431–436.

Sale, R. 2016. *Falcons*. Harper Collins. ISBN 978-0-00-751141-9.

Scott, D.A. 2014. *Airplane and Radio Setup*. 1st US R/C Flight School. ISBN 978-0-9817226-6-5.

Scott, D.A. 2014. *One Week to Solo*. 1st US R/C Flight School. ISBN 978-0-9817226-1-0.

Scott, D.A. 2014. *Park Flying 1-2-3-D*. 1st US R/C Flight School. ISBN 978-0-9817226-0-3.

Scott, D.A. 2014. *Sport Aerobatics*. 1st US R/C Flight School. ISBN 978-0-9817226-2-7.

Scott, D.A. 2014. *Precision Aerobatics*. 1st US R/C Flight School. ISBN 978-0-9817226-3-4.

Scott, D.A. 2014. *Advanced Aerobatics*. 1st US R/C Flight School. ISBN 978-0-9817226-4.1.

Stringwell, G. 1998. *A Complete Guide to Radio Control Gliders*. Nexus.

Thom, T. 2021. *The Air Pilot's Manual 1. Flying training*. Air Pilot Publisher Ltd. ISBN 9781843360643.

Thom, T. 2015. *The Air Pilot's Manual 2. Aviation Law and Meteorology*. (17th Edition). ISBN 9781843362401.

Tucker, V.A. and C. Heine. 1989. Aerodynamics of gliding flight in a Harris Hawk *Parabuteo unicinctus*. Journal of Experimental Biology 149:469-489.

Tucker, V.A. 1990. Body drag, feather drag and interference drag of the mounting strut in a peregrine falcon *Falco peregrinus*. Journal of Experimental Biology 149:449-468.

Tucker, V.A. 2000. Gliding flight: drag and torque of a hawk and a falcon with straight and turned heads, and a lower value for the parasitic drag coefficient. Journal of Experimental Biology 203:3733-3744.

Tucker, V.A. 2000. The deep fovea, sideways vision and spiral flight paths in raptors. Journal of Experimental Biology 203:3745-3754.

The goshawk binds to the pheasant before it can clear the trees.
Photo Steve Magennis

C10 Hawking in the Field

This chapter is supported by two films:
The Northumberland Crow Falcons and *Hawking in China.*

10.1 Introduction

In chapter 8 we looked at the different ways in which birds of prey hunt in the wild and the potential that each species has. Now we can look in a practical way at how these abilities can be harnessed in falconry.

Some of the methods, such as stooping, are visually more exciting than others, such as dropping onto a dead sheep. Also, some species lend themselves better to a partnership with an earth-bound falconer than do others. A goshawk sitting quietly on a branch and then dashing out 50 metres after a rabbit is easily transferable to a falconry situation whereas a harrier patrolling perhaps a kilometre of moorland to dive onto a mouse would tax the patience and fitness of the keenest falconer. Therefore many, or most, of the species of birds of prey eliminate themselves from the falconer's mews because their natural hunting methods are too unexciting or impractical.

The falconer may have some preferences over the kind of flights he is interested in. He may be so obsessed with the stoop that he regards any subsequent tail-chase if the stoop misses, as a vice. Or he may want a hawk to fly solely from his fist and kill prey in a short sprint, rather like using a shotgun. He may prefer to do all the searching for the hawk himself, only slipping the hawk once the quarry is located, or he may require the hawk to search for the prey as well as to attack it.

The hunting behaviour of raptors is geared to the terrain they inhabit, as is the escape behaviour of the prey. To expect a raptor to function well in the wrong terrain, for example a peregrine in a wood, is to handicap her. Similarly, to find prey in the wrong terrain, such as a duck over land, is to make it vulnerable.

Some hawks, such as accipiters, red-tails, Harris hawks and New Zealand falcons, take a wide variety of prey in a wide variety of habitats. It is obvious therefore that these are more likely to be successful hawking birds for the 'common man.' Waiting-on flights at game, such as partridges or grouse are much more specialized and therefore much less easy to accomplish. Access to suitably stocked land is limited and often expensive and the campaign needs to be

Harry McElroy quail hawking in Arizona.

fairly intensive if the falcons are to stay on form and not lower their pitch. If grouse or partridges are scarce or uncooperative, the game hawker is doomed to disappointment. He may spend all afternoon trying to get a proper 'set up' for his falcon: a firm point and a clean flush in the open. Most of the quarry seen during the day will remain unflown at. The goshawker on the other hand, if partridges are scarce, looks for rabbits or pheasants, woodcock, ducks, yes, even waterhens....

Provided the ingredients are all present and correct, waiting-on flights at game are not too difficult; success rates of 90 percent or more have been achieved by experienced teams although 50 percent is more normal. But when one or more of the ingredients are wrong, nothing the falconer can do will bring success; it is an intractable problem. With a goshawk, if game is scarce or cover is thick, one can hobble along one way or another, but unless a game falcon is served day after day with quarry the whole scenario is a failure.

It has been my experience that the most effective falconers are those from a shooting background. They have a knowledge of the quarry species which is invaluable in the field. Others, with no previous experience of fieldsports, progress nicely until it comes to entering the hawk at quarry; then they hit a brick wall. Many falconry books are reticent on this, often only one chapter is devoted to actual hawking. There are some people who can almost go out with a hawk which is blind in one eye and with a wooden leg and yet come home with something in the bag. Somehow the quarry always flushes in their direction, they happen to be in the right place at the right time. It is not a matter of luck, although at first it may seem so. It is a question of field experience and persistence. Others can go out with the fanciest hawk they can buy and the most they ever catch is a cold. They are not tuned in to the hawk or the quarry, or the relationship between the two.

Other authors, far more experienced than I will ever be, have described how to undertake classic flights with falcons and hawks. Many more people than ever before have access to peregrines and other powerful raptors, but do not have the lifestyle or the finances to achieve those flights which so delight the purist. The result is considerable dissatisfaction and frustration. Probably nine out of ten of the peregrines taken up for training by would-be falconers fail to take a dozen head of quarry and such people are quick to receive the derision of those in a more fortunate position. This is sad. So many flights, easy to describe in their theory, fail to work out in practice, at least on a repeatable basis.

What I would like to do in this section is first to show how, by using its potential more advantageously, the hawk can successfully take quarry in situations often thought of as impossible, and second, to enhance the falconer's appreciation of his or her bird and its abilities. There are many falconers, experienced ones too, who have never seen for example a glide attack or an indirect flying attack, who cannot recognize the field conditions which make them appropriate and who have never attempted deliberately to achieve them. This reduces both their effectiveness in the field and the variety of flights for them to enjoy. To maintain a falcon for waiting-on flights, her hunting behaviour must be carefully conditioned to remain stereotyped. But in almost all other types of hawking a hawk cannot be considered accomplished unless it is capable of taking a wide range of quarry species under a variety of field conditions and by a full repertoire of strategies.

10.2 The hawking year

Hopes and dreams - they are what falconry is built on. I'm a dreadful dreamer. Spring is the worst time, when I'm making plans for the forthcoming season and watching those embryos grow. I think to myself: "I'll hawk pipits in the mornings with a merlin, grouse in the afternoons and take the goshawk for a quick look around before it gets dark." I've done it too - for about three days! But I can't keep it up, and once the holiday is over and the working week rules, it is hard enough just keeping one hawk going.

Old falconers warn about this and young would-be falconers disbelieve them until they've tried it for themselves.

Perhaps the best way to show how hawking works out in practice, is to give figures based on my hawking records and the experiences of other falconers (Table 10.2.1). Falconry should not be undertaken in a spirit of competition but at the same time, before taking on a hawk, or attempting a particular type of quarry, the falconer needs a realistic idea of what it entails in terms of personal commitment and access to hawking grounds.

A\B. Starting dates for gamebirds are dictated by the quarry and their legal seasons. An August start benefits from the long days, good weather and inexperienced quarry and early finish gives the hawks a good settling in period for breeding or moulting. In arable country it may not be possible to start much before Christmas owing to problems of access on land. In Britain, small bird hawking licences do not start until 1st September, at the moment.

C. When grouse-hawking the falcon is normally flown on every day legally possible for a short season. With rook hawks, goshawks, and sparrowhawks, one may achieve 6 or 7 days per week early in the season but only 1-2 later in the season owing to short days or bad weather. Rook hawks require stooping to the lure or flying at roprey on non-hawking days. Buteos and Harris hawks need more flying than accipiters to maintain fitness. Once the dark evenings draw in and prevent week-day hawking, these hawks need high jumping to maintain fitness.

The hawking year - expectations

		Merlin at larks	Falcon grouse game	Falcon at rooks/ crows	Goshawk at mixed game	Sparrow hawk	Red-tail/ common buzzard	Harris hawk
A	Start hawking	1 Aug	12 Aug	10 Aug	10 Aug	10 Aug	10 Aug	10 Aug
B	Finish hawking	15 Sep	10 Dec	10 Jan	30 Jan	30 Jan	30 Jan	30 Jan
C	Number hawking days/week	7	6	3-7	3-7	3-7	5-7	2-7
D	Number hawking days/season	46	23	66	75	75	125	125
E	Flights/day	1-5	1-5	2-3	2-10	2-10	0.5-5	1-10
F	Flights/kill	9-1.5	8-1.5	5-2	10-2	10-4	6-2	6-2
G	Kills/season	5-40	5-40	15-70	30-150	15-100	10-80	30-150
H	Acres required	500+	3000+	10000 per month	1000	500	1000	1000
I	Types of ground	Open downland. Stubbles. Close-cropped moorland	Dogging grouse moorland	Very open arable or grazing land	Almost anywhere	Same as goshawk	Same as goshawk	Same as goshawk
J	Dogs required	Pointer optional	Pointer essential	None	Pointer or spaniel desirable	Flushing dog desirable	Pointer or spaniel desirable	Pointer or spaniel desirable

Table 10.2.1 is based on hawking in Britain and will obviously vary from country to country. It is instructive to jot down a few figures for your own hawking expectations and then review them again at the end of the season.

D. Quite often the hawking season will be reduced owing to a variety of hazards: prolonged bad weather, injuries, mother-in-law visits, etc. One is lucky to have a full season.

E. The first figure is a minimum for successful hawking. Beyond the second figure the hawk is possibly being overworked.

F. The first figure is for an inexperienced bird. If the hawk averages worse than this something is seriously wrong and the hawk may start refusing quarry. You should analyse your whole set up. The second figure represents a good skilled bird. If the hawk averages better than this you have setup the quarry so that it is overmatched; a happy position to be in, but is it sporting?

G. The first figure is for an inexperienced bird or handler who has had enough success to build on for the future. Below this figure for a full season one or more factors are significantly wrong. The second figure represents a good hawk well flown. In some countries there are bag limits eg. for larks and blackbirds. Taken on average throughout the season the hawk's flights: kill ratio and total score is a fairly good indication of the quality of sport enjoyed. A consistent stylish performer will be nearer the right-hand category than the left.

H. The areas quoted are for each bird flown and vary tremendously according to its game capacity and quality. With merlins and sparrowhawks it is best not to work an area more than twice a week, grouse hawks not more than once a week, partridge hawks, rook hawks, goshawks, buteos, and Harris hawks not more than once a fortnight. Six or seven kills by a rook hawk on a thousand acre patch will so educate the rooks that the falcon can no longer get near them and they are virtually unhawkable, crows are even less cooperative. A rabbit warren ferreted more than once a month or twice a season will start to show poor results. But where there is a temporary glut of game it may be best to hawk more frequently while the going is good. By November most game populations show a marked reduction; other predators have been taking their toll. In the game rich areas of Britain, particularly in the east, scores ten times larger than in the more desolate areas can be achieved. Where quarry numbers are too low to achieve the minimum number of flights in column E, hawking will be unsuccessful.

10.3 What the hawks can do

The whole essence of falconry lies in establishing the right balance between predator and prey. In this it is simply a mirror of the balance of nature, honed by millions of years of evolution acting on the players of this endless game. Weight the balance too much in favour of the hawk and she would kill too much and soon wipe out her prey resource and herself. Weight it too much in favour of the prey and the hawk would be unable to kill and would die of starvation. The prey numbers, no longer dampened by predation, would be controlled primarily by food supply and would tend to become very cyclical with population explosions followed by drastic starve outs, some of them possibly terminal to the gene pool.

In falconry we examine this balance between predator and prey, between hawk and quarry, very precisely indeed. This is one of the things which makes falconry so fascinating, and curiously, it is an aspect which seems to be scarcely covered at all in the ornithological literature.

Sage grouse have proven to be a prime quarry in USA.

When a man goes out with a gun, all he has to do is to get his quarry to expose itself within the range of his gun for long enough for him to fire the shot. The shot, whether bullet or bird shot, travels faster than the speed of sound and the prey literally does not know what has hit it. Thus all the survival skills obtained by instinct and experience to avoid predators are for nothing. How can you avoid a missile traveling so fast? Therefore the gun hunter, who might imagine he understands a lot about his prey, only actually appreciates half the story. He is at the same level as the man with a camera or field glasses.

The situation which to the gun hunter is the end of the hunt (the view and the shot) is to the falconer actually the start of the hunt. This is the point when the hawk sees its prey and starts off in pursuit, matching its own abilities with those of its prey. The prey is exposed to predators all the time and it has plenty of ways of escaping. If it is sick, handicapped, or unlucky, it may be killed. If not, it has a good chance of surviving completely unscathed and may be back feeding in the same familiar area a few minutes later. The chances of it escaping injured are probably less than 0.1 percent compared to about 25% with shotguns.

In order to understand this very fine balance, first we will examine the ways some particular species of raptors hunt using the methods described in chapter 8. Then we'll look at the ways some of the prey escape. Of course we cannot cover all hawks and all prey but just sufficient to give an idea of what goes on and enough to get you thinking about your own particular bird and its prey. Once we've painted this picture with a fairly broad brush, we'll go into some of the flights in more detail and show how they have become traditional flights because of this very fine tuning between hawk and quarry.

The Tables 10.3.1. and 2. show approximately how the more common species of falconry birds perform at various quarry species. Don't think that just because a particular quarry is abundant in your area, then that is the best quarry to hawk. The prey may not be available to the raptor because the terrain may be unsuitable for its hunting methods.

In chapter 8, I outlined the different basic techniques commonly used by wild raptors to

W = Waiting on D = Direct pursuit E = Excellent M = Moderate O = Occasional P = Poor

Table 10.3.1.	Peregrine		Gyrfalcon		Saker		Prairie		Lanner		Lugger		NZ falcon		Aplomado		Red-headed merlin		Merlin		Eurasian kestrel	
	M	F	M	F	M	F	M	F	M	F	M	F	M	F	M	F	M	F	M	F	M	F
Partridge	WE	WE	WDE	WDE	WM	WM	WE	WE	WE	WE	WM	WM	WDE	WDE	-	WDE	-	WDO	-	-	-	-
Red grouse	WE	WE	WDE	WDE	WM	WM	WO	WE	-	WO	-	-	-	WDE	-	-	-	-	-	-	-	-
Sharp-tailed grouse	WE	WE	WDE	WDE	WM	WM	WO	WE	-	WO	-	-	-	-	-	-	-	-	-	-	-	-
Sage grouse	WP	WE	WE	WE	WP	WM	-	WE	-	WE	-	-	-	-	-	-	-	-	-	-	-	-
Pheasant	WO	WE	WDE	WDE	WO	WM	WP	WM	-	WO	-	WO	-	WDM	-	-	-	-	-	-	-	-
Mallard	WO	WDE	WDE	WDE	WO	WM	WP	WE	-	WO	-	WP	-	WDE	-	-	-	-	-	-	-	-
Teal	WE	WDE	WDE	WDE	WM	WM	WE	WE	WE	WE	WO	WO	WDO	WDE	-	WDO	-	-	-	-	-	-
Woodcock/snipe	WE	WM	WDO	WDO	WO	WO	WE	WE	WM	WM	WP	WP	WDO	WDM	WDO	WDE	DO	DO	-	-	-	-
Stone curlew	DE	DM	DM	DO	DE	DE	DE	DE	DO	DM	DO	DO	DO	DE	-	-	-	-	-	-	-	-
Houbara	DP	DE	DM	DM	DO	DE	-	DE	-	DO	-	-	-	-	-	-	-	-	-	-	-	-
Magpie	WDE	WDO	WP	WP	WDM	WP	WDM	WDM	WDM	WDM	WDM	WDM	DO	DE	WDP	WDO	WDP	WDO	-	-	-	-
Crow/rook	DM	DE	DE	DE	DE	DE	DO	DE	-	DO	-	DO	-	DE	-	-	-	-	-	-	-	-
Herring gull	DE	DE	DE	DE	DE	DE	-	DE	-	DO	-	DP	-	DE	-	-	-	-	-	-	-	-
Skylark	-	-	-	-	-	-	WE	WO	WE	WO	WDM	WDM	DM	DM	DE	DP	DE	DE	-	-	-	-
Hedgerow birds	WP	-	-	-	WO	-	WE	WO	WE	WE	WM	WM	WDE	WDM	WDM	WDM	WDM	WDM	DM	DM	WO	WO
Starling	DE	DP	-	-	-	-	WE	WO	WE	WO	WO	WO	WDM	WDM	DM	DM	DM	DM	DM	DM	-	-
Rabbit	-	WP	WDO	WDO	DP	DO	-	DO	-	DO	-	DO	-	DM	-	-	-	-	-	-	-	-
Brown hare	-	WDP	WDE	WDE	-	WDE	-	DM	-	DO	-	-	-	DO	-	-	-	-	-	-	-	-
Mice/voles	-	-	WDO	WDO	WDE	WDE	WDE	WDE	WDE	WDE	WDE	WDE	WDE	WDE	-	-	-	-	-	-	WE	WE

Table 10.3.1 Falcons and their prey.

find and attack prey. Some of the techniques, such as the slow steady quartering of harriers, do not lend themselves to falconry. Others, such as the sprinting direct attack of the accipiter, are suitable for hawking in all sorts of situations. It is best if the falconer asks his trained hawk to do as similar a flight in falconry as it would in the wild. For example, if you want a really good pointing dog, train a pointer, not a terrier. You want an animal with the right build and the right instincts for the task.

Following from this have been the traditional divisions of falconry into different types of flight. On the one hand are the high flights with falcons, and on the other, the low flights with accipiters and accipitrid hawks. Both of these are pursuit flights; the sustained pursuits of the falcons are usually aerobic flights whereas the shorter sprints of the accipiters tend to be anaerobic. The third type of flight is the waiting-on flight undertaken by many falcon, buzzard, and eagle species. This is the high search followed by a stoop or dive.

Of all the flights in falconry, the traditional classic flight is the high pursuit flight, culminating at its best in the ringing flight or *la haute volée*. In days gone by, the quarry included kites, herons, some owls, and cranes. Nowadays the classical flights are obtained principally with corvids,

E = Excellent M = Moderate O = Occasional P = Poor

Table 10.3.2	Harris hawk		Red-tail		Common buzzard		Ferruginous hawk		Goshawk		Black sparrow-hawk		Cooper's hawk		Sparrow-hawk		Shikra	
	M	F	M	F	M	F	M	F	M	F	M	F	M	F	M	F	M	F
Quail	M	M	O	O	-	-	P	P	E	M	E	E	E	E	O	E	-	O
Partridge	M	M	O	O	P	P	P	P	E	E	E	E	E	E	P	E	-	P
Red grouse	P	P	P	P	-	-	P	P	E	E	E	E	-	P	-	-	-	-
Pheasant	O	M	O	O	P	P	P	P	E	E	O	E	P	M	-	-	-	-
Mallard	O	O	O	O	-	-	-	-	E	E	O	E	P	E	-	P	-	-
Teal	P	P	-	-	-	-	-	-	E	E	E	E	E	E	-	O	-	-
Snipe	P	P	-	-	-	-	-	-	P	P	P	P	O	O	-	O	-	-
Woodcock	O	O	-	-	-	-	-	-	M	O	M	M	E	E	-	O	-	-
Moorhen	E	E	M	M	M	M	M	M	E	E	E	E	E	E	-	E	-	O
Magpie	P	P	-	-	-	-	-	-	O	O	O	O	O	O	-	O	-	-
Crow/rook	O	O	-	-	-	-	-	-	E	E	-	O	-	O	-	-	-	-
Herring gull	P	P	-	-	-	-	-	-	E	E	M	E	M	M	-	-	-	-
Woodpigeon	P	P	-	-	-	-	-	-	O	O	O	O	O	O	-	O	-	-
Dove	P	P	-	-	-	-	-	-	M	O	E	E	E	E	P	E	-	O
Skylark	P	-	-	-	-	-	-	-	-	-	O	-	O	O	M	M	O	O
Hedgerow birds	O	-	-	-	-	-	-	-	O	-	M	O	M	O	E	E	M	M
Squirrel	E	E	O	O	-	-	-	-	E	E	-	E	-	M	-	-	-	-
Rabbit	E	E	M	M	O	O	M	M	E	E	-	M	-	M	-	-	-	-
Brown hare	P	M	M	E	-	-	M	M	-	E	-	M	-	-	-	-	-	-

Table 10.3.2 Accipitrid hawks and their prey.

houbara, and gulls flown by large falcons, and with larks flown by merlins. Many of these flights, and also those at some of the smaller quarry such as pigeons, stone curlew and jackdaws, tend to be long, twisting tail-chases in which the prey attempts to stay in the air and avoid the falcon without seeking cover and without necessarily climbing for height. On a windy day, this is the norm. On a still day, a crow or gull is not sufficiently fast or agile to escape the falcon and therefore seeks refuge in climbing for height. The stage is then set for a ringing flight.

In the waiting-on flight the hawk is first allowed to climb to a pitch above the falconer (high searching). The falconer then flushes the quarry within the killing cone of the hawk. The hawk is thus in an overwhelmingly dominating position and able to put in a stoop at the prey which has been 'served' to her. Buteos and eagles also like to use the high search and this is one of the most spectacular ways of flying them in falconry.

The low flights of accipitrid hawks consist mainly of direct and indirect flying attacks and perhaps some glide attacks. Whereas falcons are normally wedded to just a few, similar, prey species, the accipitrid hawks are best flown at a wide variety of species in varying circumstances.

Thus, although flights may be brief, no two are alike. Harris hawks are one of the few sociable species, hunting in family groups. Falconers sometimes exploit this by flying several together, but really this 'gang-hawking' is over-kill.

10.4 What the quarry can do

When they are attacked by a raptor, the prey respond in different ways. First of all, they can normally recognize if the raptor is attacking rather than just flying past. Then they can usually tell at least what genus the raptor is and take appropriate evasive action. If a buzzard flies over a flock of feeding pigeons, they usually flush and stay circling in the air until it has gone. If a peregrine comes over, they are more likely to stay on the ground and keep watching. If a goshawk comes over, they flush quickly and leave the vicinity or climb high. If the hawk catches one and is down on the ground eating it, the others may resume feeding unconcernedly.

Crows react in a similar way, but in addition, they will harass the raptor and rob it of any element of surprise, forcing it to quit. Even if a victim is caught, they usually gang up on the hawk and may succeed in driving it off and rescuing their companion.

Problems arise with the New Zealand falcon which sometimes looks and behaves like a falcon, and at others like an accipiter. If the crow or pigeon stays on the ground it may be caught where it stands. On the other hand, if it tries to fly off, it may be pursued aerobically and quickly beaten in the air. Thus escape tactics which are suitable for specialized raptors, may prove deficient against a versatile one, even if it does not have the peak performance of the specialist.

Most prey will seek cover if they can, either to avoid detection or to escape from attack. Game birds, rabbits and hares, and some others, will also freeze, that is, squat absolutely motionless, even away from cover, relying on cryptic colouration. This initial 'freezing' only continues as long as the victim 'thinks' it has not been detected. The moment it believes that the predator has noticed it, the prey animal will bolt. Thus it is different from tonic immobility, in which the prey has been attacked, and has either been captured or very intimidated by the predator, and 'plays dead.'For example, some horses with a genetic memory of large mammalian predators, will collapse in a heap when you saddle them up. Other prey, such as the water birds, will skulk or keep a low profile, in cover at the water's edge, but not actually remain 'frozen.'

When they flee, some prey, such as the gamebirds, rely on rapid acceleration to outsprint the predator, rather than on agility. Gamebirds and ducks usually manage a jink or two, seldom more than a 45° deviation from their flight path, whereas corvids and squirrels concentrate more on dodging than on getting away. The small gamebirds rely on anaerobic sprinting flight and soon have to land again, but the larger gamebirds can keep flying long enough to deter many raptors from continuing in pursuit. The corvids and many small birds also use their powers of flight one way or another to exhaust or discourage their pursuers. Although the gamebirds often do things which seem intelligent, most of their escape is usually instinctive or reactive; they are also often intimidated into losing heart or into tonic immobility. Corvids, and many other birds on the other hand, are strategists and will think up tricks as they go along, even helping one another and decoying the raptor away from the victim. These species are less easily intimidated and, even when caught, may still be rescued by their companions. Some of the species such as some corvids, starlings, and some waders for example, form tight knit defensive aerial flocks to intimidate or baffle an attacking raptor. This is a different type of flocking to ground flocking,

A young goshawk binds to a pheasant.
Photo Steve Magennis

such as in gamebird coveys and pigeon flocks. These feeding flocks usually have sentinel birds which reduce the risk of surprise, but once they are attacked, the flock may scatter as part of the escape strategy, rather than bunch up.

Some of the bigger gamebirds, corvids, gulls, and mammals, either by their size and strength, or by their biting or kicking, or by fluffing out their feathers, may either fight off the raptor or intimidate it so that it loses courage. Finally, when all else seems lost, many prey species have loose feathers so that the raptor is left with a foot full of feathers and the prey escapes unscathed. Often the raptor is distracted for several critical seconds by the feathers, before it realizes that what it has caught is worthless and the real meal has gone. Pigeons and doves, which are favourite prey for many raptors, have loose feathers all over, whereas the gamebirds tend to have a loose pad of feathers on the rump. Lizards shed their tails too, and falcons have been recorded with several lizard tails in their stomachs, but no whole lizards, an arrangement which seems to be a working compromise for both predator and prey!

10.5 Hawking grounds and quarry populations

If something costs you nothing, you don't value it so much. If you have a long-term commitment to something, you will look after it better. These truisms apply to many things in life, from marriage to land.

In some parts of the world there are large tracts of government land where anyone with the necessary permits can hawk. The land is managed on a broad, sustainable basis, but otherwise the hunting is a free for all. Long may this continue!

For the falconer, this gives a nice sense of freedom, but it doesn't encourage a sense of responsibility. The same applies to some of the houbara hawking grounds. Often the feeling is: if the quarry is there, go for it now before someone else does. This is the Law of the Commons. It's fine provided that everything is on an extensive basis and harvests are within yield limits. Falconry is normally self-limiting, so that falconers can seldom be accused of over-hunting, especially not in North America. On the houbara hawking grounds in North Africa and Asia, there are still

Hare-hawking in Qinghai Province, China.
Photo Alan Gates.

Suitable habitat for hunting with a red-headed merlin in Pakistan.
Photo Hyder Jaffer.

some falconers who are too keen and over-hunt. It is up to the more responsible and senior Arab falconers to encourage hawking on a sustainable basis so that there will always be quarry available for future generations.

Throughout most of Europe and eastern North America, hunting grounds are either private or more controlled. The would-be falconer first must identify suitable land and then seek access to it. Alternatively, he may get access to some land and then have to train a hawk suitable for that terrain. Obtaining access may be a simple as knocking on the door and asking permission. Or it may entail signing contracts on sporting rights or leases, or in passing hunting exams. You should also comply with the local regulations on game seasons, bag limits, display of hunting permits, and so on.

Once you have obtained permission to hawk on a particular area, the first thing to do is to buy a detailed map and mark in the exact boundaries with a coloured pen. Then you should go over the land and become acquainted with it so you can work out how best to hawk it, what hazards there are, and where the boundaries are on the ground. You should also find out if there are any other people hunting on it, possibly for different quarry. For example you might have permission to hawk rabbits, another person may release and shoot pheasants, another person may stalk the deer with a rifle, and the local hunt may draw the woods for foxes. You may have to make sure that you are not in the wrong place at the wrong time, even though you have permission.

If you are paying for the hawking on a commercial agreement, all well and good. On the other hand, if you simply have permission, then it is good manners to at least offer a small gift by way of thanks. This may be a bottle of something at Christmas, or some of the game you have taken, or a hand with some of the farming chores, such as hay making. Our crow-hawking land extends to about 100,000 acres, forming a straggling mosaic of farms, so during the season we hold a barbecue for all our landowners and their families. This gives them a chance to see the falcons and talk about things at leisure, as well as an opportunity to have a good chat and knees up with their neighbours from other isolated farms. The British Falconers' Club invite all their landowners to a dinner at the end of their major meets, and, although this is not cheap, it is a worthwhile investment in goodwill supporting the continuation of hawking in the area.

If you are just hawking pest species, such as rabbits or corvids, then the landowner is usually happy for you to kill as many as you can. But if you have taken the full sporting rights and are hawking pheasants, partridges, grouse, or ducks, then you may get involved in game management. This might entail managing the woodlands, releasing gamebirds, controlling predators or improving ponds, in other words, turning gamekeeper throughout the non-hawking season. This can

We hawk only with the kind permission of land-owners and farmers.
Photo Verity Johnson.

be a great source of satisfaction, as admirably portrayed by Dan Cover in America. The greedy person who catches as much as he can, and then leaves the place desolate at the end of the season, is obviously not popular. Particularly for people who live in a town, without roots in the country, it can be a rewarding experience to get involved with some land and follow it throughout the seasons, rather than just appearing occasionally, hawking, and then going away again. It is also a good way to get the younger generation involved and understanding the practical aspects of land management and ecology.

10.6 Rabbit hawking

Most young falconers all over the world start their hawking careers on their local version of the rabbit. The rabbit and its slightly smaller American cousin, the cottontail, have furnished sport since the beginning of hawking and no one surely can call himself an experienced falconer until he has put a few hundred in the bag.

The finest rabbit hawking I have ever enjoyed was with a goshawk and a pointing lurcher (like a small rough-haired greyhound), called *Tess*. The lurcher would point the rabbits lying out in frosted bracken or rough grass. On seeing the point, the goshawk would fly over to the dog and *Tess*, on hearing the gos's bells approaching, would flush the rabbit which would come out at a fairly high rate of knots. The goshawk would turn into the attack, often killing immediately. But if she missed first shot, the rabbit would continue at a torrid pace with *Tess* on its heels spraying snow on the turns, busting it from any cover. The hawk would throw upward briefly, look for a clear shot and smash back in. The dog's tail would come up as she braked, then she'd kill the rabbit with a quick nip and, like 'yellow-dog dingo' (grinning like a horse collar), she'd lie down beside the hawk. Such flights are the fastest thing you will see in hawking short of the quarry getting airborne. Such a powerful team should be reserved for good, strong winter hill rabbits, summer rabbits are overmatched and are mopped up unceremoniously.

Goshawking in Korea in the old days.

Where rabbits are found out in the open, there is scope for waiting-on flights with a Harris or red-tail, even, in ideal circumstances, with a goshawk. Achieving such a set up requires as rigorous attention to detail as with waiting-on flights with falcons. Groups of Harris hawks, with the males staying high and acting as spotters, and the females lower down, are probably the best form of flying Harrises are capable of. They can certainly kill a lot of rabbits this way.

Ferreting is a sport for the young and the hopeful. The ideal size of party is one or at most two. If there are two of you, then BE QUIET. Don't stamp around; communicate with signals. Otherwise, not surprisingly, the rabbits will be sticky to flush and the ferret will kill them below ground. Take two ferrets, preferably small white ones, individually marked. They should be fit and active, and given a small feed before starting. A big fat roley-poley, bleary eyed ferret which is too dozy to get out of the carrying box is no use; you will regret using it. When you open the carrying box the ferrets should pop up like jack in the boxes, and when placed at the entrance to the hole give a quick shake and beetle straight in. A good light traveling box can be made from half a plastic 5 gallon container with a strip of rubber inner tube for a shoulder strap. Remember you will have a lot of carrying to do; heavy wooden boxes are a penance. The box opening should be 10 cm square so that you can replace a ferret one handed without the other ferret getting out as you do so. The box door should bolt firmly and there should be NO WAY the ferrets can escape. If the ferret gets out of the box in the car when you are not there it will almost certainly kill your hawk if the latter is hooded or if the car is dark. For this reason, many falconers will not keep hawks and ferrets on the same premises.

On arrival at the warren, set down the box a little way away, quietly take out one ferret and place it in the bottom hole so that it works upwards through the tunnels. Softly walk round to above the warren so the hawk has a downhill slip and so that a rabbit hesitating in the mouth of the hole cannot see you. Otherwise it may well go back in, face the ferret and get killed. Particularly if the hawk is a good one, keep about 20 metres back. The rabbit should clear the warren and not double back into another hole. If the ferret sticks, put the second one in to rustle it up. If the warren is quite large you will need to work both ferrets to confuse the rabbits. If the warren is very big, leave it alone.

If the rabbits normally bolt toward a nearby set of holes it is often worth blocking them up beforehand with balls of dry grass. This can cause a hotly pursued rabbit some embarrassment. If a hard-flown rabbit escapes down a hole - leave it. If you try to ferret it again, it will usually face the ferret rather than bolt.

Most of the time the ferret will come out when called or if left for half an hour to sleep off its feed. If the ferret stays down, call at all the holes, thump the ground with your hand and rattle a dead rabbit in the hole. Don't let go of the rabbit. I remember on one occasion after half an hour in freezing conditions the ferret was still down. I wiggled my dead rabbit. Silence. I went over to look in another hole and when I came back my one and only rabbit had gone! Only after stowing it in the bowels of the earth did my muddy, bloody ferret emerge. If all else fails, tie a rabbit to the creance through a hole in its hock and put it down a hole, then stamp, yell, and rattle all over the warren. The shaking earth coming from the tunnel roof will usually stir the ferret, and when he takes the bait you can pull him out like an eel. A ferret locator and spade should ensure that you never lose a ferret, but it is all more equipment to carry.

There are a few dangerous situations when ferreting that you need to be aware of. One of these of course, is if the hawk nails the ferret. Get in fast and hold the ferret's jaws with your gloved hand. Pull out a dead rabbit and try to get the hawk on to that. Above all, get control of the ferret. It is incredible what mistreatment a ferret can take without injury. After such an incident put the ferret away, it will be extremely annoyed and will stink. Another danger is when a good hawk nails a rabbit the moment it leaves the hole because the falconer was too close. The ferret comes out after the rabbit and promptly joins in. Rabbit holes are often associated with fences - which means barbed wire and netting. Sooner or later your hawk will hit it hard, so it is worth training your hawk to it. Call it to the fist or lure so that it is likely to hit the fence at a slow speed. It must realize that wire is solid (unlike twigs) and it must learn either to go over the top or through a hole or at least put on the brakes rather than hit full tilt. A wire barb can scalp a hawk easily.

Although it is possible to ferret in open simple places in late summer, most warrens contain litters of young quite late and it is best to wait until October before starting. Avoid any warrens with extensive vegetation masking the holes; ferrets can walk a hundred metres away leaving you calling pathetically at the original hole.

Ferreting is definitely an informal activity. One soon amasses a wealth of hysterical anecdotes which would lower the tone of this book. Like Phil, whose driving amazed the traffic on the dual carriageway outside Edinburgh when a ferret got out of its traveling box and attached itself to his earlobe.

Flying the hawk at rabbits or hares after dark using a spotlight is a relatively modern development and not one, I think, to be encouraged, except as a method of pest control. Eagle owls, goshawks, and Harris hawks can be flown in this way, indeed the eagle owl can be flown with the lamp switched off, but then one doesn't see anything! An eagle owl coming in to the fist in complete darkness is an eerie and unnerving experience.

The best that can be said for lamping is that it enables rabbits to be flown in places where they seldom come out in the daytime. But the lamp beam arouses suspicion and many farmers dislike it. A well-dazzled rabbit often hardly moves before the hawk hits it, but where they have been lamped before they usually move to cover as soon as the beam hits them. As hedges and wood edges often have barbed wire fences, this exposes the hawk to a high risk of injury. Although some hawks adapt quite well to flying down a lamp beam, one needs to be fairly desperate to risk a valuable hawk in this way, particularly as the flights themselves are often poor or hard to see.

10.7 Hare-hawking

In Britain, brown and blue hares are taken by trained hawks. In North America the alternatives are the black-tailed and white-tailed jack rabbits and snowshoe hares. In the Middle East, the 'rabbit' is in fact a Desert Hare *Lepus capensis* or, in Arabic, *errneb.*

Hares are a sporting proposition only if tackled in certain ways, otherwise they can be disappointing. If a big goshawk or a red-tail, or even an eagle, is slipped from the fist at a hare on flat ground, the result is usually a slogging chase ending in a twisting turn. At this point the hawk either takes the hare or gets grounded. She may get up again and have another go, but generally as a spectacle, it is poor. If the hawk is inexperienced, the hare, instead of running fast, jinks whenever the hawk gets close and the young hawk, discouraged, gives up. An experienced hawk, on the other hand, often takes them with as little difficulty as taking the lure. Many hares are taken in Germany and eastern Europe in this way by lines of falconers with eagles or goshawks.

A white Goshawk *Accipiter gentilis albidus* takes a hare in Japan.

These flights are improved if a dog such as a slow lurcher, saluki, or similar is used. This speeds up the whole event by quite a few frames per second and keeps the hare moving if there is cover to hand. If the hawk catches the hare, the dog is close behind and can kill it quickly, preventing injury to the hawk which could kill her or put her off hares. I have used a lurcher and various pointers and spaniels for this job with good results; either the hare escapes or it is killed promptly. The dog then stands or lies nearby, panting and guarding the hawk. Frederick II gave detailed instructions for training the dogs in this role and it is a pity more falconers do not support their hawks in this way.

In Arabia the desert hare or *errneb* is hunted in two ways. One of the most skilled ways is to track it in the sand. After a windy night in the Empty Quarter, the sand has moved and it is possible to pick up tracks which are only an hour or two old. Alternatively, the trackers get up at about 4 a.m. and drive around the hunting area, towing a bush behind the car. Then, at dawn, we go out again and, picking out the fresh car tracks by the bush marks, we follow until we find hare tracks over the tire prints. Then we know that we have the tracks of a hare which was ready to find a seat for the day. We follow this trail, seeing where the hare has fed, and judging how long ago by the hardness of the droppings. When the foot prints are in pairs, very close together and soft, the hare is near its chosen seat. History is written in the sand. The braces of the falcon's hood are loosened and the last string jess or *sabook*, previously attached by a sheetbend bow to the leash or *mursel*, is untied. Eyes strain to spot the motionless hare crouching under a low camel-browsed bush. Then the cry goes up '*Dornuk! Dornuk*!' (Between us!), the hood is whipped off and the saker held high. A moment's hesitation and she is off, with the hare disappearing like a pixie through the scrub, bouncing against its shadow across the open ground. Its route is marked by the stoops of the falcon and the cries of the falconers. I jump my horse blind over a ridge, sliding down on his hocks. Just as the hare, the falcon, and their two shadows all converge at once, the hare throws itself up in the air like a rag doll, the falcon passing harmlessly through the gap between the hare and its shadow. Sometimes it looks as though it will be caught for sure, but a skidding turn sprays sand

Tracking hares in Rub Al Khali.

as it scoots into a bush. As the falcon turns over to dive in, out comes the *erraneb* throwing her out of position and making her pant in the overpowering heat. The hare's course describes a large circle but the falcon is fit and a stoop in some open ground tumbles it. In a moment all is over and the falconer runs up and kills it with his knife, murmuring the prayer *Bis'm Allah* (in the name of God). Thus it is rendered fit for our meal and we cook it with vegetables on a bed of rice and eat it with our hands. The saker gets the tongue, brain and liver; she maintains a steady weight on this ration in this climate. Curiously, she needs to be sharper for hares than for houbara, even though the *errneb* is small, like a cottontail with big ears.

Alternatively, the Arabs use salukis to go steadily through the scrub, checking likely bushes by scent. Often the hare, on hearing the approaching falconers, sneaks away and moves quietly ahead (*insellet*) so that you cannot come up to it by tracking. The saluki overcomes this problem. Once the hare is roused, the chase is on and the dog keeps the hare moving, working in partnership with the falcon. At the end of the flight the dog guards the falcon from eagles until we arrive. This is a harsh land with death always just round the corner for the unwary or unlucky.

Another exciting way of taking hares and hill rabbits is to use a waiting-on flight. I have done this with goshawks, a changeable hawk-eagle, a red-tail, and Harrises, but Ferruginous hawks and other large buteos and assorted eagles are possibly even more successful. Unfortunately, most of those flying Harris hawks and red-tails have not got experience of training falcons for waiting-on flights and seldom attempt it with their hare hawks. This is a pity because these large hawks wait on naturally and are most effective from this position. To see a golden eagle flown in this way is really impressive and a stark contrast to flights from the fist.

Blue hares are not difficult. Possibly a record best 'first' was by Cathy Blakey who, when she was working for us, slipped a female Harris out of the front door of our grouse-hawking cottage in Sutherland. Cathy was having a hot bath when the hawk took its first blue hare.

10.8 Mixed game in hedges and farmland

Let us look at strategies for hawking mixed game such as rabbits, pheasants, partridges and blackbirds in a hedgerow where vegetation is at least shoulder high, but without trees. In many parts of Europe and eastern North America, such hedges and strips form a major proportion of holding cover for game. Assume there is one falconer carrying an accipiter, a buteo or a Harris. His dog will usually be some kind of spaniel, German short-haired pointer or German wire-haired pointer. Once you start thinking about the position of the falconer, dog, and hawk, while hunting a hedgeline, then you can start to apply the same principles to other irregular patches of cover.

Figure 10.8.1 shows the arrangement when the wind is in any part of the quadrant shown. For the other three quadrants, reverse the direction of beat or the sides of the hedge accordingly. The falconer and his dog work along the hedge into or across the wind. The falconer may stand in any of the numbered positions shown according to circumstance.

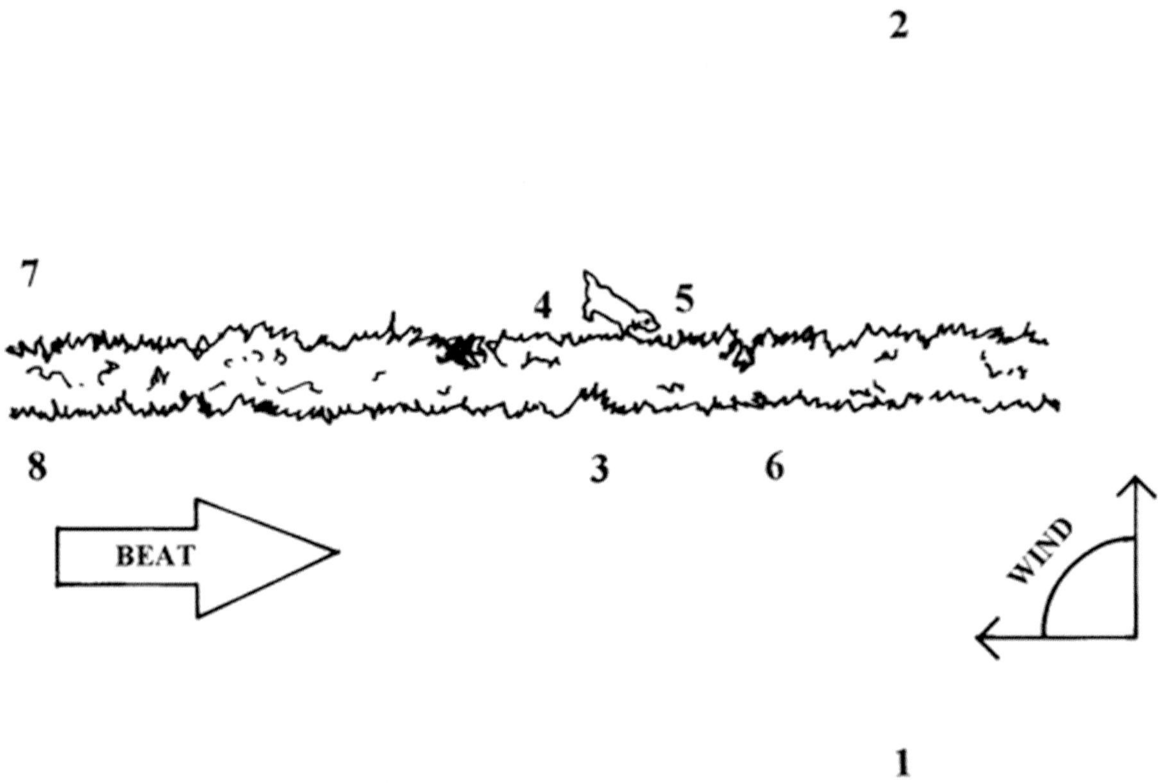

Figure 10.8.1 Hawking a hedgeline.

1. A good position for steady progress along the hedge. Staying 10 metres out and walking quietly, one can detect game running ahead or back and see the game beginning to flush before it clears cover - this gives the hawk a good start. Keeping a little off the hedge encourages the game to flush cleanly rather than run along the hedgebottom and also makes it more likely to come your way. Game flushing upwind must come toward you, game going downwind presents a nice slip. The hawk can more easily see game leaving the blind side and can more easily get over the hedge. But it is harder to see and to control the dog from this position, so it needs to be reliable.

2. A good position if the dog is unreliable but one must be resigned to losing game on the blindside or long tail-chases due to poor starts. If the dog points, or starts feathering, one can move quickly to position 4 or 5, hold the hawk up high and pray. The Harris hawk and the New Zealand falcon quickly learn to perch on a stick handle, especially if you push a hawking glove onto the stick the first few times. Then you can loft the bird high above the hedge before sending the dog in to flush.

3. A poor position for visibility as you walk along, but not a bad one to adopt if you get a firm pheasant point, or if the dog is hard to control or see.

4.You may have to adopt this position if the scent is tricky and the dog needs more careful handling, but it is not so good for the hawk.

5. From this position, the game is most likely to go out on the blind side and the hawk will be faced with a stiff climb to get over the hedge before pursuing. Some game will double back down the hedge.

6.This is a bad position because it pinches the game between the dog and the falconer. Rather than flush out cleanly, it will most likely keep to cover and go along the hedge, or even in some cases, get taken by the dog.

7/8. The falconer is too far behind so that dog and game are out of control. The hawk is faced with all slips being upwind and has a poor view

Working a hedge line.
Photo Leigh Tovey.

of the game flushing. From position 7 most game will also be on the blind side.

If there are trees in the hedge, the best place for the hawk is in the one just ahead of the dog, upwind. The falconer can then move into the side of the hedge opposite the dog and beat toward the hawk. Once he has gone past the tree, he can signal the hawk to move on to the next tree; most learn this very quickly.

Some falconers advocate flying the hawk only from the fist. I disagree with this for several reasons. First, the hawk in the tree has a much better advantage and will detect and kill more game. Secondly, the hawk can intimidate the game more easily, forcing it to put in nearer the falconer so that there are fewer hopeless long tail-chases which can easily cause a lost hawk. Thirdly, being no longer a hawk perch, the falconer is able actively to beat and participate in flushing the game so that effectively the quarry now has three adversaries rather than just two. Fourthly, flights which start from the fist show predominantly a tail view of the action, you get a poor idea of the speed of the hawk closing the gap, whereas the viewpoint when the hawk flies from a tree is usually more varied and interesting. Fifthly, a hawk flying up and down trees gets more exercise than one kept on the fist and does not face such frustrations as the falconer failing to release the jesses in time. The argument that flying a hawk from trees makes it independent simply doesn't hold water. Throughout the day, one is constantly having to move the hawk from tree to tree or call it down for one reason or another and this reinforces discipline rather than the reverse. With the slower hawks, especially the buteos, you will be lucky to catch much at all if you don't use trees.

Should she miss the quarry in an open district, she should either return to the fist or land on the ground. This is achieved by calling her from the ground during training. It is maddening if she flies off 500 metres each time, looking for a perch.

When you are hawking in the woods, the hawk should be in the trees just ahead, not lagging behind. It is surprising how much game runs

Beating for hares and bamboo-partridges in Japan.

Black Sparrowhawk *Accipiter melanoleucus* with Chinese Bamboo-Partridge *Bambusi colathoracicus* in Japan.

or flits quietly ahead unseen by the falconer. A hawk, well placed just in front, spots this game and either attacks it or intimidates it so that it stays put and is pointed properly, rather than running out ahead and escaping. On the other hand, if you go for a walk through the woods with your hawk following from tree to tree, and your 'flights' consist of the hawk simply nailing a pheasant or rabbit as it runs ahead, this can hardly be considered quality falconry.

If you are flying the hawk from the fist, you can either let her go in her own time, or cast her off, or in the case of sparrowhawks, physically throw her at the quarry. Throwing or casting off the hawk takes some skill; the reaction time of an accipiter is so much faster than a human's that they often do better by themselves. Throwing the hawk is a common method for catching quail in Turkey, where a halsband is used and in China throwing from the gloved right hand is normal for small accipiters.

Inexperienced falconers often carry on beating without reference to the hawk. If the hawk is out of position, or not ready, then STOP. Any game flushed will otherwise be wasted. Constantly check on the hawk and always endeavour to flush the game in the best way for the hawk. Similarly you must pay attention; it is no time for conversation or day-dreaming. Serving the hawk is the most serious responsibility of the falconer in the field. It is the hallmark of a good falconer and the key to a good hawk. If for some reason you cannot serve your hawk regularly and well, all sorts of problems will develop. Potentially good hawks will start self-hunting, inexperienced ones will get fist or lure-bound and may start to show undesirable aggression toward the falconer or his dog.

If the hawk does start self-hunting, cutting her weight down will not necessarily cure it; in fact it often makes it worse. Bert (1619:55) was adamant on this point. The remedy is to get the hawk more orientated toward the falconer and less to looking outward for quarry. The falconer must be the source of the quarry, and there must be little available elsewhere, so that the hawk has more hopes of obtaining quarry or food with the falconer and is less confident of her chances elsewhere. In bad cases a few days off game, simply working to fist and lure for discipline, is a good antidote. In the case of a crow hawk I do not regard self-hunting as a bad fault. A good crow hawk is expected to go out on slips of several hundred metres tackling awkward quarry unaided. If, on returning unsuccessful, she spots a few crows in a vulnerable situation, I would be worried if she ignored them.

When beating out a thick hedge or strip of woodland, game will run out ahead - pheasants and red-legged partridges (*Alectoris rufa*) may be sneaking out up to 200 metres ahead. Flying the hawk from trees or standing out a little from the hedge often allows the hawk to spot such quarry. Otherwise efforts must be made to stop off the hedge. If you believe an old pheasant is giving you the slip, do a 'P loop.' Head out from the hedge and run forwards, coming in again a 100 metres or more further on. With any luck, you have then pinched the pheasant between you and the dog and it will be forced to flush out. Where a hedge runs into a wood it is best to do this as a matter of course.

Preparing to throw our Japanese sparrowhawks in Hohhot, China.

A good goshawk, when thoroughly fit, can take both partridges and pheasants in the air in the first flight. But to achieve this level of fitness it must be flown at least six days per week with a lot of tree work and hard flights. If you fly only three days a week (as Sebright did), and fly only from the fist, you will find goshawks as disappointing as he did. Then is the time to give high jumps or roprey on non-hawking days. Goshawks improve immensely with experience, certainly for three or four seasons. A young bird's first season or 100 head are its apprenticeship for the future.

There is always a debate about whether or not to flush game for a goshawk once she has put it in. My old friend Tony Owens liked his hawk to take pheasants in the air on the first flight and if she missed and put it in to cover he would leave it and look for a fresh one, on the basis that once they learn to just sit on the pheasant's tail until it puts in, and then take stand, they get lazy. But William Barker, writing in 1853, noted: 'If a goshawk is properly trained, and given something (say the head) of everything he takes, he will never give up the pursuit until he reaches the bush wherein the quarry has taken refuge; but the dog and the falconer must be alert, and come to his assistance, and never give up the search for that particular bird, which he will distinguish from any number that may rise, and will never fail to pursue in preference to the rest, is discouraged from the sport.' There is some truth in both approaches. One of the hallmarks of Stephen Frank's falconry was his persistence in flushing game under his falcon. The hawk knows when you are trying your best to find the game for her, it is a matter of keeping

Female Cooper's Hawk.
Photo Richard Lowry.

faith with her. Her first instinct and priority is to catch the prey as quickly and as easily as she can. That may not be in as dramatic a fashion as the falconer would wish. The falconer may claim that it is not the number of kills that is important, but the quality of the flight. But the hawk doesn't see it that way. While a youthful inexperienced hawk may show some spectacular strenuous flights, an older hawk needs to be successful, confident and fit if she is to undertake challenging flights.

Woodcock come into Britain from Scandinavia normally in late November when the moon is full and the north is freezing up. During December and January, particularly when snow is on the ground, they can be found in the woods and damp scrubby corners where there is unfrozen mud or water. Often, during the daytime, they shelter in the bramble or sallow bushes around small springs in the woods. Here they can be pointed and flushed out by dogs and chased by a Cooper's hawk or goshawk waiting in a tree nearby. However, woodcock flush out strongly and once they clear the canopy, they can outstrip an accipiter. Occasionally, when the weather is very cold for several days, the woodcock lie very close and put in again after a hundred metres or more and in those circumstances a well-placed, strong goshawk can kill them. In North America, woodcock are found only in the east and the same methods of hunting them apply as in Europe.

Dogs point woodcock well, but pointers are difficult to work in thick cover, and it is often a question of pushing on as best one can. Also woodcock do not hold reliably before a point even with a hawk overhead. Even though the dog is steady, the woodcock may jump spontaneously after a few moments and time cannot be wasted when getting the hawk placed. My current dog, a Munsterlander, is very handy on woodcock.

It is very seldom indeed that woodcock can be pointed in an open enough place to fly a falcon at them from waiting-on. Given such a set up, a very sporting flight results. However, I have flown them a few times with New Zealanders starting from trees over a point and then going into a long tail-chase. So far I have always been left too far behind, stuck in a bog, to reach the falcon after she has put the woodcock in. One day I will do so, and I think in the second flight the falcon will call the tune. Modern falconers have neglected the woodcock, which is a pity. It is a fine, challenging quarry.

Rats and squirrels are sometimes taken during mixed hawking. Rats can be flushed with a dog, stick, ferret, or waterhose and are easily captured. However, the quality of the flight is usually poor, and the risk of being bitten is high, so rats are best avoided. An old common buzzard which is not of much use for anything else can show some good chaotic fun in a granary after rats. I remember a male gos of mine in Scotland out duck hawking flying across the river and picking a swimming rat out of the water like a cherry off a cake. But, generally speaking, rats are best left well alone.

Grey squirrels provide a modicum of sport for Harris hawks, goshawks, red-tails and, occasionally, Cooper's. Grey squirrels not only have a very nasty bite, but also a thick skin and a very elusive way of handling themselves. Some hawks develop a very good eye for squirrels and hunt them assiduously and effectively. Indeed at some of the goshawk nests I have visited in Sweden and Finland, red squirrels *Sciurus vulgaris* were a staple. Generally the best help the falconer can give is to keep the squirrel moving with a stick and well-aimed stones (mind the hawk with the ricochets). When the hawk catches one, unless you can quickly kill the squirrel, leave it to the hawk. Don't stand nearby distracting the hawk which may promptly get bitten. Wide protective jesses can be of some use although the bites are usually on the toes. Slitted mews jesses are definitely taboo. Do not hawk squirrels unless you are prepared to carry home a crippled hawk. If you do fly them, only use a hawk which is experienced at rabbits and which goes for the head. A fumble-footed youngster is bound to get bitten.

Mallard and teal, and to a varying extent a wide range of duck species, can be taken by an accipiter in the first 30 metres or so before they have reached full speed. The hawk flashes from the fist and binds to the duck pulling it back down. This is how most ducks are taken in Britain. This was called hawking at the 'crepe' or creep. The usual way of doing this is to use the 'P loop' method. First you walk to the river and then carefully scan up and downstream, marking the ducks. Then you walk away from the river in a loop and approach the marked ducks quietly. The reason for this is that the goshawk needs a short surprise slip to be successful at ducks. Also, the idea is that the ducks are flushed suddenly and cleanly and are properly in the air before they realize that the hawk is after them, otherwise they will dump back down, dive, and refuse to re-flush.

Where ducks are on small isolated ponds or marl pits, the simplest method is to let the hawk fly quietly into an overhanging tree while you are still some way off. This will alarm the ducks and hold them down. You then rush the pond with dogs and people together to get a sudden surprise and a quick flush. Of course, if it turns out no ducks are there, then you are the one who is surprised and flushed.

The flight with a small accipiter at hedgerow birds is very deceptive. Areas which before seemed full of birds become deserted when you appear with a sparrowhawk or sharpshin. Hedgerows which you so confidently expected would produce plenty of flights provide nothing, and amid the luxuriance of late summer suddenly seem dauntingly large, dense, and hopeless. The young falconer, perhaps with his first 'serious' hawk looking for quarry for the first time, begins to realize there is more to this hawking business than teaching the bird to come back. Much more.

Pedro Yrigoyen with his Bicoloured Hawk "Wayra" in her first season with an Ornate Tinamou (*Nothoprocta ornata*) This species inhabits in the Puna areas of Peru around 3,300 to 4,400 metres above sea level.

While it helps of course to have a good knowledge of the basic habits of the prey, the way prey behaves when hunted by a hawk can only really be discovered by experience.

For most small accipiter work, the assistance of a beater is really handy, indeed often a necessity. Even with a dog, a single falconer cannot work a hedgerow for small birds. They all fly out on the blind side or well ahead. But with a beater armed with a stick, working five metres ahead on the other side, a good proportion will come out and be flown by the hawk. Of course, after a few days, the beater will get fed up of bashing away, being yelled at, not seeing anything and then being told afterward what he missed. He will probably go on strike. But this is all good leadership training for the young falconer.

Having reached this far, he starts to find that although the hawk chases the quarry, most potential prey quickly get back into cover and refuse to cooperate further. The simple answer is that big hedges are hard in winter and impossible in summer. The falconer must look for small hedges, small isolated patches of cover, farmyards, and a hundred and one little places where birds can be found where the hawk can get a proper chance. Rushy fields, patches of stinging nettles or thistles, animal feed troughs, small ditches and streams; a bird in one of those is worth many in a big bush.

Most small birds are protected in most countries and in Britain you will need a licence to fly them. Sparrows, which are common in summer, disappear from a lot of places in the autumn and as the leaves fall, blackbirds and thrushes become more of a proposition. After a season at blackbirds you will be older, wiser and much more respectful of them as a quarry. If you've caught a few and still got a healthy hawk you have done well.

Small accipiters are frail, and in addition to being fully conversant with Mavrogordato's *A Hawk for the Bush* and Harry McElroy's *Desert Hawking*, remember the advice of the anonymous author of *The Perfect Boke for Keeping Sparrowhawks and Goshawks*. Namely, not to let the hawk stand empty in the morning after she has cast but to feed her a little to keep her 'lusty.' Especially once the frosts start, you must guard against her blood sugar levels getting low and keep her weight up. She should fly in October about 15 grams or half an ounce heavier than she did in August.

Although I flew several sparrowhawks and a shikra in my student days and acknowledge the skill required to keep such birds in flying order, I would not want to fly one now. The flights are so short and brief, the best hardly better than a poor merlin flight, and the hawks are so delicate, that I suppose I am too lazy or too spoiled to attempt it. That they have their place in falconry and provide an excellent grounding (if severe one), there can be no doubt, and certainly no falconer can consider his education complete until he has flown at least one small accipiter successfully. One of the sad aspects that occurs nowadays is that the tyro falconer soon obtains a Harris hawk. Although these are marvellous birds, they are in many ways just too easy. The falconer quickly learns bad habits and slack practices. He abandons the hood, is slack about handling the bird, and is careless around its feet and in tying knots. Out hawking, he is inattentive to the hawk's whereabouts and, when the hawk chases quarry, the falconer strolls casually after it without a care in the world. He starts to believe he knows a thing or two about hawking! But when he tries this blasé approach on an accipiter, his life suddenly becomes a nightmare in which even walking through a doorway requires careful consideration. An apprenticeship on a small accipiter helps to develop good habits and a sensitivity which are a necessity for the skilled falconer.

10.9 Waiting-on flights with falcons

Gamebirds, such as red grouse, black grouse (*Lyrurus tetrix*), willow grouse and ptarmigan (*Lagopus lagopus*), sharp-tailed grouse (*Tympanucus phasianellus*), sage grouse (*Centrocercus urophasianus*), prairie chickens (*Tympanucus cupido*), pheasants, red-legged partridges, chukar (*Alectoris chukar*), Hungarian or grey partridges (*Perdix perdix*), francolin (*Francolinus* spp), and various species of quail, are commonly taken in a waiting-on flight by falcons, although many of the smaller species which seldom fly far from cover, are more usually

Grouse-hawking at Riddlehamhope 1897. L-R Keeper with pointer, Major Charles Hawkins-Fisher, William Rutford, unknown, James Rutford.

flown by accipiters. In addition, many species of duck are flown in the same manner by falcons, where they can be flushed cleanly in open country.

The waiting-on style of flight has a long history in falconry. Shakespeare often mentioned the pitch of a falcon, for example in Macbeth: '*A falcon tow'ring in her pride of place was by a mousing owl hawked at and killed.*' Hawking 'at the brook' was usually using a falcon waiting-on overhead while beaters flushed the ducks using spaniels and drums. In the renaissance of British falconry in the nineteenth century, game hawking was a feature of the Old Hawking Club. John Anderson, Peter Ballantyne, John Pells, William Barr and others pioneered the use of eyasses for waiting-on flights, and the Scottish moors have been popular with falconers especially since the 1960s. Haggards had never been popular before telemetry, for obvious reasons. No doubt the pitches were not super high. Until telemetry became available in the early 1980s, high pitches and long flights were more trouble than they were worth, and with falcons being so scarce then, my nerves suffered.

The basic elements of game hawking are the same the world over. The falcon must go up into her 'high search' position (see 8.4). She has a natural instinct to do this, and is encouraged by the falconer's training. She understands from previous experience that the falconer below is a flushing agent who might disturb prey. She keeps him in her killing cone. It is his job, as part of this act of faith, to produce the quarry. Once it is flushed, the falcon stoops and has a good chance of killing the quarry outright. This pattern of hunting technique is repeated time and time again without variation, so that the falcon becomes 'set' in this habit. All the gamebirds fly in approximately the same way: they have a good acceleration when flushed, sprint mainly anaerobically, and then have to land. Their flight is direct, making a direct hit not too difficult, compared to agile prey such as Lapwings *Vanellus vanellus*, which seldom sustain the full impact of a stoop. Within this broad framework there are many permutations; cock sage grouse for example fly so fast and so far, and are so heavy and tough, that only the strongest falcons can kill them, and even then, only after they have

Anatum peregrine with sage grouse.
Photo Steve Chindgren.

A working Red Setter.
Photo Steve Magennis

Lochdu, Caithness. 1980. Roger Upton, Umberto Caproni di Taliedo, Eckart Schormair, Tony Walker, Simon Autie, Glennis Grant, Nick Fox and Charles de Ganay.
Photo Stephen Frank.

reviewed their ideas on speed! The medium-sized gamebirds, from partridges and red and sharp-tailed grouse through to pheasants, are tempting targets for most falcons in the 700-1200 grams (25-43 oz.) range, as are the ducks. Further down the size range, quail, teal, and snipe tend to be too nippy and agile for large falcons and this is where many male falcons come into their own.

Let's look at the stoop more closely. I will assume that the level flight speed of a red grouse when chased by a falcon is 88 kilometres per hour (55 mph). Grouse, sensibly, fly a lot faster when pursued by a falcon than when casually flushed. The grouse is the first to get its speed up and in level flight the peregrine can only catch up with it after several hundred metres. So a tail-chase is not a productive approach; the falcon needs aerial dominance.

As a falcon circles in the air, she has below her a 'killing cone' in which she can stoop (figure 10.9.1). Any quarry inside this airspace is vulnerable to a stoop from above. Outside this 45° cone is a wider 30° cone in which the falcon could make a shallower, less powerful stoop. Outside the 30° cone the falcon would more or less have to tail-chase a grouse in an angled wing-beating direct attack.

Obviously, the parameters vary tremendously and independently. We are trying here to understand the principles of what is happening rather than to arrive at exact predictions. If the falcon is 100 metres up she can cover a 200 metre

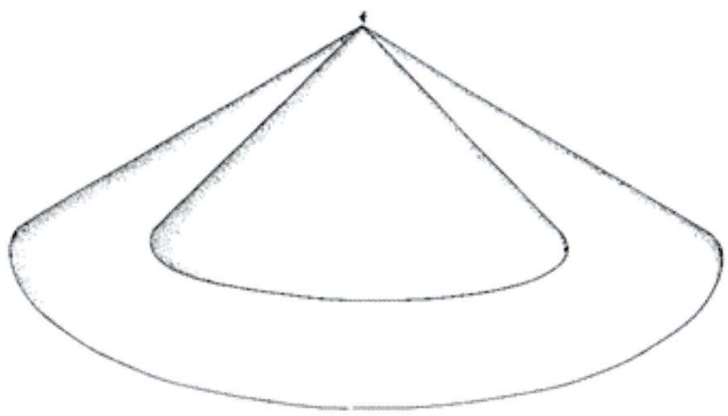

Figure 10.9.1. The 30° and 45° killing cones.

Dunbeath 2021. New generations. Mark Upton and son Robert with young peregrine.

diameter of ground (ie 100 metre radius) below, inside her 45° killing cone. In a 45° stoop she will travel down 100 metres and across 100 metres and take just over five seconds. A grouse flushed at the centre of the cone can also fly 100m and reach the edge of the cone in about the same time. But, with the grouse skimming so close to the ground, the falcon is forced to level out and lose speed. Stooping from a lower height means the falcon has no real hope and she will end up in a tail-chase. Of course she could stoop at a shallower angle, say 30°. This would take her further out, but would take longer, so the outcome is similar.

If the grouse had started, not from the centre of the killing cone but a bit nearer the edge, a tail-chase would have been inevitable. If the falconer could manage to flush the grouse from the outside of the killing cone inward, under the falcon, then the situation might be saved, but the grouse probably has its own ideas about this.

What if the falcon goes to a higher pitch, say 200 metres? The stoop would be about 60° and the grouse would only have travelled about 120 metres horizontally in about 6.2 seconds. Classic stuff. The grouse could have been 75 metres off centre and still been hit with a 45° stoop, no problem. If the falcon had started from a pitch of 300 metres, the grouse could be 150 metres off centre and yet still been hit.

The stoop itself, of course, will probably be steeper at first, then flatten out. Because grouse skim along close to the ground the falcon is forced to flatten out and so a vertical stoop is not truly vertical but hooked or spiralled. When faced with quarry high in the air, the falcon still normally flattens out before, or at, contact in order to throw up high and regain dominance. One seldom sees a falcon continue in a downward trajectory below the quarry after the strike, unless the prey is well up and easy.

So, if you count 45° or steeper as a true stoop, then the first contacts inside the 45° cone only come once the falcon is about 100 metres or more up. Less than this and she can only manage a shallow stoop or a tail-chase. There are only a few ways around this basic problem. One is to flush the grouse from the outside of the killing cone toward the centre. You could tilt the ground level so that the grouse must fly uphill. This slows down the grouse, makes the journey for the falcon shorter, and makes the grouse easier to hit. But no self-respecting grouse will put up with this for more than a few wingbeats! Round it goes and away down-hill. Now there's trouble. The tables are turned completely. The grouse hurtles away gaining speed rapidly. Wide valleys open before it. There are limitless options for escape. The falcon meanwhile is condemned to a steep tail-chase with the grouse controlling the flight. No longer so intimidated, the grouse leads the falcon a merry dance out of sight. The good news is: when the falcon gives up and comes back. The bad news is when the falcon catches the grouse a long way off and is herself killed by an eagle.

Let's say we are hawking partridges in a fairly enclosed agricultural district. The falcon must get onto them quickly before they reach cover. The birds will be holding tightly and can be flushed with reasonably accurate timing and direction. We need a falcon which waits on closely, in position and at a pitch of about 100-150 metres (300-500 ft). From there she has a chance of stooping and killing within a horizontal distance of 100 metres or so. The partridges will be so under control that they will not have opportunity to dream up spoiling tactics and with a bit of luck the falcon will kill one stone dead. From a lesser pitch, the partridges will make the hedge. Also, from a lesser pitch, the falcon is not really waiting-on at all; she is zooming round in circles. This means that the falconer is running round the field tripping over God knows what, with his head in the air, trying to watch the falcon and time his flush for the split-second when she is facing the right way. He is probably also shouting instructions to his dogs and companions, instructions which will be misunderstood and acted on five seconds too late, by which time a further diatribe of contradictory yells is already issuing forth. A very genteel partridge might die from shock at the bad language, but this is probably the limit of its risk.

And it doesn't stop there. We haven't even started to talk about wind yet. Falcons, particularly young or unfit ones, prefer to stoop upwind because they have more control and can stoop more steeply knowing they can pull out afterward more easily. The oncoming wind and the wind shear effect help lift them up away from the ground at the time of most strain on the pectoral muscles. They dislike stooping downwind because they have a faster ground speed which on low flying quarry is hazardous.

When both falcon and quarry are up in the air near together, such as in a tail-chase, wind speed is irrelevant. They are both in the same body of moving air and are thus unaffected in relation to each other, like two swimmers drifting down a river. Only their ground speed is changed by the wind direction.

But when falcon and prey are in the air and apart then they may be experiencing different wind speeds. Wind tends to be stronger higher up than near the ground. The grouse that skims off into a 20 kilometres per hour wind may make its pursuer at 150 metres (500 ft.) face a 50 kilometres per hour wind. This wind shear effect can thus aid the quarry on an upwind flight and the falcon on a downwind flight. Therefore many game falconers try for a downwind flush. The benefits come only when there is significant wind shear. In some situations, such as on windward slopes, there is a reverse wind shear; the wind low down being stronger than at height.

The effects on the killing cone of a steady wind, and wind shear are shown in figures 10.9.2 and 10.9.3. Once a falcon becomes experienced, she learns to use the wind instead of fighting it. The lessons to be learned from this are: if your falcon normally holds a pitch of 150 metres (500 ft.) it does not matter much which direction the grouse flushes in, provided that it flushes within 100 metres (330 ft.) of the centre of the 45° cone. If on the other hand the falcon has a low pitch of only 60 metres (200 ft.) the grouse is best flushed downwind, or alternatively the falcon must be placed slightly upwind. Grouse prefer to fly upwind and downhill. Most of all they have their eyes cocked to monitor the position of the falcon and will use whichever option takes them out of the killing cone most rapidly. The skilled falconer anticipates what action the grouse are most likely to take, or can be persuaded to take, and tries to place his falcon in such a position that there is no way the grouse can get outside the killing cone before the strike.

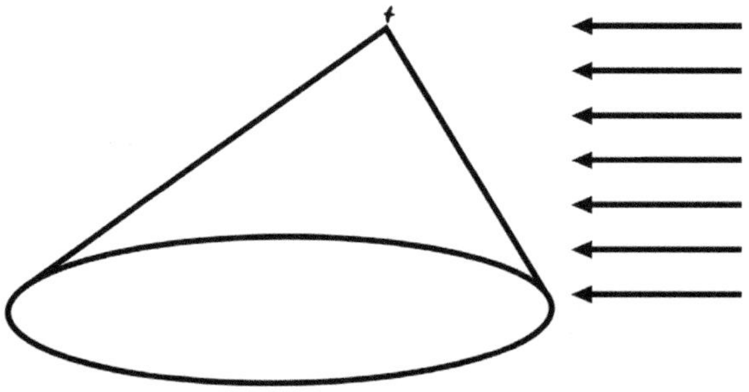

10.9.2 The effect of an even cross wind on the killing cone.

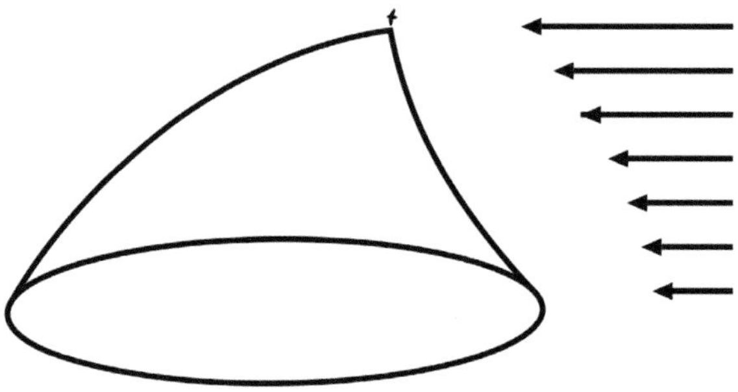

10.9.3 The effect of wind shear on the killing cone.

In the debate the question is often asked: how high is a good pitch? With modern GPS we can now measure heights accurately. The answer is that pitch should be balanced according to circumstances. For example, a high pitch for snipe on an Irish bog is counter-productive, the falcon takes too long coming down. It is important to remember that falconry, when described as an 'art' is ephemeral, like music or dance. What is the point of a flight that nobody sees? If your falcon is so high up that you must rely on gizmos to locate her, and cars to chase after the signal, what are you achieving? Certainly there is no visual art. You are simply risking your falcon for no benefit other than bragging rights. This applies to all of falconry; if nobody sees the flight, it is more or less a waste of effort. Some claim it is all about the falcon and serving her needs, but really? Why bother to do this day after day? Why not just '*let her off down the wind, to prey at fortune*'? She doesn't need you.

In Britain, especially on the high northern moors, the cloud base is often below 300m (1000 ft). The higher she goes the more likely she is to check away at pigeons, and the more likely she is to be killed by an eagle, trampled by cattle or get into some other danger. In general, if you are having to track your falcon down more than a few times per season then you are very likely to be catching prey on land where you have no permission to be and your falcon is out of control. This is poor falconry.

Usually falconers team up and hawk in pairs or trios. Where grouse are scarce and three or four hawks are to be flown, it is futile to have everybody trailing along behind one dog. Better to form a line with a dog handler at each end and two pointers running at a time working independently. Depending on the terrain each dog will take a beat 200-500 metres wide. When one gets a point, this is signalled across to the other handler who drops his dog and takes it in. The grouse are then flown. In this way everyone walks half as far and sees twice as much hawking.

In hilly terrain, the lie of the land becomes of more importance than the wind. Flushing the grouse up the slope puts them at a grave disadvantage regardless of wind direction. Grouse flushed downhill can get speed up and cover a lot of ground very quickly. On flattish land, when the wind is slight, it is a waste of time heading the point, ie for the falconer to make a circuit until he is approaching the dog head on, trying for a downwind flush. On calm days it is best to flush the grouse straight out, it is quicker, less likely to go wrong, the flush is more controlled and you have a better chance of flushing birds singly and thus getting several flights from one covey. This is particularly important for a low-flying young falcon who needs a carefully-timed flush. Also, the young hawk, coming back high after losing the first grouse, really benefits from having a second flush under her as she comes over. A few such occasions make all the difference at a critical time in a hawk's career.

One advantage of heading the point, especially when hawking alone, is that it is easier to control

Steve Chindgren's 5 year old male gyr/peregrine 'Jett'.
Photo Steve Chindgren.

the dog when it is facing you. If you are behind the dog it may, in its enthusiasm, when told to flush, run through the whole covey and spring the lot. When you face the dog, once it has flushed the first bird or birds, it is easier to drop the dog leaving you free to concentrate on the falcon.

Provided the essential ingredients are there, red grouse-hawking is one of the more realisable forms of falconry. A well-made grouse falcon enjoys a high success rate. By and large, one is free of the constant problems found hawking partridges in more enclosed areas: no power lines, roads, people, pigeons, and so on. The grouse too is a naturally tempting quarry, very predictable and incapable of serious evasive tactics in the air. On the ground it seldom reaches insuperable cover (except where bracken encroaches) or shows the intelligence of a magpie. The dogs love its scent. The whole scenario is eminently repeatable and this is the ideal way to build up the stereotyped behaviour of a good game hawk.

Black grouse, if found on a moor, are very similar to red grouse. Ptarmigan and chukar are more difficult due to the hilly terrain they inhabit. With the former it is cold, wet, and windy and with the latter, too hot for the dogs to work after mid-morning.

Of the partridges, the grey or Hungarian partridge is preferable for hawking as it flushes well whereas the red-legged and Chukar partridges run and run. No falconer who has hawked grouse or pheasants should find partridges difficult. The problems mainly arise, not from the quarry, but from the distracting influences such as other potential prey and the presence of thick cover. If there is a lot of cover you will find, as Sebright did, that nine out of ten of the partridges taken are killed by the dogs.

Although partridges themselves are not difficult, keeping a partridge falcon in work is. To find her a couple of flights a day for any length of time requires access to a lot of partridges, often in areas where game rents are at a premium. The rich coveys of September soon become scattered and unapproachable. Stubbles soon give way to bare plough. By December most partridge falcons have been grounded. In Spain, where I have had the good fortune to hawk redlegs with the Royal Spanish Falconry Club, the coto or hawking ground, is very open and partridges must be spotted a long way off from a vehicle. I detest cars and try to convince my Spanish friends to use horses, which are still used on the coto for hare coursing, but they roll their eyeballs, murmur 'Mad Ingleesh' and take me for another little snack of chopped squid! In America, a car is needed, not only to cover such large areas of ground, but to keep the falcon and falconer from freezing up in the process.

Quail can vary from very easy, if found in cultivated or arid land, to impossible where cover is thick. Harry McElroy's Desert Hawking II is still the best source of advice on them. Quail in an open landscape can be taken with a smallish falcon waiting-on. But quail love cover and a more versatile hawk, such as an Aplomado, is usually needed. I used to hawk California Quail (*Callipepla californica*) with New Zealand falcons, which were once known as 'quail hawks.' They would disappear inside thick areas of cover to reach the quail.

Various species of duck can be flushed under a large falcon waiting-on and killed in the stoop. This has been brought to perfection in parts of America where suitable ponds are available for consistent flights. Good conditions are rare for this flight in Britain; the main places are the tiny lochans of Caithness and Sutherland where one often finds a teal or two while out grouse-hawking. These little ducks know very well the distance to the next safe pool and, just when you think your falcon will be rewarded with a successful stoop, the duck disappears, showering the falcon with water. It can keep up this game until both falconer and falcon are exhausted and wet, having just completed a whistle stop tour of these tiny, hidden lochans which sprinkle the moors like jewels.

Once well on the wing, ducks leave accipiters behind, but can be flown down by a persistent falcon. The ease with which this is done depends on what water is available down below for succour. With water below, the duck will come down, but if no water is to be had, it will make a determined effort to outfly the falcon and head for distant water, making a flight of several kilometres. A big, fast falcon and a young falconer are needed

L-R: Dave Cherry, Robert Bagley (standing) David Kennedy, Joe Harmer, Dave Salem, Trevor Santochi, Eloy Carlin and Steve Chindgren at the 'House of Grouse' in Wyoming.
Photo Steve Chindgren.

for this flight. These flights usually arise when a falcon is slipped from the fist at a duck which is trading past; not always a sensible thing to do.

Occasionally, in wetland areas, high shepherding flights can be undertaken at ducks with large falcons. Beebe (1992) described these in some detail. In Spain, parts of South America, and Australia there is good duck hawking in many areas, before the land dries out too much, and high flights are possible, on a rather haphazard basis.

Woodcock, snipe, larks, and pipits can be flown successfully by small falcons waiting-on. This is an advanced flight. Snipe are not as easily intimidated as gamebirds, and are fast and agile. Usually they evade the first stoop and may then proceed to ring up rapidly, leaving the falcon behind.

There are many books on game hawking available now, giving specific details for regional conditions and quarry. If you plan game-hawking, attention to detail is everything. Do your home work, get out with some good game-hawkers, secure good ground and get your dogs steady and fit, all before starting a falcon. Game-hawking is addictive. Groups of friends hawk together year after year. Some move house to live in suitable hawking country. For them it is 99% hope, expectancy, desperation and strategy, just for the sake of the 1% of triumph when it all comes together.

10.10 Pursuit flights with falcons

This section is supported by the film *Northumberland Crow Falcons* and the book *Classical Falconry: a Treatise on Rook and Crow Hawking*.

Pursuit flights with a falcon start off as a direct flying attack and then turn to a tail-chase or pursuit as the quarry flees. The quarry has three choices: to get into cover where the falcon cannot reach it, to evade the falcon in the air using superior agility or speed, or to stay above the falcon. If the falcon is slipped in such a way that she can easily climb above the prey, then no high aerial flight will result; the prey has been dominated right from the start. It is left with only two options, and, if there is no cover nearby, it will probably be caught. The falconer may try to arrange such a situation for a young falcon to encourage her, but then will gradually make it harder and harder for the falcon, maintaining a careful balance between confidence and challenge.

Some species, such as most of the big falcons, can be trained for both waiting-on or pursuit flights, whereas some of the others, such as merlins, are better at pursuit. By and large, saker, gyr, New Zealand falcon, or merlin genes give persistence; gyr, peregrine, or merlin genes give speed; gyr, saker, New Zealand, or merlin genes give intelligence and courage; New Zealand or merlin give agility; and peregrine genes often make the bird more aerial and given to climbing above the quarry before pressing the attack. Saker and lanner genes makes the bird slower, although there is a tremendous difference in speed between different forms of saker.

The quarry needs to be of a species suited to this type of escape behaviour (see 10.4) and matched to the falcon. Larks for example, often go for height when pursued by a merlin, whereas pipits tend to try a low escape and put into cover, and are thus more easily caught. Starlings tend to be less approachable and have sorted out their escape before the merlin can get there. The starlings work in groups and use sentinels. When a falcon approaches, they either shelter in dense

A New Zealand falcon takes a mallard.
Photo Noel Hyde.

A peregrine/saker brings her rook down to us.
Photo Nigel Hawkins.

cover or take to the air in a tight flock, staying high. They can therefore pose a considerable strategic challenge.

For the larger falcons, gulls, and houbara have many similarities. It is difficult to hawk gulls regularly because they are in a different place every day and this makes it difficult to locate them and obtain permission. The other problem is that they are often in large flocks and a good falcon quickly gets in among them and takes one without much contest. It is best where they are in low numbers so that a one to one flight ensues. The same comment applies to rooks and crows. Houbara tend to be flown as singles but often prefer to face the falcon on the ground rather than in the air. If they can be got well up before slipping the falcon, there is a prospect of a long aerial chase which can last several minutes and go long distances. In Morocco, where houbara live near people, they can be approached more closely and may be dominated from the start, whereas in Pakistan they are more flighty and often produce quality flights.

Crow-hawking, first to last, is an exercise in risk management. Crow or rook-hawking on horseback must be the most exciting form of falconry available anywhere in the western world today. It is heavily addictive to those who try it! Thirty seasons ago, for example, an old friend, Tony Owens rode out with us. He had flown goshawks since childhood and dabbled with grouse-hawking with peregrines. After three days crow-hawking, his horse went down in full gallop during a flight and Tony was sent home with a suspected broken neck. The next day he was back out with his neck bound up in a neck support having arranged a new home for his seven year old gos over the telephone! Similarly, Peter Owens went down, man and beast, during a flight and nobody went to look for the pieces until afterward. Liam O'Broin found himself on the boat back to Ireland with three broken ribs, a broken shoulder blade and a broken collar bone. Two of our followers have broken their backs.

In pre-telemetry days, rook-hawking was a tricky business because it was so easy to lose the falcon. After some successes, I gave it up through

We are off hawking and we don't know when we will be back.
Photo Verity Johnson.

nervous and physical exhaustion, fortunately before losing my peregrine. Strict rules of procedure for rook hawkers, outlined by Jack Mavrogordato in *A Falcon in the Field*, while ostensibly aimed at producing quality flights, were mainly necessary to reduce the losses of trained falcons. Telemetry has had more effect on this type of hawking than on any other. Modern technology has helped overcome some of the hazards in modern falconry. We use telemetry and walkie talkies to enhance the sport, to reduce the risk of losing a falcon, and to save the horses from needless exertion. Nobody enjoys the misery of a hawk left out, or of a long hack home on a horse which has already given of its best and which will be going out again the next day.

Anyone who has trained and entered a young hawk or falcon knows the work this entails. When you have a whole team of birds to get going, the problems are immense. We judge our birds not by the best ones, but by the worst. It is easy to have a couple of star performers and the rest nowhere. The rule is quite simple, and this applies to game hawks and dog teams just as much: worst first. Always take the worst bird up and fly her first. If time runs short, the 'Ace' may not be flown. But she will not be too much the worse for it, whereas the bad bird will have improved. Of course there will be times when you have to break this rule; these are when flying conditions or quarry are too difficult for the young hawk but bearable for the good one, or when there are visitors out and you wish to guarantee sport. Many falconers put themselves under too much pressure at the beginning of the season by inviting lots of people out when the young hawks are not ready. So the old trusty ones get an airing and the youngsters get short shrift. A little willpower is required.

The system we now use has been developed in stages over the last thirty seasons or so and it works. I have described it in *Classical Falconry: A Treatise on Crow and Rook Hawking*. As in so many things in falconry, it is the details which are all important for success. We use Rocrows to train the young hawks.

The first step is to prepare the land. Few people realize just how open the land needs to be

Our hawking grounds in Northumberland.
Photo Verity Johnson.

until they have been out rook or crow-hawking properly. Most people who hawk rooks use a vehicle and make do with land which is sub-standard. The alternative is to hawk on foot. Using a vehicle, the usual method is to cruise along the farm roads until some rooks are spotted, then to drive or walk to a suitable place from which to slip, if possible. Once the hawk has been slipped, one stands there like a potted plant, watching the flight in the field-glasses and then run or drive to pick up the bits. This is as near to classical hawking as a car follower gets to fox-hunting. Only half the experience is there.

To hawk from horses, the land, first of all, needs to be both hawkable and ridable. This means open, with a minimum of 500 metres between cover. Cover includes any woodland more than a single straggly tree or bush, bracken beds, or any crop over about 50 centimetres high. Wire fences, stone walls, flocks of sheep, rushes, or bracken will ruin it for an unhacked peregrine but most hybrids will not be deterred. Crows usually need to be bettered in the air before they bale out for cover, so you may well still get a good flight. The land should preferably be gently undulating so that you are constantly exposing new vistas. Flat ground may be marvellous for the first flight but then there's a long way to go to reach undisturbed ground. If the land is at all hilly, large basins are best as they hold the flights inward. Convex domes going down to blind valleys, as in many of the Welsh hills, are disastrous. The flights always slide off into bad ground.

There are parts of Europe open enough for rook and crow-hawking, including Poland, Hungary, and Spain. Some of these areas depend on wintering rooks and offer ideal country for the flights. In North America, there is plenty of open country, but only sparse numbers of corvids, and the falconer would have to drive a lot. It remains to be seen if North American falconry will develop classical ringing flights with large falcons; the problem is suitable legal quarry.

The land must hold sufficient quarry; but not too many. Our land runs from grouse moorland and high fells along the ridges. This has very low densities of crows and is difficult riding but

acts as a good buffer zone of open country. The middle belt of 'white ground' is the best. This has a scattering of strong moor crows, the bane of the shepherd and of the gamekeeper. This white ground is permanent rough grazing for sheep and cattle. In Wales, most of the white ground is poor riding because it is either boggy, rocky, too steep or all three. In the north, the going is a lot better and in many places it can be ridden at speed. You often have to cover a fair bit of ground to find crows but the scenery is enjoyable and, when you finally come up to some crows, you have a good chance of a sporting flight, a gallop and a kill. White ground runs down to more intensively managed country; smaller enclosures cut for hay or silage, and often of re-seeded leys which are easily damaged by horses. However, they tend to hold a lot of crows and rooks. They are also more heavily interspersed with cover. The farmers will not thank you for taking a lot of horses across this land and therefore we tend to keep it for entering young falcons at the rooks, which are much easier than crows, or for days when the wind is too high to fly the higher, more exposed, ground or for days when there are only two or three horses out, and even then the falconer will usually dismount and run over sensitive ground to the kill. The fastest action occurs on the low ground if one can arrange access. The ground rides well and there is plenty of quarry; sometimes too much.

In terms of land area, we use about 10,000 acres in Wales and about 80,000 acres in the north. I believe Roger Upton, who hawked rooks in Wiltshire, used about 14,000 acres for his team of falcons. The ground is really of not much use at all if it is in blocks of less than 2,000 acres. Obviously it depends on the density of quarry, but we find that if we move off at 1pm and get back to the horse boxes around 6pm, having ridden 10-15 miles, the very least we cover is 2,000 acres and it is often double this. Therefore we manage our ground like a hunt country. It takes an hour to box across to the far end. Mounted meets three days a week requires a great deal of liaison with the farmers and a lot of work to obtain their goodwill. We therefore hawk in an organized group under the control of the Master Falconer and sometimes also, a Field Master. Green coats are worn by the members so that both the Field and the farmers can identify who is responsible for what. Except for certain meets, Fields are restricted to twelve horses or less, with no foot followers and no cars except the hawking van. From the farmers' point of view, our aim is to kill as many crows as possible with the minimum damage to land or stock. From the Field's point of view, they of course want to see high class sport, good country, good flights, and good gallops.

Our hawking grounds in Northumberland.
Photo Verity Johnson.

If the crow puts into a wall or bush, the falcon needs to follow, despite dangers such as barbed wire.

When you realize how much work is entailed in obtaining suitable land it is obvious that classical mounted crow and rook-hawking, as practised by the Old Hawking Club, is not feasible except for a limited number of falconers with the time, opportunity and finance. It is a major logistical exercise. But the individual falconer can have a lot of fun with just one or two falcons at very little expense, provided he has some land and transport.

The next piece of this jigsaw puzzle is to arrange suitable hawks. The Old Hawking Club preferred passage peregrines for rook-hawking but these are not obtainable in Britain and,

This crow put in to water and the falcon took it and dragged it ashore.

anyway, pure-bred peregrines are not first choice for crow-hawking. This is because they are aerial specialists. Some of our best performers have been peregrine/sakers, gyr/sakers and gyr/New Zealands. We need a falcon which climbs straight up without ringing, is a persistent chaser which will not give up, and which can still take the crow on the ground in light cover.

One of the annoying traits of peregrines, unlike the desert falcons, is that often they will not start at crows on the ground. This means that you either have to slip at a crow in flight which can entail some fairly fast thinking and some haphazard results, or you can flush the crows yourself, which immediately puts your falcon at a disadvantage. A good crow hawk is acutely self-aware and knows the importance of surprise as well as height. Often the falcon, once she is slipped, will fly steadily in toward the crows, keeping in close alongside the cantering horses to mask her approach, before sprinting forward into the attack.

For the basic approach to hawking we use a similar system to that described so ably by Roger Upton in his book *Falconry: Principles and Practice*. Very large falcons, birds weighing over a kilogram, can be too intimidating for rooks and crows (not to mention grouse), so that the quarry flies low or seeks cover, thus ruining the flight. Ringing flights are more frequent with smaller, less intimidating, falcons.

Strong winds are no longer quite the hazard they used to be in pre-telemetry days. The more a falcon is flown in the wind, the better she gets, until by the time she has forty or fifty head to her credit she exploits the wind as much as the crows do. When the clouds are scudding by and conversation is reduced to the utilitarian, real hawking is to be had. Slipped from a windy slope, the falcon opens her wings and, without moving a feather, simply shrinks up into the sky. Climbing along under the cloudbase over the crows, her attack - a convex hook - is effortless and uninhibited by the nearness of the ground. The straggling crows are scythed and fall about in confusion. The falcon turns and blows into one with an economy of effort that marks her as a bird worth riding your heart out for.

Strong wind magnifies everything out crow-hawking. The successes etch on the memory, the failures are more disastrous, the runs more gut-busting. A good crow hawk is really spectacular in strong wind, possibly the best hawking modern falconry has to offer. But an indifferent performer may get disheartened and will certainly cause a lot of grief. To fly crows in the stormy hills in winter in a Buchanesque landscape is to experience hope, uncertainty and despair, desperation, elation, and physical exhaustion, a privileged insight into the life and mind of a matchless falcon and her ancestral struggle for survival.

Falcons usually like crow meat provided it is still warm, once cold, they find it unappetizing; some falcons will eat it then but goshawks are reluctant. Fat hawks in aviaries normally leave cold crows. It is good policy to give crow meat to young eyasses in the breeding pens occasionally so they learn to accept it.

Rooks are more adroit than crows but do not have as much heart. Smaller falcons are better for them than big ones. When rooks billow up in great clouds, they can bewilder a young falcon, but an experienced falcon takes a rook out of a flock very easily.

Jackdaws are a more difficult proposition. It is hard to get a falcon above them, they are so wary and buoyant. The best falcons I have had for these are male gyr/New Zealands or peregrine/New Zealands.

Magpie hawking is a completely different flight to crows because the magpie has no speed and must resort to subterfuge. Magpies are flown using falcons constantly in a position of aerial dominance, and the country needs just the right type and amount of cover, not too much, not too little. There are some ideal places in North America and in Spain. A magpie will not stay in the air with a falcon unless it really has to, so there are no long tail-chases or rings. To hawk magpies well you need to be young, fit, on foot and prepared to hear quite a lot of colourful language.

When hawking on horseback, ringing flights are valued as a source of a good gallop, similar to a good run with hounds. Indeed, heron plumes were occasionally presented to those first in at the kill. When hawking on foot, one is left with the option of standing still and watching through field-glasses or running like mad, missing the spectacle and probably resorting to telemetry to determine the outcome. What action there is, occurs so high and far away that the birds, if visible, are just tiny specks. Seldom does the kill occur close at hand, and a falcon giving up on such a high flight can be tempted by fresh quarry over a very big area. Therefore to enjoy (and survive) good ringing or high flights a hawking pony is essential.

Judging the distance of a falcon in the air, either waiting-on, or during a flight, is very tricky and estimates are seldom worth the paper they are printed on. The contestants usually seem much further away than they really are. Often a flight high in the air seems a long way away, but as the birds come down to the horizon and eventually near the ground against features which are measurable on a map, it turns out that they are nearer than expected. In our terrain, with 7X or 8X field-glasses, the details of a crow flight get hard to follow at more than 1000 metres once it drops below the skyline. With better field-glasses, in good light, with a clean background, one would do better. Usually what happens is: while the falcon is in the sky and beating her wings, she is just about visible, but when she folds up and stoops, she completely disappears. If she misses and throws up, a wing may glint briefly before the second stoop and the other crows will indicate a kill. A miss at this stage almost always requires telemetry because it is hard to spot the falcon again once unsighted and in a few moments she could be anywhere. When mounted, as a rough guide, a high flight with a few rings, followed by a race for cover, usually travels about twice as fast as one can gallop. It is an amazing sensation when you get a high flight, and you gallop like mad, to have the falcon float down with her prey right in front of you, and you peel away with no brakes.

In my experience there are two types of high flight: the ringing flight and the straight climb. Birds with a low wing loading such as the heron (0.37 g/cm^2), red kite (0.31 g/cm^2), and to some extent, the rook (0.31 g/cm^2) climb most rapidly and easily in rings of about 50-100 metres diameter. The same applies to the lower wing-loaded falcons such as the saker types at about 0.60 g/cm^2. On the other hand, the crow which is more powerful than the rook, and at 0.42 g/cm^2, heavily wing-loaded for its size, is capable of standing on its tail and flying directly upwards in a steep climb. The result looks desperate and ungainly as the bird flaps hard, thrusting itself up almost vertically with hardly any horizontal travel. It will only do it when pressed by a falcon of similar capabilities, such as the peregrine/saker at 0.69 g/cm^2. Such a falcon will go up with the crow, staying only a metre or two behind. The crow cannot go down, nor can it go sideways. If it went sideways it would momentarily lose height in relation to the falcon and be immediately attacked. It can only go up. The experienced falcon will keep climbing until she is about a metre above the crow. Then she will turn over and lunge at the crow which will dive and roll to avoid the outstretched feet of the falcon. Both swoop and throw up together and the aerial contest is resumed. The purpose of this little manoeuvre is to test out the crow; will its nerve break and will it dive for the ground? Has it got the strength to avoid the falcon and the willpower to renew the upward climb? The falcon has a better measure of her prey and resolutely struggles on upwards. This time she climbs higher and higher above the crow. Two metres, three metres, five metres.... The point is reached at which both the crow and the falcon realize that the situation is now untenable. Simultaneously, both dive in a corkscrew stoop. Both are by now several hundred feet up and may even be in cloud. Down they come in the canceleer, the life or death spiral stoop. Below them the falconers are cantering gently, trying to anticipate the outcome, ready to burst into a gallop if needed, willing the falcon to bind. The two dots in free fall suddenly come together. She has him! Everyone yells and waves, the horses mill about, momentarily forgotten. Abruptly, the descent is slowed like a parachute opening. The falcon and her prey float down and drift to the ground like a sycamore seed. The riders converge from different directions, anxious to be there when she lands. On the ground, the crow contests the falcon with hoarse croaking and the falconer slips from his horse and quickly kills it, offering the brain to the breathless falcon. The horses, flanks heaving, jostle and toss their heads, their riders excitedly relive their version of the flight. The falcon, too exhausted to feed, is intent on breaking the neck of the dead crow, her face buried in the shiny black feathers, holding on grimly like a ferret and twisting and jerking with her beak. Finally convinced the crow is dead, she eats the brain and sets to plucking the body. The fluffy crow feathers stick to her eyes and she wipes them impatiently on her shoulder. The falconer's horse has set to grazing and it is time for a dram.

Not all flights have such a happy conclusion. The crow, by twisting and rolling in the descent, may manage to make it all the way down without being caught. Both birds then must lose speed rapidly and they flatten out just above the ground. The crow has been desperately seeking any refuge. If there is none then he is surely taken; on the ground, in the open with a peregrine/saker above him, his options are severely limited. But he may find a tussock or a sheep, or even a falconer's horse. If the falconers are near enough, they can rouse out the crow by riding up, shouting, and cracking whips. If the cover is small, the falcon can probably rouse out the crow herself, chasing it from sheep to sheep until finally the crow's luck runs out. But if the crow has made it to a plantation, then the flight is over; he is safe for another day. He will sulk in there, regaining his breath, his dignity, and finally his cockiness. The falcon, frustrated, is taken down to the lure, hooded up and rested while another falcon is prepared. Not until the party is well gone and all is quiet does the crow emerge, wiser but unharmed.

A high flight like this may take five to ten minutes. If there is little or no wind and the flight starts in open ground there is a good chance of a high flight without the crow reaching cover. This is achieved by the falconer standing off from the crow by 200 metres or so. The falcon is unhooded and slipped, either when the crow is in the air (in which case timing is critical), or at a crow on the ground. As soon as the falcon starts, the falconer shouts 'Au volée!' to alert the other members of the Field and to flush the crow so that it can start its climb. Of course it may not climb, but go for some nearby unforeseen escape route or it may try to outmanoeuvre the falcon. In these cases, in still air, with an experienced falcon, his outlook is bleak. But if he is a strong crow who is not so confident that he thinks he can outface the falcon on the ground (as he could a wild peregrine) or too intimidated because the falcon is too large, he will take to the air, and the falconer will immediately ride underneath to persuade him that he has taken the best option. If, on the other hand, the falconer slips the falcon in such a way that the falcon is above the crow when it reaches it, no high flight will ensue. The crow, dominated from the outset, will easily be driven down. Thus the game falcon, given to checking at crows when waiting-on, knows that she can easily master them from such a height advantage; they are like moorhens then.

So, to obtain the high flight, the falconer is trying to slip his falcon in such a way that, when they come together, the falcon is below the crow, cutting it off from the ground and forcing it into an aerial contest. If such a contest lasts five minutes and the day is windy, the flight will drift. In a 12 kilometres per hour wind the flight will go a kilometre, in a 12 miles per hour wind, a mile. Without even trying, the crow thus has access to a kilometre or two of downwind ground. If this is completely open, all well and good. But few of us are fortunate enough to have such plain ground and it is more likely that somewhere in this distance, perhaps slightly to the side, is some sanctuary for the crow. The crow will keep his head. He knows his ground intimately. Although too busy avoiding the falcon to have time to look around him, he knows where he is and he calculates his chances. When he sees his moment, he begins his dive for safety, flinging himself about wildly to avoid the last desperate attempt of the falcon. The experienced falcon too is not stupid. She knows his game. She will keep herself between the crow and the cover, shepherding him away and anticipating his moves. When he does make a break for it she will exert all her energy into one final effort, scything down to skim him off the fringe of the trees. And as she bears him down to some safe spot she will hear above the noise of the trees the cries of the Field which have turned from despair to triumph.

In late summer there are days when a breezy day goes quiet toward evening. The falconer keeps back a strong climber and some good land for this time in the hope of setting up a good ringing flight. Each season we get just a few really high classic flights that none of us will ever forget.

10.11 Field meets

Whenever two or more falconers hawk together, they need to agree certain ground rules. This will ensure the safety of their hawks, the success of the day's sport, and the longevity of their friendships. Field meets thus vary from an informal mini-meet, to an international meeting lasting several days. The British Falconers' Club International Field Meeting, held every four years, lasts four days and organizes 67 hawking parties over 240,000 acres. To achieve this requires major organization by dedicated local falconers approaching landowners, mapping boundaries, arranging dates and meeting places, which falconers will be coming, with what hawks, dogs, and visitors, inviting honoured guests, arranging control of Press coverage, accommodating all participants, arranging evening entertainment, and providing a formal Landowners' Dinner.

NAFA's annual meet usually attracts 350 or more falconers, not to mention visitors and families. Again, all arrangements must be made, although the hawking grounds have to be

A goshawk field meet.
Photo Leigh Tovey.

A crow-hawking Meet in Northumberland.

sorted out by the individual falconers. This can mean that small groups go off and do their own thing, whereas in Europe the groups are more organized.

If you plan to attend such a meet, then it is up to you to ensure that you have organized yourself properly. All legal details and permits should be attended to well in advance, as well as travel plans and room bookings. Your hawk should be an experienced hunter, not a raw novice, fit, and in good flying order. An accipiter which is

A wild saker takes a young crow.
Photo Stig Olsen

liable to be unsettled by the change of scene will need to be on the sharp side of her flying weight, but with enough condition to stay strong despite long days in bad weather. Your gear, from your transmitter mount to your car tool kit, should be clean, checked, and in good order and you should have with you whatever you need to cope with eventualities, including a detailed map of the area.

Most organized field meets have a Field Master who is a senior falconer who knows the ground and is responsible for the whole group's activities for the day. You should report to him and do what you are told. If you follow in convoy, then help the car behind to keep in touch. I remember once in north Wales, after hawking, we were a convoy of about 16 cars trying to find our way back to the hotel in the dark. We drove and drove down more and more tortuous country lanes, and ended up all blocked into a dead-end farm track. It turned out the leading car was a young couple trying to find a quiet place for a cuddle, and extremely embarrassed to find a whole convoy of muddy falconers crowding round wondering what was going on up front.

A few simple rules are needed when hawking in groups: keep spectators together and away from the hawks and game. Avoid what Jack Mavrogordato used to call, somewhat rudely, 'inconsequential female chatter'. Out hawking or hunting, stay quiet, avoid 'coffee-housing'. Post markers in blind areas to mark flights. Shout 'Hawk free,' 'Hawk secure' when flying accipiters. Agree who is to slip next. If you slip out of turn and catch someone else's hawk then don't expect much sympathy. If you are hawking in a line then stay in line and keep an eye on everyone else. The object is to go slowly and find game, not to race to the other end. The best treatment for a persistent 'racer' is to load him down with dead hares and rabbits to carry.

If there are slow hawks, such as buzzards, present, the easy slips, such as moorhens, will be reserved for them. Field Masters should aim to give every hawk opportunities within their capabilities so that everyone gets a kill or a good flight. When a falcon is waiting-on and the cover is thick, such as a big field of crops, don't leave the falconer and his dog on their own pathetically

Hawking in Japan we were graced by a young lady in traditional dress. Notice the *Kumatakaor* Mountain Hawk-eagle *Nizaetus nipalensis.*

running round trying to flush game for the tiring falcon, get in there and help him serve her. While some senior visitors and ladies may be permitted to watch 'from the gallery' and even be excused for a little 'coffee housing,' able-bodied people should do their best to help.

Falconers should declare the state of their hawk beforehand to the Field Master. Tell him if your hawk is poor on feather, good on fur; a bit high, better flown later; flies free but has never taken quarry before, takes other people's hawks off the fist and so on. Similarly, they should declare their dogs. Big field meets are not a place for training or even for birds which are not 'spot on.' It is infuriating to work hard for a good flush for a bird, only to be told afterward by the falconer that his bird has only ever been flown to the lure before. It is also amazing how many falconers go to field meets without dogs and tag along, expecting you to find all their slips for them. What do they do at home? When you eventually serve their hawk, they muff the flight and then blame it on the dog frightening their bird. One's smile becomes glassy.

A non-hawking member of the field should be allocated to assist each falconer by marking his hawk and accompanying him on long flights, so that if the hawk is lost he can relay back for more help. One soon gets to recognize certain individuals whose birds spend a lot of time Away Without Leave. If they don't return within about quarter of an hour, then carry on hawking, but try to ensure that the next bird slipped is as big or bigger than the missing one. Some Harris hawks are very crabby with strange birds, even on the fist. When faced with bad behaviour, the Field Master may send the miscreant home or cross him off the list for subsequent days. He may then be unable to find any party prepared to take him.

Similar escorts should be provided for visiting guests and overseas visitors as a matter of courtesy and hospitality. We always appreciate visitors, and making return visits to them, and this is one of the charms and attractions of big meets. Similar efforts should be made for young and inexperienced participants, so that they feel welcome and not cold-shouldered. Many are too shy or intimidated to approach people, and it is up to everyone to ensure that they are not left out. Otherwise first impressions become last impressions.

All falconers present should check their transmitters to avoid duplication of signals, and receivers should be carried in the field, if not by the falconer, then by an assistant. Losing a hawk on strange ground at a field meet is not good for relations with landowners, and at the very least wastes everyone else's precious hawking time. Transmitters should be switched off when not in use, particularly at the meet site, otherwise a lost hawk is hard to recognize from all the other signals.

If a flight takes the falconer away from his party, it is permissible for him to attempt an immediate re-flush of game which the hawk has put in, but not to seek fresh game. He should rejoin the party as soon as possible and not go hawking on his own. Permission from the landowner will have been obtained for that specific group on that specific day. Under no circumstance should a falconer return later and hawk the ground. This will arouse the wrath of the landowners and of the local falconers and will jeopardize future hawking in that area.

Field Masters should do their best to avoid random flushes caused by uncontrolled dogs or beaters. Random flushes increase the chances of two hawks being slipped and crabbing and also cause long unsuccessful tail-chases which disrupt the party and delay the hawking. Beaters or people working dogs should say to themselves 'If I flush quarry now, what hawk is there in a position to have a good chance of catching it?'

Some quarry such as pheasants, partridges, grouse and quail, will run ahead of the flush when they hear a party approaching. Field Masters should ensure that a falcon is put into the air or a hawk into the trees ahead to hold the quarry down or, if this is not feasible, to work the quarry away from holding cover so that it runs into a flush point rather than into thick cover.

Field Masters should warn falconers of prohibited areas or dangers such as hidden roads, electricity lines, deep water, swamps, or other impassable places. There are four main classes of water that interest the falconer: water that is too shallow for quarry to dive into and escape, water that can be waded in or across, water that is clear to swim, and water that is clogged with floating weed, swamp, or ice which will support a hawk on a kill and yet which is impassable to falconer or dog. This last water is the most dangerous. If the kill is out of reach of branches or poles, it may be possible to throw a creance across the area with a weight on it, so that it forms a V. By walking around the pond one can sometimes wind up or snag the hawk or its kill sufficiently to drag it to the bank or to make it lift off with the kill. Alternatively, if an opened previous kill is attached to the middle of a creance, by walking around the pool, it can be dragged to the hawk who may be coaxed to transfer its attention to it and then be slowly dragged ashore.

While clubs in some countries impose a semi-formal bag limit of four per hawk per day, the bag really depends on a variety of circumstances, ranging from prey density and ease of capture, to the experience and fitness of the hawk and to future hawking plans. What is inexcusable is to waste quarry that has been killed; almost all quarry is suitable for consumption by hawks or humans. In European countries, the best of the quarry is normally donated back to the host landowner. If you live in a state where there is a 'let lie' clause on some quarry, such as waterfowl, you must ask yourself whether the wastage of game justifies multiple kills.

Falconry is an exercise in cooperation between hawk, man, and dog. Field meets are no place for a competitive spirit; there is no need to give a prize to the person whose hawk caught the most or flew the highest. A good falconer who has served his hawk successfully will stop flying to help those in the party who have been less fortunate.

Since the advent of Harris hawks in falconry we now have the phenomenon of 'gang hawking' - several hawks being flown loose together. Under most circumstances, more than two hawks are not very sporting.

10.12 Hawking dogs

Over many thousands of years in association with man, dogs have been bred for different tasks. Of the hunting dogs, we have flushing dogs, such as spaniels and terriers, pointing dogs, coursing dogs or gazehounds, and scenting

A hare-hawking team in Cang County, China, with hooded TuHu (a passage saker).
Photo Alan Gates.

Teamwork.
Photo Steve Magennis

hounds. For sheep and cattle, we have heading dogs, huntaways and heelers. For guarding, we have Alsatians and mastiffs. There are also of course companion dogs, fighting dogs, hairless hot water bottle dogs and edible dogs, of which no more needs be said!

Of the hunting dogs, the gazehounds and scenting hounds were bred to catch and kill the quarry themselves, the flushing dogs were bred to find and flush game for men armed with bows, nets, spears, clubs, slings, or hawks. In recent years guns have been added to the list of weapons and it has become fashionable to refer to the pointing and flushing dogs as 'gun' dogs. But do not let the dictates of vile saltpetre blinker your thinking - falconers were there first!

My own experience of dog work includes springer, cocker and Brittany spaniels, German short haired and wire haired pointers, English pointers, Irish setters, Munsterlanders, lurchers, greyhounds, and salukis; scenting hounds for hares, foxes, coyotes, and deer, and huntaways, heading and eye dogs for stockwork. Normally we have 2-4 working dogs here at any one time. Through days shared with friends one gets to see other breeds and other handlers working and there is always plenty new to learn in this complex and, by falconers, often neglected subject.

Although a few flights, such as crow-hawking, do not require a dog, dog work is important for most types of falconry. Even if you are lucky enough to live in an area so rich in game that you can get slips regularly without a dog, you will not easily be able to serve your hawk 'in the retrove' (ie re-find game which she has flown in and flush it for her to kill on the second flight, a common occurrence in pheasant and partridge hawking).

I remember my first springer, 'Smokey,' whom I trained 'by the book.' He would drop to flushed game, retrieve from the other sides of rivers - everything. But I soon found that when you're hawking with a goshawk the last thing you concentrate on when a pheasant flushes is the dog - and he knows it. He'll follow the bells to the end of the earth and between him and the goshawk the pheasant will have a hard time. I soon learned that a dog sitting on its backside is not a very useful piece of apparatus and that while we were going through the gundog rule book together, the pheasant would be legging it for the nearest rabbit hole.... So nowadays I expect my dog, on flushing game, to be there with the hawk. On the few occasions when I want a careful flush, such as when flushing a covey of grouse one bird at a time, with a pointer, I drop the dog verbally or by whistle or eye, and, if the dog is in its first season or two, put on a checkcord for the flush.

You need to be clear in your own mind what your policy is with ground game. When the dog is young you must be able to stop him and this can be linked in with his steadying to sheep. But it is a mistake to be too firm with a dog on rabbits. How can the dog be expected to divine that moment between legitimately working a rabbit and what you consider to be rioting? One of the critical things in hawking is the timing of the flush. When the dog is on point and you say'Get him up!' you want the dog to go in with conviction and flush immediately, not in fifteen seconds' time. A dog that prissies about instead of flushing cleanly (which it will start to do if you keep checking it) can quickly ruin a young falcon or even cause it to drift away and be lost. Secondly, when hawking rabbits, few flushing dogs can move faster than a hawk. I prefer it if the dog stays hot on the rabbit and keeps it moving. When the hawk catches the rabbit, the dog should lie down beside the hawk and honour her. If the hawk is having trouble with the quarry the dog should quickly kill it by nipping the back end and then leave it for the hawk. The dog shouldn't shake the quarry or disturb the hawk. When the hawk is down on a kill, especially on bird flights which finish some way from the falconer, the dog should stay beside the hawk and guard her from interference by wild predators, especially eagles and buzzards, and, in farming areas, curious cattle. All this requires that the dog should be reared with hawks and be taught to treat them with total respect as an extension of you, the falconer. I have trained springers, wirehairs, and a lurcher in this way, working with goshawks and various falcons. For further details see Frederick II:267.

When using a pointer for grouse, it is often possible to get several flights at the same covey, if the dog continues to hold the point during each flight. However, if you live in a place like Wyoming, which Bob Berry tells me holds many thousands of golden eagles in winter, you may well prefer to train your dog to run and find the falcon and stay guarding her.

Usually, toward the end of its life, a spaniel develops the habit of coming out from the cover and running round the blind side to see if the quarry has flushed yet. This is annoying and should be stopped. Always shout 'Ho!' clearly when the quarry flushes and rate the dog if he comes out before the shout. Of course there are times when he has to come out as the quarry may not be in there after all - but then it's up to you to read the situation right. The shout of 'Ho!' should be his signal, not the sound of hawk bells.

The hawking dog's main role is to find game, mark and flush it to order. For waiting-on flights this usually entails a pointer or setter quartering the ground and holding point. If you have enough work to warrant such a specialist dog, an English pointer is hard to beat. If, on the other hand, you expect waiting-on flights for only 3 or 4 weeks and have more varied terrain to cover later in the season, a GWP, GSP or Munsterlander is a better all round choice. Whatever it is, it should get well out, at least 200 metres each side of you, more if the visibility is good. Under good conditions the dog may cover up to 800 metres (about 1/2 mi.) between turns and so the falconer needs to walk only slowly. If a car is being used, the dog can be allowed to range out further but a telemetry collar is sensible. A dog lost in American sagebrush country runs the risk of coyote and mountain lion traps, quite apart from natural hazards. Under good scenting conditions, the dog may draw scent from birds up to 200 metres upwind, but in calm weather, he may be able to detect the birds from only a few metres away. In good conditions then, the dog may take in 200 metres at each sweep and range well out, but in poor conditions he will need to work more closely and carefully for fear of missing or bumping birds. The birds themselves vary in their reactions. On some days they may be very wild and run ahead and flush off before the dog can get near enough for a solid point, on other days they will often flush spontaneously after a minute or two and it is easy to think it is the dog's fault when it isn't; best to put the

An English Pointer at a British Falconers' Club Fieldmeet at Grinton.
Photo Steve Magennis

falcon overhead without delay when the birds are getting wild or cover is sparse.

It is a pity that many falconers tend to train their dogs according to shooting field trials precepts rather than to their own needs in the hawking field. We don't have field trials for hawking dogs, mainly because no two falconers can agree on what their dogs should or should not do. Long may it continue! For my own part I expect a minimum of handling. Once the dog is more than a few metres away from me I use a whistle or hand signals. Nothing frightens game more, or ruins my afternoon so quickly, as someone who roars at his dog for some minor or imagined misdemeanour. The mark of good dog handling is calmness and unhurriedness and enough relaxation to be able also to concentrate on the hawk, the quarry and on appreciating the scenery. Foresight, preparation, and avoidance will prevent many problems in dog handling together with the clear understanding that one is hawking for ENJOYMENT, not for peak efficiency. I do not expect the dog to work every single scrap of ground, to sit every time I stop for a pee, or to quarter like a metronome. I hawk to escape the rat race, not to create a new one. If your top priority is to kill game you may as well use a heat-seeking missile or a good pesticide.

There are however, a few times in falconry when the dog needs to do the right thing. One of these is when obtaining a good point for a waiting-on falcon, one is when working with an accipiter which is frightened of the dog and one is when working ferrets and the dog must sit quietly. A dog should be trained and prepared methodically beforehand. Never expect to train a dog when you are working a hawk; train the dog first and then work them together.

The kingpin of dog training is the 'drop'. Over long distances I use one long blast on the whistle; close to, I click my fingers quietly. If you can drop your dog any time, any place, then at least you can 'switch it off' and sort out whatever disaster has occurred. Get this one command certain and the rest is just icing. The drop is a positive command, whereas 'No' is an open-ended command which leaves the dog in limbo.

A plea here to those with pointers on the moor. Train your dogs to heel and walk on a lead properly. It is no fun for a member of the field (often of the fair sex) to be towed around a hot grouse moor by a headstrong dog while you swan ahead in the distance.

Where there is heavy cover, a general-purpose pointer is of more use, particularly if it has a good coat, as has a GWP, Munsterlander or a Brittany spaniel. It is better, in my experience, to have one or two dogs that you know well, to do a variety of jobs rather than have several specialists. A wirehair may not be as suitable for grouse-hawking as an English pointer, but you can get more solid points over a good dual-purpose breed than an over-eager English pointer with no brakes.

Often one finds that while the initial point was in the open, such as stubble or a root crop, the quarry puts in to such thick cover that a thin-skinned pointer cannot evict it at the 'retrove' and dances around on the outside. In these circumstances you need a dog with a dense coat and a certain amount of railway train in its breeding. Dogs soon learn to alter tactics according to the situation and will benefit greatly from the extra experience and from the rapport that you have with it. I don't think it is possible to over-emphasize the importance of frequent and regular work for both hawk and dogs. Behind every good performer is a background of solid work.

For very dense cover there is nothing to beat a good English springer. They will go through the brambles like a dose of salts and if they give a couple of barks when the scent gets red hot, I regard that as no fault; it sets the goshawk up wonderfully. The cocker is good too, but he cannot take full days for long and maintain condition. Teams of 6-8 spaniels used to be worked in the woods with a goshawk and such an arrangement has terrific flushing power. Even two or three dogs will confuse and flush game which would give one dog the run around. For really dense cover and holes, a terrier or wirehaired dachshund is useful but they cannot keep up if you have a lot of running, and they are easily unsighted in long cover. They are not known for their obedience and I worry about their testicles in the snow.

Some recommend using spaniels to flush game for pointers. I think grouse-hawking is tense enough without introducing a spaniel into the focal point. I prefer to flush the grouse myself with a stick, sending it back over the pointer, rather than use a spaniel which can make a pointer as unsteady through jealousy as it might become through flushing the game itself. With a stick, one can flush the grouse with better timing and better directionality than with a spaniel, particularly if you can read your pointer well. If you hawk rabbits or hares a lot in open country, then a lurcher will improve the quality of the sport. A rough-haired greyhound/collie cross, about 22 inches high, is large enough for both quarries and has plenty of brains. A good lurcher can be taught to point and, when unsighted, can follow by scent at a useful pace. Only one running dog should be slipped with a hawk, otherwise the quarry is over-matched and the hawk risks being injured through jealousy. This type of hawking has a long history in the Middle East, Asia and parts of Europe but has been neglected by many falconers, particularly in America where it has great potential.

To work well, dogs need to be fit, preferably with plenty of roadwork behind a horse, bicycle, quad or slow car. Our pointers get at least five kilometres at 20 kilometres per hour most days of the year except work days. This keeps their pads hard and their muscles firm. It also extends their working lives, often by two or three seasons, and remember, an extra season at the end of a wise, experienced dog's life, is worth much more than trying to cram in work too early in a young dog's life, through impatience.

Of course, there will be times when the dogs must be confined in soundly-made pens. But these, like stables for horses and blocks for tethering falcons, are necessary evils for the benefit and convenience of the owner. No animal is improved by being confined for long periods. When you are working with a lot of animals, ones which need to be able to 'read' you and you to 'read' them, at a subliminal level, one of the hardest things is to maximize contact time, just being together.

When choosing a dog, first see plenty of specimens of the breed actually working and also check on similar breeds which may have more

With HPR breeds, such as the Munsterlander, the same dog can be used both for lowland game and for upland game.

potential. Then see one or preferably, both parents working, with particular attention to physical conformation and well-being, intelligence, drive, and nose. Do not bother too much about the parent's training - this is not inheritable! Having bred several litters of pups myself and followed their progress, some to be champions, some to be straight workers, I know that most, if not all, have the potential to be champions, but it depends whose hands they fall into. The ones with that extra, are often the least likely to shine in field trials.

Also for slow developers, such as the wirehaired pointers and those breeds required for multipurpose work, it is essential to build a firm foundation of discipline and fieldwork. Few are ready for serious work before three years old and should certainly be kept steady until about four years old. Never, at your peril, put a wirehair loose in a car, always use a dog box. A wirehair can remove the seats, side panels, roof lining, trim, sun visors, and all handles in fifteen minutes. Or less.

I had a cocker spaniel called 'Sue' in New Zealand whom I used for hawking California quail. The quail, on putting in, would flutter up inside the 'Bush Lawyers' a particularly nasty type of three dimensional crucifix. Sue would struggle to climb up after them in the thorns. Once she climbed about 25 feet up a Macrocarpa tree after a possum. But, with her short legs, she'd get tired on long days, and I taught her to jump against the side of the horse so that I could pull her up to travel home in front of the saddle. Like many spaniels, Sue was virtually impervious to punishment (I'd received her as an old dog) and I never succeeded in devising a treatment which had any effect on her, short of causing actual bodily harm.

Nothing cements the bond of companionship more than the hunt, and one's dog is the ultimate companion, always eager when you reach for the hawking bag, always cheerful when you're low, never asking questions. Many a man who treats his dog sternly in public is soft as fudge when they're alone. A certain late president of the Irish Hawking Club had one of our pups, and I remember hearing his voice change when he thought nobody was listening. Yes, a hawking dog is something special.

10.13 Further Reading.

Allen, M. 1980. *Falconry in Arabia*. Orbis, London. ISBN 0-8289-0481-2.

Anon. 15C. *A Perfect Booke for Kepinge of Sparhawkes or Goshawkes*. Quaritch 1886.

Beebe, F. L. and H. M. Webster. 1964. *North American Falconry and Hunting Hawks* 7th edition. Denver, CO.

Beebe, F. L. 1992. *The Compleat Falconer*. Hancock House, Surrey, BC. ISBN 0 88839 253 2.

Bert, Edmund. 1619. *An Approved Treatise of Hawkes and Hawking*.

Blaine, G. 1936. *Falconry*. 1970 Reprint, Neville, Spearman Ltd. ISBN 84435 320 8.

Blome, R. 1686. *Hawking or Faulconry*. Extracted from *A Gentleman's Recreation*. Facsimile Edition Thames Valley Press. ISBN 0-900294-04-3.

Brander, M. 1963. *Gundogs: their care and training*. A. & C. Black. London.

Brewer, G. 1995. *Buteos and Bushytails*. GLB Publications, Texas.

British Falconers' Club: Journals. 1937-present.

British Falconers' Club. 2003. *Celebrating 75 Years of The British Falconers' Club*. Privately Printed.

Brodie, I. 1978. *Ferrets and ferreting*. Blandford Press, Poole. ISBN 0 7137 0903 0.

Carlisle, G. L. 1983. *Grouse and gun*. Stanley Paul, London. ISBN 0 09 153360 0.

Ceballos, J. 2002. *Conocimiento y Practica de la Cetreria*. Cairel Ediciones. ISBN 84-85707-37-0.

Chindgren, S. 2018. The Art of Hawking Sage Grouse. New Holland.

Compendium of authors. 2006. *One Thousand Years of falconry*. Eryr Press.

Cover, D.A. 1996. *Oh, Those Wonderful Hawks!* Reef Graphics.

Cummins, J. 1988. *The Hound and the Hawk: the art of medieval hunting*. Weidenfield and Nicolson, London. ISBN 0 297 79459 0.

Davis, H. P. 1970. *Training and Hunting your own Bird Dog*. G. P. Putnam's Sons.

De Chamerlat, C. A. 1987. *Falconry and Art*. Sotheby's, London. ISBN 0 85667 338 2.

De La Fuente, F. R. 1970. *El Arte de Cetreria*. Libreria Noriega, Mexico. ISBN 968 18 2226 9.

Derbyshire, M. 2017. *Out of the Hood*. Coch-y-Bonddu Books. ISBN 978-1-904784-82-1.

Durman-Walters, D. 1994. *The Modern Falconer*. Swan Hill Press, Shrewsbury, England. ISBN 1 85310 368 3.

Fox, N.C. 2004. *Classical Falconry: A Treatise on Crow and Rook Hawking*. Hancock House. ISBN 0-88839-548-5.

Frederick II, 1248. *De Arte Venandi cum Avibus*. 1943 translation by C.A. Wood and F.M. Fyfe. Stanford University Press.

Freeman, G. E. 1869. *Practical Falconry*. Horace Cox, London.

Glasier, P. 1978. *Falconry and Hawking*. Batsford. ISBN 0 7134 0232 6.

Harting, J. E. 1898. *The Rabbit*. Longmans, London.

Hiebler, J. 2000. *Der Steinadler in der Falknerei*. Panhofer Verlag. ISBN 3-9501123-9-1.

Hollinshead, M. 1993. *Hawking Ground Quarry*. Hancock House, Surrey, BC. ISBN 0 88839 320 2.

Hollinshead, M. 1995. *Hawking with Golden Eagles*. Hancock House. ISBN 0-88839-343-1.

Humphreys, J. 1990. *The Complete Gundog*. David and Charles, Newton Abbot. ISBN 0 7153 9412 6.

Irving, J. 1983. *Gundogs: Their learning chain*. Loreburn, Dumfries. ISBN 0 9506670 1 3.

Jackson, T. 1989. *Hunter, Pointer, Retriever*. Ashford, Southampton. ISBN 1 85253 189 4.

Jameson, E. W., Jr. 1962. *The Hawking of Japan*. Privately Printed, Davis, California.

Jameson, E.W. 1993. *Shortwinged Hawks: A review of Ancient and Modern Hawking*. Privately Printed, Davis, California.

Lascelles, G. 1892. *The Art of Falconry*. Re-printed 1985, Neville Spearman. ISBN 85435-031-4.

Latham, Symon, 1615. *Latham's Faulconry, or The Faulcon's Lure and Cure*.

Latham, Symon, 1618. *Latham's New and Second Booke of Faulconry*. 2012.Published in two volumes with a Commentary by David Horobin. Coch-y-Bonddu Books. ISBN 978-1-904784-51-7.

Loft, J. 1999. *A Merlin for Me*. Privately printed. ISBN 0-9535881-0-6.

Loft, J. and R.W. Radcliffe (Eds) 2018. The Complete Merlin and Red-headed Falcon in Falconry. Western Sporting. ISBN 10: 188835707X.

Martin, B. P. 1990. *The Glorious Grouse*. David and Charles, Newton Abbot. ISBN 0 7153 9237 9.

Mavrogordato, J. G. 1960. *A Hawk for the Bush*. H. F. & G. Witherby Ltd.

Mavrogordato, J. G. 1966. *A Falcon in the Field.* Knightly Vernon Ltd.

McKelvie, C. L. 1990. *The Book of the Woodcock*. Swan Hill Press, Shrewsbury. ISBN 1 85310 113 3.

McElroy, H. 1977. *Desert Hawking II*. Privately printed.

Michell, E. B. 1900. *The Art and Practice of Hawking*. 1959 Edition, Holland Press.

Mirza, Taymur. 1868. *The Baz Nama Yi Nasiri*. Translated by D. C. Phillott, 1908, Bernard Quaritch, London.

Moxon, P. R. A. 1967. *Gundogs: training and fieldtrials*. Popular Dogs, London. ISBN 09 029762 8.

North American Falconer's Association: Journal 1961-present.

Mullenix, M. 1996. *American kestrels in Modern Falconry*. Eagle Wing Publishing. ISBN 1-885054-02-5.

O'Broin, L. 1992. *A Manual for Hawking*. Privately printed.

Osman, S.M. 2001. *Falconry in the Land of the Sun: The Memoirs of an Afghan Falconer*. Western Sporting. ISBN 1-888357-12-6.

Osman, S.M. 2005. *Musings of an Afghan Falconer.* The Eryr Press.

Phillott, D.C. 1906. *Observations on Eastern Falconry*. 2001 facsimile reprint Eryr Press.

Remple, D. and C. Gross. 1993. *Falconry and Birds of Prey in the Gulf.* Motivate Publishing, Dubai. ISBN 1 873544 39 1.

Roy, J. 2004. *Duck hawking and the art of falconry*. Hancock House. ISBN 0-88839-553-1.

Schuster, J. 2002. *Die Krahenbeize* (Crow Hawking, in German). Verlag Peter N. Kluh. ISBN 3-933459-09-5.

Sebright, J. S. 1826. *Observations upon Hawking*. J. Harding, London.

Stevens, R. 1957. *Observations on Modern Falconry*. Privately Printed.

Timimi, F. A. 1987. *Falcons and Falconry in Qatar*. Privately printed, Doha.

Turner, R, and A. Haslen. 1991. *Gamehawk.* Gallery Press, Lavenham.

Tyers, S. 2020. *The Specialist Falcon*. Privately printed.

Upton, R. 1980. *A Bird in the Hand*. Debrett.

Upton, R. 1987. *O, for a falconer's voice*. The Crowood Press, Marlborough. ISBN 1 85223 015 0.

Upton, R. 1991. *Falconry, Principles and Practice*. A. & C. Black ISBN 0 7136 3262 3.

Van de Wall, J. W. M. 1986. *De Valkerij Op Het Loo. The Royal Loo Hawking Club 1839-1855*. Joh Enschedéen Zonen. Haarlem. ISBN 90 70024 42X

Waller, R. 1987. *Der Wilde Falk istmein Gesell*. J. Neumann Neudamm, Melsungen.

Webster, H. M. 1988. *Game Hawking at its very best*. Windsong Press, Denver. ISBN 0 912510 04 X.

Whitehead, S. 2008. *Ferreting: An Essential Guide*. The Crowood Press Ltd. ISBN 978-1847970367.

Woodford, M. 1966. *A Manual of Falconry*. A. & C. Black. ISBN 0 7136 0736X.

C11 Raptors and Humans

Mongolian sakers, electrocuted on power lines.
Photo Andrew Dixon.

11.1 Introduction

Falconry depends on five pillars: the falconers, the raptors, the land, the quarry and a place in society. In this section I examine these by first taking an overall perspective of world falconry in the 20th century, our cultural heritage and then the ways in which we contribute - positively or negatively - to raptors, their habitats and their prey, both as populations and as welfare of individuals. The issues are complex, so of necessity this is just an introductory insight.

11.2 Falconry as an Intangible Cultural Heritage

This section is supported by a film produced by Dr Javier Caballos as part of our submission to UNESCO and by Cordi and Jo Atkinson's film on the *Second International Festival of Falconry* in Abu Dhabi in 2011.

- ICH is culture that we have inherited from previous generations of practitioners.
- It is intangible (as distinct from the artefacts) and transmitted from one generation to the next.
- It is still alive, rather than re-created or re-enacted.
- It can continue to evolve rather than be 'frozen'.
- It is not about dead history, it is about passing on this ICH to FUTURE generations.
- So, paradoxically, it is not about the past, but about the future!

In 2003 the United Nations Education, Scientific and Cultural Organisation (UNESCO), recognising that there is more to heritage than just sites and ruins, created a new Convention for Intangible Cultural Heritage (ICH):

UNESCO text is in blue and our submission responses are in green.

Article 1 – Purposes of the Convention

The purposes of this Convention are:

a. to safeguard the intangible cultural heritage;

b. to ensure respect for the intangible cultural heritage of the communities, groups and individuals concerned;

c. to raise awareness at the local, national and international levels of the importance of the intangible cultural heritage, and of ensuring mutual appreciation thereof;

d. to provide for international cooperation and assistance.

Dr Monica Reuter in Austria quickly realised that this was an opportunity for Falconry to be accepted as part of our global Heritage. In 2005 HE Mohammed Al Bowardi, on behalf of the government of the United Arab Emirates, asked me to prepare a submission. There followed six years of meetings and preparations.

Of the 80 or so countries in which falconry is practised, we had first to find out which governments had signed up to the new Convention, or could be persuaded to sign up. This was quite instructive. USA, Canada, Australia and New Zealand, have not signed. They all have minority indigenous peoples who of course would be the first to benefit from the new Convention possibly to the detriment of the 'new' colonisers. UK too is becoming a multi-ethnic society and finds itself in a quandary. On the one hand its policy is integration, on the other it is supposed to support minorities. I approached the Ministers for Culture in UK and in New Zealand and both claimed that they could not define 'Intangible Cultural Heritage'. But UNESCO has already defined it:

Article 2 – Definitions

For the purposes of this Convention,

1. The 'intangible cultural heritage' means the practices, representations, expressions, knowledge, skills – as well as the instruments, objects, artefacts and cultural spaces associated therewith – that communities, groups and, in some cases, individuals recognize as part of their cultural heritage. This intangible cultural heritage, transmitted from generation to generation, is constantly recreated by communities and groups in response to their environment, their interaction with nature and their history, and provides them with a sense of identity and continuity, thus promoting respect for cultural diversity and human creativity. For the purposes of this Convention, consideration will be given solely to such intangible cultural heritage as is compatible with existing international human rights instruments, as well as with the requirements of mutual respect among communities, groups and individuals, and of sustainable development.

2. *The 'intangible cultural heritage', as defined in paragraph 1 above, is manifested inter alia in the following domains:*

a. *oral traditions and expressions, including language as a vehicle of the intangible cultural heritage;*

b. *performing arts;*

c. *social practices, rituals and festive events;*

d. *knowledge and practices concerning nature and the universe;*

e. *traditional craftsmanship.*

We held a meeting in Paris in 2008 of falconry representatives from countries that had signed, or might sign the Convention. If a country signs, then its government accepts that it has to safeguard the heritage:

3. *'Safeguarding' means measures aimed at ensuring the viability of the intangible cultural heritage, including the identification, documentation, research, preservation, protection, promotion, enhancement, transmission, particularly through formal and nonformal education, as well as the revitalization of the various aspects of such heritage.*

Each State Party shall:

a. *take the necessary measures to ensure the safeguarding of the intangible cultural heritage present in its territory;*

b. *among the safeguarding measures referred to in Article 2, paragraph 3, identify and*

Preparing our UNESCO submission, Paris 2008.
Photo Frank Bond.

define the various elements of the intangible cultural heritage present in its territory, with the participation of communities, groups and relevant nongovernmental organizations.

When a country signs the Convention, its first task is to draw up an Inventory List of cultural elements of initial concern for consideration:

Article 12 – Inventories

To ensure identification with a view to safeguarding, each State Party shall draw up, in a manner geared to its own situation, one or more inventories of the intangible cultural heritage present in its territory. These inventories shall be regularly updated.

1. *When each State Party periodically submits its report to the Committee, in accordance with Article 29, it shall provide relevant information on such inventories.*

So once a country had joined the Convention, our task was to persuade it to add Falconry to its national Inventory List of ICH. This meant meeting various Ministers and representatives, and sometimes taking them out hawking, or showing them falconry in some way. Many had not even heard of falconry. As well as this we had to establish relations between falconry organisations and UNESCO governmental representatives, sometimes establishing formal bodies to undertake safeguarding activities:

Article 13 – Other measures for safeguarding.

To ensure the safeguarding, development and promotion of the intangible cultural heritage present in its territory, each State Party shall endeavour to:

a. *adopt a general policy aimed at promoting the function of the intangible cultural heritage in society, and at integrating the safeguarding of such heritage into planning programmes;*

b. *designate or establish one or more competent bodies for the safeguarding of the intangible cultural heritage present in its territory;*

c. *foster scientific, technical and artistic studies, as well as research methodologies, with a view to effective safeguarding of the intangible cultural heritage, in particular the intangible cultural heritage in danger;*

d. *adopt appropriate legal, technical, administrative and financial measures aimed at:*

i. *fostering the creation or strengthening of institutions for training in the management of the intangible cultural heritage and the transmission of such heritage through forums and spaces intended for the performance or expression thereof;*

ii. *ensuring access to the intangible cultural heritage while respecting customary practices governing access to specific aspects of such heritage;*

iii. *establishing documentation institutions for the intangible cultural heritage and facilitating access to them.*

Article 14 – Education, awareness-raising and capacity-building

Each State Party shall endeavour, by all appropriate means, to:

a. *ensure recognition of, respect for, and enhancement of the intangible cultural heritage in society, in particular through:*

i. *educational, awareness-raising and information programmes, aimed at the general public, in particular young people;*

ii. *specific educational and training programmes within the communities and groups concerned;*

iii. *capacity-building activities for the safeguarding of the intangible cultural heritage, in particular management and scientific research; and*

iv. *non-formal means of transmitting knowledge;*

b. *keep the public informed of the dangers threatening such heritage, and of the activities carried out in pursuance of this Convention;*

c. promote education for the protection of natural spaces and places of memorym whose existence is necessary for expressing the intangible cultural heritage.

Article 15 – Participation of communities, groups and individuals

Within the framework of its safeguarding activities of the intangible cultural heritage, each State Party shall endeavour to ensure the widest possible participation of communities, groups and, where appropriate, individuals that create, maintain and transmit such heritage, and to involve them actively in its management.

In order for each State to fulfil these obligations, we had to prepare Action Plans to advise each Government on how to achieve them. We started off with about thirty countries which gradually got whittled down to eleven. The International Association of Falconry became an accredited advisory NGO, entitled to attend the UNESCO meetings as an Observer.

The Cultural Heritage of Falconry in UAE

The geographical and political history of the region in relation to Falconry and the human inhabitants.

Known history of Falconry in the region from manuscripts and artefacts.

Traditional craftsmanship in falconry equipment.

The falcons traditionally used.

Traditional trapping and handling methods.

Traditional training methods.

Traditional quarry hunted.

Traditional Hunting Areas in the region.

Traditional knowledge and practice such as tracking methods.

Traditional hunting assistants: the Saluki, the Arab horse and the Camel.

The International Association of Falconry and Conservation.

Recording historic falconry medicines evolving into modern methods.

The vocabulary of falconry.

Falconry in poetry.

Falconry in Islam.

Falconry artefacts in coins, stamps and other usages.

Some well-known Falconers in the region.

The Action Plan

1. *Defining managed Hunting Areas with safeguarded status, possibly as Heritage Sites, to conserve natural habitats.*

2. *Steps to safeguard the quarry species both in UAE and in Range States.*
   The Houbara
   The Hare.
   The Karrowan.

3. *Steps to safeguard the falcon species both in UAE and in Range States:*
   The Saker Falcon.
   The Peregrine Falcon.
   The Gyrfalcon.
   Hybrid Falcons (preventing genetic pollution).

4. *Steps to safeguard the Assistants in Falconry:*
   The Saluki.
   The Arab Horse.
   The Camel.

5. *Education of young Falconers through the Emirates Falconers Club.*

6. *Disseminate information about Falconry through a magazine such as Al Saggar.*

7. *Education of the public through schools, summer camps and a Falconry Heritage Centre.*

8. *Encouragement of Falconry Poetry through an annual Event at the Hunting Exhibition or Festival.*

9. *Promoting a national and international legal framework necessary for Falconry, such as the acquisition of falcons, their registration, sale and international movement, legal hunting areas, seasons and access to quarry at home and abroad.*

10. *Promoting the good husbandry, welfare and medical treatment of animals used in falconry through education and through vets and falcon hospitals.*

11. *Promoting the international exchange of ideas and companionship with Falconers from other countries through the Emirates Falconers Club.*

12. *Promoting one or more festivals or events such as the Hunting Exhibition.*

13. *Promoting an awareness of Falconry as part of the national culture to tourists and visitors to the UAE.*

14. *Promoting Falconry as part of daily life and culture including weekend hunting trips to the desert where fathers can introduce sons to Falconry and their desert heritage, tracking their quarry, hunting it, preparing and cooking it and living outdoors.*

15. *Making a demographic survey of falconers; their numbers, ages, experience, species used etc as a waypoint in maintaining Falconry.*

16. *Undertake research programmes where necessary, particularly for items 1-5.*

17. *Disseminate research information to Falconers and researchers through a journal such as Falco.*

18. *Promote Falconry on the world stage amongst politicians, NGOs and other interested parties and to make International Agreements where necessary.*

19. *Maintain an Archive of information about falconry, perhaps at a Falconry Heritage Centre.*

Abu Dhabi Falconers at the First International Festival at Englefield 2007.
Photo Linda Wright.

While we were preparing the UAE submission, and helping other countries prepare their own, we also had to create the overall group submission. In support of this, we created a 12 minute film portraying falconry in the 11 countries, we produced an illustrated book *Falconry: Celebrating a Living Heritage* and we organised the first International Festival of Falconry in 2007 at Englefield, UK, inviting participants from 60 countries. We went on to do another in 2009 in UK, then two more in Abu Dhabi in 2011 and 2014, with 80 nations participating. These brought together about 700 falconers for several days of hawking, conferences and demonstrations.

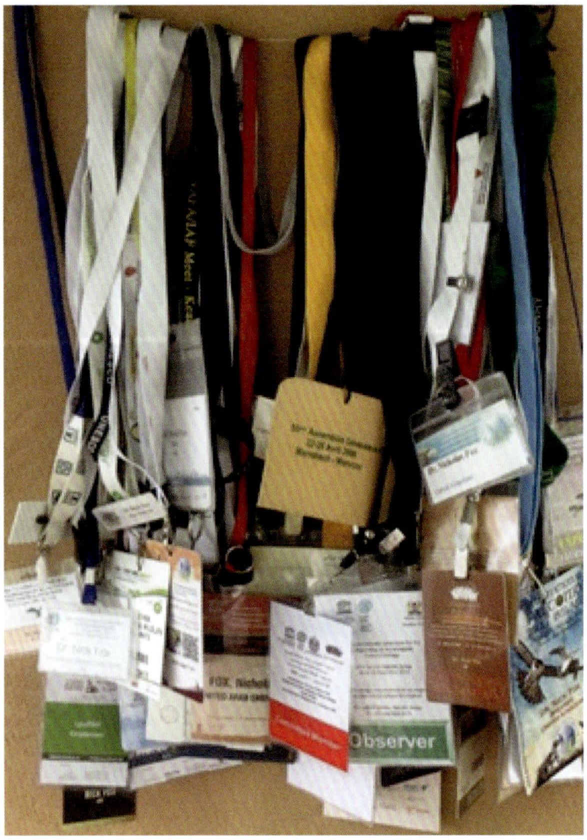

Conferences!

On the UAE side, our active supporters included HE Mohammed Al Bowardi, HE the late Mohammed Khalaf Al Mazrouie, Prof Sulayman Khalaf and our energetic diplomat, Dr Awadh Saleh. Our IWC team included Jevgeni Shergalin, Jo Oliver, Nick Kester and many others. Internationally, Jose-Manuel Rodriguez Villa, Dr Bohumil Straka, Dr Javier Ceballos, Kris Ulens and Tula Stapert were very active. As well as the national meetings that we organised or attended, Dr Awadh and I attended many UNESCO conferences in which the finer details of the Convention were finalised and the first submissions administered.

A. State(s) Party(ies)

For multi-national nominations, States Parties should be listed in the order on which they have mutually agreed.

United Arab Emirates, Austria, Belgium, Czech Republic, France, Germany, Hungary, Italy, Kazakhstan, Republic of Korea, Mongolia, Morocco, Pakistan, Portugal, Qatar, Saudi Arabia, Spain and Syrian Arab Republic

B. Name of the element

B.1. Name of the element in English or French

Indicate the official name of the element that will appear in published material.

Falconry, a living human heritage

B.2. Name of the element in the language and script of the community concerned, if applicable. Indicate the official name of the element in the vernacular language corresponding to the official name in English or French (point B.1).

الصقارة، الصيد بالطير الحر، القنص، البيزرة، تبيازت

Falconry, Hawking, Chasse au vol, Falconaria, Sokolnictví, solymászat, Sokoliarstvo, Cetrería, Cetraria, Halconería, valkerij, Falknerei, 매사냥, ҚҰСБЕГІЛІК (kaz.), Beizjagd, Falconeria, Шувуучлахуй یرادزاب

B.3. Other name(s) of the element, if any. In addition to the official name(s) of the element (point B.1) mention alternate name(s), if any, by which the element is known.

In this submission 'falconry', 'falconer' and 'falcon' covers the use of all species of birds of prey (raptors), such as falcons, eagles, hawks, buzzards, etc.

C. Name of the communities, groups or, if applicable, individuals concerned

Identify clearly one or several communities, groups or, if applicable, individuals concerned with the nominated element.

Communities involved in falconry are villages and kinship groups, tribes, families and individuals, as well as organized falconry clubs, falconry heritage trusts and institutions. In addition, there are supporting agencies and associations such as falcon hospitals, breeding centres, conservation agencies, traditional falconry equipment makers, artists, poets, and professional falconers.

Examples of specific communities are as follows:

Austria: Österreichischer Falknerbund, ÖFB and Zentralverband Österreichischer Falkner, ZÖF

Belgium: Belgian Federation for Falconry 'Valkeniers.be', which includes the following:

- *Belgische Vereniging voor Valkeniers en Havikeniers, BVVVH vzw*
- *Beoefenaars Vluchtbedrijf de Valkenier, BVdV vzw*
- *Flanders Valkerij Academie, FVA vzw*
- *Studiegroep Behoud Valkerij, SBV vzw*
- *Association Belge de Fauconnerie 'Club Marie de Bourgogne'*

Czech Republic: Klub Sokolníků (Falconry Club of the Czech-Moravian Hunting Union)

France : Association Nationale des Fauconniers et Autoursiers

Germany: Most falconers are organised into associations. Approximately 2,000 falconers are distributed among the three largest organisations, the Deutscher Falkenorden e. V. (DFO), the Orden Deutscher Falkoniere e. V. (ODF) and the Verband Deutscher Falkner e. V. (VDF)

Hungary: Magyar Solymász Egyesület (Hungarian Falconry Club)

Italy:Circolo dei Falconieri d'Italia per lo studio e la protezione dei rapaci, founded in 1955;

FIF (Federazione Italiana Falconieri) founded in 1992

UNCF (Unione Nazionale Cacciatori con il falco), founded in 2003

OFI (Ordine Falconieri d'Italia), founded in 2006

CPF (Associazione Conservazione Progresso della Falconeria), founded in 2014

Republic of Kazakhstan: 'Burkitshi' and 'Kyran' Federations, 'Zhalayir Shore' Centre for Falconry, 'Sunkar' Falcon Nursery Centre

Republic of Korea: Korea Falconers' Association and Traditional Falconry Association.

Mongolia: Falconry is practised primarily by the Khazakh minorities in Bayan Ulgii Province, Mongolian Falconry Association, Mongolian Burkut (Eagle) Association.

Morocco : Kwassems Oulad Frej Tribes in Abda-Doukkala Region, Association Marocaine des Fauconniers (Al Noubala), Association des Fauconniers d'Ouled Fraj pour la Chasse au Vol, Association de l'Éducation Environnementale et de la Protection des Oiseaux au Maroc

Pakistan: Pakistan Falconry Association

Portugal: Portuguese Association for Falconry (APF); Municipal Division of Social and Cultural Action Salvaterra de Magos; UNESCO Chair 'Intangible Heritage and Traditional Know-how: Linking Heritage'

Qatar: Qatar Falconry Society

Saudi Arabia: Villages and towns in the Eastern Region, Festival Nuaireyah Province Spring Falconry Competition, National Authority for the Protection of Wildlife and its Development.

Spain: Spanish Association for Falconry and Conservation of Raptors (AECCA), Real Gremio de Halconeros de España

Syria: Al Rehaiba Falconry community

United Arab Emirates: Tribal families and communities in the Western and Eastern regions of the United Arab Emirates, Emirates Falconers' Club, Abu Dhabi Falconer's Club, and the Emirates Heritage Club.

D. Geographical location and range of the element.

Provide information on the distribution of the element within the territory(ies) of the submitting State(s), indicating if possible the location(s) in which it is centred.Nominations should concentrate on the situation of the element within the territories of the submitting States, while acknowledging the existence of same or similar elements outside their territories, and submitting States should not refer to the viability of such intangible cultural heritage outside their territories or characterize the safeguarding efforts of other States.

Falconry is practised along the migration routes used by falcons for thousands of years. Thus falconry is mainly found within these traditional migration flyways and corridors that run from north and east Asia and north Europe through Mediterranean Europe, the Middle East and the Caspian Sea countries to North Africa, and from North America, south to Central and South America.

Falconry is found in more than sixty countries. It depends on open terrain so that the falconer can follow the bird. Thus, habitat dictates the practicality of falconry and shapes its particular local variations of traditional styles. For example, in the deserts of Arabia, the open terrain allows the use of falcons that fly long distances, and the steppes of Asia permit the flying of both falcons and large eagles. However, in forested areas and mixed farmland, as found in much of Europe as well as Japan, parts of China, Republic of Korea and Pakistan, short-range birds, such as goshawks and sparrow hawks, are preferred. Spain and Italy provide opportunities to fly both short and long-wing birds of prey according to varying habitat.

Falconry retains an unbroken tradition in central, south and east Asia, the Middle East, North Africa and most parts of Europe. Following a brief decline in 18-19th century Europe, it is recovering and provides a link to the countryside by increasingly urban populations. When settlers from Europe reached the Americas, southern Africa and Australasia, they took many of their traditions along with them, including falconry. Falconry is even present in the Azores islands.

In most countries falconry is a stable minority activity. However, in some areas rapid urbanisation has restricted the opportunity to practise falconry, leading to a decline. The migration from the countryside to towns is a major threat to rural-based traditions.

Provide a brief summary description of the element that can introduce it to readers who have never seen or experienced it.

Falconry is the traditional art and practice of keeping, training and flying falcons to take quarry in its natural state, and has been practised for more than 4000 years. It is one of the oldest relationships between human and bird. It probably evolved in the steppes of Asia, and spread via cultural and trade links to other countries, first to Europe, North Africa and East Asia and later in the 16th century to the rest of the world. The practice of falconry in early and medieval periods of history is documented in many parts of the world. The falcon and her prey have evolved together over millions of years; their interaction is an age-old drama. The falconer's task is to bring the actors together on nature's stage.

Falconry as intangible cultural heritage is integrated into communities as a social recreational practice and a means of connecting with nature. Originally a way of obtaining food, falconry has acquired other values over the centuries, and is identified with camaradèrie, sharing, and expressions of freedom. Nowadays falconry is practised by people of all ages in local communities in more than eighty countries.

Falconry expresses itself in multiple cultural domains, in particular 'social practices, rituals and festive events', as well as 'knowledge and practices concerning nature and the universe'. These include traditional skills, traditional knowledge about the biology and behaviour of birds and the environment, the making of traditional falconry equipment, as well as linguistic and artistic expressions found in idioms, painting, sculpture, poetry, rituals and music.

Falconry has its own set of culturally-shaped traditions and ethics; while falconers come from different backgrounds, they share universal values, traditions and practices. Falconry is transmitted from generation to generation as a cultural tradition by a variety of means, such as mentoring, learning within families, or formalized training in clubs and schools. The process of transmitting falconry heritage activates not only the historical memory of the falconry community but also enriches its cultural identity. In recent times falconry has seen a renaissance.

Falconry provides modern man with links to nature and strengthens local identities. It is an important cultural symbol in many countries. The modern practice of falconry aims at safeguarding not only falcons, quarry, and habitats but also the practice itself as a living cultural tradition.

Who are the bearers and practitioners of the element? Are there any specific roles or categories of persons with special responsibilities for the practice and transmission of the element? If yes, who are they and what are their responsibilities?

Falconry is practised by people of all ages, men and women, amateurs and professionals. Falconers develop a strong relationship and spiritual bond with their birds; commitment is required to breed, train, handle and fly falcons. Falconers come from all walks of life, societies, cultures, linguistic and religious backgrounds. Falconry shapes the lives of practitioners, as individuals, in families, villages, and clubs, and practitioners are the essential leaders for the conservation of birds of prey. Falconers are considered to be responsible for protecting falconry heritage by their own practice, and many are also involved in transmission through mentoring or more formal teaching.

Some countries have also developed a system of master practitioners, who have the responsibility of transmitting knowledge about falconry. For example, in Germany each state association has its own 'master falconer', whose main role is to assist in the guidance and training of new falconers. In Kazakhstan, falconry is mainly transmitted from officially registered Burkytshi/ Kusbegi to their apprentices, and in Korea, falconers are designated as Living Human Treasures by the state, and they are responsible for transmission.

Falconers regard themselves as a group, thus reinforcing the cultural values of sharing, interdependence and mutual support. Falconry has its own set of culturally-shaped traditions and ethics, including how to cooperate as a group. A bond of 'almkhuwa' (brotherhood, UAE) or 'al-shareek' (partner ethics, Saudi Arabia) governs the behaviour of falconers. Almkhuwa signifies a bond of companionship and equal sharing of tasks, costs and responsibilities during the trip. There are usually groups of six to twelve falconers on an excursion of one to three weeks. After a day of falconry, they sit around the camp fire and recount stories and anecdotes of the day, often composing poems. For the almkhuwa group, falconry is the means to enjoy a cultural tradition that celebrates the camaraderie of falconers in the open desert.

European falconers often join together in special national and international social events, such as field meets. The camaraderie experienced by the falconers can be summed up by the Czech statement, 'We do not hunt for the quarry itself, but for all the other nice experiences'. At the end of the day the prey may be honoured by speeches and the blowing of horns. In many European countries, such as Austria, Belgium, Czech Republic, Hungary and Spain, it is a tradition to hold mass in churches for certain patron saints of falconry to pray for successful hunts. In Arabia and Pakistan, falconers utter God's name over the bird or animal caught by the falcon.

Falconry creates a sense of pride and identity within the larger community. For example, the Kwassem tribe of central western Morocco gained prestige and recognition of their practice of falconry over centuries, and the receipt of letters of praise from the Sultans distinguishes them from other tribal communities. Some families of practitioners are so well known that they carry the family name of Biaz (falconer).

How are the knowledge and skills related to the element transmitted today?

Transmission is vital to all falconry communities and it includes many activities related to falconry practice, knowledge of the environment, the proper management of the falcons and cultural values. Falconry as a cultural tradition is transmitted between generations by a variety of means, such as mentoring, learning within families, or formalized training in clubs and schools.

Because falconry is essentially a practical activity, the main method of non-formal education is mentoring, where an experienced falconer instructs the beginner and demonstrates techniques. Informal methods of transferring knowledge often occur within families, which is typical in Mongolia, Morocco, Pakistan, Qatar, Saudi Arabia, and UAE. Falconers give their children training in how to handle and build a relationship of trust with the bird. This is a long process that develops skills such as feeding and holding the falcon on the fist, and calling the falcon to the lure. These centuries-old methods are equally effective for transmitting cultural values and traditions. In many Middle Eastern countries, where most nomadic Bedouin have settled in towns within one generation, falconry is one of the few remaining links with the desert and their traditional culture and ways of life. As stated by Sheikh Hamdan bin Zayed, Chairman of Emirates Falconers Club, 'Falconry lets our children experience the magic of the desert and instil in them the virtues of patience'.

More formal systems of learning have been developed in many falconry groups and clubs, and this has led to apprenticeship schemes and courses for nationally recognized certificates. For example, in Austria, the Czech Republic, Germany, Hungary, Italy and Portugal the candidate must pass an official state examination to become a falconer.

It is especially important to educate the youth about the importance of intangible cultural heritage of falconry. For example, school curricula in several countries refer to falconry and specific learning materials have been produced. Examples are in Belgium, where children use a book on falconry to learn to read Flemish and in Kazakhstan, the Burkitshi and Kyran Federations hold training sessions with sports clubs and secondary schools. The UAE has introduced falconry competitions which include junior categories to encourage the youth, including Fakhr Al Ajyal Championship (Abu Dhabi) and Fazza Championship (Dubai).

The practical knowledge of falconry has also been documented over the centuries; the 13th century book, 'De Arte Venandi cum Avibus (The Art of Hunting with Birds)', by Frederick II is still widely referred to today. A more recent example is from Spain, where a book about falconry with a code of ethics has been put online to help beginners learn about the practical aspects of this element.

What social functions and cultural meanings does the element have today for its community?

Falconry is integrated into communities as a social recreational practice and a means of connecting with nature. Originally a way of obtaining food, falconry has acquired other values over the centuries, and is identified with camaraderie, sharing, and expressions of freedom. Nowadays falconry is practised by men and women in their communities in more than sixty countries, and therefore has a wide diversity of cultural traditions.

While falconers come from different backgrounds, they share universal values, traditions and practices. For example, the methods of training and caring for birds, the equipment used and the bonding between falconer and the bird are similar throughout the world. Falconers can understand each other by simple gestures; it is these shared traditions and knowledge that make falconry universal and keep it alive.

Falconry is a dynamic tradition, which is constantly being recreated by the communities in response to changes in their environment. While

Kazakhi berkutchi Abdelkhak Turlybaev at Englefield.
Photo Linda Wright

adapting to changing times by using modern technologies and equipment, such as telemetry, some groups still keep their local traditional costumes. For example, Austrian, Belgian, Korean and Mongolian falconers indicate identities with special hats, buttons, headbands and jackets respectively.

Falconry has also inspired artistic creativity, with a rich heritage of books, manuscripts, poetry, painting and historical buildings. The falcon is an important cultural symbol in many countries, and is used widely in postage stamps, coins, and coats of arms; it is the official state emblem in several Arab and European nations. The gift of falcons has been used in diplomacy throughout history. Falcons have been used to name children and human settlements. In Mongolia, during the national festival competition, wrestlers perform the 'eagle dance' showing the power and beauty of its flight.

Is there any part of the element that is not compatible with existing international human rights instruments or with the requirement of mutual respect among communities, groups and individuals, or with sustainable development?

Falconry is in harmony with existing international human rights instruments and with the requirements of mutual respect among communities, groups and individuals.

Falconry depends on the sustainable use of wildlife resources, and falconers are deeply involved in looking after those resources. Since its inception, falconry depends on the sustainable use of natural resources, birds of prey, the quarry and the environment itself. Falconers are personally involved in the care of these resources. Falconry by its very nature has a very low environmental impact; since the beginning falconers have endorsed the concept of 'sustainable use' and the traditions and ethics of falconry follow this principle.

Falconry is subject to statutory regulations and is accessible to everyone, regardless of their gender, education, religion, or nationality. Limitations to practice falconry in some countries only arise due to the requirement to prove the relevant expertise, which is ensured by means of the examinations. This limitation is prescribed by law and takes into account the legitimate interests of animal welfare and conservation.

For the procurement of the birds themselves, numerous regulations apply at national and international level. Currently the practice of falconry in most countries takes into consideration the 1973 Convention on International Trade in Endangered Species of Wild Fauna and Flora (CITES) and the 1992 Convention of Biological Diversity (CBD). Falconers endeavour to preserve the living conditions of birds of prey and their prey, and to protect the birds. Many rehabilitation centres for injured birds of prey are led by falconers.

As countries become increasingly urbanized, falconers are committed about the lands where they fly their birds and concerned to keep them in their natural state to ensure the survival of all plant and animal species. Sheikh Zayed commented on the sustainability of nature, '...it is not what you catch that is important; it is what you leave behind.'

2. Contribution to ensuring visibility and awareness and to encouraging dialogue.

For Criterion R.2, the States shall demonstrate that 'Inscription of the element will contribute to ensuring visibility and awareness of the significance of the intangible cultural heritage and to encouraging dialogue, thus reflecting cultural diversity worldwide and testifying to human creativity'. This criterion will only be considered to be satisfied if the nomination demonstrates how the possible inscription will contribute to ensuring visibility and awareness of the significance of the intangible cultural heritage in general, and not only of the inscribed element itself, and to encouraging dialogue which respects cultural diversity.

How can inscription of the element on the Representative List contribute to the visibility of the intangible cultural heritage in general and raise awareness of its importance at the local, national and international levels?

Inscriptions of Falconry on the Representative List (2010, 2012) have been successful in raising the visibility of ICH in general in many communities worldwide; especially as from a global perspective falconry is not restricted to a particular season or locale. Inscription has led to increased visibility of ICH through many ways, including media, internet and festivals. Practitioners and the public have become better aware of the wide variety and diversity of intangible heritage, as well as UNESCO's role in promoting mutual respect and knowledge of different cultural expressions.

For example, in France falconers were asked to explain ICH within their communities, and specialized magazines have explained the definition, advantages and obligations of ICH to 1.3 million readers. In Czech Republic the falconry community played a major role in encouraging Parliament to ratify the UNESCO ICH Convention.

Communities which practise falconry ensure through their clubs, societies and communal festivals that falconry is viewed within the context of its wider cultural heritage, such as traditional handicrafts, costumes, food, music, poetry, communal rituals and celebrations. International falconry gatherings, such as the 3rd Falconry Festival in 2014, celebrate many aspects of the cultural heritage of falconry communities. UNESCO recognition has contributed to the safeguarding and sustainability of these cultural traditions in all countries that practise falconry. It has also highlighted the use of many important cultural spaces and artefacts, and contributed to public recognition of their significance. Falconry captures the interest of the young, and has raised awareness of ICH among this generation.

The addition of five States Parties to this multi-national submission file will lead to further visibility. International NGOs are important cultural agents in this process, and will continue to organise meetings, forums, seminars, and festivals to encourage raising awareness of ICH in general and to promote the importance of cultural values.

How can inscription encourage dialogue among communities, groups and individuals?

The inscription of falconry as a multi-national file has led to increasing cooperation and collaboration among the participating states on many aspects of intangible cultural heritage. Inscription has encouraged five additional States Parties to join this nomination, a reflection of increased dialogue and cooperation.

The falconry submission was presented at a South-East Asian ICH conference (Korea, 2009) to illustrate how an ICH element can bring about international participation and cooperation. Falconry was also selected in a conference in December 2011 (University of Bordeaux, France) as the best example of a multi-national submission, as it reflected greater cooperation among nations and raised awareness of ICH in many different communities.

Since inscription, there has been an increase in international conferences and festivals. The International Falconry Festival (December 2014, UAE) brought together 800 falconers from 80 participating countries, and a further festival is planned for 2017. Qatar and Saudi Arabia hold falconry festivals every year and Qatar hosted conferences in 2013 and 2014.

The International Falconry Association (IAF) organises the annual World Falconry Day on 16th November, the date of first inscription. Falconers in more than 40 countries organise activities to raise awareness of falconry ICH to the general public, and to share the sense of belonging to the falconry community world-wide.

Cooperation among falconers through websites, forums and international meetings has led to increased dialogue.

Inscription has helped promote the interface between falconry and the general public and media. Falconers have made active efforts to promote this rich cultural heritage by conducting public displays with their birds at a wide variety of venues, exhibitions, heritage events, festivals and country fairs. These have led to increased dialogue between falconers and the general public, encouraging a spirit of cooperation to the benefit of society and nature. Continued inscription will further promote this cultural dialogue.

How can inscription promote respect for cultural diversity and human creativity?

Falconry embodies knowledge, a practice, a craft and also a way of life. Although countries share the cultural heritage of falconry, the individual manifestations of the element (handling, instruments, equipment etc.) are diverse and follow centuries-old local traditions. The resultant tradition of each community displays a unique cultural character and therefore represents cultural diversity, as well as demonstrating the extraordinary creativity of humanity.

Inscription has highlighted the fact that falconry is an art form that varies based on distinctive local conditions and traditions. National and international festivals have brought together communities to share their falconry practices, and increased respect for cultural diversity. As a national example, a large traditional meet is held in the Czech Republic in Opočno castle, as well as in other countries. These communal displays of rich and diverse cultures and heritage promote dialogue and mutual respect for cultural diversity, to the benefit of both falconers and the general public.

Inscription has also led to increased interaction among falconry groups to celebrate their common cultural values as well as knowledge about and respect for the diversity of expressions of this ICH element. Events such as international festivals are particular occasions to express human creativity with the sense of sharing personal and collective experiences. Inscription has led to an increase in the number of these events, as well as increases in the participating communities and a greater attendance of the general public of all ages.

Falconry has also inspired a great diversity of artistic expression, such as medieval tapestries from Belgium, France, Germany, Hungary, Italy, Portugal and Spain, a rich heritage of the written word in books, manuscripts and poetry, and modern paintings, urban sculptures and symbols.

3. Safeguarding measures.

For Criterion R.3, the States shall demonstrate that 'safeguarding measures are elaborated that may protect and promote the element'.

3.a. Past and current efforts to safeguard the element

How is the viability of the element being ensured by the concerned communities, groups or, if applicable, individuals? What past and current initiatives have they taken in this regard?

Falconry communities are actively involved in a wide variety of safeguarding measures concerning transmission, promotion and interaction with the environment. Many safeguarding initiatives have brought together falconry communities who collaborate with NGOs at national and international levels. Specific examples are:

Transmission

Some falconry communities have formalized transmission by setting up mentoring and apprenticeship schemes. They have established clubs, teams ('equipages') and groups that meet regularly and host visiting speakers and events. At the national level, falconers have formed themselves into national clubs and federations.

- *The Austrian Falconry Association (ÖFB) has established a Falconry Academy which runs theoretical and practical courses several times a year.*

- *In Kazakhstan the Federation Berkutshi runs annual competitions and trains young falconers in sports schools with funding 'Samuryk-Kazyna' of US$500,000.*

Promotion

There has been an increase in the number of books, magazines, websites, artwork and films related to falconry, helping to raise public awareness.

- *Czech Falconry Club publishes an annual newsletter and a wide range of promotional materials, information brochures, CDs and DVDs.*

- *Two falconry associations in Korea present falconry to the general public each spring.*

- *In Portugal, falconers established a national association in 1991 to provide information and training.*

- *Pakistani falconers presented falconry to different communities.*

- *Saudi Arabia holds annual falconry festivals and has established a Committee of Friends of Falconry.*

Respect for nature:

- *Falconers in Czech Republic and Hungary provide technical support for wildlife conservation.*

- *Since 1971 French falconers have collaborated with ecologists for the official protection of falcons.*

- *In Germany, over 1,000 peregrine falcons bred by falconers have been reintroduced into the wild since 1977.*

- *In Hungary, a research programme involving the Hungarian Academy of Sciences was initiated by Hungarian Falconry Club with support of Ministry of Environment and Water.*

- *In Italy, the CRAS (Wildlife Rescue Centre) is run by UNCF falconers in Bologna.*

- *Spanish falconers are leading 'breeding in captivity' programmes for endangered species. An ethical code, written by falconers, is on the Ministry of Culture webpage.*

- *The Pro-Falcon Centre (UAE), established in 1996, encourages the use of captive-bred falcons.*

International safeguarding measures:

National clubs have linked together to form the International Association for Falconry and Conservation of Birds of Prey (IAF), with 40,500 members from 67 countries. The Federation of Associations for Hunting and Conservation of the EU (FACE) and the International Council for Game and Wildlife Conservation (CIC) represent the interests of falconers in matters of international legislation.

- *Various internet forums have been set up for falconers to promote international communication and cultural dialogue.*

- *Falconers from many countries, including Austria and Belgium, participate in the apprentice scheme of the Falconry Heritage Trust, which is continuously upgrading the World Falconry Inventory.*

- *At national and international levels, falconers have held many events and field meets open to the public, such as those in Austria (2010), UAE (2011, 2014), Qatar (2014), and Spain (2013, 2014).*

- *A unique falconry exhibition was shown at European Parliament in Brussels in November 2012 to introduce international and cultural aspects of falconry to members and officials.*

- Tick one or more boxes to identify the safeguarding measures that have been and are currently being taken by the communities, groups or individuals concerned:

- transmission, particularly through formal and non-formal education
- identification, documentation, research
- preservation, protection
- promotion, enhancement
- revitalization

How have the concerned States Parties safeguarded the element? Specify external or internal constraints, such as limited resources. What are its past and current efforts in this regard?

As safeguarding measures, States Parties such as Austria, Belgium, Czech Republic, France, Italy, Pakistan and UAE have established museums and heritage research institutions to record and protect falconry artefacts, original documents and art. In many countries, efforts were made to collect and safeguard falconry heritage, promote cultural values, and encourage production of falcon-related handicrafts.

- *Austria has provided the legal framework for falconry (Tierschutzgesetz, Landesjagdgesetze) and has made falconry exams mandatory in certain provinces.*
- *Since 2003 Belgium has an official training centre for falconry, subsidised by the government. This intensive training takes 2 years, and successful participants receive an official diploma recognised by the Flemish Community of Belgium.*
- *Czech Republic offer state subsidies for the breeding of birds in falconry, and also support for the international meeting in Opočno each year.*
- *In France, periods for falconry have been studied with the Ministry of the Environment in charge of regulations, the National Federation of Hunters and ANFA, to enable a larger practice of falconry. Thanks to the participation of falconers, the raptors are completely protected (the authorizations for owning a raptor are given only for scientific aims or the practice of falconry).*
- *In Germany, a falconry license was implemented 80 years ago, and the Deutscher Falkenorden has been recognised at its request by the Federal Ministry for Environment, Nature Conservation and Nuclear Safety as a conservation association.*
- *In Hungary, to ensure sustainable falconry, falconer and nature conservation courses are offered by the Hungarian Falconry Club with final state exams.*
- *In Italy, both national and regional legislation has been introduced to recognise falconry and protect falcons.*
- *Kazakhstan adopted a programme of conservation, restoration and use of predatory birds in 1997, with priority given to endangered species. The falcon nursery centre 'Sunkar' maintains the collaborative work in rehabilitation and breeding of birds of prey together with the ornithologists of the Institute of Zoology of the Ministry of Science and Education of Kazakhstan. The falconers participate in bird research and population control projects in remote areas of Kazakhstan.*
- *In Mongolia the Ministry of Environment, Green Development and Tourism organises an 'Eagle Festival' every year in the capital.*
- *The Capital Development Authority (CDA) in Pakistan has included falconry in the annual Margalla festival, a sports and cultural event in Islamabad.*
- *Qatar has established a heritage area (open air museum) in which falconry is displayed.*
- *Saudi Arabia has established two falconry centres: Prince Sultan bin Abdulaziz and Prince Fahad bin Sultan bin Abdulaziz.*
- *In Spain, the Ministry of Culture gave financial support of 26,350 Euros in 2014 for a National School competition about*

falconry (photography, painting, literature and research).

- *The UAE has established several falcon hospitals to provide comprehensive veterinary services for falcons. The Abu Dhabi Falcon Hospital is staffed by about 20 veterinarians, and offers internship programmes. The UAE has also established falcon passports to facilitate travel, and has initiatives for conservation of migratory birds of prey with an MOU under the Convention on Migratory Species (CMS).*

Tick one or more boxes to identify the safeguarding measures that have been and are currently being taken by the State(s) Party(ies) with regard to the element:

- transmission, particularly through formal and non-formal education
- identification, documentation, research
- preservation, protection
- promotion, enhancement
- revitalization

3.b. Safeguarding measures proposed

This section should identify and describe safeguarding measures that will be implemented, especially those intended to protect and promote the element. The safeguarding measures should be described in terms of concrete engagements of the States Parties and communities and not only in terms of possibilities and potentialities.

What measures are proposed to help to ensure that the element's viability is not jeopardized in the future, especially as an unintended result of inscription and the resulting visibility and public attention?

The viability of falconry depends on five essential pillars for its survival: falconers, falcons, habitats, quarry species and place in society. Future plans therefore hinge on safeguarding measures for each of these.

1. The Falconers

It is important to encourage the youth and urban dwellers to take up falconry. Proposed measures include expanding falconry and environmental awareness programmes in schools and establishing falconry clubs for youth. Falconers also need a legislative framework to enable them to continue practising falconry. The tradition protects itself against an uncontrolled rise in new practitioners, due to factors such as strict legal conditions, the time-consuming work involved with birds of prey, and the limitations of natural resources.

Examples of specific safeguarding measures include:

- *Austrian Falconers Association, in conjunction with the IAF and international experts, is currently developing a new code for falconers, detailing expected conduct with regard to ethics, animal welfare, cultural tradition and heritage. They are also working on unifying the falconry examination structure throughout the country.*
- *The French Association (ANFA) through a convention with the National Federation of Hunters (FNC) will continue the training programme for technicians on methods to teach young people how to obtain falconry permits.*
- *In Italy there is a proposal by the National Commission for Falconry supported by the community of falconers for the definition of ethical guidelines and an internal code of ethics that recognizes and defines the profile of falconers, with specific competences for breeders, trainers, craftsmen and procedures for free flight display.*
- *The local governments in the Republic of Korea will continue their support of falconry by giving monthly allowances of $700 to the two state designated practitioners.*
- *Mongolia will introduce support measures to protect intangible heritage in general and falconry in particular, including the nomination of some falconers as 'living human*

treasures' and the establishment of a heritage museum.

- *The Portuguese Association for Falconry and national partners will work on a comprehensive proposal for a national preparation system for those who wish to be legally entitled to practice falconry.*

- *Saudi Arabia is proposing to establish a Falconry Club, and Prince Sultan will continue to support falconry with a fund of US$135,000.*

- *At the international level, the Falconry Heritage Trust has raised an endowment fund of $1,630,000 and is providing scholarships for the study of falconry heritage, to be recorded on an international electronic archive available to all.*

2. The Falcons.

Nowadays wild falcons face hazards, such as pesticides, habitat loss and degradation, which reduce their populations. Falconers are making efforts to ensure the sustainability of these populations at national and international levels as follows:

- *Falconers in the Czech Republic will continue to support wild-life conservation.*

- *Germany will continue to share its expertise and work with other countries on programmes to re-establish peregrine falcons into the wild.*

- *Kazakhstan will continue to conduct research and support a programme of conservation, restoration and use of predatory birds, aimed at the preservation and sustainable development of birds of prey, with special priority given to endangered species. This will be implemented by 'Burkitshi' and 'Kyran' Federations, falconry centre 'Zhalair Shore', and the Association of the National Sports.*

- *In the Republic of Korea, goshawks will continue to be protected under the Cultural Heritage Protection Act to ensure future stable*

Turkmeni falconer Annaaman Movlamov.
Photo Linda Wright.

populations.

- *In Mongolia, a large-scale artificial nest programme for saker falcons, which will also support local herding communities, is being pioneered with the support of the UAE and the agreement of CITES. Italian children in middle-schools are also supporting this project and Belgium is in the process of connecting five schools to Mongolian schools. The purpose is to have children from the age of 10 to 12 communicate with Mongolian children in the exchange of information of birds of prey in their local areas. Ref: http://www.schoollinksprogramme.org/*

- *Qatar will continue to support a falconry hospital and falcon breeding programme.*

- *In Saudi Arabia, the National Committee for Conservation of the Environment will continue supporting breeding programmes for falcons and houbara with an annual budget of US$6 million.*

- *The Sheikh Zayed Falcon Release Programme (UAE) will continue to re-establish falcon populations back into their breeding grounds in Asia.*

- *IAF is currently working on the development of an international database with information regarding birds of prey that have been electrocuted with the support of all its members. This can subsequently serve as a basis for further research.*

3. The Habitats.

- *Falconry has a direct relationship to nature and is dependent on a cultural landscape which is characterised by its biodiversity. Efforts to preserve falconry areas will be supported at both national and international levels, and by greater awareness in the media.*

- *In France, the Foundation for the Safeguarding of Habitats will continue to obtain and conserve suitable grounds.*

- *In Hungary, a nine year research programme initiated by the Hungarian Falconry Club in 2010 and involving the Hungarian Academy of Sciences and other research institutions will continue.*

- *In Portugal, national habitats and important bird areas will continue to be protected by national law, programmes and specific measures.*

- *The Environment Agency Abu Dhabi (UAE) will continue its plan to develop a system of managed protected areas.*

4. The Quarry.

Many falconers breed and reintroduce quarry species in order to maintain sustainable populations. In many European countries, there are on-going agriculture and environmental measures to provide suitable habitats for quarry species, such as grey partridge and brown hare.

- *In Korea, the local government of Geollabuk-do Province has designated Jinan region, a central place for falconry, as a protected area to prevent illegal catching of wild rabbits and pheasants, and this will continue.*

- *Morocco will continue collaborating with the Environment Agency Abu Dhabi and the Falconers Club to release more than 5000 houbara per year into the eastern desert.*

5. Safeguarding falconry's place in society

Falconry will continue to be promoted by activities to raise awareness, promote intercultural dialogue, disseminate information, and share experiences, as well as the dedication of cultural spaces. This will involve a wide range of activities such as rural events, traditional country fairs, conferences, workshops and publications.

Specific examples are:

- *The Austrian Archives of Falconry, located in Castle Waldreichs, will continue its work on consolidating knowledge on falconry practice.*

- *The citizens of Ghent (Belgium) will continue their tradition every October of celebrating the patron saint of falconers, Saint Bavo*

(~600AD), in the cathedral. Falconers in their traditional costumes take part carrying falcons. This is a very popular event and is followed by a procession to the market square.

- *France has created a Foundation that will offer all the private archives held by French falconers to the International Museum of Falconry in Gien, so that this heritage will be accessible to the public.*

- *In July 2015, Italy is planning to run a 3 day public event in Rome on the theme of 'Falconry Wings of Peace among the Nations', to disseminate information and raising awareness about falconry and ICH in general.*

- *In Morocco there is a project to establish a museum for falconry in Al Jedida town, with the support of the local government and falconry associations.*

- *The Pakistan Falconry Association (PFA) is working with Lok Virsa to develop a diorama of falconry in the National Ethological Museum. This will present falconry as part of the intangible culture of the country, and include portrayal of the living culture, tradition and life style of Pakistani falconers.*

- *Qatar has approved a project to carry out a field study of falconry heritage traditions, and will produce an annual report on falconry published by the Ministry of Culture.*

- *Saudi Arabia will continue the annual traditional falconry skills event under the name of Saudi Arabian Festival to support local falconers and their culture, as well as annual festivals at Al Noaireeh and other locations.*

- *In Spain the royal group of falconers, established for more than 1000 years, will continue its programme of public engagements. The citizens of Madrid (Spain) will celebrate the patron saint of Spanish falconers Saint Francisco de Borja (XVIth century) every October 3rd in their church.*

- *In Syria, there are plans to re-establish the falconry festival in al Rhaiba town, with cultural, artistic and performance events.*

- *The UAE will open the Zayed National Museum in 2016. This will contain a section on falconry.*

- *IAF shall, through its channels, continuously reflect the diversity of falconry practices and raise awareness about the importance of falconry as intangible cultural heritage. IAF will continue to allocate an annual sum in the order of $15,000 for its awareness-raising, information and publication activities aimed at the general public. It will also continue to organise World Falconry day on November 16th each year.*

The group of falconry countries involved in this UNESCO submission will continue to cooperate and support falconry heritage issues.

How will the States Parties concerned support the implementation of the proposed safeguarding measures?

State Parties, by working together on the largest multi-national submission, have demonstrated their strong commitment to safeguarding falconry and the principles of the Convention.

- *Austrian government will continue to provide the legal framework for falconry and support infrastructure necessary for transmission of falconer's knowledge and skills. Federal falconry organizations will carry out the legally-required falconry exam and improve legal instruments.*

- *Belgium will support and subsidize ICH umbrella organizations, including falconry. A databank will enable heritage communities to describe their element and present safeguarding measures. Falconers are able to contact organizations and fellow communities on this forum and exchange experience and knowledge.*

- *Ministry of Agriculture in the Czech Republic will continue subsidizing the breeding of rare and indigenous species of birds of prey.*

- *France will safeguard falconry through*

elaborating and sustaining agreements and conventions with Ministry of Ecology with the Office for Hunting and Wild Game and the Ministry of the Environment for falconry regulations. The ICH documentation centre will continue organizing meetings for communities with an inscribed element, including falconers, to exchange experiences and define future safeguarding measures.

- *'Greening' measures of the Common Agricultural Policy of the European Union and its implementation by German authorities will be supported to promote species diversity in agricultural landscapes, including wildlife such as partridge and hare.*
- *In Hungary, the state continues to support final exams for falconer courses. There is an undertaking to clarify the legislative background of falconry practice, including coordinating existing and future regulations focusing on the safeguarding and survival of falconry.*
- *In Italy, the state continues to support legislation to protect falconry, and there are plans to establish a falconry exhibit in a museum or historical location.*
- *Kazakhstan will continue the implementation of safeguarding measures in cooperation with the IAF. Falconry is officially recognised by the Agency of Tourism and Sports, and will be promoted throughout the country.*
- *• Korea will continue to provide annual support (currently around US$10,000) to falconry associations for programmes to enhance visibility.*
- *Mongolian government and local authorities will continue to assist the minorities which hunt with eagles by promoting ecotourism, which generates income for communities and enables them to continue practising falconry and remain in this harsh environment.*
- *Moroccan government and local authorities are providing support to establish a museum for falconry in Al Jedida town. They will continue to collaborate with the Environment Agency Abu Dhabi in the houbara breeding and release programme.*
- *In Pakistan, the government will provide assistance to falconers in conservation projects.*
- *In Portugal, the government welcomes proposals for law improvements regarding the practice of falconry and a register of birds of prey.*
- *Qatar will continue support for an open air museum which includes falconry, and new premises for the Falconers' Club.*
- *Saudi Government is continuing to support the National Commission for Wildlife Conservation and Development, with 16 large natural reserves, Houbara bustard breeding programmes, and a satellite monitoring system for falcons. It will support cultural festivals for falconry heritage, and run media campaigns to preserve and protect wildlife. The latest allocated budget is around US$6 million*
- *In Spain, falconry is included in Castile-Leon's Historical Heritage Plan, with documentation, archives and heritage dissemination. Historical Heritage Council will continue to support the 'National Plan to Safeguard Intangible Cultural Heritage', sponsored by the Ministry of Culture in conjunction with Autonomous Communities, with actions to safeguard, transmit, and promote heritage. Priority is given to inscribed ICH elements, including falconry.*
- *Ministry of Culture in Syria has plans to support falconers through a comprehensive survey to identify the needs and number of falconers.*
- *The UAE will host future International Falconry Festivals every three years (budget of US$5 million). There are plans to launch a Mohamed bin Zayed Falconry and Desert School.*

How have communities, groups or individuals been involved in planning the proposed safeguarding measures and how will they be involved in their implementation?

- *Falconers are determined to maintain and pass on their traditions to future generations. Hundreds of individual falconers have been actively involved in planning and formulating proposed safeguarding measures within their respective countries. They have submitted photographs and materials, held club meetings, and funded delegates to attend preparatory meetings, demonstrating their strong interest and commitment to maintain falconry as an intangible cultural heritage.*

- *Safeguarding measures have also been planned and supported by local, national and international communities, with the participation of NGOs, institutions, experts and different cultural agents. This collaboration extended to include communities in the five States Parties joining this multi-national submission.*

  Examples:

- *In Austria, community organizations ÖFB and ZÖF have initiated, planned and implemented the majority of falconry safeguarding measures, including several breeding, scientific research and bird release programmes. Practical implementation will continue to be carried out by members of the falconry communities.*

- *In Czech Republic, the local communities, Czech Falconry Club, Ministry of Agriculture and the State Forest Enterprise will continue to support the cost of large falconry events, e.g. Opočno castle, with US$20,000 every year. This will be extended to include private and public sponsorship schemes. Czech Falconry Club has been very active in planning safeguarding measures, including promoting falconry.*

- *French falconers have been and will continue to be associated with every measure concerning falconry, through the presence of representatives in the signing of conventions or treaties under the auspices of the local and national authorities.*

- *In Germany, the DFO has made a significant contribution with its representatives to the husbandry survey of the Federal Ministry of Food and Agriculture for the keeping of birds of prey and owls. The falconry associations also provide examiners for the state falconry exams.*

- *In Hungary, the community-based Hungarian Falconry Club met to plan, discuss and promote future safeguarding measures. On March 7, 2011, a Community Forum dealing with the legal background and issues pertaining to ICH safeguarding was held in Mohács, attended by bearer communities, experts and ministry representatives. Falconers were able to raise concerns and present recommendations. Community proposals are taken into fullest account in developing safeguarding measures, as their active cooperation is required to implement these measures.*

- *In Italy, the Lanner Foundation was founded by experts and falconry enthusiasts to promote awareness and run reintroduction campaigns for the Lanner Falcon, a rare protected species in slow decline (around the Mediterranean there are around 250 breeding pairs).*

- *In Kazakhstan the 'Burkitshi' and 'Kyran' Federations, 'Zhalayir Shore' Centre of Falconry and the 'Sunkar' Falcon Nursery Centre have been actively involved in delineating and implementing safeguarding activities, including training volunteers.*

- *In Korea, members of falconry associations continue to practise falconry and transmit to future generations with the help of financial support and government policies, planned to reflect the wishes and aspirations of the practitioners.*

Árpád Ferencz from Hungary at the 2007 festival.
Photo Linda Wright

- *In Mongolia, the Community Association for Preservation and Conservation of Golden Eagles have planned and will continue to organize an annual falconry meet. Through community efforts, their culture of flying eagles from horseback in the steppes is becoming well known internationally.*

- *The Pakistan Falconry Association is developing a breeding programme for birds of prey and game birds. It holds an annual meeting for its members, and raises public awareness through falconry presentations and local media.*

- *The Portuguese Association for Falconry initiated a survey concerning safeguarding measures, which formed the basis of a position paper on proposals for future safeguarding measures. Falconers continue to make recommendations regarding falconry law and heritage safeguarding.*

- *The Qatari Society of Algannas regularly meets to discuss safeguarding measures and is actively involved in their implementation.*

- *In Saudi Arabia, the community of falconers continues to provide information and documentation and give suggestions as to how to improve safeguarding, including improving methods of transmission.*

- *Falconers in Spain initiated and planned a successful breeding programme for birds of prey, and they will continue to support this safeguarding measure.*

- *Emirates Falconers Club (UAE) helps organise the International Falconry Festival and various falconry competitions.*

Falconry communities worldwide plan and will continue to support national and international festivals. As representative bodies of the falconry community, IAF, FACE and CIC will continue to safeguard all aspects of falconry at the international level. The IAF will compile and update regularly an inventory of international cultural elements of falconry.

2010, our Falconry submission is inscribed by UNESCO in Nairobi. L-R: Author, HE the late Mohammed Khalaf Al Mazrouie, Dr Awadh Ali Saleh, Prof Suleyman Khalaf.

Falconers in Serbia.
Photo Stavros Athanasiou.

Finally on 16 November 2010, in Nairobi, our submission was accepted on behalf of eleven countries and Falconry was inscribed on the Representative List of the Intangible Cultural Heritage of Mankind. UNESCO said it was the largest and best prepared submission they had ever had.

Currently 21 falconry countries are inscribed by UNESCO: Austria, Belgium, Croatia, Czechia, France, Germany, Hungary Republic of Ireland, Italy, Kazakhstan, Republic of Korea, Kyrgystan, Mongolia, Morocco, Netherlands, Pakistan, Poland, Portugal, Qatar, Saudi Arabia, Slovakia, Spain, Syrian Arab Republic and United Arab Emirates.

In writing the core submission I had to consider first the total dimension of world falconry heritage in all its different forms, bearing in mind that additional countries might join in future and should not be constrained by tight wording. I also had to bear in mind the future and how falconry might change in the next few hundred years. Would it last that long or still be recognisable? Different countries are at different stages of their cultural journey. Some, such as China, were deeply traditional and used the cockerel feather on their goshawks' tails because they did not yet have radio-transmitters. Of course, when they came to the festivals and saw some of these modern developments, they wanted them too. And so, in organising a festival to celebrate and support cultural heritage, we were unwittingly leading to its demise. Not only this, but international cultural transmission was leading to homogenisation so that we all ended up pretty much the same. Grey people in grey suits….

This is all 'progress'. But not all traditions stand up to modern scrutiny. Most Western falconers find the practice of seeling a falcon's eyes by stitching them shut quite horrid. And many traditional medicines and nostrums are worse than harmless. Most falconers use modern medicines if they are available.

It was clear also that the smaller nations were much stronger in their cultural identity than the large ones. I was particularly struck by the Americans. Whereas the South Americans came to the festivals with beautifully made national dress and customs, those from USA and Canada bemoaned that they had no national dress. The dress and traditions of native North Americans are not appropriate for the majority of Americans who had only settled there in recent generations. Their only offering was the baseball cap, which is a sorry state of affairs. Until then, they hadn't really understood how important having your own culture is to many peoples.

Another strong lesson I learned is that Heritage is not the same as History. History is about the past. It's dead and gone. Heritage on the other hand is a baton that is passed on from one generation to the next, which is something very different. It is 'alive' and, like Chinese Whispers, is constantly changing as it passes from one person to the next. UNESCO not only recognises this; it encourages it:

Article 2.3 says 'as well as the revitalization of the various aspects of such heritage'. In other words, it acknowledges that Falconry will change, and indeed has to change, if it is to survive into the future. The world is changing and the only certain thing is change. So while it is nice to remember and acknowledge our own traditions and history, we mustn't be afraid to embrace the future and do whatever it takes to adapt.

Many falconry organisations emphasise their definition of Falconry as '*The art of taking wild prey in its natural state by means of trained birds of prey*'. This bald statement is fine as far as it goes, but it misses the point: Why do it? At the end of the day, it is not about the pursuit of prey, but of enjoyment. That is the actual aim of the exercise. The days of hawking primarily for food are long gone. Those same organisations take the definition a step further. They claim that anything that is not hunting is not Falconry. 'If you cannot hunt with your bird every day then you should not be doing it at all.' This is where idealism departs from reality. Many prominent people in these organisations, (and I could name plenty) do not fly their birds at prey for more than a short period each year, if at all. For the rest of the time the birds do very little. The reality is that most people who keep birds of prey either do not hunt at all, or at best catch less than a dozen prey per year. But, one way or another, they clearly enjoy whatever it is they are doing, call it 'hawk-keeping' if you like. To the general public, someone with a hawk on hand is a 'Falconer', they know no different and care less.

To try to understand and come to terms with this reality, Jim Chick and I undertook an Audit of Hawk-keeping in UK in 2007. At the time we were Chair and Vice-Chair of the British Hawk Board. Amongst other things we estimated that there are about 25,000 people keeping about 70,000 raptors in UK of which only about 5000 people are active Falconers who hunt. Almost all captive raptors in UK are captive-bred and UK produces sufficient to supply other countries, especially the Middle East. As a follow up to this audit, Jevgeni Shergalin and I decided to go global and dig up some facts and figures about current falconry around the world. It makes for interesting reading, a sort of Domesday Book audit of the current situation.This audit is a work in progress as more information comes in from different countries.

We have the perception amongst some traditional Falconers that only hunting with their birds is approved of, and those who just fly their birds to the lure or in demonstrations do not deserve to call themselves Falconers. But amongst the general public the perceptions are the opposite. For many non-falconry people, hunting in any shape or form is abhorrent and the only 'falconry' they have ever seen is in public demonstrations, which are relatively innocuous and thus not opposed. The public image of falconry is thus based on these centres and displays, while the falconers who hunt go relatively unseen. The moral high ground is a tricky place, and sometimes over-crowded with wannabees busily virtue signalling.

In recent years, Arab falconers, lacking opportunities to go hunting, have developed racing their falcons as a substitute for hunting. Now there are falconers training birds solely for racing, and breeders developing bloodlines for racing. We see parallels with both horses and dogs. In UK, with the Inclosure Acts from 1604, open 'common' land became partitioned into fields enclosed by hedges. By 1914, 6.8 million acres had been fenced in. This ruined a lot of land for falconry. Fox-hunting, still in its infancy, also had to adapt. Hugo Meynell, who became Master of the Quorn Hunt in Leicestershire in 1753, developed the art of jumping hedges at a flying pace, to the extent that for many followers jumping hedges at speed became what hunting was all about. Some began to challenge each other to races across country, using church steeples as markers. And so the sport of Steeplechasing began and is carried on to this day as Point-to-Point races for horses that have been 'fairly hunted' during the season. These developed into events overseen by the Jockey Club, including races 'over the sticks' such as the Grand National at Aintree, and Flat Racing. So we see parallels in falconry with the development of Flat Racing to the *Telwah* or lure and Hunt Racing chasing robotic prey.

For some people, like myself who hates rules, racing is a poor substitute for real falconry. But the sad reality is that in many countries there is not enough suitable land or quarry for falconry with falcons, whether for pursuit flights or for waiting-on. So the choice is either to carry on doing it badly, with falcons checking away at pigeons and wandering away onto private land, or to take up racing. There is no doubt that racing does improve the standards of management and training, and for many people the only chance they will ever have to see the drama of powerful falcons performing high in the sky is at the races.

It has all gone a long way from trapping your hawk in the wild, using fieldcraft to find the prey and hunt, and all those things that bring us close to nature. The whole world is changing rapidly. There are many places in Britain where I have hawked rabbits or pheasants, that are now housing estates, or areas of desert in the Gulf where I have hawked karrowan or hares, that are now unrecognisable as cities. My own hawking grounds in Northumberland, that have been hunted for thirteen centuries, deteriorate each year by a process of attrition. A new wire fence here, a forestry block there, or a new tourist pathway and 'open access' where all was once quiet for the curlews and black grouse. How long can Falconry carry on? Will its dying embers be seen in some vestigial computer game in which you can become a virtual reality raptor and chase down aliens?

Prince Philip, Duke of Edinburgh, with a gyrfalcon trained by Roger Upton in 1967 given to HRH Sh Zayed Al Nahyan.
Photo Roger Upton.

Falconry also preserves its tangible heritage in archives and museums in UK, France, Holland, Portugal, Kazakhstan, Russia, USA, Czech Republic, Abu Dhabi, Dubai, and Qatar. These are listed on the IAF website. In addition, the Falconry Heritage Trust is an online archive of falconry curated by Jevgeni Shergalin documenting falconry internationally and through the ages. These all provide interesting records showing how falconry has evolved over the centuries. Often called the Sport of Kings, falcons are still given as diplomatic gifts. We breed the falcons belonging to the British Royal Family and these originated as gifts from the Middle East.

11.3 The international contribution of falconers to raptor conservation and welfare

Man's first personal contact with raptors came through falconry and religion sometime in prehistory. In some countries, raptors were worshipped as gods, such as Horus in Egypt; in others they were esteemed, more prosaically, as hunting companions. Falconry probably originated in Asia, reaching Japan by 247 A.D. and Europe and North Africa by the fourth century. Although North and South American Indians revered raptors, falconry itself was introduced there by Europeans.

Falconers have traditionally conserved raptors in the wild as a sustainable source of hunting birds. In 1501, in Burgundy, anyone stealing a hawk was heavily fined, or if unable to pay, six ounces of flesh was cut from his breast and fed to the bird. In Britain, during Henry III's reign, anyone stealing a hawk from a nest was imprisoned for a year and a day. If he stole another person's hawk, he was executed. It was common in Britain to turn hawks free at the end of the season. Merlins were released in late autumn in the same way that sakers and peregrines are still released now in the Middle East. Goshawks were often turned out into a local wood to nest, to supply young for the next season.

Falconry became deeply entwined in the cultures of nations, and people wanted to learn more about the birds. Frederick II's 1250 *The Art of Hunting with Birds* examined all aspects of raptor biology and on some topics is still unrivalled today. In 1379, Henri de Ferrières wrote in a French manuscript: 'He who wishes to learn falconry must possess three qualities: he must be very fond of birds, must be interested in them, and be good and kind to them.'

So falconry, right from the start, grew up with the basic premise that wildlife is a renewable resource, sustainable indefinitely if managed carefully. Surprisingly, even as I write now, in the mid-2020s, there are still many people and organizations who cannot or will not grasp this fundamental fact. This is what lies at the root of many of today's wildlife problems.

As firearms become widespread, people could kill animals more quickly and easily and falconry fell from favour. This is the way with technology. Children now play war games on their computers without even leaving their rooms. Wildlife was given a human morality. There were the good guys - game animals, and the bad guys - vermin and pests. The former were preserved for hunting, and the vermin were eliminated. Gradually the big mammalian predators were exterminated in areas populated by humans. Raptors, without a role in falconry, were now seen as competitors for game and therefore vermin. Massive campaigns in Europe (documented by M. Bijleveld) eradicated whole populations, and migrating raptors were shot for sport, as they still are in Malta. The Victorian era was a century of shame for raptors in Britain, with systematic slaughter from gamekeepers and greedy exploitation from egg collectors. Fortunately for raptors, the two World Wars put an end to the 'golden era' of game shooting. With the reduction of income from the British Empire, the big estates went into recession, and most of the keepers went to the Front, never to return. Raptor populations had a respite and a recovery. But for some it was too late: the goshawk had been exterminated, as had the white-tailed sea-eagle and osprey. The red kite had just a few pairs left in Wales, and the common buzzard and golden eagle hung on in places with few gamekeepers. Similar scenes occurred in Europe, and in North America, where it is even now legally easier to obtain a gun than a hawk, people were still busy winning the West and shooting every raptor they saw on sight. As Fran Hamerstrom pointed out: in the 1960s you could shoot an eagle but weren't allowed to keep a live one.

After the Second World War, many farmers returned to the land, and chemicals were seen as a way of improving productivity by winning the struggle against those pests. DDT was the wonder chemical, to be used on crops, in orchards, on livestock and around the home. The sad impact on raptors was unravelled by Derek Ratcliffe in Britain, followed by a spate of further investigations throughout the world. Whole ecosystems were poisoned and even on remote places like the Auckland Islands, 1000 kilometers south of New Zealand toward Antarctica, the New Zealand falcons were contaminated through feeding on fish-eating petrels.

The Americans, always ready to do everything bigger and better than everyone else, almost wiped out their peregrines completely before discovering the folly of pesticides. In Germany and Poland, the tree-nesting populations of peregrines were exterminated. Raptors were in big trouble and, as biological indicators, became the flagship species for clean-up campaigns. Suddenly, they were endangered national heritages. Anything which might threaten them was stamped on hard, including falconers.

A few simple facts were overlooked. Falconers have never, in any country at any period in human history, endangered any population of raptors or any other species. The same cannot be said for agriculture, pesticides, shooters, or domestic cats, all of which have exterminated species completely, in the case of cats, 63 species (40 birds, 21 mammals and two reptiles, representing 26% of all known contemporary vertebrate extinctions). Cats endanger a further 367 species. Wherever falconry has become popular, falconers have organized systems of self-control either in law or through codes of conduct, recognizing their dependency on sustainable use. And falconers, with their millennia of experience of hands-on contact with raptors have pioneered their welfare and breeding in the wild and in captivity. Finally, falconers are sitting targets for new-found protectionists. The falconer does not want to kill raptors; he tends his few birds alive and thus is easily detected and targeted. The shooter who kills or poisons raptors is almost undetectable, as is the farmer using pesticides. We have had gamekeepers in Wales who, on their own admission, kill over a hundred raptors a year. Nothing comes to court. The effort by protectionists fell disproportionately on falconers as 'soft' targets, rather than on those who made a significant impact. In the U.S.A., where raptors are heavily regulated for falconry, they are still frequently shot illegally and few North American Falconry Association field meets go by without one of the falconers' birds being shot at by casual shooters.

Although falconers were prosecuted for keeping hawks, to my knowledge, no chemical companies in Britain or Europe were ever prosecuted for poisoning them. Indeed, pressure from large companies resulted, not in the chemicals being outlawed but in a voluntary ban, a gentlemen's agreement. This enabled them to continue exporting these chemicals to Third World countries which, despite knowing their ecological impact, had become dependent on them. We had major problems with pesticides being used in Mongolia.

In Britain, sufficient peregrines survived in non-agricultural areas to make a slow unaided recovery. But in North America and parts of Europe, they were virtually extirpated. Other bird-eating raptors were similarly affected. What was to be done? In North America, Britain and West Germany, some of the prominent falconers pooled their resources and placed their own peregrines into joint breeding programs. On both sides of the Atlantic major milestone successes were made in breeding raptors in captivity. As the feasibility and potential were realized, public funds were raised and some of the projects became formalized. In the U.S.A., The Peregrine Fund was set-up under the astute leadership of a falconer-biologist, Prof. Tom Cade, heading a team of a new breed of applied biologists pioneering breeding techniques, undertaking field research, and then reintroducing young peregrines back into their former haunts.

A yearling goshawk hunting at one of our beaver ponds. 2017.
Photo David Woodfall.

In Canada, Richard Fyfe's team at Edmonton and David Bird in Montreal set up programs, while in Europe, the Deutscher Falkenorden under Prof. Christian Saar and Dr Trommer set up their breed and release scheme for peregrines. The British Falconers' Club meanwhile concentrated on the goshawk and reintroduced it back to Britain in a quiet success which has restored it once more as an integral part of the British fauna. Led by Dr Robert Kenward, our team hacked out eyasses in Argyll and I released a trained one on Speyside with Doug Weir. The simplest route though was simply to trap passage gosses, bring them to UK and turn them straight out.

When the kite builds, look to lesser linen. Shakespeare: A Winter's Tale.

Kites were common scavengers in Shakespeare's time and they line their nests with wool and old cloth, sometimes snatched off the washing line. But, in Victorian times, kites were decimated by gamekeepers and died out in England and Scotland. In 1985, the British population was down to about 25 breeding pairs in mid-Wales. Their productivity was seldom more than one chick each, and many nests were robbed by egg-collectors, despite intensive wardening, even by the Ghurkhas. We were breeding various raptors at our farm in Wales at that time, so I approached the Kite Committee and suggested that the Nature Conservation Council (as it was then) took eggs from threatened nests and replaced them with dummies, bringing the real eggs to us for hatching. In 1986 they visited our facility and we showed them what was possible with artificial incubation and rearing. They agreed, and in the 1987 season the first kite eggs were delivered by field-workers Tony Cross, Iolo Williams and Dee Doody. Some of the eggs were pretty filthy!

Over the next seven years we hatched or reared 53 Red Kites and all were released to the wild. The project was so successful that soon we ran out of foster nests in Wales and that led to the next stage, which was to

Baby kites and falcons.

Peter Walters Davies MBE collects kite chicks ready to place in foster nests.

establish them in England and Scotland.

Various incidents occurred along the way. One year the field team, using a car wing mirror on a telescopic pole, spotted a very strange egg in the nest of a new pair. The pattern looked familiar. Tony Cross climbed up and retrieved – a tennis ball! The young female, presumably not laying herself, had brought it to the nest and started to incubate it. So Tony replaced it with a Buzzard egg, which eventually hatched. Then we swapped the buzzard chick with a red kite chick from us, so eventually the pair succeeded in fledging a kite chick of their own.

The general plan had been to take one egg from a clutch of two, replacing it with a dummy, so that the kites could hatch their remaining egg, then place a second chick in the nest from the egg that had been artificially incubated. We had a foster-Buzzard to rear the chicks. She was an imprint called 'Beast-that-lurks'. For some reason BTL was nice to me but horrid to anyone else who so much as approached her door. She would wait on the floor to grab them if they opened it. BTL could rear twelve kite chicks a season on a roll-on roll-off basis. There were no Cain and Abel squabbles. BTL would feed all the chicks to repletion and then sit on them until next feed time. The result of this was that our chicks were fed well and grew rapidly. But when we came to return a chick to its natural parents at 3-4 weeks old, we found that the original chick was barely half the size. Our chick would have killed it. So we had to do 'musical chairs' with the chicks to get them into matching pairs.

It was clear from this that the kites in mid-Wales were under food stress. That is why they could only rear one, poor quality chick. Often they were reduced to feeding rotten lambs' tails that had fallen off the lambs, complete with the rubber ring. Of course these would impact in the gizzard and kill the chick. Although the kites had hung on in mid-Wales, where there were no gamekeepers to kill them, the habitat was far from ideal. So it seemed better to increase their range, either from the edges, or to start again in a new area where there was more food available.

As the project became more successful, more organisations became interested. The RSPB made a film in 1990 and we hatched chicks for the film in replica nests in our lab. Did we get a free copy of the film? No, we had to buy our own for £14!

There was a bit of politics going on around this time. The iconic Red Kite had become very emblematic of Wales. The Welsh were not keen to supply kites to the English. Also, as the Welsh kites had been through a genetic bottleneck and were very in-bred, fresh blood was needed. So we cast around to see if any neighbouring countries would donate kites and as a result, a plan was made to import kites from Spain for the English releases, and from Sweden for the Scottish releases, starting in the Black Isle. Chicks bought in from the continent still had to be quarantined before release and I designed a hack pen which could

The buzzard 'BTL' rears some kite chicks.

serve both purposes. When the first pen was ready on an estate near Stokenchurch, I took two Welsh Kite chicks which were ready for release but had no foster nest available, in our old Cortina and put them into the release pen. So they were the first kites to fly again in England since extermination in Victorian times.

Feeding stations were set up in Wales to help kites through the winter months because there was a clamp-down on dead farm stock and farmers had to remove all dead animals and incinerate them. This drastically reduced food availability for scavengers during the hard months. Nowadays the feeding stations have to use butchers meat, which is not a balanced diet for kites and has further implications. But the kites are compensating by attending rubbish dumps!

Of course the releases in England and Scotland were spectacularly successful. With more food around, they often reared three chicks per nest. Soon these chicks were moved on to new release areas. We had kites flying all over the place and the new populations started to link up. By then there were plenty of people and organisations jumping on the band wagon, so I retired from the UK Kite Committee. The project is a success and the work was done. The kites became a multi-million pound jobs gravy train for tourism and conservationists. Now I can look out of my office window across the valley and sometimes see a dozen kites, not to mention buzzards, and occasionally a goshawk as well!

Falconers have always been determined individualists, and when they applied themselves to restoring raptors a whole night sky of individual stars started to twinkle around the globe. Experienced old-timers like John and Frank Craighead and Fran and Frederick Hamerstrom set a powerful example of what could be achieved when the practical skills of falconry were linked to the discipline of the biological sciences. This inspired further generations of up and coming falconer-biologists, myself amongst them. Gradually they became more organized, and the Raptor Research Foundation and Peregrine Fund were formed in North America, and the Hawk and Owl Trust in U.K. Others, such as the Durrell Institute for Conservation and Ecology tackled a wide variety of species, including Carl Jones's efforts for Mauritius kestrels *Falco punctatus.*

In Poland teams lead by Janus and Slawomir Sielicki have released captive-bred peregrines, gradually restoring the tree-nesting population since 1990. In Russia, Dr Alexander Sorokin has established peregrines breeding on the Moscow State University building. In Bulgaria we have supported the Green Balkans project which has released 98 sakers since 2011, resulting in the first wild nestings since they became extinct. In Spain, Fernando Feás had the first captive breeding success with Bonellis Eagles *Aquilla fasciata* and Luis Jimenez undertook the Lesser kestrel *Falco naumanni* project at GREFA in Andalucia.

In South America raptors are threatened by habitat loss, widespread shooting and pesticides. There is little governmental interest in conservation. In Brazil, Gustavo Trainini runs projects at the Vale dos Falcoes on Harpy Eagles *Harpia harpyja* and Orange-breasted falcons *Falco deiroleucos*. In Peru, Jose Antonio Otero is working on Andean Condors *Vultur gryphus*, and, with Fernando Feás, on White Hawks *Leucopternis albicollis*. In Paraguay at the Proyecto Taguató, Fernando Feás is breeding Bat falcons *Falco rufigularis*. Manual Maier in Uruguay runs a raptor rescue centre as does Cristian Gonzalez Bulo in Chile, working on Pallid falcons *Falco peregrinus kreyenborgi*. In Columbia, Alex Ospina works on Black-and-Chestnut eagles *Spizaetus isidori*.

Australian falconers, such as Mike Calvin, rehabilitate raptors, while in New Zealand Debbie Stewart and Noel Hyde have both won awards for their work at Wingspan.

Similarly, there are projects in Africa, the Middle East and Asia, some large and well supported, but also many other, more modest, individual efforts. Virtually every falconer at some time or another rehabilitates raptors, if only because people arrive with injured birds at their doorstep.

Nowadays, there are raptor biologists working all over the globe. Wherever there is hands-on work to be done you will find that the biologist either has a falconry background, or is using techniques developed by someone else who has. It is all very well shouting about the plight of raptors, but when it comes actually to doing something about it, then falconry, through the dedication of hundreds of individuals, has made the international contribution which has turned the tide. Of course, with climate change and all the increasing threats to biodiversity, the word 'conservation' has become virtue signalling. Dozens of organisations and businesses have slotted 'Conservation' into their titles and their 'projects' have been more about PR image than results.

Falconers learned from all this. With the increasing human pressure it is obvious that few species of wildlife will ever again be secure. Any international event could wipe out a species very quickly. The future of falconry as a field sport would be jeopardized if it relied totally on the sustainability of wild populations. Activated by the alarm of pesticides, we set up breeding projects to become self-sufficient in falconry birds, and in Britain, Germany, and North America we achieved this aim in a few short years.

Falconry continues all over the world, bringing with it a back-to-basics missing from the crazed modern urban lifestyle. In North America, there are many regional clubs all under the umbrella of NAFA, the North American Falconers' Association. Founded in 1961, it now has several 'senior statesmen' and this growing maturity shines through in their publications such as *A Bond with the Wild.* In Mexico, falconry is popular and the standard is improving. In South

Connecting schools together.

Dr Andrew Dixon explains a nest barrel to local children.

America, there are numerous small groups of falconers scattered in different places. In Britain, there are around 25,000 people keeping birds of prey in a plethora of clubs which are jointly represented by the Hawk Board. The main club, the British Falconers' Club, traces its origins back through the Old Hawking Club to the heydays of falconry, and this history is admirably documented in Roger Upton's books.

In Europe, there are clubs in Ireland, Britain, Denmark, France, Belgium, Holland, Germany, Switzerland, Austria, Italy, Spain, Portugal, Hungary, Czechoslovakia, Poland, and across to Russia, Georgia, Kazakhstan, and central Asia. There are also clubs flourishing in Japan, India, Tunisia, Zimbabwe, and South Africa. In the Middle Eastern countries clubs are beginning to form. Falconry is still an integral part of society in most of the Arabian countries. Falcon competitions are becoming very popular and a focus for people who have limited access to hunting. In many other countries there are small clubs and groups of individuals, some relatively new, and some who have hawked for many generations.

Most of these clubs are represented internationally by the IAF, the International Association of Falconry and Conservation. The IAF fights for falconry on international issues, promotes the international Code of Conduct and helps maintain contacts between the clubs to the benefit of the sport and of the birds themselves.

For conservation to work it is important to reach the up-coming generations. For our Mongolian artificial nest project to work, we needed the support of local people, but they were scattered over hundreds of kilometres of steppe. So Nicola Dixon started visiting Soum leaders and schools, explaining about the project, and the children relayed this to their parents. As technology improved we were able to link the schools together by internet, first nationally, then internationally, using falcon conservation as a talking point for school lessons and cultural exchange. We tied this in with field trips to visit nearby nest barrels. The School Links project is now sponsored by the International Association of Falconry and covers 37 schools in eight countries.

11.4 The ethics of rehabilitation

There are three elements we must consider in rehabilitation: the individual bird itself, the state of its wild population, and the situation of the person giving aid.

Surprisingly, little effort is made by welfare societies to *prevent* the main causes of wildlife casualties, thus reducing the need for rehabilitation. Changes in the design of roads, cars, electricity poles and wires, and the lay-out of windows, and the control of cats, would do far more for wildlife than all the rehabilitation programs. Although a proportion of animals come in for treatment because of starvation in the autumn, the majority of animals needing treatment appear to result from man-made causes, rather than from natural selection. Even as I write this, one of my best gyr/peregrines has just broken his wing on a wire while chasing a crow, and his sister last year was killed by a wind generator while chasing a crow. Possibly the main use of rehabilitation is simply to document and amass data on wildlife injuries with a view to preventing them, along the lines of the pioneer work on electricity pole design by Morlan Nelson.

Although some raptors have moved into cities, the reality is not as rosy as it seems. Urban goshawks living on a rich diet of pigeons, breed better than their rural counterparts. But the pigeons are infected with *Trichomonas gallinae* and cities are full of plate glass, so more goshawks die of disease and collision injuries than are produced. The city populations rely on a constant supply of fresh goshawks from the surrounding areas.

When a bird comes into human hands it may be in the form of an egg, or as a chick, or as an older bird which has been through legal procedures and is now ready for return to the wild. All of this group have originated in the wild and are potentially physically and mentally healthy, but are naive about survival in the wild. From the legal point of view they belong in the wild, not in captivity, even if this requirement over-rules biological considerations. Some of these may have been hand-reared and imprinted on man, rendering them unsuitable for release.

A Japanese hand puppet for preventing imprinting on humans.

The second group of birds includes those which come into human hands because, although they are physically uninjured, they have been unable to survive in the wild. These are mainly starving young birds in the autumn and winter which are part of natural attrition and selection. They are the failed surplus population which has been lucky enough to be found before actually dying. It is possible to fatten them up, get them fit again, and give them a second chance in the spring.

The third group is those which have been physically injured, often through man-made hazards, but which, with proper care, could potentially make a full recovery and return to an independent life in the wild.

The fourth group includes the injured ones which will obviously never be able to return to the wild but which could still live in captivity with a reasonable quality of life. Such a life would need to be moderately low in stress, and might include a 'useful' element such as breeding or educational purposes which could justify the cost of keeping it. This group includes psychologically damaged birds (malimprints) as well as physically injured ones. Unfortunately, many of these crippled birds are kept alive with no real quality of life because the rehabilitator does not want to face the issue. On the other hand, some governments insist that any rehabs that cannot be returned to the wild are euthanised. The IUCN has created a decision pathway to triage animal patients.

The fifth group are the injured which will not recover sufficiently to join the previous group. These have poor prospects. Even badly injured birds can contribute in that they provide veterinary information and experience on wild birds. This may give insights into disease aspects of the wild populations and may enable treatments to be developed which help other birds in the future.

So the first step is to assess the bird carefully and assign it to one of these groups. This may require the services of a vet and in many cases will only gradually emerge as the success of treatment becomes apparent. In the case of 'attacking' species (see chapter 8) it is essential that they achieve full function and fitness before return to the wild.

So far, so good. But now we must consider this bird as it relates to its wild population. The bird is two things: it is an individual animal with resource needs, and it is a parcel of genes which will survive only if passed on to another generation before the bird dies. The wild population too is a quantity of individual animals with collective resource requirements. The individuals need a prey base, hunting areas, nest sites and territories, and so on. The resources available to the wild population are finite and limit the size of the population. Every year a healthy population outstrips its resources and is cut back to its original size when the resources become tighter during the winter. A healthy population is controlled by *mortality,* not by *productivity.* Thus, if a population is healthy, adding further individuals to it will not improve its health, or its numbers; it will only add to its mortality. Either the added individuals will die, or they will compete with others who will die instead. The only way to increase a healthy population is to increase its resources by increasing the carrying capacity of the habitat in one way or another.

Therefore, before releasing a bird into a wild population, one must know whether or not the population is already at the limit of its resources. We will examine simple population dynamics in 11.5. and criteria for reintroductions in 11.8. There are examples of productivity-limited populations below carrying capacity, such as the red kite and the sea-eagle in Britain, but they are not common.

The wild population is also a gene pool which includes a variety of genes. In the case of a common, widespread species in genetic contact with neighbouring populations, such as the common kestrel in Europe, there is a wide variety of genes in the pool and thus a high degree of unrelatedness. On the other hand, if the population has been reduced in numbers, such as with the Californian condor *Gymnogyps californianus* and the Mauritius kestrel, the variety of genes is also reduced.

In a case like this, imagine there are ten original birds left in the wild and that the genes are evenly spread between them (unlikely, but bear with me). Each of these ten carries 10 percent of the quantitative responsibility for the population and also 10 percent of the genetic responsibility. Along comes a conservationist and traps two of the birds (20 percent of the genetic responsibility) and breeds 90 young ones. Meanwhile none of the birds die (you can tell this is a model and not real life!). We now have a population of 100 birds. In quantitative terms the population is now ten times better off. In genetic terms it cannot be better off, in so far that, although the existing genes and their alleles can recombine into different variations, no new genes are created (for practical purposes). But are the birds genetically equally responsible? No. Eight birds carry 80 percent of the genetic responsibility and 92 birds carry 20 percent of the genetic responsibility. The population is now genetically unbalanced. Can you see where we're going? What if one of the first eight birds fell ill? It would be important to save it and get some genes out of it and into further individuals. Thus, by keeping it alive you have not actually 'saved' it. Only once the genes have been transferred to others is it saved and, bearing in mind that it only gives a random 50 percent of its genes to each offspring then, if it only has one chick you have only genetically saved half of it. It would need to have at least five young ones in order to save about 97 percent of its genes.

All of this applies to any genetically isolated population. It might be the last few birds in the whole world, or it may be an isolated population like the Welsh red kite which had become in-bred but which is now getting a genetic shot in the arm from European gene pools. It also applies to genetically isolated captive populations, including closed stud books of domestic species.

Thus rehabilitation to the wild of individual birds may have a negative or a positive effect on the wild population both in terms of competition for resources, and in terms of genes. You can assess for yourself when the impact is likely to be negative and when it is likely to be positive. It is quite clear that for most species which are already numerically common and genetically varied, the effect on the wild population is deleterious. This is all quite apart from the risk of introducing disease into the wild population. Thus, most of the releases are futile.

This can be a bitter pill to swallow. The law, which is based on only a primitive understanding and application of population dynamics, assumes that the more birds go back to the wild the better.

The law is geared to prevent uncontrolled taking from the wild and automatically assumes that, if a bird has come from the wild, the best thing to do is put it back there. In this it is often wrong.

We are gradually moving away from the protectionist viewpoint toward a conservationist viewpoint. This means that populations must be managed both quantitatively and genetically on a sustainable basis. There is a better understanding of how habitat limits populations and that healthy populations produce excess young for their habitat, a surplus which is harvestable on a sustainable basis if the species has some value as a resource to man. Many of the common raptors fall into this classification, as do game birds and mammals, trees and so on. The problem with pest species takes us one step further; with these, man has first to kill off all the annual harvestable surplus and then get stuck into the breeding population before he can hope to reduce the population below the carrying capacity of the habitat. Thus with pests, it is a never-ending battle to suppress populations and a good insight into the effect of self-regulating mechanisms of populations. Unfortunately, because raptors were hit so badly by pesticides, they became subjects of a massive media campaign which gradually changed public attitudes, and which resulted in over-legislation. Everyone viewed raptors as endangered and rare. Now that many populations have recovered, people have difficulty in readjusting their attitudes and in targeting their conservation effort into aspects giving genuine benefit.

So finally we come back to people. Most people rehabilitating raptors are private individuals or small groups. Most big organizations do not do it because it is of little conservation benefit, which is their main criterion. Others do it by obtaining funding from non-specialist sources who believe that it does have conservation benefit or who wish to be associated with what they see as the moral high ground. The global economic recession and the increasing raptor numbers are gradually reducing this support. Meanwhile, the public expect to be able to bring these casualties to someone, somewhere, who will care for them.

In Britain, thousands of orphaned or injured raptors and owls are rehabilitated by volunteers and about 47% of the birds are released back to the wild. Many of these had been referred to them by the Royal Society for the Protection of Birds or the Royal Society for the Prevention of Cruelty to Animals. Some must be euthanised, some are nursed back to health and are returned to the wild, and some live to old age in retirement aviaries, some even becoming parents. Unfortunately, many of these birds are kept alive in unsuitable conditions by people who mean well, when euthanasia would have been more appropriate.

Most people rehabilitate animals for humanitarian reasons. They might just as quickly come to the aid of a little old lady who fell down in the street. This is genuine altruism which commands respect. It is all very well to look at cold scientific reasoning, as I have done above, or to comply with laws which are often nonsensical. It is something else to go out of your way to help a fellow creature in need. People often denigrate rehabilitators on the grounds that it is a waste of time. Critics fail to appreciate that, while *they* may consider it to be a waste of time, others obviously don't. It depends by what criteria you run your life. Rehabilitation work evidently gives some people a sense of satisfaction and achievement, and this in itself is worthwhile. The hypocrisy I would point out is the different morality expressed toward old or injured humans compared to the other life forms. With humans the prevailing morality seems to be 'keep them alive at all costs' (although many old people would not agree), with other species it seems to be 'if it has no breeding future, kill it.' My own personal view is that first, decisions must take into account the resources available, and second, if a human or any other animal is terminally ill, *and does not want to carry on*, we should try to allow it to die painlessly and with dignity. With terminally ill animals, we obviously cannot ask them, we can only do our best to assess the situation and reach a decision based on the animal's needs, rather than our own sentiments.

The humanitarian reaction springs from the heart rather than from the mind, and often the

Mike Calvin doing early fitness training of an Australian *macropus* peregrine.
Photo David Palmer.

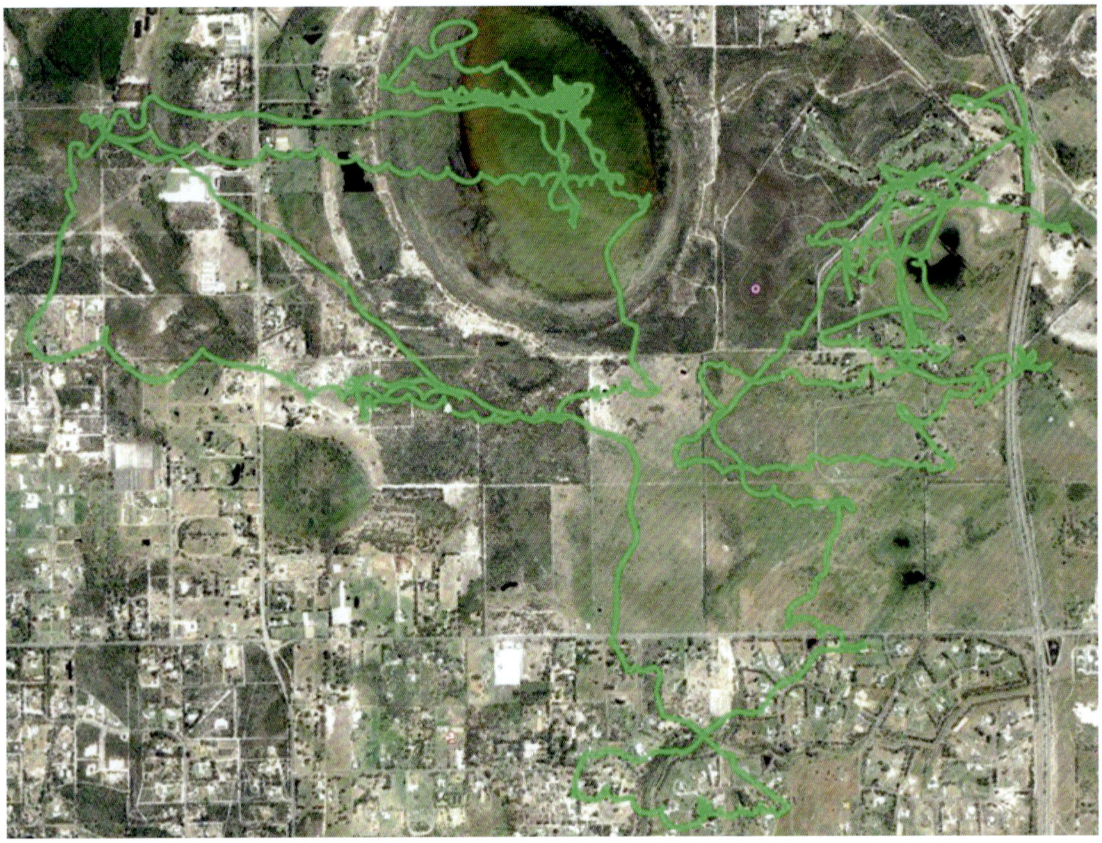

Six weeks into her free flight programme this peregrine flew for 1 hour and 52 minutes before finally being safely called in. She flew a total of 92km and climbed over 500 metres three times, stooping unsuccessfully at over 200 kph.
Photo Mike Calvin.

resource implications are neglected. To look after injured animals takes time, skill, facilities, and funds. If these are overlooked, animals will suffer. These aspects must be considered and be properly attended to, and some rehabilitation organizations have codes of conduct and support systems for this. Also there is a great deal of difference in commitment between *rescuing* an animal, often with attendant media publicity, and *rehabilitation* which may tie up resources for months. Often it takes one organization to rescue, another to rehabilitate and a third to assess habitat and supervise the return to the wild.

Often rehabilitation is used as a justification for hands-on contact with wild animals. There should be no need for this justification because there is no reason why many wild animals should not be kept as pets. This is a perfectly valid and sound recreational use of a resource, provided that welfare issues are attended to. There is a growing attitude amongst hard-line protectionists that the only justification for hands-on contact with wild animals should be for the conservation benefit of wild populations. I disagree with this approach, and I cringe at some of the misunderstandings which arise because of it. Much of my own understanding of wild animals comes, not from science, but from being myself brought up in constant contact with animals. My approach to them is not anthropomorphic, but it is not strictly scientific either, because science is an inadequate tool. The more academic an institution is, the less suited it tends to be for hands-on work with wild animals.

I don't want to go into the finer details of assessing habitat, preparations for and release into the wild, and post-release monitoring. Most of the raptor handling techniques have already been covered in detail in this book. Having myself surveyed habitat and monitored its use by wild Mauritius kestrels in detail, I believe that a human can reach only a very gross assessment of habitat. Prey *availability* (and I mean that in its full technical sense), is a very hard parameter to

A saker falcon being released in the Sh Zayed Falcon Release Scheme.

predict, bearing in mind that it is dependent on the hunting skills of the release candidate. For all practical purposes it is a question of trying it and seeing what happens. I used to spend days trailing around west Wales looking for a place with no resident peregrines. As soon as I put my release bird on the wing, a resident falcon would show up. My own assessment was futile; the birds told me the answer.

Getting rehabilitated raptors fit can be a major difficulty. There are still some rehab facilities which eschew any techniques to do with falconry and so do not allow the birds free-flying before release. Hack pens are fine for initial flying practice but certainly cannot get the birds half fit. We have ten 25x25 metre hack pens here which we use for the fledging falcons, but this is just the start of their fitness journey. We can put 30 same age young falcons in each pen without problems, but rehab birds need to be segregated to protect them from one another. This effectively means that the whole pen is being used for just one bird, otherwise there is a risk of bullying or disease transmission. In the later stages of fitness training, release birds need a lot more flying and practising of skills than could ever be achieved in a flight pen.

When one investigates rehabilitation, one of the first things to become apparent is the lack of data on 'success,' whether measured in terms of survival of released individuals, or as positive impact on wild populations or some other way, such as the effect on the person doing it. These parameters are very difficult to assess to a scientific level of satisfaction. The result is a lot of claims and counterclaims, all of which are inconclusive. Only when you have analysed all the *real* reasons for rehabilitation can the real criteria for success be evaluated.

Isn't there a paradox that a falconer might spend all afternoon trying to catch a prey animal with his hawk and then all evening tending to some poor injured bird? Or that he might go out to kill a rabbit in order to feed the injured bird? Life is seldom straightforward. I think if you are fundamentally against killing, you must stay well away from the world of animals because it is a mayhem of births and deaths with no individual justice. Best to hide your head in the

sand and pretend that by eating plants you are not competing with other life forms for resources.

Over 2,000 wild falcons have been released in Central Asia in the Sheikh Zayed Falcon Release Scheme since 1995. How many survived to breed? Satellite tags give some feedback but also impede the falcons.

11.5 The effects of falconry on wild raptor populations

This section is supported by the film *Mongolian Artificial Nest Project* and the film *The International Trade in Falcons* which we made for the Environment Agency in Abu Dhabi and the CITES Secretariat.

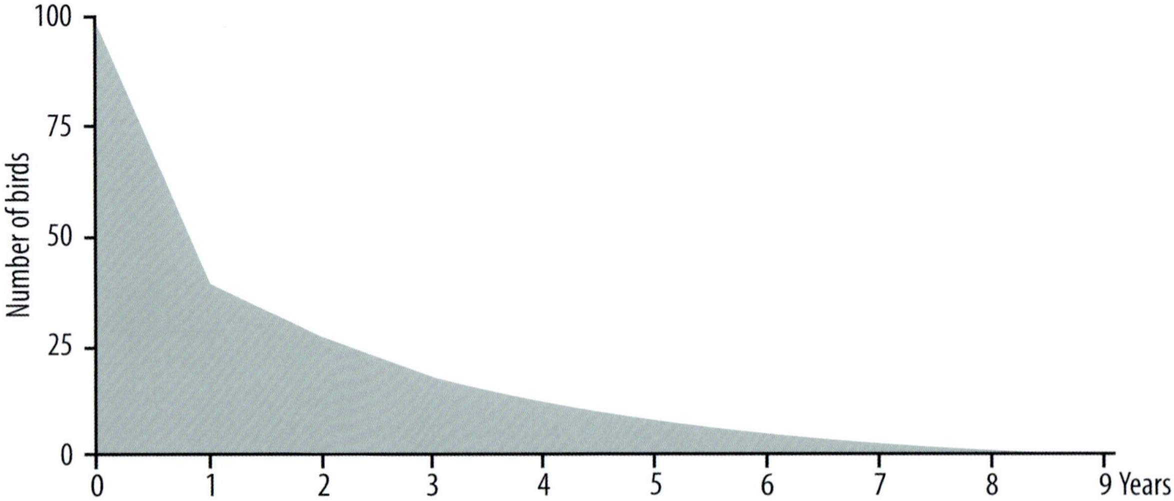

Figure 11.6.1 The average survival for a cohort of a hundred juvenile European sparrowhawks.

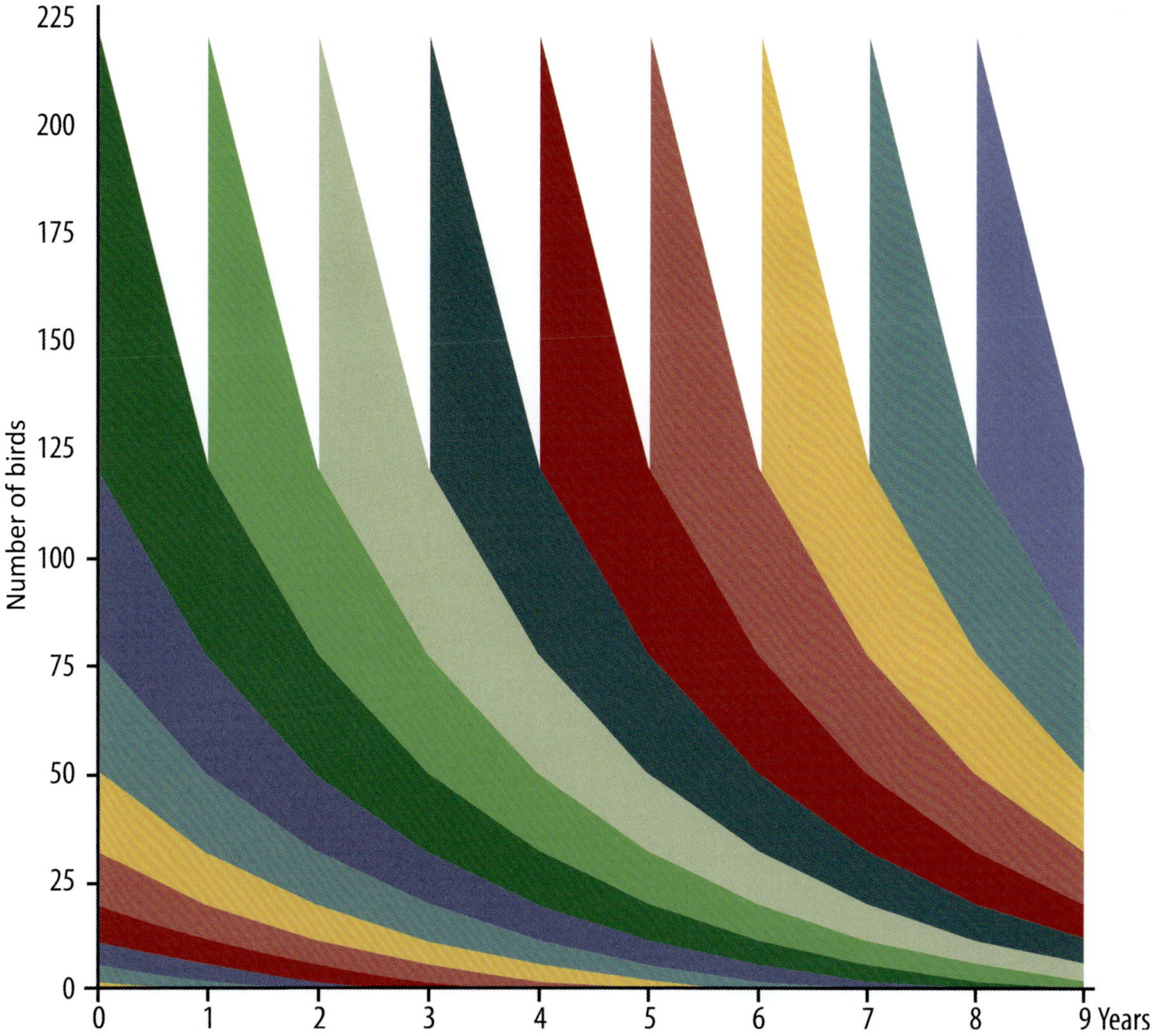

Figure 11.6.2 The age structure of a population of European sparrowhawks.

In most countries where falconry is practiced, falconers are 'consumers' of hawks. Britain, Spain and Germany are exceptions in that they are now probably self-sustaining in domestic raptors. Most other countries rely on taking a 'harvest' of hawks from the wild each year. Many of these birds are returned to the wild the following spring. Within limits, the yield is sustainable long-term and this can be demonstrated both on a scientific basis and on a historical basis. To understand why this is so requires a brief excursion into population biology.

Using Ian Newton's excellent data in his monograph on the sparrowhawk as an example, we can visualize the survival of sparrowhawks as in figure 11.6.1. Of a sample of a hundred juveniles, only about 40 percent survive after their first year. This is because many fail to become skilled independent hunters in their first few weeks of life and die. After their first year the hawks do better; about 68 percent survive each year. Only the occasional individual survives more than nine years.

Put this cohort together with successive years and we get figure 11.6.2. You can see that each year the population, on average, is made up of diminishing proportions of successive age groups. Of these groups, those between the numbers 25 and 75 on the vertical axis are the most effective breeders. One way or another, only about half the adults produce the year's crop of youngsters. By the third year, only about 26 of the original hundred are left alive to breed. In other words, it takes about four juveniles to fledge to produce one breeding adult. So if a falconer comes along and traps an adult, he is having about the same impact as if he had trapped four juveniles. This is why conscientious falconers restrict themselves to trapping passage birds (juveniles in their first autumn). In terms of minimizing the impact of a harvest, the earlier in the life cycle the birds are taken, the less effect it has.

Now let's throw a few more variables into the basic model. Imagine what the figure would look like for a peregrine, or for an eagle. The eagle has a longer immature period, a longer adult life, a lower productivity rate and a lower mortality rate. Some of the eagles might still be alive after forty years and the whole figure would stretch across four pages of this book.

When we look at productivity, many different factors can affect how many young are produced. For example, there may not be enough nest sites available for all the adults to breed. They are forced to play musical chairs. Peregrines in Britain, kestrels in Holland and Wisconsin, sakers on the plains of Asia; many populations are restricted by availability of nest sites. When artificial nest sites were provided, the populations increased to new levels. When nest sites are in short supply, only a proportion of the available

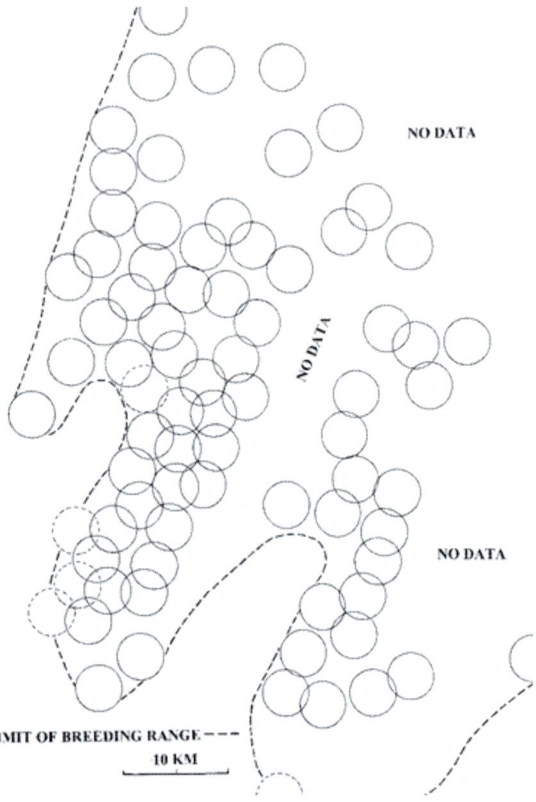

Figure 11.6.3 Evenly spaced pairs of New Zealand falcons.

Not exactly a cragg, but for this eagle on the Mongolian steppe, it's better than nothing.

An old tyre is enough for this upland buzzard.

An old water tank for this saker.

Or a custom-built, palatial, artificial nest.

adults can breed. If one of the nesting adults is killed, its place may be taken by a newcomer within hours. There have even been cases in which both parents have been shot and the young ones have ended up being reared by replacement parents. This is a sure indication that nest sites are limiting and that surplus adults are present which are unable to breed through lack of a nesting territory. Another indication is to look at a map of all the nests (figure 11.6.3). A nice even distribution indicates that breeders have spaced themselves out away from their neighbours. A clumpy and irregular distribution tends to point to a shortage of nest sites, or of birds. Falcons often show this distribution, whereas species which build their own nests are able to take a more constructive approach to the problem and make a nest where it suits them.

Prey is a common limiting factor. In early spring, if the male cannot provide enough food, the female may not lay. If she does, she may rear less young. And the young themselves will have difficulty surviving and more will die off. The hawks themselves must compete for the available prey. Imagine hacking out extra birds into this situation. Even more would die. Conversely, removing some of the young may actually increase the number of survivors. Thus, what we might subjectively perceive as 'good' (hacking out extra birds) or 'bad' (trapping up surplus birds), may objectively be the opposite of what was expected. Prey availability thus affects both productivity and mortality. Of course, prey availability is not the same as prey density, although it may depend on it.

Weighing saker chicks from artificial nests in Mongolia.

Although raptors can locally affect prey numbers (goshawks can wreak havoc on concentrated pheasant areas), by and large it is the prey numbers which control the raptor numbers. This is most obvious with raptors which specialize on mammals, such as voles and susliks, which cannot travel far. Once the voles have eaten all the local food, their numbers crash. The vegetation regrows, and the mammals slowly increase again. Thus there are cyclical 'vole years' and the population graph is like saw-teeth. The predators which rely on these voles tend also to be cyclical, following the prey cycle. It is a different matter for avian prey. If food gets short, they can fly somewhere else. The hawks too can fly somewhere else. Therefore after a vole or hare crash in the Arctic, numbers of goshawks, gyrfalcons, and owls head south. These raptors are surplus and most are doomed. Removing them from the wild population will have minimal effects on future numbers.

Of course, to measure whether any factor is impacting a population, one first must be able to survey the population accurately, and this can be fiendishly difficult. You can prove an animal is present, but you cannot easily prove there are none there. For example, when we surveyed wild Mauritius Kestrels in 1984/5 we were told that there were only 5-7 individuals left. But after four months of surveying and radio-tracking, we concluded that in fact the population was around 50 individuals. Similarly, although the Saker falcon has disappeared from many parts of its range, our field surveys concluded that in Mongolia at least, it was doing well, despite heavy trapping for falconry. Yet people still declared it was almost gone, so we decided to ask it directly. After three years of design trials, we put out 5,000 nest barrels on poles at 2 km intervals on the open steppe. Previously we had occasionally found sakers nesting on the ground, so we knew they must be nest-site limited. Sure

The sakers use old stick nests built by upland buzzards or ravens.

An electrocuted Mongolian saker.

Our power line surveys collected over 600 electrocuted sakers.
Photo Andrew Dixon.

enough, there were plenty of non-nesting sakers around only too pleased to find our nest barrels. By 2014 we had 760 pairs of sakers producing 2,500 chicks, plus 1,122 pairs of European kestrels, 348 pairs of ravens and 422 pairs of upland buzzards (*Buteo hemilasius*).

The Minister for the Environment was delighted because these raptors were feeding mainly on Brandt's Vole *Lasiopodomys brandtii* which eats the scarce grass on the steppe. These steppe voles were destroying large areas for grazing, so the Mongolians started a poisoning campaign. The predators then ate the dead and dying voles so they died too. So we started a public awareness campaign to ask the government to stop buying poisons from China. But, before this could get going properly, an election campaign started and the government spent its money on this instead. Meanwhile the Mongolians started an electrification programme, erecting new power lines across the steppe. Although the Mongolians, as Buddhists, leave wildlife alone, the Chinese workers on the new roads and

An electrocuted New Zealand falcon with toes burnt off.

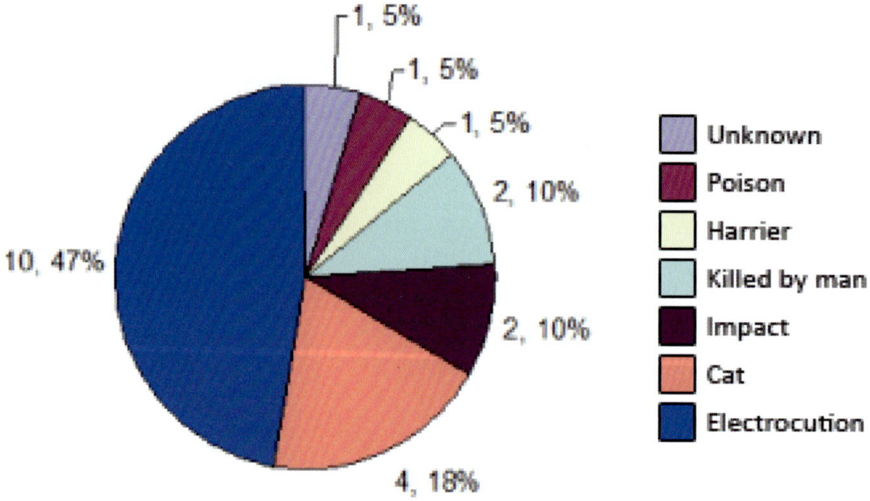

Causes of death of 21 falcons in our Marlborough, New Zealand study.

This New Zealand pole electrocuted a falcon. Subsequently the power company has retro-fitted red insulators on the transformer.
Photo Colin Wynn.

It looks scary for this goshawk, but this pole in Japan is well-insulated.

lines would eat any raptor eggs or chicks within 20 km of the construction lines. Soon raptors were being electrocuted massively, so we surveyed some sample lines using two people on motorbikes, travelling the lines daily to find the carcasses before scavengers took them. We found that, besides the other raptor species, about one saker falcon was killed per week for every 10 km of power line. So then we had to organise meetings and conferences with the ministry and the power companies, to get designs altered or retro-fitted. We found a similar thing in New Zealand where we established a new breeding population of falcons in the Marlborough vineyards to reduce pest bird damage. Most trees were cut down to discourage pest birds, leaving only the power poles.

When you consider that birds have been taken from the wild for falconry for thousands of years,

This was one of our crow falcons in Northumberland. She caught fire and we had to throw stones to get her down.

coupled with the fact that falconry has never contributed to the extinction or near extinction of any raptor or its prey, it seems extraordinary how unbalanced the pressure has been on falconry by protectionists, compared to the pressure on the major factors in the fluctuations of raptor populations. Shooting by game preservers, for example, has exterminated raptors in some countries, and yet it is allowed in all the countries where falconry is banned. The use of pesticides known to affect raptors is permitted in most countries where falconry is banned. One must look further for the real reasons behind the differing attitudes to the use of raptors as a resource, and we will do that in some of the following sections.

11.6 The ethics of hybrids

Hello. Are you a hybrid? If you are African you are probably relatively pure *Homo sapiens*, but westerners and Asians have around 2-5% Neanderthal or Denisovan genes. We share a lot of our genome in common. So roughly speaking, you could be 31/32 *sapiens* and 1/32 *neanderthalis*. That's the same ratio as some of our gyr/saker hybrids we produce!

So, first of all what is a hybrid? A hybrid is a cross between different species. It is not a standard commodity. Some hybrids are very outcrossed, such as crosses between separate genera, for example redtail X Harris, or Cooper's hawk X Harris. Some of these species and genera have completely different chromosome complements, and their hybrids are probably infertile. The genus *Falco*, as we know it today, has in the past been allocated into separate genera or subgenera such as the peregrine *group (Rhyncodon*) and the great or desert falcons *(Hierofalco*). Hybrids between these subgenera are more outcrossed and usually less productive than hybrids within each subgenus. Thus a saker X peregrine is more outcrossed than say, a saker X gyr or a shaheen X peregrine. Research is continuing on the relationships within the gyr/saker complex, and it is likely that genetically they are a single 'superspecies.' They are fully fertile over several generations. It is thus debatable whether or not these should be termed hybrids at all.

Are hybrids likely to interbreed with wild raptors and thus alter their genetic structure? Again, there is no blanket answer. There are two aspects: *could* they breed (are they sufficiently genetically compatible) and *would* they pair up (would they recognize a wild raptor as a potential mate)? For example, the only large falcon in Britain is the peregrine. A shaheen/peregrine hybrid might breed normally with a wild peregrine, but a peregrine X desert falcon hybrid paired to a peregrine (in captivity at least) shows a reduced hatchability. A desert falcon X desert falcon hybrid, such as a gyr/saker, would be no more likely to breed with a peregrine in Britain than would a pure desert falcon such as a gyr, saker, or lanner. The reasons why individual birds pair or fail to pair have already been discussed in 2.6 and 4.11 and there is growing evidence from falcons, and from some non-raptorial species, that imprinting on the parents critically affects future mate selection. Thus, for example, a saker reared by a peregrine may be more likely to pair up with a wild peregrine than would a peregrine reared by a saker. Even here there is a natural block because, although the saker might try to pair with a wild peregrine, the wild peregrine would repel its advances, recognizing it as an alien species. The implication for managing hybrid raptors is either to imprint them on man, or on

a species which is not found in the wild in the country where the bird will be flown. Precocial and social species, such as ducks, have different systems of imprinting and mate selection and we cannot make valid comparisons between them and solitary, altricial raptors.

The only example in Britain in which there is a serious risk of crossing is not a hybrid at all but red-tail X common buzzard. Even here there have been very few successful inter-breedings in the wild. But in the event that that happened on a sufficient scale to warrant concern, we would have to sterilize or malimprint the red-tails, but there would be no need to stop flying other exotics, such as Harrises.

Basically the problem is not one of hybrids, as such, but one of genetic pollution of a wild gene pool. This is happening all over the world and with increased movements of plants and animals is likely to continue. In Britain for example, the domestic cat has polluted the Scottish wild cat, the ferret has polluted the polecat, and examples among the ducks and pigeons are rife. To argue that wild gene pools must remain pristine is nice, but it implies that the habitat for that population also remains pristine so that the two remain well adapted in evolutionary terms. This is not so easy. Most of the species are living in very different and fast-changing habitats to those in which they originally evolved. Those which can adapt in the space of a few generations have a chance to survive. The specialists go. In replacing the *anatum* peregrine in North America, The Peregrine Fund released several races of peregrine, on the principle that natural selection would gradually choose the best-adapted birds for modern conditions.

Curiously, our attitude to this problem is very dichotomous. With plants and animals other than man, the moral high ground is to maintain original gene pools. On the other hand, with the human races, the morality is against discrimination and pro-introductions of non-indigenous genetic races. Morality is a wonderful thing.

The wild gene pool is seldom what one thinks it is. Studies by Dr Wink and others have shown that the falcon species share most of their genetic material, just as man has about 95 percent of his genetic material in common with the other apes. Our whole concept of species is changing, and taxonomic edifices are tumbling. So it is foolish to jump to conclusions. But, as a rough and ready guideline, the most likely individuals to interbreed with a wild population are subspecies or very close relatives, non-indigenous to the area, for example, Peales peregrines, shaheens or barbaries flown in central U.S.A. or Europe.

Genetic pollution is not necessarily a bad thing, in fact the very term is a misnomer. The introduction of a few genes from a related gene pool can inject some vigour into ailing or maladapted populations. The red kite in Wales probably received genes from a stray German red kite several generations back, which seems to have given the population a much-needed kick in the genetic pants. Some of these local populations which have recently become genetically isolated from the main population may benefit from a flow of genes to prevent them becoming inbred.

If lost, could hybrids survive in the wild? Certainly the hybrids between gyrs, sakers, or peregrines kill quarry in falconry as well or better than their pure-bred counterparts, and one must therefore assume that they would continue to do so in the wild. Thus, they could survive as individuals for their lifetime, even though their genes might die out. Unentered falconry birds of whatever parentage have a poor chance of surviving in the wild, and it is essential that the falconer takes all possible precautions to prevent the loss of an unentered bird.

There is a last point on the 'con' side of the argument, and that is that hybrids are in some way unnatural or unsporting. I cannot see that hybrids are less 'natural' than say flying a non-indigenous species, or using telemetry, or transporting hawks in cars with hoods on their heads, or using a spaniel instead of a wolf. On the point about being unsporting, we have already discussed in 10.4 the balance between predator and prey, and it is up to the falconer to maintain this balance.

Spitfire, a male peregrine/NZ falcon flying at 454 g (16oz).

The trained peregrine is only suitable for certain types of land. In Britain, these are grouse moors and in certain very limited lowland areas. Most falconers do not have legal access to suitable land in lowland Britain to fly a peregrine. Does this stop them? No, instead they try to hawk land which is unsuitable for peregrines, and they come unstuck. The peregrine is a superb aerial specialist and in the right situation, such as a grouse moor, can scarcely be equalled. But she has three faults which make her unsuitable for pursuit flights in most of lowland Britain: first, she chases in a series of stoops, a method which although spectacular to watch, means the flight tends to cover a lot of distance, and most prey can reach cover first. Secondly, the peregrine is seldom willing to enter the flimsiest of cover. If the prey just lands on the ground, or under a sheep, the peregrine will stoop ineffectually and then, if not taken down promptly, is likely to drift away and get into mischief. Thirdly the peregrine tends to lack persistence; if things aren't exactly to her liking she often refuses or checks at something easier. Peregrine/sakers on the other hand get onto prey more quickly than a peregrine and, even when stooping, do not go so wide that the prey has so much time to make toward cover. They can kill in thin trees which means that large areas are not rendered unhawkable by walls, fences, sheep and so on, as they are for a peregrine. Finally, on the ground, they have the confidence and determination to kill a crow quickly, whereas a peregrine, although physically capable of doing so, often hesitates or refuses. Thus, for hawking with trained birds in managed habitats, hybrids are sometimes better adapted than pure-breds.

It may well be that some hybrids, such as those between red-tails, Harrises and ferruginous hawks, have no significant advantages, in which case their production will soon be abandoned. So-called 'developed' countries are suffering increasing obstacle pollution in the form of roads and traffic, barbed wired, and overhead cables. This has had a major impact on wildlife and has closed huge areas of country to all but the most basic types of hawking. If some designer hybrids enable falconry to adapt and continue despite these massive changes then they will have a role to play in the future.

11.7 The past and future of domestic breeding

Once upon a time, about 2.8 million years ago, before the first 'humans' in our genus *Homo* evolved, there were already predators and prey, hunting and escaping from one another. Recognisable falcons were hunting 10 million years ago. When a prey animal detected a hunting falcon, it took avoidance action. In other words, in relation to the falcon, the prey was 'wild'. The prey didn't run away from flowers, it was only wild in respect of animals that might hunt it. When humans came on the scene as predators, prey became wild towards them too. Being 'wild' described the relationship between the prey and humans.

But humans did not reach all parts of the world. The wildlife of New Zealand only met humans about 700 years ago, and then only in small numbers. The indigenous species were not 'wild', nor were they 'tame'. They simply had no cognisance of humans. Humans quickly took advantage of this and hunted many of the flightless birds to extinction. But to this day the surviving indigenous species are still relatively unafraid of humans. There have been too few generations and too little evolutionary pressure for them to become 'wild' with respect to humans.

Elsewhere, many species of animals were hunted by humans for hundreds of thousands of years. Weapons of increasing sophistication were used, enabling humans to kill at a distance. Only the fearful ones survived. Some ran away, some became nocturnal. So our first effects on animals were not to domesticate them, but to 'wild' them. We called them 'wild' animals, and we extended this description to apply to <u>all</u> animals outside our control.

With so many animals afraid of us, hunting became harder. Prey became scarcer. The alternative was to reverse the wilding process and to domesticate animals. We learnt to rear them alongside us so that we could eat them when we wanted, or we could use them for transport or to help us hunt. Similarly we discovered that we could plant seeds from plants and grow more seeds, and we could select the most plump and nutritious seeds to grow varieties of plants that

These Wekas *Gallirallus australis* love to catch my dummy mouse lure. Who's wild?

were better for us than the original ones. We modified them genetically and these changes were reflected in their phenotypes (what they looked like) and their ethotypes (their suites of instincts). Some were noticeably tame with respect to humans, but not all; plants don't have any behaviour to show tameness with.

Now our relationships with wild organisms included 'wild' ones and 'domestic' ones, as well as most other species which are, if you like, 'neutral'. These held no immediate interest or significance for us. The domestic ones were animals and plants under our control. Some escaped from our control and became 'feral'. The places they escaped to were outside our control: 'the wild'. This is a somewhat nebulous concept. We thus had 'the wild' as a place, and 'wild' as in a state of mind, either ethotypic or learned avoidance, and wild as in untampered with by humans.

Other species had relationships with one another. Some are commensal, such as cleaner fish or cleaner birds removing parasites from bigger hosts. Some are parasitic, and some are various forms of associations, such as lichens or corals. The relationship may have altered the two genetically, resulting in changes in form or behaviour, and the engine for this change was mainly evolutionary rather than consciously intentional. In the case of humans, it is unlikely that someone sat down and thought "I'm going to catch a wild sheep and breed it and select it until I've developed a Merino". More likely a lamb survived as an orphan when its mother was hunted and the lamb was taken back to camp as a pet for the children. And one thing led to another... Only later did selective breeding feature in domestication.

Thus our words 'wild', 'domestic', 'feral' and so on are a reflection of our human relationship with other species. They say as much about us as they say about them. The species concept is a man-made construct too. It is our habit to want to organise things and put them in boxes, and the species concept is our attempt to make sense of taxonomic relationships. But the more we research it, the more we realise that it has huge failings as a working model. Everything is fuzzy at the edges. A point was reached when having meddled with Wolves for so long, we no longer called them *Canis lupus* but re-named them as Dogs *Canis familiaris*. We had created a new species! Or should we hedge our bets and call them a sub-species, *Canis lupus familiaris?* Oh dear, it is so hard to decide!

So we can take a species and 'evolve' it into a different species. Or we could somehow give our limited 'domesticated' population a different common name such as Cow (*Bos taurus*) to differentiate it from its wild ancestor, the Auroch (*Bos primogenius*) which did not survive our hunting it to extinction.

Other domestic animals, such as geese and chickens, still have extant relatives in the wild of what we are prepared to label as the same 'species'. We point to differences in them but consider the differences insufficient to name them as different species.

Other times two species of animals, or of plants, may interbreed and produce a hybrid that may be fertile and effectively form a new group, such as the Altai falcon. This is evolutionary braiding, where two species diverge and differentiate, then come back together again. In medieval times sakers, as well as lanners, still bred in Europe and may well have inter-bred with peregrines, producing what was called a 'Bastard' falcon. This equally could have been a peregrine x lanner hybrid. As Sir Thomas Sherley noted in 1603 : '*…indeed diverse times very unlikely hawkes, both for plume and shape, do prove excellent hawkes, especially such as are called Bastard hawkes, which is, when one kind doth engender with another kind.*' We know that both peregrine/sakers and peregrine/lanners have occurred in the wild and also in captivity and that the males are fertile, and so are some of the females. This year an 11 year old peregrine/saker, one of our crow team, laid 9 eggs without pulling.Thus in raptors, a species can be modified progressively through captive breeding to become a domestic falcon, and/or it can be modified in one generation by hybridising two parent species.

Although lanners at least were probably bred in captivity in Britain in the sixteenth century, there were few serious attempts to breed hawks because it was easier to obtain them from the wild. Strict laws were passed to protect the nesting falcons and often goshawks were put out at the end of the season to nest in a local wood. Their youngsters were then available later for hawking.

When hawking went out of favour, raptors had a very hard time at the hands of shooters, and in Britain some became extinct. In North America, and in many other countries, hawks were often shot on sight. Then came organochlorine pesticides. Hawk numbers crashed. Suddenly raptors became 'magnificent, rare, and endangered.' Many falconers were prosecuted or persecuted just for keeping a live hawk. While falconers were hounded and blamed for the demise of raptors, the pesticide producers and users, which had a much bigger impact, were persuaded to limit their activities voluntarily over a period of thirty years. I was working in New Zealand at the time and found egg-shell thinning as well as pesticides in the raptors there.

In Britain, falconry was stuck between a rock and a hard place. On the one hand it was getting almost impossible to obtain licences to take hawks from the wild in any significant numbers owing to the population declines. On the other, there was a burgeoning interest in raptors and in people who wanted to take up falconry. Despite the gloomy predictions of some old-timers and the loudly-voiced scepticism of most protectionists, some falconers had made their first tentative attempts to breed hawks in captivity. It wasn't easy, but success followed success, and a body of expertise was established. Ronald Stevens and Johnny Morris in Ireland crossed a peregrine with a saker. In America, Tom Cade, backed by many prominent NAFA members, set up The Peregrine Fund to restore U.S. peregrine numbers through captive breeding. It was a long haul, but it was an outstanding success and a monument to what can be achieved by hard work and dedication, pioneering techniques of breeding and reintroduction. In Britain, the British Falconers' Club held breeding seminars and conferences. Many projects sprang up in Europe.

By the late 1970s, although the screws were being tightened on the supplies of wild hawks, we gradually realized that falconry had a future if it could be entirely self-sustaining by breeding its own birds. Not only did it have a future, but also it could control its supply, and escape from the yoke of legislation surrounding wild birds. In 1981, Britain implemented the Wildlife and Countryside Act in response to the EC Birds Directive. During the same period CITES, controlling international trade in endangered species, started to come into operation. These were new and far-ranging controls aimed at protecting wild birds. Because few raptors were being bred in captivity, the legislation was worded confusing wild-source birds with domestic ones. Domestic-bred raptors had to be registered and ringed.

By 1994, there were around 16,000 diurnal raptors registered to about 8,000 hawk-keepers in Britain. Domestic raptor production was booming-and so were wild raptors. Much of the legislation was now unnecessary and no longer cost-effective. With free trade throughout a growing European Community it was impossible to maintain such high levels of regulation out of step with the other countries. So now falconry has access to as many birds as it wants and there is virtually no limit to its potential growth.

Of course during this period, some wild raptors were taken illegally by egg collectors, game preservers, pigeon racing enthusiasts, falconers and people posing as breeders. I do not have reliable data on the total illegal take of each species, however it was insufficient to stem the increase in most of the wild raptor populations, some of which are now at saturation. These illegal birds probably were significant in boosting the captive population sizes of some indigenous species during the early years of domestic breeding. However, the domestic productivity rates of indigenous and non-indigenous species are very similar and therefore it is probably safe to assume that most of the declared breeding is genuine. The main exception to this is the goshawk, which is difficult to breed, is a desirable falconry bird, and is under great pressure from game preservers who often try to persuade falconers to take them rather than kill them illegally. But nowadays, thanks to progress in voluntary inseminations, even goshawks are in over-supply.

During the period 1981-1993 when full registration was in force, it was possible to get statistics which can never be repeated. During 1980-1991, 23,804 diurnal raptors were registered as bred in captivity in U.K. alone, including 16 species to the second generation. By 1992, the following species had been bred worldwide for two generations in captivity ie F2, and have internationally recognized domestic populations, many of which are numerically and genetically self-sustaining:

Vultur gryphus Andean condor
Milvus migrans Black kite
Milvus milvus Red kite
Haliaeetus albicilla White-tailed sea eagle
Gyps bengalensis Indian White-backed vulture
Gyps fulvus Griffon vulture
Accipiter gentilis Northern goshawk
Accipiter nisus European sparrowhawk
Accipiter cooperii Cooper's hawk
Parabuteo unicinctus Harris hawk
Buteo jamaicensis Red-tailed hawk
Buteo buteo Common buzzard
Buteo regalis Ferruginous hawk
Aquila chrysaetos Golden eagle
Aquila rapax Tawny or steppe eagle
Polihierax semitorquatus African pygmy falcon
Falco sparverius American kestrel
Falco tinnunculus Common kestrel
Falco punctatus Mauritius kestrel
Falco chicquera Red-headed falcon
Falco columbarius Merlin
Falco novaeseelandiae New Zealand falcon
Falco subbuteo European hobby
Falco biarmicus Lanner falcon
Falco mexicanus Prairie falcon
Falco jugger Lugger falcon
Falco cherrug Saker falcon
Falco rusticolus Gyr falcon
Falco peregrinus Peregrine falcon
Falco pelegrinoides Barbary falcon

The process continues and the list gets longer. In the 1970s, falconers already kept numbers of indigenous species because they had been able to obtain them under licence, and many were rehabilitating injured hawks. It was easy too, without quarantine and without CITES, to import raptors. By 1980, these initial populations were already being bred in small numbers. Everyone was excited about captive breeding, and every available bird was paired up. As a result the commoner species, such as the kestrel, sparrowhawk, and buzzard, increased exponentially in captivity. In 1987, over 1000 young kestrels were produced. Suddenly nobody wanted them. They were hardly worth the cost of registration. Pairs were separated or died, and production fell drastically within three years. By 1993, it was half of peak level and still falling. In 1988, sparrowhawk production peaked, followed the next year by buzzards. These three species were only in limited demand by serious falconers, and nobody was prepared to pay high prices for them. People began to realize that it costs money to breed hawks, and kestrels do not pay the bills.

Falconers, on the other hand, were prepared to pay for the species which were the backbone of the sport: peregrine and merlin, Harris hawk, goshawk, and red-tail. These in 1980 were in lower numbers and also were slower to come to breed. The goshawk and the merlin posed difficulties for breeders. By 1985 their production was increasing slightly. People who had gained experience on kestrels now set their sights on more expensive birds. Everybody wanted a pair or two for breeding. Prices rocketed. In 1987, Harrises reached a peak price of 1,600 pounds ($2,560) for a female. The falconry demand was totally swamped by the demands for breeding stock. Three years later this generation came into production; output soared. Britain meanwhile went into a deep economic recession. Prices halved. The demand for breeding stock fell back, leaving falconers as the main market. But the Harris gained popularity, being tough and undemanding, and easily flown near towns, it does not have the divorce-making propensities of the goshawk. It is also easy and cost-effective to breed and will no doubt dominate the falconry world in Britain for the foreseeable future.

The peregrine, red-tail, goshawk, and merlin similarly increased, but output will never reach the same levels as the Harris, because they do not have the same market volume potential as the Harris and are not so cheap to breed. The peregrine suffered from a lot of price hype, the worst of which is promulgated by protectionists in sensationalist media. Prices of more than 10,000 to 15,000 pounds ($16,000-$24,000) were still being quoted in 1993. In reality, peregrine prices in U.K. peaked in 1988 when the best birds fetched up to 1,200 pounds ($1,920). In 1993, similar birds fetched 500-600 pounds ($800-$960). In 1994, the German protectionists claimed that sakers cost 30,000 pounds ($48,000), whereas adverts by saker breeders in U.K.'s *Cage and Aviary* newspaper averaged around 400 pounds ($640) for a female saker.

By 2018 prices for some species or bloodlines that performed well in the falcon races increased, whereas falcons not used in races decreased. The value generator for big falcons has always been the Middle East falconers who traditionally buy passage sakers and peregrines, usually at a ratio of about 10:1. The commonest peregrines are the arctic *calidus* (nebli), trapped on autumn passage. Many of these falcons are released in March of the following year so they have a chance to return to the breeding grounds. One of our projects at the National Avian Research Center in Abu Dhabi marked sakers at nests in Asia using microchips. Some of these birds were later trapped and found their way to falconers in the Middle East. We organized the falcon hospitals so that when a falcon was brought in for health checks the microchip was detected and recorded on our central computer. Similarly, all released birds were chipped and some were detected back on the nesting grounds. Additionally, a proportion carried either conventional radio tags or satellite tags which enabled biologists to trace their migration routes. We thus slowly built up a picture of the turnover and sustainability of falcons in Arab falconry.

Unfortunately, many protectionists do not understand that there is a world of difference between a young aviary-bred peregrine or saker and a big passage falcon which is already a proven hunter. As already discussed in earlier parts of this book, there is a lot of skilled work, and attrition, in bringing on a young domestic falcon to the standard of a passage falcon. Thus falcons, like horses, can have very different values.Two animals might, to the untutored eye, look alike, but one animal could be worth ten times more than the other. Many breeders have been swept up in the idea of breeding falcons for the Middle East market, only to find that their birds were either physically or mentally unsuitable for this market, or that their prospective buyers did not know how to manage them, as they were only used to passage birds. Having now bred, hacked, and trained falcons for the Middle East for 35years, I know that it can be done on a repeatable basis.

Nowadays we train the birds using robotic prey instead of hacking. and Arab falconry could survive on domestic falcons if the wild falcons face problems on their nesting grounds. There is plenty of scope in this situation for education - of breeders, of falconers and of protectionists.

The British have always had a liking for variety and in the late 1960s and early 1970s imported many different species of raptors. Some of these became sustainable through domestic breeding; others were imported later from other collections. As a result, there are quite a lot of species producing under 30 or so young per year, the prairie falcon and American kestrel being typical. Many of these species are held by just a few breeders who are slowly breeding them up. Over the last ten years, the ferruginous hawk has slowly increased to the point where it is ready to double or treble output quickly. But for many of these species, the demand (reflected in prices) is not there. Therefore many of the birds are not actually paired for breeding and it is likely that, although they may increase slowly, they are probably at market saturation below 50 birds per year.

The big danger with these small populations is inbreeding and a population crash. The black sparrowhawk, the red-headed merlin, and the New Zealand falcon have all started from very limited gene pools in Britain. It only takes the deaths of a few key birds in these small projects to endanger the future of the whole captive population. With the New Zealand falcon, of which all the birds in the world outside New Zealand derive from my original stock of six individuals, we have kept back or loaned out almost all the progeny to breeding projects so that numbers can be built up. We really need to see an annual production of about 50 birds per year to be self-sustaining.

Other species are even less fortunate. Many of these are poor or slow breeders, such as eagles, and many are represented by just a few individuals. They are not sustainable long term at these levels.

Thus in Britain, falconry is now self-sustaining for all the main falconry species and quite a few fringe species as well. If one pieces together the different figures shown, starting with the small output species, and going through to the peak ones, it is possible to discern a general pattern. Imagine starting with say, two pairs of a medium-sized raptor which breeds at two or three years old and which breeds tolerably well. In the first few years, the loss of any of these four birds could be critical. Slowly numbers increase, but they may be all of one sex, or all from one pair. Gradually further pairs are made up. Genetics then become a major headache, and maybe a few more unrelated birds are obtained to make further outcrosses. From that point it could take 10-15 years to reach an output of 15 birds a year. Once this point is reached the species should start a steady climb, potentially going exponential. This could take 10-15 more years. This climb will flatten out as market saturation is reached, which could be at 50 birds or could be 500 birds.

Although this model gives an idea of the time-scale, reality is seldom this straightforward. In practice there are other little twists and turns. For example, the founder stock may be owned by four different people, none of whom are speaking to each other. To try to overcome this problem and maximize production from these small populations, zoos have organized TAGs (Taxon Advisory Groups). Breeders specializing in a particular taxon get together to breed on a co-operative and planned basis. This is a very sensible approach but it hasn't had much traction.

The British populations are not of course totally genetically isolated. It is easier now to move birds around Europe and to obtain birds from North America. Weird, CITES-inspired, regulations come and go every few years. What is government policy in one country is illegal in another and vice-versa. The nub of the matter, which governments are having to come to terms with, is the clear differentiation in law between domestic birds and wild-source ones.

What does 'domestic' mean in law? Many of our domestic animals can and do interbreed with wild ones. For how many generations do you have to breed an animal for it to become domestic? Should it be genetically or physically distinguishable from the wild-type? Should it show 'tame' rather than 'wild' behaviour? Should it have some sort of agricultural connection? Should its counter-part wild-type be extinct?

None of these questions really get to the point of domestication. Legislators tie themselves into knots without solving the problem. A classic example in the British law is *Columba livia*, the rock pigeon, feral town pigeon, racing pigeon, fancy pigeon, and edible pigeon. All are one single species. In law, as a feral pigeon, it may be killed or sold dead by authorized people at all times. As a rock dove, it is fully protected. As a racing pigeon, it is not included under Schedule 3 allowing it to be sold alive. Otherwise, it is regarded as domestic poultry and thus escaping the strictures of wildlife law. A footnote below each schedule mentions that common names are for guidance only, the law being based on scientific species. The law, as they say, is an ass!

The British 1981 Wildlife and Countryside Act was intended to help preserve *wild* life forms, which, in fairness, it does quite well. It does not really come to terms with the question of domesticated forms. It is based on the premise that certain species are 'wild' and certain other ones are 'domestic poultry.' Birds, such as pigeons, which didn't fit this mould, are fudged. Since then, among raptor breeders alone, we have bred many thousands of raptors in captivity. Even here we have bred over 5,000, and some to 15 generations. The law didn't plan on this happening. It has been an unprecedented and successful large-scale domestication of wildlife.

You may have noticed that I've still avoided defining 'domestic.' In reality I think the word defies definition and is not really helpful. It has so many connotations. But for legal and management reasons one must come to a working definition. CITES for some time regarded life forms which were second generation captive-bred onwards as different from wild-source life forms. But this leaves a huge group of the most contentious animals, the first generation and the half generations, in a legal limbo. With this definition you could actually breed a wild animal from a domestic one. Therefore I think that in legal terms the only clean definitive cut-off point is the first generation. In other words either something was born (or laid, not hatched) in the wild, or it was born in captivity. This leaves us two clear groups: wild-source animals and captive-source domestic ones. No other 'definition' is definitive. Laws in many different countries are now being challenged and changed in line with this. Once there is a clear international acceptance of 'wild' and 'domestic' and an appreciation that the law refers not to a whole species but to the individual animal in contention, then it is possible to make progress. The full weight of international law can be brought to bear to help *wild* life forms and separate legislation can cover domestic ones.

A further aspect of domestic breeding which is still under scrutiny in several countries is that of ownership. In some countries wild animals are owned by the state. If you shoot a wild bird (legally) then usually, you own the body. You may or may not be allowed to sell it. I'm not too sure what happens if you eat it and then the state claims it back! If you trap the bird alive under a permit, you may be allowed to possess it, but the state may own it. If you breed it, the state may still claim ownership of the youngsters. Or you may own them. The state may control the sale. The state may also give you a permit to export the bird. This means that the bird will pass into another jurisdiction. The new possessor will almost certainly legally own it. Can you see what a jungle we've got into? If the new possessor sends it, or its progeny, back to you, that is a personal ownership transaction. You have thus obtained ownership from the state. Of course you did not need to export it to another person in the first place. You could export it to yourself. Many people own birds in several different countries. This whole question is an international mess. Again, the only solution can be that animals born in captivity are privately owned.

The final area of confusion is the definition of 'trade.' Some legislators regard trade as a movement across a frontier. So if you take a couple of falcons hawking for a week or two in a neighbouring country and then come home, the official statistics show that four falcons

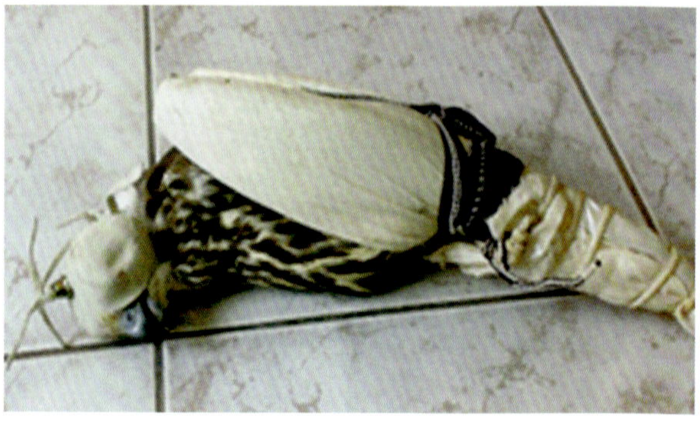

Illegal trapping in Pakistan. When a falcon is sighted, a lure bird, such as a lugger falcon, is thrown into the air trailing a bundle of nooses. The wild falcon, thinking the lugger has prey, comes down to rob it and gets caught. The falcon is trussed up and may be smuggled hidden in white goods, such as washing machines.

have been traded. It's amazing what you can do with statistics! Other legislators include some financial element. This may be a 'value for trade' as defined by Customs and Excise; this figure is usually a work of fiction in itself. You may take your home-bred hawk abroad for hawking, only to be charged import tax in each direction on a notional value which the Customs official has just established by telephoning a protectionist organization which has given him an astronomical figure. Other regulators distinguish somehow between commercial and non-commercial trade. The idea is that price equals production costs plus profit margin. Some regulators allow commercial trade but not non-commercial trade. If you keep birds in two separate countries and move them regularly, as we do, you find yourself faced with all sorts of fees and taxes which are completely unwarranted, and with laws designed for wild birds which are entirely inappropriate and achieve nothing in conservation terms.

Breeders who are trying to set up and run a program are faced with all these issues. I'm not trying to be mischievous; I have had personal experience with every single point mentioned in this section, and my peer reviewers think I have understated the issues!

The basic problem is that birds of prey, mainly falcons,were being trapped illegally in several countries, such Syria, Iraq, Afghanistan, Pakistan, Kazakhstan, Mongolia, Russia and China, and smuggled into the Middle East. It is impossible to stop because of the strong financial incentives and porous borders.

In 1995, we produced an identification guide in Russian for customs officials but it had little effect. Practically all the borders had corrupt officials.We tried microchipping all the young falcons in our field surveys and organised the vet hospitals in the Gulf to scan for the microchips. We held many conferences and published a magazine *Falco* for 25 years, linking up vets, field biologists and falconers.

This is a common problem with wildlife legislation – the laws are unenforceable. The only way to reduce the trapping was to reduce the demand by producing cheaper or better alternatives. When I first took captive-bred sakers to Saudi in 1987, they told me that they were *wahesh* – wild-caught. I explained to them about captive-breeding and how the closed rings are fitted when the chick is still small. But they couldn't get their heads around it. Falconers there had never seen a saker nest and they still believed that the larger bird must be male. Over the next fifteen years or so, some falconers, mainly in the UAE, learned how to train captive-bred falcons that had never hunted before. It was a slow process, but now about 95% of the falcons in UAE are captive-bred. Saudi still uses a lot of wild-caught falcons, but it too is slowly discovering the benefits of captive-bred birds, especially in the racing scene.

At the 2007 Conference of the Parties for the Convention on Migratory Species (CMS) we proposed, and it was adopted in the Action Plan (5.7), to '*Seek to promote appropriate programmes of captive breeding so as to alleviate the pressure of wild harvests on populations of birds of prey.'* In other words, captive breeding became officially recognised as an offset conservation method for raptors.

11.8 Breeding for conservation and gene banks

Having looked at the progress of projects motivated primarily by falconry, let's move on to projects intended primarily for the conservation of species in the wild.

The first point to consider is money. Most people consider it last. They run out of funds and their project goes to the wall. You must look at ways in which money can be brought into a project. Bills must be met. In the case of birds which have a falconry interest it is possible to sell surplus young and cover costs. Some birds can earn their keep in displays at zoos and centres. Others, with a reputation for being endangered, such as the Mauritius kestrel, can help swell the gate and people tend to dip into their pockets if they feel it will help save a species in some way. Rarity is always a crowd puller even if the object of attention has no other special features. Rarity itself is a marketable commodity, even without any birds being involved. Paradoxically, many conservationist organizations rely on certain species remaining 'endangered' or 'threatened' in order to attract funds. This often results in major inconsistencies between true, biological endangerment, and perceived endangerment reported to the public. The peregrine has become a conservation icon and a political pawn.

Some projects have sponsors, some have endowments, some run on the backs of more lucrative projects. All projects require stability, long-term commitment, and financial security. If a project is under-capitalized the birds will suffer. To provide this continuity usually requires more than a private individual can supply; it needs a team effort of many individuals and organizations.

The next question to consider is: what exactly is the object of the project? The best projects last no more than a day and don't involve any breeding, just a translocation. This is how long it takes to move eggs from one wild nest to another. Egg manipulation is the quickest and most cost-effective way of boosting a raptor population. The next step on from this is to take threatened eggs, hatch them out, and then return older chicks to nests in the wild, as we have done with red kites in Wales. The next step is a much bigger commitment, to maintain a gene bank for a protracted period.

Given that egg and chick manipulation is by far the simplest and most cost-effective way of increasing wild raptor populations and, once a policy decision has been taken, can be achieved on a somewhat opportunistic basis, I will leave it on one side. The techniques are reasonably straightforward.

If a 'breed for release' scheme is contemplated, then there needs first to be a clear management decision. Will the progeny be used to supplement an existing population? If so what are the factors causing the population to be below capacity? Are these factors known? Have they been addressed? Are they tractable - will they respond to remedial treatment? Are they essentially irreversible? If the factors are not known, a release scheme may be useful as a method of investigating them. If you put out all your young birds with radio-tags or some other identification and, for example, you find that they all get poisoned, you then have a pretty clear idea why the original population died back. In other words the release scheme was not intended to boost the wild numbers but was being used as an investigative method. The next stage is to address the factors causing the attrition and reverse them. The wild population may well then reassert itself unaided.

For these reasons, releasing stock into an existing population is seldom necessary. Two circumstances spring to mind where supplementation can be useful. One is where a population has died back and now only occupies a small part of its former range and, although the historical factors causing the decline no longer exist, the remnant population has failed to expand geographically. This is what happened to the red kite population in Britain; it needed a little push to encourage it to expand its breeding range. Another reason for supplementation is when a population has died back in numbers right across its range, leaving scattered pairs which are no longer in contact with one another. This happened with the *anatum* peregrine in North America. The subspecies was extirpated by pesticides over most of the continent, but a few scattered, poorly reproductive pairs and individuals hung on. The Peregrine Fund's strategy was to release captive-bred young to support these local foci and maintain traditions of use. These then rebuilt local populations which gradually linked up again.

The other reason for breed and release is for reintroductions. There are seven IUCN-recognized criteria for reintroductions:

a. There should be good historical evidence of former natural occurrence.

b. There should be a clear understanding of why the species was lost to the area. In general, only those lost through human agency and unlikely to recolonize naturally should be regarded as suitable candidates for reintroduction.

c. The factors causing extinction should have been rectified.

d. There should be sufficient suitable habitat available.

e. The released individuals should be as close as possible genetically to the original population. (Because habitat is changing so rapidly, what most populations need is a wider variation in genetic variation, rather than genetic 'purity.' Wider variation gives more chance of surviving fast-changing selection pressures.)

f. Obtaining individuals for a release program should not prejudice the survival of the donor population.

g. There will be increasing examples in the future of original range being no longer suitable, so that the only option if the species is to survive in the wild, is an introduction or transplant to a completely new area, as distinct from a reintroduction. The impact the introduced species may have on the new ecosystem may well outweigh any advantages gained for the species itself, so this course requires very careful consideration. A gene bank in captivity may be the only prudent alternative.

In practice, however much field data are gathered and however much theoretical modelling takes place, there is always an element of trial and error. Ask the birds; often they come up with some surprises. When red kites were reintroduced by hacking, into England and Scotland, the survivability was phenomenal - about 80-90 percent. Obviously the habitat was crying out for colonization but the Welsh population was simply not producing sufficient excess to spread into England. Similar results occurred with the reintroductions of peregrines in Germany.

These are the cold, clinical management

A hack site for our Saker re-introduction project with the Green Balkans in Bulgaria.
Photo Andrew Dixon.

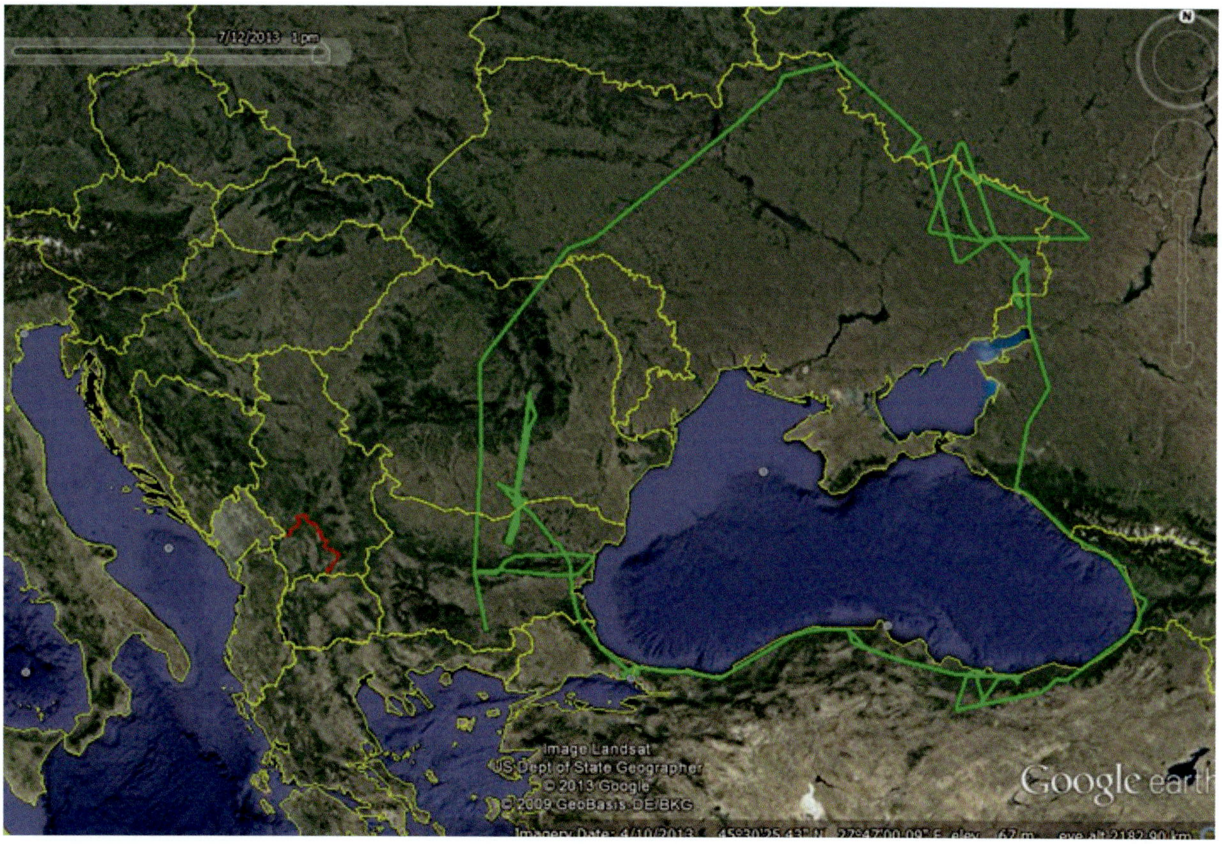

The breeding and release for this young saker in Bulgaria is just the start of an international journey.

considerations to be assessed before a release scheme is contemplated. They are often overlooked by amateurs who are so keen to get involved in an active conservation project that they fail to see an overall perspective. Professionals quite rightly castigate them for rushing in 'where angels fear to tread.' At best the professionals consider it a waste of resources.

On the other hand, professionals can be curiously blinkered in their approach to wildlife management. In many countries they assume proprietorial rights over wildlife which are not theirs and for which they cannot actually handle the responsibility. Most wildlife managers nowadays are tightly restricted on budget and yet do not appreciate that an action plan which for them might be a waste of resources, for some dedicated amateurs is a good use for resources which are not transferable anyway to another action plan. The typical clash is when amateurs are told 'not to breed a species in captivity, that it is a waste of time, that what is needed is more field work.' The fact that the amateur may have a job, a family and be agoraphobic, escapes the professional. The professional too easily just sees the immediate wildlife problem and is unable to handle the people problem and the people resources. The result is a stalemate. The professional doesn't get his fieldwork done because he hasn't got the funds. The amateur is prevented or opposed in his contribution because the professional thinks it is a waste of time.

On a national scale this leads to friction and frustration. On an international scale it amounts to neglect. In many Third World countries there are many species crying out for active conservation work. Local and expatriate professional biologists do not have the resources to do anything significant. Specimens could have been collected for captive breeding as a gene bank or as a source of study material. This is resisted by professionals invoking CITES. The situation is thus left to deteriorate further until the population becomes 'endangered.' This puts it in a position where a professional has some prospect of raising sponsorship money for fieldwork. He does a survey. He thinks that is 'conservation.' It isn't, it's just studying the animal. The animal is no better off; in fact it may be worse off carrying various transmitters and other research detritus. The actual conservation only starts with management, as far as the animal is concerned. Management requires different skills and different thinking to those employed by most field biologists. Some field researchers resent management because it devalues their datasets! Eventually the crisis point is reached, as it was with the Californian condor and the Mauritius kestrel. Captive breeding becomes the last available option. But by then the gene pool is into a severe bottle neck, and the person or organization designated to breed the remnant birds is under grim pressure to be successful. In North America, hard-line protectionists wanted the California condor to become extinct 'with dignity' rather than be taken into captivity and bred up for reintroduction.

This is not the way to tackle these situations. Specimens should be made available to dedicated amateur breeders much earlier on in the process. Without depriving field-scientists of resources, they can then develop breeding techniques for the species and establish a captive breeding population on a broad and representative genetic base.The field scientist should spend some time studying the captive birds so that he starts to understand what makes the species 'tick' and so that he can pose questions to the breeder which may require several breeding seasons to answer. The captive birds can also be used for filming and publicity campaigns to inform people about the total project and to raise further funds. During this process the qualified scientist will start to gain some respect for the experienced aviculturalist and vice versa. I am quite sure that the experience I have gained in breeding and handling captive raptors is easily equivalent to the nine years of university degrees and to the fieldwork I have done. I also know many qualified biologists whom I would not trust to so much as open the door of an incubator.

When immediate release of progeny is not contemplated, the breeding project takes on the complexion of a gene bank. Genes can either be stored as whole live birds, or as frozen semen. Whole birds have a nasty habit of dying and so it is necessary to keep breeding, generation after generation. The important thing to concentrate on is the genes, not the birds.

The first step is to establish the extent of genetic variation in the wild population. This is what we have been doing with the New Zealand falcons. We trapped falcons and obtained blood samples from as many birds from different places as possible. We achieved about 35 birds. By DNA profiling it is possible to assess the coefficients of relatedness (see 2.2). The founder birds should reflect this full range of genetic and geographic variation.A minimum of about eight 'unrelated' birds should represent about 90 percent of the total current genetic variation. I put 'unrelated' in apostrophes because although two birds have come from the wild doesn't mean that they are unrelated, far from it. We have trapped two resident falcons 300 kilometers apart and found that their coefficient of relatedness was over 50 percent. In other words, they were as close genetically as brother and sister or mother and child. Obviously all populations are related to some extent and each has its own characteristic background relatedness. A species, such as the Mauritius kestrel, which has been through at least one severe genetic bottleneck, will be highly inbred and may show a background coefficient of relatedness of 40 percent. This is not necessarily a bad thing. Deleterious genes will also have been heavily weeded out during this process; if they hadn't, the bird would not be here now. Other species, such as the peregrine, found all over the world in big and varied populations with lots of forms or subspecies, intrinsically show more genetic variation and thus a background relatedness of perhaps only 15 percent. We don't know these figures yet; we've also started on the saker and gyrfalcons, but it will be some years before we can put everything together.

Assuming the founder birds carry 90 percent of the genetic variation, the next task is to preserve all these genes as long as is anticipated. For all practical purposes, no new genes are created. But genetic variation can be lost. Every time you breed, you risk losing genetic variation. If you don't breed, you risk losing birds *and* their genes. So the trick is to increase the interval between generations. If your generation interval is four years, ten generations will only last forty years. Therefore, as breeding stock for the next generation, you pick birds which were the last to be produced by the previous generation. Youngsters produced earlier by those birds can be

Mongolian Minister for the Environment Luimed Gansukh signs an MOU with HE Mohammed Al Bowardi at the Environment Agency in Abu Dhabi for 5,000 artificial saker nests.

used for other purposes.

As well as making efforts to preserve genes, it is important to be ever on the scrounge for additional genetic material. CITES has been a major barrier to sources of breeding stock. Sad to say there are many extremist conservationists who would rather see a crippled animal killed than used for breeding.

Unfortunately, conservation is seen as a luxury. When a country's economy goes down, conservation is marginalized. It is very important not to separate ordinary people from hands-on work with real wildlife. Time and time again, the committed amateur achieves more than teams of paid professionals who, when they stop getting paid, stop working. It is simply not safe to rely entirely on professionals, nor is it safe to have a public whose interest in wildlife has been obtained solely through television rather than real life. People who have been reared on an edited view of reality are not in a position to make judgements on wildlife issues; they have become puppets of the media men.

TAG groups on the other hand, can easily become introverted. Many participants may not have even seen their study animal in the wild. It is important for them to build a symbiotic relationship with field workers. Field workers also often have access to wild genes in the form of deserted eggs, or injured birds which would be invaluable to a breeding project and which would otherwise be wasted. Most breeders, by the nature of their work, are rooted to one particular location. They would like to contribute to the conservation of species in the wild, but they have to stay at home. Far from being avaricious for birds for themselves, many dream about providing animals for reintroduction projects. Most of these projects are impractical and do not fulfil the biological criteria, but some do. The barn owl situation in Britain is a classic example of an animal hyped up as threatened, followed by an amazing response from people thus motivated to 'do' something about it. The owl's real problems, such as habitat, are rather intractable, but people felt they could breed them up and release them. And they did. Thousands upon thousands were bred, released, and died. Protectionists then turned on

We take the Minister out to visit the project, while Nicola Dixon (centre) engages with the local communities.

these people and castigated them for their well-intentioned efforts. The point to come out of the whole fiasco is that there are many ordinary people who, pressured by tales of endangered wildlife, want to do something practical to help. They feel that they can't preserve a rain forest in their back garden, but they can breed owls.

Now they are realizing that this approach was simplistic. Proper research, guidance, and coordination is essential. But this is not to deny that there is an amazing human resource there if only it could be tapped. Perhaps over the next decade, some of these TAG groups will be able to set up gene banks for threatened endemics in a coordinated program with field workers. It may not be particularly glamorous or high profile, but then the best work is often steady and low-key.

The alternative to keeping captive birds in a breeding project is to support their breeding in the wild, using artificial nests. This is much simpler and cheaper. It may entail simply putting up nest boxes in a local wood, but if done on a large scale, such as our artificial nest programme in Mongolia, which covered 20,000 km^2, you need support from local communities and governments.

11.9 Managed hunting areas

There are essentially three kinds of hunting areas:

- Private lands
- Areas managed by syndicates or clubs
- State open hunting areas.

Private and managed areas are capable of providing sustainable hunting but open areas are subject to the 'tragedy of the Commons'. There is no incentive to leave any game behind for the future.

The decline of falconry in the Middle East was due to loss of habitat and over-hunting. Once oil was discovered, off-road vehicles and water bowsers became available. This meant that herders could penetrate further into the already marginal desert areas with flocks of goats and herds of camels. The sparse vegetation could not withstand this grazing pressure and the habitat which provided wintering grounds for houbara deteriorated. After thirty years or so, the 'moving baseline syndrome' meant that the next generation of people thought that this was normal and they forgot what the desert used to be like. We interviewed old Bedouins from that period and their tapes are stored at the Abu Dhabi Tourism and Cultural Authority, consigned to history now. On top of this, roads, fences and whole cities sprang up. The falconers now had SUVs, radios, GPS. They had the mobility and technology to far out-gun any wildlife.

The supply of falcons was dependent on local trapping and on wild-caught birds coming in from Pakistan and Syria, some trapped further north in China, Mongolia and Russia. But through captive breeding, mainly in Europe and North America, supplies of good falcons became available and eased the pressure on wild falcons. Starting in 1989, we bred falcons for UAE and within 20 years 98% of the falcons there were captive-bred. Other countries such as Saudi Arabia, still depended on wild falcons, but now they too are switching to sustainably-bred farmed falcons and setting up restoration programmes for their indigenous Lanner and local *jebelia* shaheens which have been trapped out.

Also, in 1989, HRH Sh Khalifa Bin Zayed set up a houbara breeding project and I was on the Advisory Committee for many years. Its initial aim was to breed 500 houbara a year and was spear-headed by HE Mohammed Al Bowardi.

Houbara hawking. We stop for lunch.
Photo Tom Spink.

Cars fan out to search for houbara.
Photo Tom Spink.

After many difficulties and false starts, techniques were developed to breed houbara and gradually projects and production built up. A big project and managed hunting area was set up at Missour in Morocco for the North African Houbara *Chlamydotis undulata undulata* and at Sweihan in Abu Dhabi for the MacQueen's Bustard *Chlamydotis undulata macqueenii.* I designed the research centre there in 1990.

Further centres have now been set up in Kazakhstan, Uzbekistan and Morocco under various initiatives. Production has increased year on year, co-ordinated by the International Fund for Houbara Conservation, managed by Dr Mohammed Saleh Hasan Baidani. The total to date exceeds 600,000 birds.

There had been a big demand for houbara needed for training the young falcons each autumn. Huge numbers were trapped in Pakistan and smuggled across to the Gulf. Faris Al Timimi has documented some of the appalling losses in this illegal trade, with piles of birds dying in transit. And of course, each houbara trapped for training meant one less available for breeding or hunting.

So supplies dwindled. Arab falconers who wrote lyrical poems about the Houbara Moon when migrating Houbara would arrive in the Gulf overnight, now found that very few came. Those that made it found the habitat too poor and had to move on further south. Any that lingered were spotted, reported and cars quickly arrived and hunted them. Houbara scientists in Abu Dhabi set up international programmes to study all aspects of wild houbara ecology, and we used trained falcons to catch the wild houbara, to attach satellite tags and monitor migration routes.

Experiments were done to release the farmed houbara, with some success. A few pairs bred in quiet reserves. But these were nothing like enough to provide sustainable hunting. One of the main problems was that except in very heavily guarded areas, it proved impossible to stop falconers going anywhere they liked to hunt houbara. Although they also hunted hares and karrowan, houbara were the ultimate prize for the falconers. In 2011,

Reaching the kill. *Photo Tom Spink.*

At the end of the day, time for a game of cards.
Photo Tom Spink.

The Houbara Research Centre, Sweihan, Abu Dhabi.

Farmed houbara chicks come to be fed.

for the International Festival of Falconry held in Abu Dhabi, we managed to obtain permission to use an area at Al Khatim, fenced 15 years previously by HRH Sheikh Zayed. It was about 7x5 km and some of the habitat was regenerating, but recovery is very slow. We organised three days of hunting for visiting international falconers and locals and I insisted that only camels, horses or foot could be used. No vehicles. We released houbara and hares in the days leading up to hunting. The hunting was successful except that the prey were very naïve. They had no experience of raptors and were easily caught. Any that flew over the boundary fence were quickly mopped up by local falconers. This reserve, although tiny, still exists and now houses the Sh Mohammed Bin Zayed Falconry School, but sadly it is also used

Houbara love mealworms.

Wild Houbara trapped in Pakistan to be smuggled into the Gulf.

for trophy hunting using geriatric Arabian Oryx *Oryx leucoryx* destined anyway for slaughter.

At the next International Festival in 2014 we fenced a new 25x20 km area in the Western Region, on the road to the Liwa oasis. The fence was only just completed in time and the farm camels removed, so the habitat was still minimal. Again, I banned vehicles. So we had 80 camels and six horses with about 500 falconers wanting to hunt. During mornings and evenings in December we gave everyone the opportunity

A saker equipped to catch wild houbara unharmed so that we can fit satellite tags. She has beads on her claws and her beak is covered.

Once the satellite tag has been fitted, the houbara is released again.

to go hawking. The farmed houbara were tracked and flown, but many were disappointing fliers. This area was subsequently retained for managed hunting but then vehicles were permitted. Such a small area can be hunted out by cars in a single day. So they resorted to putting farmed houbara out hours before each hunt. Basically you could pre-order how many houbara you wanted to hunt. So these houbara behaved more like chickens and the organisers asked me how to make them fly properly. Obviously they need more time living in the wild, which means they have to adapt to wild food, of which there needs to be plenty or they

Hawking houbara on horses is fun!
Photo Mark Williams

will simply fly away. And they need to be reared to be wary and then get experience of predators. And this is where you hit a brick wall.

Do the maths. I had to do it recently for the Saudis. There are about 20,000 falconers in KSA, with say, two falcons each. Each falconer wants to hunt maybe one day per week during the season November to February. That's about 18 hunting days, a very conservative estimate. If he hopes to catch one houbara per trip, that's 18 x 20,000 = 360,000 houbara. When game is released, if allowed to acclimatise to the area, there will be high attrition. Some will starve to death, some will be caught by foxes and eagle owls. Some will fly away. A return rate of 50% is very good going. So you'd need to release 360,000 x 2 = 720,000 Houbara each year. This may not seem very much compared to the 53 million pheasants and partridges released in UK each year. But pheasants and partridges are easily bred on game farms, costing only about £5 each to produce. Breeding houbara is usually done using artificial insemination and is hugely labour intensive. Each one costs about £400 and most are heavily government subsidised. So to enable the falconers to go hunting just 18 days per winter season could cost each person 18 x 2 x 400 = £14,400 simply for the houbara, and human nature being what it is, many will try to cut corners. It is impossible to properly warden and police the managed hunting areas. People simply cut the fences. Community-led projects are the best hope but these are breaking down as more people move to the cities. Hares can be farmed more easily and at least do not fly away. But they have a high attrition rate in poor habitat, and they do not have the same social cachet that houbara have. We set up a hare farm in Abu Dhabi, but really people want houbara.

Thus, despite huge investments and falconry being inscribed by UNESCO, the reality is that real falconry is on its knees. Recently falconers have flocked to Azerbaijan to hunt Little Bustards (*Tetrax tetrax*) but quickly, with habitat loss, shooting and over-hunting, the great flocks that were there have diminished. Similarly in USA, the deterioration of the great plains habitats has led to the slow attrition of the upland grouse populations as foreseen by Aldo Leopold. Falconers there are supporting the Sage Grouse Initiative. In Britain, the red grouse populations have been maintained by driven shooting, which pays big money, allowing a few falconers to hawk on the fringes and poorer moors. Moves to prevent driven grouse shooting will mean these areas are no longer keepered, and as fox and crow numbers increase, ground nesting birds will disappear. I have seen it in my own lifetime on a local moor here in Wales. In the 1920s it was a driven grouse moor. But then, with the decline of farming, commercial forestry was planted around it and, without proper keepering, the predators flourished. The graziers burnt off the old heather and the last grouse I saw there was in 1995. The hares too, lacking cover and lamped by people with lurchers, disappeared at the same time.

Houbara cannot survive this level of shooting.

Migrating Demoiselle Cranes *Grus virgo* shot in Saudi Arabia.

Dismayed at all this, in 1999 I discussed the situation with HRH Sheikh Mohammed Bin Zayed and HE Mohammed Al Bowardi, subsequently publishing a book with Helen Macdonald and Dr Fred Launay. We used this as a policy guide at meetings for many years but the human and cultural difficulties are immense, and this was one of the things that motivated me to develop roprey.

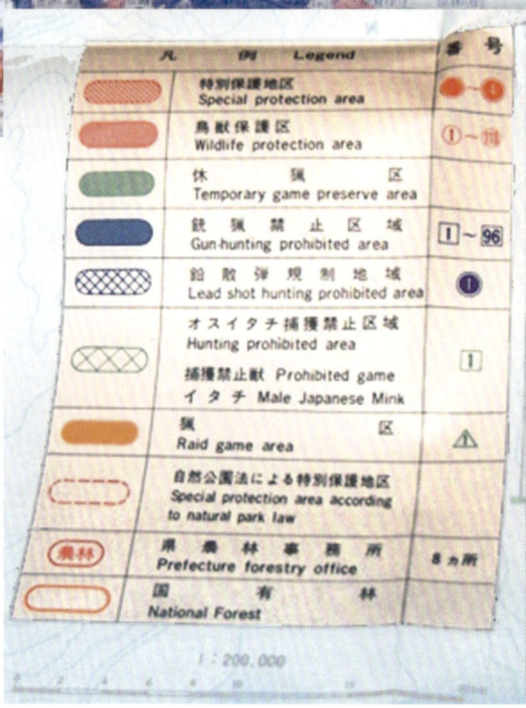

In Japan, hunting is strictly controlled, with designated areas and quarry species.

Access can be an issue.
Photo Pippa Scott-Harden.

Hawking Hadrian's Wall country.
Photo Pippa Scott-Harden

We hold a party for our farmers and do some country dancing in the mews.

Similar stories are repeated all over the world. The 'tragedy of the commons' means that open hunting areas get hunted out and only exclusive areas can support game. In Morocco the large hunting area established by Sh Zayed is successful because it is huge and the hunting is tightly controlled. But in UK, if you want daily flights at game, you need to lease suitable land and release your own game, as so ably described by Simon Tyers in 'The Specialist Falcon'.

11.10 Sky trials and bagged game

In this section I want to tackle the ethics of using prey animals which have been previously held or captured by man. *This of course is not falconry, which is the taking of wild prey in its natural state by means of trained hawks,* but it is legal in many countries with Britain, Germany and the Netherlands as notable exceptions.

What do we mean by 'bagged?' The commonest bagged game is a domestic pigeon which is handled and released in the same way as when releasing pigeons in a pigeon race. Another way is to use the same method as that employed by dog-trainers, namely to 'dizzy' the pigeon and place it on the ground in rough grass. It will stay there for up to half an hour, allowing the person to come back with a dog and obtain a point, which presents a realistic set-up for a young falcon learning to wait on. 'Dizzying' is just a form of predator-induced tonic immobility. A further way is to use a commercial spring trap beloved of the American shooters. This can be triggered from a distance and will throw birds up to pheasant-size into the air.

Less innocuous forms of release include handicapping the animal in some way, for example by trimming the primaries, or by attaching a string to it. Thus, quite apart from the hawk, the use of bagged game in itself may entail some degree of suffering or discomfort.

Bagged game is used under these circumstances:

a. To enter a young raptor, in the same way that its parents do in the wild.

b. To help teach a falcon to wait on.

c. To exercise a falcon.

d. To make competitive sky trials for falcons.

e. To call a bird back with, for example, a live pigeon on a string.

The use of captured game to teach youngsters to hunt is widespread in wild raptors and in many mammalian predators too, such as the cat family. Falcon parents tend first to fly away from the pursuing youngster and then drop dead prey for it to catch and then later, as the youngster gets more proficient, to release live prey high in the air where the young falcon has a good chance of catching it. To this extent bagged game is natural. But now roprey are available, these obviate this role for bagged game.

There are some ethical dilemmas in entering young raptors. The first is the practice of flying raptors which are unentered, that have never caught prey before. This is common (for very good reasons) among display birds and among people who keep raptors as pets rather than for hawking. If an unentered raptor becomes lost, she has only a small chance to learn to catch prey in the few days before she becomes too weak to fly. This is especially true for older birds which have passed their youthful learning period and become 'lure bound.' For the hawk's benefit therefore, if not for the falconer's, it is desirable that the hawk should have at least taken one or two prey for herself so that she knows that prey mean food. Then, if she gets lost and gets hungry, she will at least try to catch her own food. Otherwise she is likely to starve to death without so much as launching an attack on prey which may be right in front of her.

Rehabilitators are often faced with the task of returning raptors which have never hunted before back to the wild. We have had to do several peregrines like this. Usually they have been taken from a nest illegally and then been the subject of a court case. By the time the legal procedures are finished it is well into the autumn. The bird must become proficient at flying, and become fit, at a time of short winter days, bad weather and low prey numbers. Being a single, older bird, she cannot be hacked properly because she would quickly wander off. She cannot be allowed to

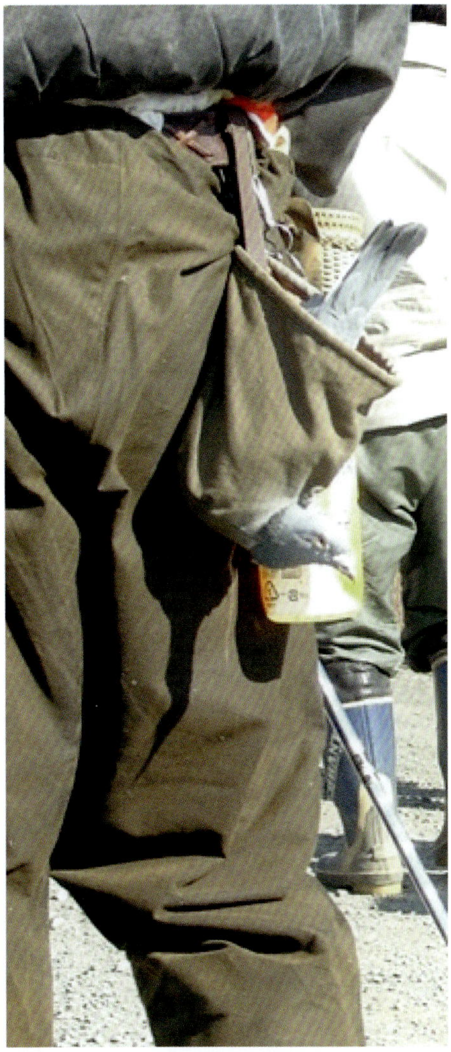

A bagged pigeon being carried in Japan.

stray too far because of the risk of losing her in an unentered state. To get such a bird, at such a late stage, to catch prey reliably, without using bagged game, is at best a major work commitment, at worst, impossible. The current healthy state of wild peregrine populations in U.K. no longer justifies such an expenditure of resources; indeed, releasing such a bird is probably counter-productive. In such a scenario it is more ethical to flush some dizzied pigeons under such a bird than to release her unentered to almost certain starvation. Alternatively, if one is against killing, it might be better to kill the peregrine and thereby save the lives of her future prey…

b), c), and d): Training to wait on, exercise, and sky trials, can all be lumped together. Essentially the falcon is taught to wait on, as she does in the wild and a racing pigeon is released. The falcon then stoops at the pigeon and may or may not catch it. The method is described in Beebe and Webster in detail. This is a common practice in the Americas, Spain, Italy, and the Middle East. Most pigeons escape. This is not a particularly cruel practice but it is the lazy falconer's way of training and exercising his bird. When time is short, it is obviously easier to give your falcon a flight at a pigeon than to go hawking. Many falconers in North America stoutly maintain that it is impossible to train a high mounting falcon without using pigeons. They are wrong! Certainly it is a lot easier to use pigeons, but it is disappointing when experienced falconers find themselves unable to cope without pigeons. It is one thing to use some pigeons to teach a young falcon how to hunt, another when the falconer relies on using pigeons on a regular basis for the rest of the falcon's career.

Sky trials are when several falconers meet to fly their falcons at pigeons on a competitive basis. The method has been described by Gerald Richards in *Game Hawking...at its very best.* Only nine out of 202 pigeons were killed. Thus, in itself, it is no more cruel than falconry. My objections to it are that it debases falconry into a competition between people rather than a hunting experience. To me, when you hawk together, you all work as a team to help the hawk, whoever's it is. It is sad to see a fine old sport like falconry reduced to a slick televised competition. If you want to do competitions you can go to the races, which are not falconry and do not require killing live prey.

And this brings me onto my second objection. Once you hold a sky trials, impressionable youngsters attend and go away with the idea that this is falconry. Worse still, the media will be there (in Spain they are actively promoted) and before long it all appears on television. The public then think that this is falconry and a bad image is created. In Spain, where attitudes to animals are different, falconers find my comments bewildering. But when films of sky trials are sold abroad, the damage is done. Interestingly, when I last went to Spain three years ago, the organisers of the Spanish Hunting Federation approached me to discuss the feasibility of replacing live pigeons with robotic prey, because they were getting so many public objections to sky trials.

e) The use of a pigeon on a string to call a falcon back is still prevalent in the countries mentioned above. Again, it is a sign of a lazy or inexperienced falconer. Many game falcons are flown in such a high condition that they won't come down to a dead lure and the falconer has become slack on his discipline and weight control. Often the lure used is a sad little leather pouch on a short line. Many have never even tried the far more attractive pole-lure.

In America the pigeons are usually carefully home-bred but in many other countries, especially the Middle East, the pigeons and other species, such as sea gulls, are caught at industrial and landfill sites. They are caged together so that any diseases spread throughout the whole group. I was talking with a falconer recently who proudly told me of his falcons chasing bagged sea gulls. Then he mentioned how all his falcons had died suddenly. Whereas some bird species have some resistance to diseases, such as avian influenza, falcons have none.

With all this in mind we developed robotic prey as a humane and reliable alternative, and using this we can now bring a young falcon from her first day loose to catching wild houbara high in the sky in six weeks from her first day flying free.

Thus, with a few exceptions, I condemn the use of bagged game. Most of it is unnecessary to the sport of falconry and brings it into disrepute. In most cases, with a little thought, there are humane alternatives.

11.11 Legislation issues

Lawyers, politicians, and civil servants, for all their faults, have a job to do. Most of them are not specialists in wildlife biology, and few have any real personal freedom of decision. Their decisions are often made for them by panels of 'experts' and by pressure groups. As a result, national and international laws concerning wildlife show little underlying logic or science, they merely reflect popular levels of concern on different issues. Therefore they are not coordinated, and hypocrisy is rife.

I don't want to list a catalogue of criticism. This would be easy, but is a negative and polarizing approach. But I'll give a few examples to show you the kind of galloping hypocrisy we are talking about.

In Britain, the Wildlife and Countryside Act 1981, Part I: 8 (1) strays onto welfare issues and makes it illegal to confine any bird in a cage too small for it to stretch its wings freely. The very next subsection promptly exempts poultry! Thus a falconer who leaves his hawk in its car box overnight when away from home (a very safe and comfortable arrangement for all concerned) is liable to prosecution entailing a fine of up to £1000 and a ban on keeping birds for five years. The chicken farmer meanwhile, who is cramming millions of birds into cages, gets away with it. How is this? Simply due to pressure from the agricultural lobby on politicians on 'economic' grounds.

Another classic example which permeates legislation in America, Europe and internationally through CITES, is the failure to differentiate between three different categories. The first category is the *species*. This encompasses all individuals genetically included in an internationally recognized taxon, the species (including all *subspecies*), whatever their immediate origins. Obviously, if a species is not internationally agreed, then it can have no legal definition. Similarly subspecies are taxa which cannot be perfectly differentiated and therefore cannot be subjected to separate legislation. If one could clearly differentiate between all individuals of two subspecies then they could be classed as two full species. It is the very fact that there is a gray area which means they are given subspecies status. Geographical origin is not a valid criterion in legislation because it is seldom known with certainty.

The second category is the differentiation between *wild* and *domestic*. We have already discussed this in sections 11.3. International legislation must take into account the substantial populations of animals which, although genetically and physically identical to free-living wild animals, are of domestic origin. These populations are going to increase in the future and will be traded and moved internationally more and more. Legislation needs to identify these and cater for them. Many species are now more numerous in domesticity than in the wild, and in the long-term many species will depend on domestic populations for survival as a gene pool. Paradoxically, although CITES is intended to support wild populations, some authorities and NGOs use it as a tool to hamper or prevent trade in captive-bred animals, which is a sustainable process, recognised by the Convention on Migratory Species as a form of offset conservation. All the unenforceable regulations in the world did not prevent the trade in wild falcons from Asia to the Middle East, what did impact this trade was the supply of domestic falcons.

The third category is the differentiation between *species* and *individuals*. Many laws are framed to encompass whole species, whereas when the law is actually applied, the item in court is not a species but an individual animal. This individual animal may have been born in the wild, or in captivity. This has ownership implications. For example, the Mauritius government claims ownership of all Mauritius kestrels. Individuals shipped abroad still 'belong' to the Mauritius government although now outside its jurisdiction and subject to the laws of their new host country. As these individuals breed, the breeders consider that they own the progeny. Birds may be sold or exchanged with other collections. Everytime birds are shipped somebody must pay import taxes on a 'value' for customs. The whole thing is a mess.

Protectionist organizations are not slow to cash in on this confusion. For example, the Royal Society for the Protection of Birds launched a very commendable campaign to stop the international trade in wild birds. This trade is undesirable if it is an unsustainable drain on wild populations or if there are welfare implications with birds suffering or dying in transit. But because the documentation on bird movements does not include the source of the birds, the monitoring organizations count all movements of non-farmyard domestic birds as 'trade in wild birds.' Thus, in the 1980s, I found at the top of their list of 637 species which are 'near threatened and which are currently traded,' the New Zealand falcon. As it happened, only six birds have been exported from New Zealand, all to me. The trade referred to was based entirely on movements of my domestic-bred progeny from these founder birds. Nor were most of the birds sold, most were sent to reputable breeders to establish a gene bank. But again, monitoring records do not cover monetary matters; all movements between two parties are automatically recorded as commercial trade. Given that the whole thrust of the protectionist argument is to reduce supplies of wild-caught birds in favour of domestic ones, it hardly promotes goodwill and understanding by using domestic birds as scapegoats.

CITES has been a disaster in that it has failed to protect wild populations, failed to cater properly for the requirements of domestic populations, failed to design a system that is internationally workable, and left the door open for the crook at the expense of law-abiding people. Although the overall aims of CITES are commendable, it has suffered at the hands of idealists and theorists with insufficient experience or understanding of real-life situations. There is so much disparity between the priorities of the different nations, and imbalance in their representation, that I despair that CITES will ever achieve its goals.

Recently there have been efforts to include animal welfare within the remit of CITES and from there, ban any trade with welfare issues. There are rafts of other legislation dealing with animal welfare, for example in UK we have The Animal Welfare Act 2006. The target of these efforts is to use animal welfare to ban trade. This is a nihilist approach. The target should be to improve welfare.

Gradually though, we are seeing a change of attitude away from old-fashioned, total protectionism toward sustainable management. While it is obviously essential to maintain wildlife habitat, particularly in critical areas, there is a growing awareness that man and wildlife have to live *together,* not apart. All the best areas of the world for living things are dominated by man. For wildlife to survive there, it can only be by careful management. This cannot be by total protectionism (except in very limited cases) but by balancing the needs of species with resources. Wildlife is itself a resource, whether in a managed 'wild' or in closer association with man. Decisions must be made on aspects such as carrying capacity, over-production, culls, harvestable surpluses, funding habitat maintenance, and so on. Some of these decisions are hard. The approach must be pragmatic rather than idealistic. We need balanced international legislation to support sustained management. And organizations need to depolarize and bring their members toward realistic attitudes for the survival and welfare of all life forms. Some of these attitudes toward the relationship between man and raptors are described in the next three sections.

11.12 Falconry as a 'green' field-sport

Falconry is the most environmentally friendly field-sport and method of pest control. In recent years there has been an astonishing revolution in falconry. At least 50 species of diurnal birds of prey have now been bred domestically for two or more generations and most of these are available on a sustainable basis. This has almost been like condensing about 10,000 years of domesticating farm animals into just twenty years. The significance for falconers has been immense because at last their hunting birds are available on a secure long-term basis.

No raptors have been taken under licence from the wild for falconry in Britain since 1988. British falconry is now self-sufficient for birds of prey and has no negative direct impact at all on wild raptors. Over 2,500 injured wild raptors are tended by hawk-keepers in U.K. every year, and about half of these birds are returned to the wild (admittedly with unknown success), so the only impact falconry has on wild populations is a positive one.

But what is true falconry? Falconry is the hunting of wild quarry in its natural state by means of trained birds of prey. As such it is really just a controlled form of natural predation. This is what I find so fascinating about it. I remember on one occasion radiotracking a wild Mauritius kestrel in a last remnant of the forest there. I ran along behind him in the trees while he searched for and attacked tree geckos *Phelsuma guimbeaui*. At the end of the day he went back to his nest area. If he had been a falconry bird I would have taken him home safe to his mews, but this was the only difference; otherwise it could well have just been a sequence in a normal day's hawking.

This close intimacy enables the falconer to see at first hand the way in which predators function and also to understand just how closely the prey have adapted to this predation. Hawks are remarkably quick at identifying weakness in a prey. Any animal which is weak, moulting, diseased, pricked by shot, or simply an odd colour, is immediately targeted. I would say, from watching goshawks attacking flushing wood pigeons and ducks, that an experienced hawk recognizes a weak bird in a group within the first half a second of them rising. By the time the birds are at waist height any odd bird has been singled out, and the goshawk will fly it with much more conviction, almost as if it knows the outcome is inevitable.

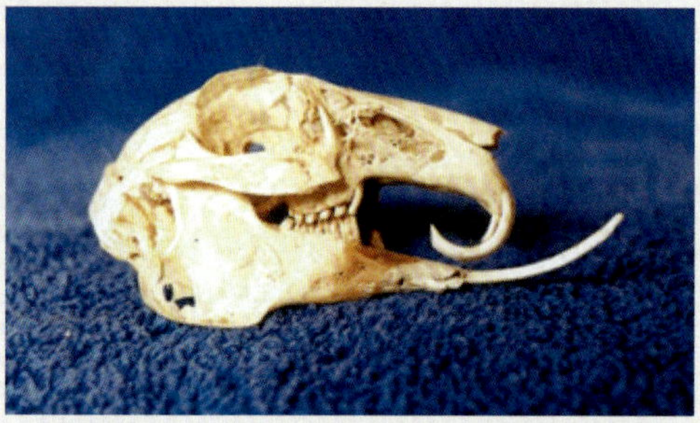

This rabbit was easily taken by my male gos. The right lower incisor had an abscess and had broken off. The other teeth had nothing to wear against and kept growing. The rabbit was starving.

Falcons, such as peregrines, tend to be much less impulsive. They like to be in control of the situation. But studies by Rudebeck in Sweden and by German falconers hawking corvids, have shown that falcons too are selective. Initial attacks at flocks are often feints, designed to try out the prey and reveal any weakness. If the whole flock is strong the falcon may well lose interest, but if one of the individuals appears slower than the rest, or loses its nerve, then the falcon attacks in earnest. Of course raptors are not selective all the time; they are perfectly capable of taking healthy prey. But studies reveal that the proportion of weak or handicapped prey in their tally is considerably higher than the proportion in a shot sample.

By watching the way in which a raptor hunts its prey, one comes to a much deeper understanding of the prey themselves, their cryptic colouring, the way they go about their daily business, always alert, always near an escape route or in flocks with sentinels, their whole way of life finely tuned to predators. But when the predator actually arrives, a whole new kaleidoscope of behaviour is revealed, and one suddenly realizes the prey have got a lot more tricks up their sleeves than at first appeared (see 10.4). The falconer and his bird together discover that in certain situations prey are safe from attack and, as they get more experienced, they learn to recognize situations where an attack could be successful. This breeds a respect for the prey to the extent that an animal, such as the crow, which to others may be regarded as a pest to be killed at every opportunity, to a falconer poses intriguing intellectual possibilities and a physical challenge. To hawk it successfully may entail the use of four-wheel drive vehicles or horses, radiotracking, specially bred and trained falcons, and access to large areas of open terrain, together with a daily commitment for several weeks. Not (you might say) the most cost-effective way of going about killing crows. Quite possibly not! But wait a minute, what are we actually trying to catch in a field-sport? Not, as a superficial look might suggest, the quarry. That is a necessary adjunct. No, our real quarry is a much more elusive customer. His name is enjoyment. Just when you think you've got him, he slips away. And then, in a quiet moment, you suddenly realized you've glimpsed him again. This is what falconry is all about.

I remember a day a few decades ago. A falconer called Graham was working for me and it was his first day out hawking red grouse. We were hawking in Sutherland in Scotland with important guests. Graham had helped hack and train the young falcon, called '*Snake*,' and when my dog locked on point in a broad basin, Graham was told it was his flight. He unhooded *Snake*, and with a shake of her bell she climbed effortlessly to a good pitch above us. I faced the dog and with a quick rush flushed a single grouse under the falcon. She dipped a moment and then ignored it. Unbelievable! What could be up with this bird? And then we realized. Twice as high as *Snake*, in an unmarked sky, appeared the searing crossbow-shape of a wild peregrine. In tight rings *Snake* caught an updraft and climbed to do battle. Not daring to take our glasses off them for fear of losing them, we watched the aerial dogfight, silenced by distance. Alternately stooping and locking talons, they drifted out of sight over the hill. I could hear the radio signal fade as *Snake* disappeared over the ridge. Graham ran uphill through the long heather, whistling and swinging his lure. After ten minutes he was almost exhausted, and we lost sight of him too. We were left, actors in an unresolved drama, waiting for a prompter.

Then I heard some very soft bleeps on the radio. Like the faint pulse on a recovering heart victim, we sensed that *Snake* was returning, coming back on stage. Graham, suddenly recharged with energy, came bounding down the hill, stumbling and sprawling with fatigue. We were ready. The pointer, after twenty minutes, was still absolutely motionless on the remainder of the covey, which in turn had stayed tight due to the falcons overhead. *Snake* came back over and resumed her position of authority. The moment we were in her killing cone I flushed the grouse back up the hill over the dog's back. Like a dark bee, one skimmed up toward the burn. *Snake* peeled over and hooked down and as the two dots merged they disappeared from view into a small gully. The falcon did not come up again, and I knew she had got it. When I arrived I found Graham lying beside the falcon under a bank. Steve Frank took a photo of them. Graham's first grouse had left him totally physically and emotionally drained. He'd gone from the heights to the depths and back again and was now being eaten alive by midges. Such is falconry!

The point of this story is not to show how inefficient falconry is, but to show how *efficient* it is. Many, many man-hours of interest, involvement and enjoyment are obtained per head of game taken. Indeed some falconers go several years before they manage to catch anything at all, so absorbing is the sport.

This is not to say that falconry cannot be effective at killing game. Many professional falconers are engaged in pest control, particularly in places, such as in towns and parks, airfields, and rubbish dumps, where to discharge a gun would be dangerous. Goshawks and Harris hawks are

Hare hawking in Japan.

quite capable of accounting for a dozen rabbits in an afternoon without making a lot of noise or endangering anybody. Hawks and falcons used in pest control are also used, not primarily to kill the pest, but to frighten it off. And as it doesn't pollute the environment. It is a very 'green' method of pest control.

Finally, as well as being self-sustaining, selective, natural, and non-polluting, falconry seldom leaves prey injured. Normally it is either caught or it escapes unscathed. The kill, especially by experienced falcons, is often instantaneous. Hawks on the other hand often hold their prey until the falconer arrives. He may then decide to kill it humanely or to release it. But whatever happens the falconer leaves the field with a clear conscience that no animals are left behind wounded and that the prey have faced only a situation to which they are naturally adapted.

In the old days falconry had a huge following across Europe and Asia. Nowadays, despite increasing urbanization and huge pressures on the countryside, it is a splendid way to renew contact with nature and to come to understand how natural systems work. In Britain 83 percent of people live in towns. Many now have a very distorted view of nature from being reared on 'Bambi' TV programs. Despite computer games based on killing one another, there is a big taboo on real death. But many strive, consciously or unconsciously, to renew their links with nature. Maybe they take up rambling or birdwatching. Others feel that this is more like voyeurism and want to be actually involved in nature. Many people, especially boys and men, still feel a strong instinct to hunt. Some play computer games or football, but others want to be out hunting real quarry, not an artificial one. Falconry enables people to get out into the fresh air and hunt real quarry without having a major impact on the environment.

Some falconers feel that falconry should just be reserved for an 'elite', that somehow it is too difficult or demanding for the masses. They call for increasing legal restrictions on falconry so that they themselves could continue but others would be excluded. The problem with this is that the numbers of falconers in a country are then tiny. They have no significant political influence, nor public support. So they are easily swept away, often unintentionally by legislation intended for other purposes. And it is often a process of attrition, death by a thousand cuts. For example in Britain, Starlings *Sturnus vulgaris* and House Sparrows *Passer domesticus* have been protected because their populations, although massive, show a decline. This has meant they are no longer available as falconry quarry. And so it goes on.

Falconers have formed clubs to represent their interests, or other NGOs such as the UK Hawk Board that represents all the clubs. But many falconers are not in clubs. Clubs in all spheres are struggling because young people now link up through social media instead, and these often do not carry the same ethos of social responsibility that clubs do. At the international level, the International Association for Falconry and Conservation faces major difficulties because there are huge cultural differences between the 80 or so falconry nations. Even on basics such as animal welfare, there is little shared opinion.

The reality is that politicians are swayed not by science, not by facts, but by public opinion and that is shaped by what the public sees. The major interface between falconers and the public are not with the guys quietly hawking rabbits, but with the centres and flying displays that attract thousands of spectators. While falconers plaintively plead

UK	Legal	How many participants	Impact on public awareness	How organised	Number of organisations
Centres and mobile flying displays	Yes	6000	70.0%	commercial	150 businesses
Courses/education	Yes	7500	18.0%	commercial	35 centres
Corporate days	Yes	5300	7.0%	commercial	20 businesses
Pet-keeping	Yes	20000	1.0%	hobby	
Recreational flying free	Yes	15000	1.0%	hobby	
General Zoos	Licence	50	1.0%	commercial	
Flying at wild prey	Yes	5000	0.5%	hobby	23 clubs
Breeding	Yes	2000	0.5%	hobby/commercial	20 businesses
Pest control	Yes	200	0.5%	commercial	21 companies
Re-habilitation	Yes	50	0.5%	hobby/charity	
Raptor related businesses	Yes	50	0.0%	commercial	40 businesses
Research	Licence	5	0.0%	commercial	

General notes

Total hawk-keepers	25000
Total visitor/spectator days	64 million
Total raptors in captivity	70000
Total full-time job equivalents	3000
Falconers in clubs (approx)	2500
Falconers in IAF clubs	1500

Figure 11.12.1 In a 2007 survey we did for the UK Hawk Board it was clear that the major influencers of public opinion are centres and flying displays, educational courses, and corporate days.

that this is not 'real' falconry (according to their own definition), the public enjoy the spectacle and end up, at a superficial level, supporting falconry. Like it or not, this is falconry's window and the hunting element slips by scarcely noticed.

A falconer can hunt through a path of countryside and leave no trace. No bangs, no empty cartridges, no wounded game, no risk to other people. He doesn't need to modify the countryside by building sports grounds or golf courses, or by killing vermin, rearing large numbers of game birds, and restricting public access. His demands on prey populations are small; his bag for the whole season is often less than a shooter's bag in one day, and his prized quarry is often a pest species.

Thus falconry is a natural, low-impact field-sport, self-sustaining and well suited to the needs of modern man. It helps us to understand natural ecosystems and to realize that we are part of them.

11.13 Ways to learn falconry and hawk-keeping

When I was a boy I hawked for three seasons before I ever met another living falconer. The books available then were few to someone of my limited means. Jack Mavrogordato had published his first book *A Hawk for the Bush* and Michael Woodford had just published *A Manual of Falconry*. Most of the other books were too historical to have much relevance for me. Growing up in the country, without others to help, my attitude has always been 'Ask the birds.' They will tell you what they need.

This is a very hard, slow way to learn something as unforgiving as falconry. I treasure those years very much. They were simple, natural years when I hacked my birds without really being aware of the deep intellectual pool of knowledge whose surface I was swimming on. In factual terms I could probably teach somebody in a couple of days all the information I learned during my childhood, how to put on jesses and all that kind of thing. But I could never easily teach someone how to read a hawk's mind, how to interpret the glance of a hawk which betrays what she is thinking.

Making hoods with Philip Glasier, 1968.

When I started meeting and corresponding with other falconers, all the complications of modern falconry surfaced. In the twenty years after the war the small groups of falconers started to come to a focus. Pesticides decimated the hawks which suddenly had special status. Rumours circulated that hawks could be bred in captivity.....

More and more people wanted to learn falconry and to meet this demand Philip Glasier started to run courses. In return for working at his centre in the summer it opened, he taught me to block hoods, to exercise falcons, and a host of other practical things which cannot easily be picked up from books. Many of these details had been passed on, person to person, from the Old Hawking Club and back into the mists of time.

In 1971, I was approached to run three courses under contract during the summer. Although I had only taken my first head of game scarcely five years earlier, I took the opportunity and all went well. For the first time I realized what a diversity of approaches there are to falconry. The British Falconers' Club's attitude to beginners then was that you would not be accepted into the Club until you were taking quarry regularly with a trained hawk. This rather left beginners out in the cold; by the time they were taking quarry regularly many of them had lost all desire to join!

The North American Falconers' Association and some of the clubs in other countries have their own apprenticeship schemes. This is an admirable way of getting beginners off on the right foot. Of course it depends how good the tutor is, both at falconry and at teaching, but it is the next best thing to actually being reared in hawking. In Zimbabwe, Ron Hartley, ran a club at Falcon College which not only gets the youngsters off on the right foot in falconry but also instils in them a deep ethos of responsibility toward raptor conservation. As a result, Zimbabwe is one of the few African countries with established local

One of the basic skills to learn is how to carry a hawk comfortably. First lessons include carrying a glass of water without spilling it. It's good training for the horse too! 1970.

interested people caring for their raptors.

Mentoring apprentices is fine and does help to uphold and maintain falconry traditions and heritage. Japan used to have 'schools' of falconry dating back to the Edo period in the 1600-1800s. Traditional techniques were passed down from master to pupil. Similar but less formal systems exist in Arabia, Asia and Europe. This helps to maintain traditional ways of doing things, but at the same time may stifle innovation and evolution of the sport to adjust to changing modern conditions. It can also encourage the pupil to go running to the master for plug-in answers rather than developing independently. The difficulty in many countries is that there are not enough falconers willing or able to mentor apprentices.

Learning from the birds themselves, and from real life, can be tough for all parties. Some falconers rely on intuition. Falconry is after all about developing working relationships between humans and other species. But under-pinning all of this is biology, some of which has been covered in this book. To learn from the animals this way requires careful note-taking and record-keeping. If you do this, and check back from time to time, and sit and think through situations, you can carry on learning almost indefinitely. This is part of the intellectual challenge of falconry.

Some people maintain that falconry is an art, suited only for a limited elite. Of course, falconry has a deep and varied culture and some people make it their life's work. But it is much more than an art. For those of us trained as scientists, falconry is a rich field of applied sciences, from taxonomy, anatomy, ethology, lots more -ologies, right through to anthrozoology which examines our own human relationships with the animals. Beyond science, falconry is undoubtedly a sport and we are seeing more and more applied sports science, fitness techniques, nutrition and so on. From the elitist point of view, even if you define falconry in its narrowest sense, falconry is egalitarian, encompassing most social, demographic and cultural groups. Just a cursory examination of the statistics shows that large numbers of people enjoy having some kind of active relationship with birds of prey. They may not all be hard-core hunters, but there is nothing wrong with that. In the Middle East, where real hunting is becoming almost impossible, racing has given new impetus and motivation to the falconers.

The problem in Europe was that up until the 1980s, hawks were in short supply. Tyros were discouraged because they represented future competition for scarce birds. This is not so any longer. Beginners now do not represent competition for hawks but they do represent competition for hawking grounds and, worse still, they represent an increasing threat to the future of falconry through potentially bringing the sport into disrepute by doing something stupid. When there were only a hundred falconers in Britain keeping a low profile, there was no problem. Now that there are perhaps 25,000 hawk-keepers in Britain alone, many in populated areas and many actively seeking publicity, then there is great potential for a bad press.

Through the early eighties we taught courses for three or four weeks in the summer and learned more about the process of learning. The first thing we learned was the amazing capacity of the human race for messing people about. The person who could book a course, send a deposit, arrive

Classes of 1984/5. Matt Gage, Professor of Genetics. Helen MacDonald, author of H is for Hawk, Tom Bailey, Falcon Veterinarian.

on time without bringing their dog/ferret/mother/children, without losing their flight ticket/hearing aid, able to both speak and write English, capable of getting up in the mornings and eating normal meals, was a rare gem.

Apart from their idiosyncracies, the beginners fell into three groups: those who had no natural aptitude or empathy with hawks and who should not be allowed anywhere near a live one. These we weaned off the idea by whatever means we could. The second group really wanted a summer break doing something special, a taster. They had no serious intention of continuing, but supported falconry and enjoyed actively participating for a few days. We tried to make their week enjoyable, to show them some of the fascination of falconry and we didn't worry too much if they didn't take in all the technical details. The third group were the ones who meant business. We were more fussy with them, picking them up on small errors which could have been fatal to a bird. At the end of the three weeks we lent them the buzzards which had been used on the course. They took them away for a year. Some caught quite a lot of rabbits with them, some gave up falconry when they sent the bird back. But they had had a chance to get started without great expense using birds that were tough enough to weather callow treatment. With dogs, horses and hawks I find it is always better if one of the partnership is already trained.

We found that the best courses consisted of four to six people. This made the group 'gel' with fertile discussions and a happy atmosphere. You can only teach intensively in short bursts, then there must be time to consolidate. Then coursers can talk amongst themselves and discuss different viewpoints. There's always a wise guy who wants to show the others how to fit hood braces! In the mornings we covered theory and indoor practical work such as fitting anklets and repairing feathers. In the afternoons we did outdoor practical work such as calling off hawks and, on the advanced course, hawking, either crow or magpie hawking or rabbit -hawking.

When people first arrived on their course, on their first morning, I would ask them one question: 'Why are you here?' Few of them had thought about this and we would return to the point several times during the course. Everyone has different motives for taking up hawk-keeping and/or falconry. Some of these motives are not even conscious. We tried to get people to come to terms with their real motives and to act on them, either to give up or to go forward.

One of the main problems is that most of the people coming into falconry have urban backgrounds. They have not been brought up in the country and do not understand it. They do not understand farming or farmers' problems. They do not understand how country folk live and work. And most of all they do not have the field craft skills necessary to produce good slips for their birds. Usually they have come into it through a fascination with a hawk on the fist. Thus we found that some of the keen rough-shooters who came on a beginners' summer course had taken over fifty head by Christmas whereas many with no country background caught nothing within three years. These are the kind of people who become 'muddlers' who never really get it together. A lot of people never really get it together in falconry.

More and more courses have sprung up. There are also now 'taster' days, corporate entertainment days, weekends, and hunting days. These are situations in which falconry interfaces with the public on very intimate terms. Anyone can ring up and go on one of these days. They can take a camera or video. That material could be used in the media against falconry. It is thus important to ensure that all these activities are properly conducted and do not bring falconry into disrepute. It is also important that beginners are properly taught and receive value for money. Courses can help avoid many of the elementary pitfalls. They are almost always better than nothing at all. But they are only a very small beginning down a very long and interesting path. The good falconer, when he hits a problem, will not immediately go running for advice. He will try to puzzle it out for himself, he will ask his bird, and he may come up with some very original solutions.

Much better than a course is to become a working pupil for several months or a year. There are various establishments involved professionally in some aspect of hawk-keeping, such as commercial and private breeding projects, some research and conservation institutions, zoos and 'falconry' display centres, rehabilitation centres, pest destruction and clearance contractors and genuine falconers. Anyone hoping to be involved with hawks as a career would do well to work at one of these places on a voluntary or semi-voluntary basis in order to get a grounding in the subject and to get a realistic view of it as a potential career. We have had over 300 interns staying usually for six months each, similar to Fran Hamerstrom's 'gaboons'. Most find the work harder than they had expected, and less glamorous. Unfortunately there are very few openings for genuine professional falconers. Most falconers nowadays are amateurs; those who do employ a professional look for a good person and are understandably reluctant to let someone inexperienced anywhere near their precious hawks. Having said that, it is often better to bring on a young inexperienced person and teach them good habits from the start than to employ someone who has flown hawks for a few years and reckons he knows it all.

In North America and Germany there are falconry tests, linked to possession licensing. Some of the questions in the multiple-choice tests have several possible answers and raise a smile or two, but in the main they are a genuine attempt to ensure that beginners have a basic grounding before they obtain a hawk. In Germany, they have a practical exam, and a verbal test, on raptor biology, falconry and management, hawking dogs, and wildlife law, and 30-50 percent fail. In Britain, where anyone can buy a hawk, ignorance is widespread and humility is rare. But a balance has to be struck in this, as in all things. You don't have to pass any tests to have a baby human, and a child is a much greater commitment and responsibility.

Many young people nowadays are not good readers and find it hard to learn from books. Videos are a good way of teaching these people some aspects and more and more films of variable quality are becoming available. They can demonstrate things which cannot easily be described in books, on the other hand they can encourage an 'instant expert' attitude based on a very shallow approach. The films that accompany this book will help bring to life some of the things I have tried to explain in the text. From the teaching point of view, with hands-on tuition, one can only reach a handful of people per year, but with books and films one can reach many thousands, and although tutorial feedback is not feasible, the course work can be better prepared, and can be re-visited repeatedly.

It is important to achieve and maintain certain standards, not just for the beginner, but also for experienced hawk-keepers. The Hawk Board in Britain, and now the International Association of Falconry, have produced guidelines on hawk-keeping covering the welfare of both hawks and quarry, and intending to prevent disreputable activities. In Britain, Germany, and North America standards of welfare are among the highest in the world. Part of the intent of this book is to help maintain those standards and to encourage them elsewhere.

These young ladies have got up early, groomed their horses, made a picnic to offer round, and ridden for five hours out hawking. What kind of world are we leaving them?
Photo Verity Johnson.

11.14 A protected belief

Despite fox-hunting being banned in 2005 in UK, it continued unabated throughout the land. Young and old several times per week through the season continue to meet and hunt as they have done for thousands of years. It is the largest and longest example of civil disobedience in British history. As such, it has brought politicians and the law into disrepute. Why haven't the hunters just meekly killed all their hounds and gone home? The reason is that hunting is not some relatively trivial pursuit, like tennis or golf. It goes deep into our human psyche. It pre-dates all of the received religions by hundreds of thousands of years. Hunting is so deep in our ancestry that we take it for granted until some politician comes along and tries to ban it.

Recently Judge Robin Postle, in an employment tribunal hearing in UK for Mr Jordi Casamitjana, ruled that 'Ethical Veganism' satisfied the tests required for it to be a 'philosophical belief' protected under the Equality Act 2010. Falconry is also capable of protection as a philosophical belief. To qualify as a protected belief, it must:

1. Be genuinely held;
2. Not just be an opinion or viewpoint;
3. Relate to a weighty and substantial aspect of human life and behaviour;
4. Attain a certain level of cogency, seriousness, cohesion and importance; and
5. Be worthy of respect in a democratic society and not be incompatible with human dignity or conflict with the fundamental rights of others.

Examples include:

- Ethical veganism is capable of being a protected philosophical belief, but vegetarianism is not.
- A 'gender-critical' belief that sex is biologically

immutable and that sex rather than gender identity is fundamentally important could be a philosophical belief.

- Other philosophical beliefs that are capable of protection include a belief in anti-fox hunting, Brexit, Scottish independence, stoicism, and climate change / environmentalism.

Other countries, such as USA, have similar legislation. Protected beliefs give protection to individuals against discrimination, for example in employment law. They may also entitle you to kill animals in ways that would otherwise be banned. Over 90 million animals are killed in UK byhalal or kosher methods due to protected beliefs. Falconry has been practised since pre-history, in over 80 countries, and is inscribed by UNESCO. It is not something trivial to be written off by the stroke of a politician's pen.

11.15 The philosophy of the falconer

Throughout the course of history, and recurring in many cultures, falconry has been seen as a training for princes, leaders, and gentlemen. It is the most senior of the hunting sports because it is the most difficult and the most challenging. Falconry teaches humility and fortitude, it enhances the spirit, it improves fine judgement and gives the falconer a sense of oneness with nature.

The falcon is a solitary animal, not a social one. She has no instincts to obey a leader, to follow the commands of a society. She cannot therefore be punished or ordered to obey; she can only be asked and rewarded. Pauper or prince, she treats all men as equals, and there are times when even princes crawl on their knees before a falcon.

She is never servile or sycophantic. To a man accustomed to being obeyed, this is a lesson in humility. His horses, salukis, and people may obey his commands but to his falcon he is just a hunting partner.

HRH Sh Mohammed Bin Zayed, President of Abu Dhabi, kneels before his saker on a houbara in Baluchistan.

And what a poor one! Weak of eyesight and slow of foot, the falconer is more a hindrance than a help to his falcon. He must resort to cars and radios to help him keep up with and find his falcon. The nerves at the bases of her feathers enable her to feel the air with the sensitivity of a concert pianist's fingers. Her world is an interplay of wind, gravity, and speed. She sees the plodding falconer on his flat earth below and it is only the invisible trust between hunting partners which stops her raking off downwind to prey on fortune alone. Even that imagined trust is based on her instinct to follow anything that might flush prey for her.

Because he is such a poor hunting partner, the falconer must learn fortitude. When he wakes in the morning he must think of his falcon's well-being before his own. Out hunting he must be prepared to go through great physical hardship, trying to find prey for his falcon, following her flight, helping her when things go wrong, searching for her when she is lost. Sometimes his body will be aching with fatigue but his spirit will be soaring in the clouds with her, willing her success in the flight. A moment later an eagle may appear and the falconer's emotions turn to anxiety and desperation. Forgetting his tiredness he will make every effort to save his falcon. If all goes well he will end the day contented, his falcon beside him on her perch with a good crop. But if she is still out there in the dark, lost, he can think of nothing else. He does not notice what he is eating. He gets no sleep.

His heart is out there with his falcon. Where is she? Is she safe? What plan can he make to find her again?

This is the difference between the falconer and a man with a gun. At the end of the day the gun is cleaned and put away, a mere machine, a tool. The falconer never puts away his falcon. She is always in his thoughts. As HRH Sheikh Mohamed Bin Zayed Al Nayhan said recently, 'We value our sakers like our children.'

When the falconer flies his falcon, part of his heart flies with her. All the time and effort he has spent on her have forged a bond in his heart. When she does well his heart almost bursts with joy. When she does badly, he is disappointed and sad. But he must learn not to show it, to bear his sadness without burdening it onto others. Although his emotions are taken on a roller-coaster ride by his falcon, and although his body may be brought to the limit of physical endurance, the falconer must cope with it all with fortitude. If he is not prepared to do this, if he is not prepared for the downside as well as the upside, he should never have flown his falcon. He should have stayed at home.

When he takes out his falcon, and slips her, he does not know what fate has in store. But he is prepared for it. He will have prepared in advance as best he can and he is ready to meet the challenge. He is ready for life.

The wise falconer can feel the spirit of a falcon, like an aura. The spirit of a horse or dog is different. They are social animals with a desire to please. They are dominated by man. The spirit of the falcon is independent, self-reliant and usually stronger than man's. Daily contact with such spirit enhances, inspires, and strengthens our own, through example.

Is this why we admire the falcon so much? She may be frightened of you, she may be angry at you, she may be pleased to see you. She will never lie to you, although you may not understand her. She has no self-pity and, even on the point of death from some consuming disease, her spirit is unbroken. It is a privilege to be in the presence of such a spirit, to draw strength from it and to learn from it. The gaze of a dying falcon is the distillation of life and a denial of triviality.

The wise man learns to listen when no words are spoken. How does he do that? He learns from his animals who communicate with more subtle signs than mere words. These signs are the truth. The falconer learns judgement.

Sometimes out hawking the falconer finds a set-up that looks perfect. Wide open terrain, quarry well placed, wind direction right. Everything fine except for one bush out in the middle. That one bush is sufficient to alter the course of the whole flight. Instead of trying to outfly the falcon in aerial combat, the quarry makes an extra effort for the bush. It is the falconer's job to know his ground in advance. He should know about that

If the crow comes to this bush, we are ready.

Enjoyment isn't just about flying the birds. After hawking we squeeze into the Hawking Van to go and collect the horse trailers.
Photo Verity Johnson.

bush and be prepared for it. Perhaps he will be able to manoeuvre the quarry far enough away from the bush to get a fair flight. Or perhaps he will select a falcon which can take quarry inside such a bush. He must use his judgement and balance the risks. And so it is in the boardroom. A strategy may look perfect, but the good tactician will look for that small bush and will evaluate the risks in just the same way as he does out hawking.

Similarly, the falcon teaches us the importance of time and timing. When we are breeding falcons we are carefully considering plans up to five years ahead, trying to predict our needs and envisage all the factors. When we are training the falcon we are making preparations for hawking trips some weeks or months ahead, carefully ensuring that when the date comes everything will be ready. The falcons must be fit and keen and in good health. They are not machines; if something is wrong you can't get a mechanic to fit a replacement part. And out hawking there will be moments when two seconds will make the difference between success and failure. A brief fumbling with the hood is all it takes to delay the falcon long enough for the quarry to become badly placed and the whole flight ruined. Particularly when the quarry is on passage and the falconer is trying to judge the right moment to slip his falcon, timing is everything. And so the falconer learns to manage time and to handle timing. He learns to treat time almost as a physical entity. You can't see it, like distance. You can't touch it, like food. But nevertheless it is a major factor and taskmaster. A falcon which seems dull now, may simply with the passage of two hours of time, fly like an arrow. In politics and in everyday life, time and timing are crucial.

A good falconer is measured not by his best bird, but by his worst one. It is not hard to have one or two top class birds, but not easy to make a whole team of good ones. If there are only one or two good ones and they are flown first, then they get better and better but perhaps become overworked and too exposed to accident or disaster. If something happens to them the falconer is left with a very poor team, unable to show sport. But by bringing on his other birds he

In China we enjoy cooking and eating our pheasants.

can work from strength and if some cannot be flown he still has others he can rely on. It is just the same in life. The most effective team has all its members consistently and reliably good, not one or two good people carrying too much load and the rest a poor lot.

Everyone looks for excitement. But it must be balanced according to each person's needs. When we are young we risk our necks. As we grow older and the body begins to fail, we value our health more and turn more to mental challenges and risks. The businessman gets excitement from making big business decisions. Others just gamble at cards or get second-hand excitement through films or computer games, living in a world of Haraway's 'cyborgs'. Falconry offers a blend of physical hardship with risk management. Falconry is such a deep intellectual pool that you can never reach the bottom of it. People looking at falconry from the outside think that it is all about killing things, pitting one animal against another. This is not so. The quarry which the falconer is actually hunting is enjoyment. And that, as everyone knows, is a most elusive prey. Enjoyment cannot be bought, bottled, organized, or preserved. Often you don't appreciate it is there until it has gone. It comes in all shapes and forms from a long contentment, to a fleeting glimpse, to a surging thrill. The falconer realizes this and understands that all the ingredients must be there for it to work. This is why falconers are keen conservationists for their falcons, their quarry, and their hunting grounds. What use is the most prized falcon if she has no quarry to match her powers? Yes, the falconer wishes to catch his quarry because falconry is about life and death. But he wishes to pursue the quarry in a way which enables it to use all its own natural abilities.

The widespread definition of falconry is 'Hunting wild prey in its natural state by means of trained hawks' and indeed this is the one I used when writing the UNESCO submission. By this definition, other activities such as displays to the public, lure flying, Sky Trials, racing and so on are not 'falconry'. But the reality is, most falconers spend more time with their birds doing 'non-falconry' activities than they do actually hunting. Some spend day after day training with only a few days or weeks hunting each year, if at all. Some of these activities are enjoyable in themselves. Many people enjoy tame hacking or social training evenings. Some enjoy a close relationship with their birds. Others are happy just lure flying and not hunting at all. We should not denigrate these aspects because, at the end of the day, falconry is about enjoyment, both for the falconer and (one hopes) for the bird. One of the reasons we developed robotic prey was to enhance the enjoyment element and to enable the raptors to do more 'hunting', fitting into a semi-urban landscape and limited free time after work hours. We cannot all live on a grouse moor for several months.

The prey has no natural defence against a gun. The bullet or pellets travel too fast for it to react. It may fly up into the air, be hit, and fall into a crumpled heap, or run away, winged. With the falcon it is different. Millions of years of evolution and a lifetime of experience has taught her prey to deal with this eventuality. Some fly high, some fly low. Some creep and hide. Some disappear down a hole. Some twist and turn and outfly the falcon. The strong are more likely to survive than the weak. The falconer thus gets an insight into natural systems. He knows how each species has different tactics to escape. He works out his own strategy with this in mind. And he has respect for his quarry, not as a mere target for a shot, but as a worthy opponent for his falcon. None escape injured to die later.

The man who has no personal contact with animals is like a man who has lived all his life in the same village, unaware of the wide world outside. The man whose only contact with animals is at a distance, such as at a zoo or through television, is like a traveller in a foreign country who never gets to speak to the inhabitants. When people say 'animals should be left in the wild' they are saying: 'keep them separate from us, we are of a different world.' Falconry is an interface between humans and the wild, a chance to be involved in the wild and not just a voyeur. You are not just looking in the window; you are actually in the party.

When you have understood, whether we like it or not, that we are all in this together, man, animals, plants, all stuck on Planet Earth, all dependent on one another, you become less arrogant and less confident of man's ability to control everything around him. We have not begun properly to understand complex natural systems yet. We cannot even predict the weather or earthquakes. We are using up finite resources. We call it all 'progress'. But we are running in the dark, faster and faster. We may just be hastening toward a cliff when we should be quietly feeling our way along, step by step. Falconry is a way of feeling our way along, of comprehending the complex life systems which we depend on and are part of, and falconry is a preparation and training for life.

11.16 Further Reading

Abuzinada, A.H. 2005. *The National Strategy for Conservation of Biodiversity in the Kingdom of Saudi Arabia.* National Commission for Wildlife Conservation and Development Dul Al-Hijja 1425 H.

Allsen, T.T. 2006. *The Royal Hunt in Eurasian History*. University of Pennsylvania Pres. ISBN 978-0-8122-3926-3.

Amirsadeghi, H. 2008. *Sky Hunters*. Thames and Hudson. ISBN 978-0-500-97687-6.

Anderson, S.H. and J.R. Squires. 1997. *The Prairie Falcon*. University of Texas Press. ISBN 0-292-70473-9.

Anderson, D.L., C.J.W. McClure, and A. Franke, editors. 2017. *Applied raptor ecology: essentials from Gyrfalcon research.* The Peregrine Fund, Boise, Idaho, USA.

Arnall, L. and I. F. Keymer. 1975. *Bird diseases.* TFH Publications Inc., New Jersey. ISBN 087666950X.

Azafzaf, H., E. Sande, S.W. Evans, M. Smart, N.J. Collar. 2005. International action plan for Houbara Bustard Chlamydotis undulata undulata in North Africa. BirdLife International, Nairobi, Kenya.

Azar, J.F., P. Rautureau, M. Lawrence, G. Calabuig, Y. Hingrat. 2016. Survival of reintroduced Asian houbara in United Arab Emirates' reserves.The Journal of Wildlife Management, 80 :1031-1039

Bailey, T.A., Samour, J.H., and T.C.Bailey. 1998. Hunted by falcons, protected by falconry: Can the houbara bustard (*Chlamydotis undulata macqueenii*) fly into the 21st century? Journal of Avian Medicine and Surgery. 12: 190–201.

Blaine, D.P. 1840. *An Encyclopaedia of Rural Sports*. Longman, London.

Blűchel, K.G. *Game and Hunting.* KönemannVerlagsgesellschaft. ISBN 3-89508-471-9.

Bijleveld, M. 1974. *Birds of Prey in Europe.* MacMillan, London.

Bogyai, F. 2021. *Hungarian Falconry Almanack.* Privately printed. ISBN 978-615-01-2553-4.

Bradshaw, J. W. S. 1992. *The Behaviour of the Domestic Cat.* C. A. B. International, U.K. ISBN 085198715X.

Broom, D. 2014. *Sentience and animal welfare.* Cambridge University.

Burnham, B. *A Fascination with Falcons.* Hancock House. ISBN 0-88839-415-2.

Burnside, R.J., N.J. Collar, P. M. Dolman. 2018. Dataset on the numbers and proportion of mortality attributable to hunting, trapping, and powerlines in wild and captive-bred migratory Asian houbara *Chlamydotis macqueenii.* Data in Brief, Volume 21: 1848-1852

Burr, E. W. 1987. *Companion Bird Medicine.* Iowa State University Press. ISBN 0813803624.

Cade,T.J., J. H. Enderson, C. G. Thelander, and C. M. White. 1988. *Peregrine Falcon Populations:Their Management and Recovery.* The Peregrine Fund Inc. ISBN 0961983906.

Cade, T.J., and W. Burnham (Eds). 2003. *Return of the Peregrine.* The Peregrine Fund. ISBN 0-9619839-3-0.

Cavé, A.J. 1968. 'The breeding of the Kestrel *Falco tinnunculus L.*, in the reclaimed area Oostelijk Flevoland.' *Netherlands Journal of Zoology* 18 (30): 313407.

de Chamerlat, C. A. 1987. *Falconry and Art.* Sotheby's Publications, ACR Edition Internationale. ISBN 0856673382.

Clarke, N. 2016. Sentience and animal welfare: Affirming the science and addressing the skepticism. Animal Sentience, 2016, 049.

Coles, B. H. 1985. *Avian Medicine and Surgery.* Blackwell Scientific Publications, Oxford. ISBN 0632014032.

Cooper, J. E., and J. T. Eley. 1979. *First aid and care of wild birds.* David and Charles, London. ISBN 0715376640.

Cooper, J. E., and A. G. Greenwood (eds.). 1981. *Recent Advances in the Study of Raptor Diseases.* Chiron Publications, Keighley, England. ISBN 0950771600.

Cox, J. 1985. *Observations on Falconry and Pakistan.* Pilot Publishing, Karachi.

Craighead, J. J., and F. C. Craighead. 1969. *Hawks, Owls and Wildlife.* Dover Publications. ISBN 486221377.

Craighead, J. J., and F. C. Craighead. 2001. *Life with an Indian Prince.* Archives of American Falconry. ISBN 0-9619839-2-2.

Cromie, R and M. Nicholls. 1995. The Welfare and Conservation Aspects of Keeping Birds of Prey in Captivity. A report to the RSPCA. Durrell Institute of Conservation and Ecology, University of Kent.

Crowley, S.L, M. Cecchetti and R. A. McDonald. 2019. Hunting behaviour in domestic cats: An exploratory study of risk and responsibility among cat owners. People and Nature 1(1): 18-30.

Czudek, R. 2006. Wildlife Issues and Development Prospects in West and Central Asia. FOWECA.

Davis, J. W., R. C. Anderson, L. Karstad, and D. O. Trainer. (eds.). 1971. *Infectious and parasitic diseases of wild birds.* Iowa State University Press.

Dawkins, M.S. 1980. Animal suffering: the science of animal welfare. Chapman and Hall, London.

Dawkins, R. 1989. *The Selfish Gene.* Oxford University Press. 0192860925.

Dickson, B,. J. Hutton and W.M. Adams (Eds). 2009. *Recreational Hunting, Conservation and Rural Livelihoods: Science and Practice.* Wiley-Blackwell. ISBN 978-1-4051-9142-5.

Dixon, A., Batbayar, N., Purev-Ochir, G and N. C. Fox. 2011. Developing a sustainable harvest of Saker Falcons (*Falco cherrug*) for falconry in Mongolia. In *Gyrfalcons and Ptarmigan in a Changing World*: 363-372. The Peregrine Fund, Boise.

Dixon. A. 2012. Conservation of the Saker Falcon *Falco cherrug* and the use of hybrids for falconry. Aquila 119:9-19.

Dixon. A. 2016. Commodification of the Saker Falcon *Falco cherrug*: Conservation Problem or Opportunity? Problematic Wildlife pp68-89. Springer.

Dixon, A., R.Maming , A. Gunga, G. Purev-Ochir, N. Batbayar. 2013. The problem of raptor electrocution in Asia: Case studies from Mongolia and China. Bird Conservation International 23:520–529.

Dixon. A., Bold. B., Tsolmonjav. P., Galtbalt. B. and N. Batbayar. 2018. Efficacy of a mitigation method to reduce raptor electrocution at an electricity distribution line in Mongolia. Conservation Evidence 15:50-53.

Dolman, P.M., N. J. Collar, & R.J. Burnside, 2018. Captive breeding cannot sustain migratory Asian houbara *Chlamydotis macqueenii* without hunting controls. Biological Conservation. DOI: 10.1016/j.biocon.2018.10.001

Dolman, P.M., R.J. Burnside, K.M. Scotland and N. J. Collar. 2021. Conservation and the captive breeding of the threatened houbara bustards. Endangered Species Research 46:161-173.

Dolman, P.M., K.M. Scotland, J. Burnside and N.J. Collar. 2021. Sustainable hunting and the conservation of the threatened houbara bustards. Journal for Nature Conservation 61.

Döttlinger, H. 2002. *The Black Shaheen Falcon (Falco peregrinus peregrinator* SUNDEVALL 1837): *Its morphology, geographic variation and the history and ecology of the Sri Lanka (Ceylon) population*. PhD thesis, University of Kent. ISBN 3-8311-3626-2.

Eastham, C.P., J.L. Quinn and N.C. Fox. 2000. Saker *Falco cherrug* and peregrine *Falco peregrinus* falcons in Asia; determining migration routes and trapping pressure. pp. 247-258 *In* R.D. Chancellor and B.-U. Meyburg [eds.]. *Raptors at risk; proceedings of the V World Conference on Birds of Prey and Owls*. Midrand, Johannesburg, South Africa, 4-11 August 1998. Hancock House; WWGBP, Blaine, Washington; Berlin, Germany.

Erikson, W.A., R.E. Marsh and T.P. Salmon. 1990. A review of falconry as a bird-hazing technique. Proceedings of the Fourteenth Vertebrate Pest Conference, University of Nebraska.

Fenech, N. 1992. *Fatal flight: the Maltese obsession with killing birds*. Quiller Press, London. ISBN 187094853X.

Ferrer, M and G.F.E. Janss (Eds). 1999. *Birds and powerlines: collision, electrocution and breeding*. Quercus 239pp.

Ferrieria, D.F. 1616. *Arte de Caca de Altaneria*, translated as *Ferriera's Falconry* by Anthony Jack, 1996, The Signet Press. ISBN 09526906-16.

Filkins, K (ed). 1993. *A Bond with the Wild: A Celebration of American Falconry*. North American Falconers' Association. LCCCN 9385241.

Forbes, N. A., and G. N. Simpson. 1994. *Emergencies and First Aid: A course for rehabilitators and falconers*. Clockhouse Veterinary Hospital, Wallbridge, Stroud U.K.

Fox, N.C .and J.W. Locke. 1978. Organochlorine residues in New Zealand birds of prey. New Zealand Journal of Ecology 1:118-125.

Fox, N. C. 1981. Deutsche Falknerprufung – The German Falconry Test. Journal of the British Falconers' Club VII (5): 331-334.

Fox, N.C., B.I.M. Fox and T. Bailey. 1985. The Predation Ecology of the Mauritius Kestrel (*Falco punctatus*). Unpublished research report, The Peregrine Fund.

Fox, N. C. 1987. People problems in endangered species conservation. In 'Reproduction and Management of Birds of Prey'. Conference Proceedings, University of Bristol.

Fox, N. C. 1987. The artificial incubation and rearing of wild Welsh Red Kites (*Milvus milvus*). Report to the Kite Committee: Nature Conservancy Council and Royal Society for the Protection of Birds.

Fox, N.C. & C.P. Eastham. 1996. The conservation programme on the Saker Falcon (*Falco cherrug*) in Kazakhstan, Mongolia, Pakistan and the Middle East. Abstracts of 2nd International Conference on Raptors. Urbino, Italy 2-5 October 1996. RRF, University of Urbino. 129.

Fox, N. C., and H. Macdonald. 1996. An investigation on behalf of the Hawk Board into the nature and extent of suffering caused by current methods of pest control and field sports. The Hawk Board 14pp. https://www.falcons.co.uk/wp-content/uploads/2019/12/An-investigation-on-behalf-of-the-Hawk-Board-into-the-nature-and-extent-of-suffering.pdf.

Fox, N.C., M. Ahmed, C.P. Eastham. 1996. Research and conservation of the Lugger Falcon (*Falco jugger*) in Pakistan. Abstracts of 2nd International Conference on Raptors. Urbino, Italy 2-5 October 1996. RRF, University of Urbino. 131-132. [Abstracts].

Fox, N. C., M. A. Bowardi, H. Macdonald and F. Launay. 2000. A Global Strategy Plan for the Conservation of Falcons and Houbara. ERWDA, UAE. 74p [In English and Arabic]

Fox, N.C. 2001. Introduction to the International Programme on the Saker Falcon. Saker Falcon in Mongolia: Research and Conservation. Proceedings of the 2[nd]International Conference of the Middle East Falcon Research Group on Saker Falcon and Houbara Bustard, Ulanbaatar 1-4 July 2000. Potapov, E., Banzragch, S., Fox, N., and Barton, N. (Eds). Ulanbaatar. 2001. 20-24. [In Mongolian, Russian and English].

Fox, N. C., N. Blay, A. G. Greenwood, D. Wise and E. Potapov. 2005. Wounding rates in shooting foxes (*Vulpes vulpes*). Journal of Animal Welfare 14: 93-102.

Fox, N.C. and C. Wynn. 2010.The impact of electrocution on the New Zealand falcon (*Falco novaeseelandiae*). *Notornis, 2010, Vol. 57: 71-74*

Francione, G. 1995. *Animals, property and the law.* Temple University Press.

Fraser, D. 2010. Toward a synthesis of conservation and animal welfare science. Animal Welfare, 19(2), 121–124.

Freeman, G. A. and F. H. Salvin. 1859. *Falconry: Its Claims, History and Practice.* Longmans, London.

Gersmann, K-H and O. Grimm (Eds) 2018. *Raptor and Human - falconry and bird symbolism throughout the millennia on a global scale.* Wachholtz Verlag. ISBN 978-3-529-01490-1.

Grenet, S. 2019. Heritage of display: Interculturality and cultural diplomacy in the 2010 UNESCO falconry file. International Journal of Heritage Studies 25(26):553-55-64.

Haak, B.A. 1992. *The Hunting Falcon.* Hancock House.

Haas, D., M. Nipkow, G. Fiedler, R. Schnieder, W. Haas and B. Schurenberg. 2005. Protecting birds from powerlines. Convention on the Conservation of European Wildlife and Habitats (Bern Convention). Nature and Environment No. 140. Council of Europe Publishing.

Hamerstrom, F. 1970. An Eagle to the Sky. The Crossing Press, New York.

Haraway, D. 2016. A Cyborg Manifesto. University of Minnesota Press.

HMSO. 1981. *The Wildlife and Countryside Act 1981.* Her Majesty's Stationery Office, London. ISBN 0105469815.

Harrison, G. J., and L. R. Harrison. 1986. *Clinical Avian Medicine and Surgery, including aviculture.* W. B. Saunders Co, Philadelphia. ISBN 0721612415.

Ingold, T. 2000. The Perception of the Environment: Essays on Livelihood, dwelling and skill. Routledge, London. ISBN 0–415–22831–X.

Higbee, D. and D. Bruzina (Eds). 2018. *Hunting and the Ivory Tower: Essays by Scholars Who Hunt.* University of South Carolina Press. ISBN 978-1-61117-849-4.

IUCN. 2007. *IUCN Guidelines for the Placement of Confiscated Animals.* IUCN ISBN 2-8317-0622-x

Ivanov, K.P, and E.V. Sedunova.1993. Action of wind-power plants (WWP) on ornithofauna. Russian J. Ecology 24:315-320.

Jefferies, R. 1882. *Bevis: The Story of a Boy.* Sampson Low, London. ISBN 0192822292.

Jefferies, R. 1883. *The Story of my Heart: An Autobiography.* Longmans, London. ISBN 1548001759.

Khattak, K.K. 2014. *The Book of Falconry.* Translated by S.U. Rahman. Pangraphics, Islamabad.

Knecht, C. D., A. R. Allen, D. J. Williams, and J. H. Johnson. 1987. *Fundamental techniques in Veterinary Surgery.* W. B. Saunders Co. Philadelphia. ISBN 0721613977.

Kowalsky, N. (Ed). 2010. *Hunting: In Search of the WildLife.* Blackwell. ISBN 978-1-4443-3569-9.

Lacy, R. 1988. 'A report on population genetics in conservation.' *Conservation Biology* 2 (3):245247.

Lack, D. 1954. *The natural regulation of animal numbers.* Oxford University Press.

Lack, D. 1966. *Population studies of birds.* Oxford University Press. ISBN 0198573350.

Lehman, R.N., P.L. Kennedy and J.A. Savidge. 2007. The state of the art in raptor electrocution research: A global review. Biological Conservation136:159-174.

Leopold, A. 1949. *A Sand County Almanac: And Sketches Here and There.* Oxford University Press. ISBN 0197500269.

Llewellyn, P. J. and Brain, P. F. 1983. 'Guidelines for the rehabilitation of injured raptors.' In Olney, P. J. (ed). *International Zoo Yearbook* 23: 121125. Zoological Society of London.

Loft, J. 2003. *D'Arcussia's Falconry 1643.* (English translation). Privately printed. ISBN 0-9535-8811-4.

Lovegrove, R. 1990. *The Kite's Tale: The story of the Red Kite in Wales.* RSPB, U.K. ISBN 0903138379.

Macdonald, H. 2006. *Falcon.* Reaktion Books. ISBN 1-86189-238-1.

Mace, G. 1986. 'Genetic management of small populations.' *International Zoo Year Book* 24/25: 167174.

Madden, D.H. 1897. *The Diary of Master William Silence: A study of Shakespeare and of Elizabethan sport.* Longmans, Green and co.

Malacarne, G. 2011. *Lords of the Sky: Falconry in Mantua at the time of the Gonzagas*. ArtiglioEditore. ISBN 978-88-88752-10-5.

Merling de Chapa, M. *et al.* 2020. Phantom of the forest or successful citizen? Analysing how Northern Goshawks (*Accipiter gentilis*) cope with the urban environment. R. Soc. open sci.7:201356201356.

Marvin, G. 2000, 'Natural Instincts and Cultural Passions: Transformations and Performances in Foxhunting', Performance Research, 5(2): 108–115.

Marvin, G. 2003. A Passionate Pursuit: Fox Hunting as Performance. Sociological Review.

Medway, Lord. 1980. *Report of the Panel of Enquiry into Shooting and Angling (197679).* c/o The Causeway, Horsham RH12 1HG, U.K.

Meijer, T. 1988. *Reproductive decisions in the Kestrel: a study in physiological ecology.* Drukkererij Van Denderen B. V., Groningen.

Mellor, J.E.M. 1949. *Notes on Falconry*. Privately printed.

Al Nahayan, Zaid Bin Sultan. *Our Arab Heritage*. Privately printed.

Morris, D. *The Animal Contract*. Virgin Books. ISBN 1-85227-392-5.

Newton, I. 1979. *Population Ecology of Raptors.* T. & A. D. Poyser Ltd. ISBN 0856610232.

Newton, I. 1986. *The Sparrowhawk.* T. & A. D. Poyser Ltd. ISBN 0856610410.

New York State Department of Environmental Conservation. Undated. *New York State Falconry Examination Manual.* New York State Department of Environmental Conservation, 50 Wolf Rd., Albany, NY 12233.

Oakes, W. C. 1993. *Topical Index of North American Falconry: The 1940's, 1950's, 1970's, 1980's.* Eaglewing Publishing. Roy, UT.

Oggins, R.S. 2004. *The kings and their hawks: Falconry in medieval England.* Yale. Sheridan Books. ISBN 0-300-10058-2.

Olney, P. 1990. Proceedings of a Raptor Rehabilitation Workshop, London Zoo 1990. The Falconry Centre, Newent, England.

Page, R. 1977. *The Hunter and the Hunted.* Readers' Union, Devon.

Parkes, C., and J. Thornley. 1994. *Fair Game: The Law of Country Sports and the Protection of Wildlife.* Pelham Books. ISBN 0-7207-2030-3.

Pitcher, E and R. Velarde. 2009. *The Flying of Falcons*. Delfin Seral, Madrid.

Potter, C. (ed.). 1991. 'The Impact of cats on Native Wildlife.' Proceedings of a Workshop 89 May 1991, The Endangered Species Unit, Australian National Parks and Wildlife Service.

Rahman, M.L., Purev-Ochir,G., Etheridge, M., Batbayer, N and A. Dixon. 2014.The potential use of artificial nests for the management and sustainable utilization of saker falcons (*Falco cherrug)*. Journal of Ornithology 155(3):649-656.

Rahman, S.U. 2017. *The Art of Falconry in the Mughal Empire*. Rainbow Graphics, Islamabad.

Ratcliffe, D. 1980. *The Peregrine Falcon.* T. & A. D. Poyser Ltd. ISBN 0856610267.

Redig, P. T. 1993. *Medical Management of Birds of Prey.* 2nd Edition. The Raptor Center, University of Minnesota.

Redig, P. T., J. E. Cooper, J. D. Remple, and D. B. Hunter (eds.). 1993. *Raptor Biomedicine.* Chiron Publications Ltd, Keighley, U.K.

Rollin, B. E. 2006. *Animal rights and human morality*. Prometheus Books.

Rollin, B.E. 2016. *A new basis for Animal Ethics.* University of Missouri Press. ISBN 978-0-8262—2101-8.

Royal Society for the Protection of Birds. 1992. *The International Trade in Wild Birds.* RSPB, The Lodge, Sandy, Beds, U.K.

Rudebeck, G. 1950. The Choice of Prey and Modes of Hunting of Predatory Birds with Special Reference to their Selective Effects. Oikos 2(1):65-88.

Russ, T., and J. Foster. 2010. *Law of Field Sports.* Wildy, Simmonds and Hill Publishing. ISBN 9780854900695.

Ryman, N. and L. Laikre. 1991. 'Effects of supportive breeding on the genetically effective population size.' *Conservation Biology* 5 (3): 325328.

Scruton, R. 2001. *On Hunting*. St Augustine Printing. ISBN 1587316005.

Sielicki, J. 2017. Falconry as a Biodiversity Conservation Tool. In U. Szymak and P. Sianko (Eds) Falconry – Its influence on Biodiversity and Cultural Heritage.

Smith, J. A. and K. M. Boyd (eds.). 1991. *Lives in the Balance.* Oxford University Press.

Al-Timimi, F. 2007. *Falconry and Hunting in Arabia*. Gulfosaker, Canada. ISBN 978-0-9738853-0-9.

Szymak,U. (Ed.) 2016. *Falconry – Its Influence on Biodiversity and Cultgural Heritage in Poland and across Europe.*Muzeum Podlaskie w Bialymstoku. ISBN 978-83-87026-59-2.

Travis, A. S. (ed). 1980. *Angling in Britain 1980: Report of the Travis Commission.* The Angling Foundation.

Trouwborst, A., P.C. McCormack and E.M. Camacho. 2020. Domestic cats and their impacts on biodiversity: A blind spot in the application of nature conservation law. People and Nature 2(1) 235-250.

Tubbs, C. R. 1974. *The Buzzard.* David & Charles, London. ISBN 0715363239.

Turner, D. and P. Bateson (eds.). 1988. *The Domestic Cat: the Biology of its Behaviour.* Cambridge University Press.

Upton, R.C. 1980. *A Bird in the Hand.* Debrett's Peerage Ltd. U.K. ISBN 0905649346.

Upton, R.C. 1987. *O for a Falconer's Voice.*Crowood Press, Marlborough, Wilts. ISBN 1852230150.

Upton, R.C. 2004. *Hood, Leash and Lure*. Privately printed. ISBN 0-9548104-0-6.

Upton, R.C. 2010. *Arab Falconry: History of a way of life.* Abu Dhabi Authority for Culture and Heritage. ISBN 978-9948-01-730-1.

Van den Abeele, B. 1990. *La Fauconnerie dans les lettres Francaises du XIIe au XIVe siècle*. Leuven University Press. ISBN 90-6186-400-3.

Van de Wall, J.W.M. 2004. *The Loo Falconry.* Hancock House. ISBN 0-88839-576-0.

Village, A. 1990. *The Kestrel.* T. & A. D. Poyser Ltd. ISBN 0856610542.

Walter, H. 1979. *Eleonora's Falcon.* University of Chicago Press. ISBN 0226872297.

Watson, J. N. P. 1991. *A Green Guide to Country Sports.* The Sportsman's Press, London. ISBN 0948253525.

Webster, H. M. and J. Enderson. 1988. *Game Hawking...at its very best.* Windsong Press. ISBN 091251004X.

Winkelman, J.E. 1995. Bird/wind turbine investigations in Europe. Proc. National Avian-Wind Power Meetings, Denver Colorado, 20-21 July 1994. Pp. 43-48.

Wyatt, T. 2014. Non-Human Animal Abuse and Wildlife Trade: Harm in the Fur and Falcon Trades, *Society & Animals*, *22*(2), 194-210. doi: https://doi.org/10.1163/15685306-12341323

Wyatt, T., J. Maher, D. Allen, N. Clarke and D, Rook. 2022. The welfare of wildlife: an interdisciplinary analysis of harm in the legal and illegal wildlife trades and possible ways forward. Crime, Law and Social Change 77:69–89.

Ye Xiaodi, Wan Ziming, Bai Yanxia, Zhou Lizhi, Nick Fox. 2001. Saker Falcon (*Falco cherrug*) resources and its sustainable utilization. Aridnyeekosistemy [Arid ecosystems]. Editor-in-chief Z.G.Zalibekov. Vol.7. N14-15: 93-99. In English with Russian summary.

Index

A

B

C

D

E

F

G

H

I

J

K

L

M

Q

R

S

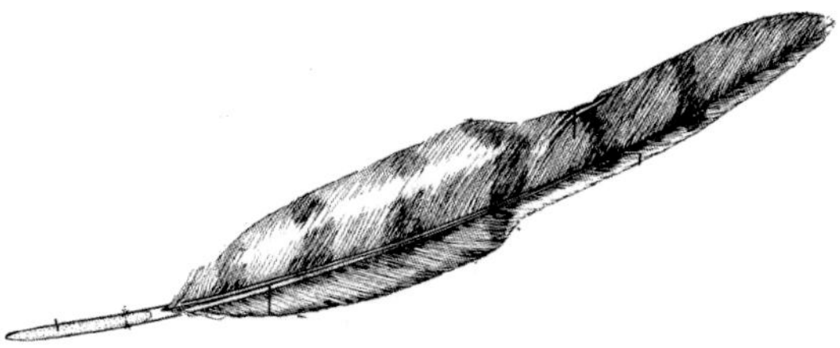